The Physics of Free Electron Lasers

Springer
Berlin
Heidelberg
New York
Barcelona
Hong Kong
London
Milan
Paris
Singapore
Tokyo

Advanced Texts in Physics

This program of advanced texts covers a broad spectrum of topics which are of current and emerging interest in physics. Each book provides a comprehensive and yet accessible introduction to a field at the forefront of modern research. As such, these texts are intended for senior undergraduate and graduate students at the MS and PhD level; however, research scientists seeking an introduction to particular areas of physics will also benefit from the titles in this collection.

E. L. Saldin
E. A. Schneidmiller
M. V. Yurkov

The Physics of Free Electron Lasers

With 231 Figures

Springer

Evgeny L. Saldin
Evgeny A. Schneidmiller
Automatic Systems Corporation
443050 Samara, Russia

Mikhail V. Yurkov
Joint Institute for Nuclear Research
141980 Dubna, Moscow Region, Russia

Library of Congress Cataloging-in-Publication Data

Saldin, E. L. (Evgeny L.), 1951-
  The physics of free electron lasers / E.L. Saldin, E.A. Schneidmiller, M.V. Yurkov.
    p. cm.
  Includes bibliographical references (p.
  ISBN 3-540-66266-9 (alk. paper)
  1. Lasers. I. Schneidmiller, E. A. (Evgeny A.), 1963- .
  II. Yurkov, M. V. (Mikhail V.), 1957- . III. Title.
  QC688.S25 1999
  621.36'6--dc21   99-39000
                                                                       CIP

ISSN 1439-2674

ISBN 3-540-66266-9 Springer-Verlag Berlin Heidelberg New York

This work is subject to copyright. All rights are reserved, whether the whole or part of the material is concerned, specifically the rights of translation, reprinting, reuse of illustrations, recitation, broadcasting, reproduction on microfilm or in any other way, and storage in data banks. Duplication of this publication or parts thereof is permitted only under the provisions of the German Copyright Law of September 9, 1965, in its current version, and permission for use must always be obtained from Springer-Verlag. Violations are liable for prosecution under the German Copyright Law.

© Springer-Verlag Berlin Heidelberg 2000
Printed in Germany

The use of general descriptive names, registered names, trademarks, etc. in this publication does not imply, even in the absence of a specific statement, that such names are exempt from the relevant protective laws and regulations and therefore free for general use.

Typesetting: Camera ready copy from the authors using a Springer TeX macro package
Cover design: *design & production* GmbH, Heidelberg

SPIN: 10569967   56/3144/mf - 5 4 3 2 1 0 – Printed on acid-free paper

Preface

This book contains a systematic treatment of the basic principles of free electron laser (FEL) physics. It is primarily intended for physicists specializing in FEL physics and related fields: laser physics, microwave electronics, particle accelerator physics, etc. At the same time it might be useful for those who use the FEL as a research or industrial tool.

The treatment requires that the reader has a knowledge of classical mechanics and electrodynamics. It is assumed that the reader is familiar with the kinetic theory of charged particle beams, in particular the Vlasov equation. All the results presented here are derived from "first principles", and all steps involving physical principles are given. To preserve a self-consistent style, we place the derivation of auxiliary results in appendices.

Theoretical study is performed with an extensive use of similarity techniques, so the results obtained are simultaneously highly general and completely specified. The use of similarity techniques involves a particular way of thinking and leads to a deeper insight into FEL physics.

We use a synthetic approach to present the material: some simple models are studied first, and more complicated ones are introduced gradually. We start with the one-dimensional theory of the FEL amplifier and FEL oscillator. Then we move on the analysis of diffraction effects and waveguide effects. Finally, we introduce the reader to the part of FEL theory dealing with the start-up from shot noise in the electron beam.

To help readers form their own opinion on the topics discussed, the end of each chapter has a suggested bibliography together with relevant remarks. The list of references includes only the papers we have consulted directly. A lot of papers remain unmentioned, and for this we apologize.

We have also included a list of symbols. When a symbol appears for the first time (or is not used for a long time), we provide its definition in the body of the text. In this book we use Gaussian units for electromagnetic quantities, since they are more convenient for theoretical study.

The book has evolved over several years in parallel with our scientific work on FELs at the Automatic Systems Corporation (Samara), the Joint Institute for Nuclear Research (Dubna) and DESY Hamburg (Deutsches Elektronen Synchrotron). We thank all our colleagues from these institutes for fruitful collaboration and for creating a friendly atmosphere for scientific work. We are particularly indebted to V.P. Sarantsev, B.H. Wiik and A.N. Lebedev. Without their support and encouragement, the appearance of this book would have been problematic.

Samara, Dubna and Hamburg
October 1999

Evgeny L. Saldin
Evgeny A. Schneidmiller
Mikhail V. Yurkov

Contents

1. **Introduction** 1
   1.1 Advantages of FELs Over Quantum Lasers 3
   1.2 Principle of FEL Operation 4
   1.3 Suggested Bibliography 11

2. **One-Dimensional Theory of the FEL Amplifier** 13
   2.1 Linear Mode of Operation 15
       2.1.1 Effective Hamiltonian 15
       2.1.2 Self-Consistent Equations 18
       2.1.3 Solution of the Initial-Value Problem
             by the Laplace Technique 22
       2.1.4 General Solution of the Initial-Value Problem 38
       2.1.5 Linear Theory of the FEL Amplifier
             with a Planar Undulator 41
   2.2 Saturation Effects 48
       2.2.1 Self-Consistent Equations 49
       2.2.2 Numerical Simulation Algorithm 51
       2.2.3 Power Balance 51
       2.2.4 Saturation in the High-Gain FEL Amplifier 52
       2.2.5 Space Charge Effects 59
       2.2.6 Energy Spread Effects 61
       2.2.7 FEL Amplifier with a Planar Undulator 63
   2.3 FEL Amplifier with Tapered Undulator 65
       2.3.1 Low-Efficiency Approximation 66
       2.3.2 The High-Efficiency FEL Amplifier 76
       2.3.3 Some Generalizations 79
   2.4 Concluding Remarks 82
   2.5 Suggested Bibliography 86

3. **One-Dimensional Theory of the FEL Oscillator** 87
   3.1 Small-Signal Gain 89
       3.1.1 Basic Relations 89
       3.1.2 Cold Electron Beam 91
       3.1.3 Gaussian Energy Spread 92

           3.1.4 Space Charge Effects 94
      3.2 Saturation Effects in the FEL Oscillator 96
           3.2.1 Self-Consistent Equations 97
           3.2.2 Nonlinear Simulation Algorithm 98
           3.2.3 Resonator Losses and Efficiency Optimization 99
           3.2.4 Space Charge and FEL Efficiency 104
           3.2.5 Energy Spread and FEL Efficiency 106
           3.2.6 Some Generalizations 107
      3.3 FEL Oscillator with Nonuniform Undulator 111
           3.3.1 Basic Equations 113
           3.3.2 Optical Klystron 118
           3.3.3 FEL Oscillator with a Prebuncher
                  and a Tapered Main Undulator 126
      3.4 Start-Up from Shot Noise in the FEL Oscillator 134
           3.4.1 Basic Equations 135
           3.4.2 General Results 139
           3.4.3 Operation Below Threshold 144
           3.4.4 Operation Above Threshold 146
      3.5 Concluding Remarks 149
      3.6 Suggested Bibliography 152

4. **Diffraction Effects in the FEL Amplifier** 155
      4.1 Self-Consistent Equations 159
      4.2 Power Balance 164
      4.3 Linear Theory of the FEL Amplifier
           with a Sheet Electron Beam 168
           4.3.1 Eigenvalue Problem for a Stepped Profile 170
           4.3.2 Analysis of the Beam Radiation Modes 178
           4.3.3 Initial-Value Problem for a Stepped Profile 187
           4.3.4 Epstein Profile 194
           4.3.5 Parabolic Profile 205
           4.3.6 Arbitrary Gradient Profile 209
      4.4 Linear Theory of the FEL Amplifier
           with an Axisymmetric Electron Beam 214
           4.4.1 Eigenvalue Problem for a Stepped Profile 214
           4.4.2 Analysis of the Solutions 216
           4.4.3 Initial-Value Problem for a Stepped Profile 225
           4.4.4 Parabolic Profile 234
           4.4.5 Arbitrary Gradient Profile 238
           4.4.6 Numerical Solution of Initial-Value Problem 243
      4.5 Nonlinear Mode of Operation 245
           4.5.1 Nonlinear Simulation Algorithm 246
           4.5.2 Some Results of Numerical Simulations 248
           4.5.3 Planar Undulator 258
      4.6 Concluding Remarks 259

	4.7	Suggested Bibliography 261

5. Waveguide FELs 263
5.1 Self-Consistent Equations 265
5.1.1 Integro-Differential Equation for the Field 266
5.1.2 Integro-Differential Equation for the Beam Modulation 268
5.2 Power Balance 272
5.3 Beam Radiation Modes in a Circular Waveguide 276
5.3.1 Stepped Profile of Electron Beam 278
5.3.2 Parabolic Profile 303
5.3.3 Arbitrary Gradient Profile 307
5.4 Initial-Value Problem 309
5.4.1 Analytical Solution 309
5.4.2 Effective Potential for a Circular Waveguide 320
5.4.3 Numerical Solution 328
5.5 Nonlinear Mode of Operation 330
5.6 Rectangular Waveguide 340
5.7 Wall Resistance Effects 342
5.8 Concluding Remarks 350
5.9 Suggested Bibliography 351

6. FEL Amplifier Start-up from Shot Noise 353
6.1 Shot Noise in the Electron Beam 357
6.2 One-Dimensional Theory of SASE FEL 360
6.2.1 Analytical Description of the Linear Regime 360
6.2.2 Numerical Simulation Algorithm 380
6.2.3 Numerical Simulations of the Main Characteristics of a SASE FEL 385
6.3 Three-Dimensional Simulations of SASE FEL 402
6.3.1 Numerical Simulation Algorithm 402
6.3.2 Transverse Coherence 407
6.4 SASE FEL: Experiment and Theory 414
6.4.1 Region of Physical Parameters 415
6.4.2 Numerical Analysis of the Experiment 419
6.5 Suggested Bibliography 424

Appendices 425
A.1 The Extended Hamiltonian Formalism 425
A.2 Longitudinal Space Charge Field of a Modulated Electron Beam with Finite Transverse Size 428
A.3 Green's Function for a Homogeneous Waveguide 429
A.4 Eigenfunctions of a Passive Circular Waveguide 435
A.5 Calculation of the Sums in (5.119) 437
A.6 List of Symbols 441

Suggested Further Reading 449

References 453

Index 457

1. Introduction

In the middle of the 1970s John Madey and colleagues constructed the first free electron laser operating in the infrared wavelength range. Since that time tremendous progress has been made in the FEL technique and free electron lasers now occupy an appropriate place among other sources of coherent radiation. When the first operating free electron laser was constructed, it seemed that FELs would form only a small supplement to a long list of available at that time quantum laser. Actually, the appearance of FELs led to a new direction for sources of coherent radiation: they embodied the type of coherent source to which all experimenters have aspired since the invention of the laser. It is relevant to note that despite the FEL being referred to as a laser, the principle of its operation is similar to that of conventional vacuum-tube devices: it is based on the interaction of electron beams with radiation in vacuum. From this point of view FELs form a separate class of vacuum-tube devices capable of generating powerful coherent radiation at any wavelength from the millimeter to the X-ray part of the spectrum similar to the vacuum-tube devices which generate coherent radiation at any wavelength, from the kilometer to the millimeter range. Also, free electron lasers possess all the attractive features of vacuum-tube devices. FEL radiation is always totally polarized and has ideal, i.e. diffraction, dispersion. FELs are capable of providing a high efficiency of transformation of the electron beam power into radiation power. Remembering that electron accelerators of driving beams for FELs can provide high average and peak power with an effective transformation of electric power into electron beam power, one can expect to reach a high level of total FEL efficiency and high peak and average output radiation power.

Despite strong competition from conventional lasers, the FEL is recognized nowadays as a unique tool for scientific applications requiring tunable coherent radiation in the far-infrared or VUV ranges. Taking into account the future perspectives of the FEL, many industrial firms undertake intensive investigations into FEL technology, aiming at constructing powerful UV FELs for industrial applications such as material processing, lithography, isotope separation, and chemical applications.

Significant effort of scientists and engineers working in the field of conventional quantum lasers are directed towards the construction of X-ray lasers.

Nevertheless, this problem is still unsolved: progress in this field is rather moderate and we cannot expect a significant breakthrough in the near future. A similar problem can be formulated also for the free electron laser. Pioneering investigations performed at the beginning of the 1980s have shown that this problem can be solved by FELs with appropriate development of accelerator and FEL techniques. During the last decade there has been extremely rapid progress in linear accelerators, new developments in low-emittance, high-current electron guns, and successful operation of high-precision undulators. As a result, at present there exists a technological base for the construction of free electron lasers operating in the X-ray wavelength range. Recently X-ray FEL projects have been initiated at several laboratories around the world. The unique properties of these facilities will open up a multitude of new scientific and technical opportunities.

This book contains a systematic treatment of the basic principles of free electron laser physics. It might be useful for physicists specializing in FEL physics as well as in related fields: laser physics, microwave electronics, particle accelerator physics, etc. It will be of help also for those who use the FEL as a research or industrial tool in their work.

All the results presented in this book are derived from "first principles" and the reader can follow the whole derivation process from beginning to end. The present treatment requires from the reader only a knowledge of classical mechanics, electrodynamics, and a moderate knowledge of higher mathematics (differential equations, the Laplace transform, and special functions). Both the simplicity and the strictness of the treatment are explained by the features of the object under study (FELs are described rather well in the framework of classical physics) and by the relative simplicity of the accepted FEL models.

In traditional microwave electronics similarity techniques have been widely used and have served a good deal to clarify the basic ideas of this science. But in the physics of free electron lasers these techniques still did not achieve the popularity they might have, due to the variety and generality of their possible applications. At present, there is an urgent need for a handbook to effectively help researchers in using similarity techniques applied to FEL theory. This book is an attempt to satisfy the needs of specialists for such a tool.

The advantages of the application of similarity techniques are evident. The dimensional analysis of any problem, performed prior to its analytic or numerical investigation, not only reduces the number of independent terms but also allows one to classify the grouping of dimensional variables in a way that is most suitable for subsequent study. Furthermore, the solution of the dimensionless equations produces final results in such a form that the information contained in them will be simultaneously of a high degree of generality and completely specified. The use of similarity techniques forms a style of physical thinking and leads to a deeper insight into FEL physics.

To make the book useful for practical calculations, we have included numerous universal graphs illustrating various modes of FEL operation, together with reduced design formulae which will show the reader how to find FEL characteristics using simple dimensional analysis only. It may be useful for FEL physicists, especially at the design stage of an experiment.

1.1 Advantages of FELs Over Quantum Lasers

We mentioned that FELs form a separate class of vacuum-tube devices. From this point of view it is easy to appreciate the origin of the advantages of the FEL over quantum lasers mentioned above. The main advantage is the tunability of the radiation. In the quantum laser, the lasing wavelength is defined by discrete energy transitions between the quantum levels of atoms or molecules of an active medium. Despite the variety of the active media types discovered so far, the number of quantum levels is and will remain finite. As for the FEL (or, in a wider sense, for vacuum-tube devices), their operating frequency is defined by their design, namely by the electron beam parameters, the characteristics of the electrodynamic structure (waveguide walls, resonator mirrors, etc.), and by characteristics of the electrical and magnetic fields in the interaction region. For these reasons, the FEL can be tuned, in principle, to any desired operating frequency.

Another important feature of the FEL is that its radiation is always coherent and has ideal, i.e. diffraction, dispersion. In other words, the FEL radiation can always be focused on to a spot whose size is defined totally by diffraction effects. This feature of the FEL is a consequence of the fact that the process of electromagnetic field amplification develops in vacuum. It reveals a wide range of possibilities for FEL applications in the transportation of the radiation over long distances and in obtaining high intensity. Contrary to the FEL, the dispersion of radiation of powerful lasers usually exceeds significantly the value of the diffraction limit. The main effects which determine the growth of the radiation dispersion are fluctuations of the refractive index of the active medium due to thermal effects, and nonlinear effects in the active medium.

FELs predominate significantly over conventional lasers in their ability to attain a high level of average output power. In the conventional laser, the unused fraction of the pumping power (which significantly exceeds the output radiation power) is dissipated in the active medium. So, the possibility of increasing the average output power is limited by the heat elimination problem. By contrast, in the FEL amplifier the process of electron beam energy conversion to radiation takes place in vacuum. Utilization of the electron beam is a routine problem in the accelerator technique.

The FEL, as a device for converting the net electrical power to the radiation power, can provide, in principle, a high efficiency close to unity. There are no principal physical limitations which prohibit attaining such a high efficiency and it may be achieved with appropriate development of FEL technology. In this sense the situation with the FEL is similar to that with vacuum-tube devices. For instance, the development of the theory and technology of vacuum-tube devices during the last fifty years has offered the possibility of constructing powerful klystrons with an efficiency of about 80%.

1.2 Principle of FEL Operation

We begin with a short introduction to the principle of free electron laser operation. It is mainly addressed to readers with limited knowledge of FEL physics. Fortunately, the principle of FEL operation does not require specific knowledge of physics and can be explained in a very simple way.

As with vacuum-tube devices, FEL devices can be divided into two classes: amplifiers and oscillators (see Fig. 1.1). FEL amplifiers amplify the input electromagnetic wave from an external master oscillator. There is no feedback between the output and the input of the FEL amplifier. The FEL oscillator can be considered as an FEL amplifier with feedback. The radiation in an FEL oscillator grows from fluctuations of the electron beam density. For an FEL oscillator in the optical wavelength range the feedback is carried out by means of an optical resonator which also defines the radiation modes which can be excited in the resonator. When the gain of the radiation per pass exceeds the radiation losses in the resonator, the lasing process occurs.

The key element of a free electron laser is the undulator (or wiggler) which forces the electrons to move along curved periodical trajectories. There are two popular undulator configurations: helical and planar. The helical

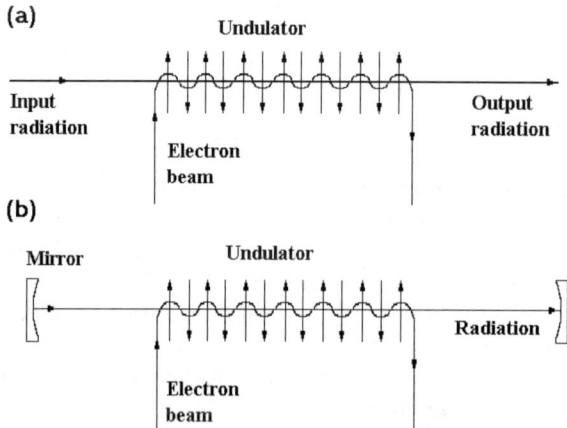

Fig. 1.1. Free electron laser configurations: (**a**) amplifier, (**b**) oscillator

1.2 Principle of FEL Operation

undulator is formed, for instance, by bifilar winding and produces a rotating transverse magnetic field which forces the electrons to move along helical trajectories. The planar undulator is formed by a sequence of dipole magnets of the opposite polarity and produces a linearly polarized transverse magnetic field which forces the electrons to move along sinusoidal trajectories.

To understand the basic principles of FEL operation, let us consider the helical undulator. The magnetic field on the axis of the helical undulator is given by (we neglect the transverse variation of the magnetic field):

$$\boldsymbol{H}_\mathrm{w} = \boldsymbol{e}_x H_\mathrm{w} \cos(k_\mathrm{w} z) - \boldsymbol{e}_y H_\mathrm{w} \sin(k_\mathrm{w} z) \;, \tag{1.1}$$

where $k_\mathrm{w} = 2\pi/\lambda_\mathrm{w}$ is the undulator wavenumber and $\boldsymbol{e}_{x,y}$ are unit vectors directed along the x and y axes of the Cartesian coordinate system (x, y, z). The Lorentz force

$$\boldsymbol{F} = -\frac{e}{c}\boldsymbol{v} \times \boldsymbol{H}_\mathrm{w}$$

is used to derive the equations of motion of electrons with charge $(-e)$ and mass m_e in the presence of the magnetic field

$$m_\mathrm{e}\gamma\frac{\mathrm{d}v_x}{\mathrm{d}t} = \frac{e}{c}v_z H_y = -\frac{e}{c}v_z H_\mathrm{w}\sin(k_\mathrm{w} z) \;, \tag{1.2}$$

$$m_\mathrm{e}\gamma\frac{\mathrm{d}v_y}{\mathrm{d}t} = -\frac{e}{c}v_z H_x = -\frac{e}{c}v_z H_\mathrm{w}\cos(k_\mathrm{w} z) \;, \tag{1.3}$$

where $\gamma = (1 - v^2/c^2)^{-1/2}$ is the relativistic factor and $v^2 = v_x^2 + v_y^2 + v_z^2$. Introducing $\tilde{v} = v_x + iv_y$, $\mathrm{d}z = v_z \mathrm{d}t$, we obtain from (1.2) and (1.3)

$$m_\mathrm{e}\gamma\frac{\mathrm{d}\tilde{v}}{\mathrm{d}z} = -\mathrm{i}\frac{e}{c}(H_x + \mathrm{i}H_y) = -\mathrm{i}\frac{e}{c}H_\mathrm{w}\mathrm{e}^{-\mathrm{i}k_\mathrm{w} z} \;. \tag{1.4}$$

Integration of the latter equation gives

$$\frac{\tilde{v}}{c} = \frac{K}{\gamma}\mathrm{e}^{-\mathrm{i}k_\mathrm{w} z} \;, \tag{1.5}$$

where

$$K = \frac{\lambda_\mathrm{w} e H_\mathrm{w}}{2\pi m_\mathrm{e} c^2}$$

is the undulator parameter. It is useful to present another form of this expression convenient for numerical calculations:

$$K = 0.0934 \times H_\mathrm{w} \text{ (kGs)} \times \lambda_\mathrm{w} \text{ (cm)} \;.$$

The explicit expression for the electron velocity in the field of the helical undulator (1.5) has the form:

$$\boldsymbol{v}_\perp(z) = c\theta_\mathrm{s}[\boldsymbol{e}_x\cos(k_\mathrm{w} z) - \boldsymbol{e}_y\sin(k_\mathrm{w} z)] \;, \tag{1.6}$$

which means that the electron in the undulator moves along the constrained helical trajectory parallel to the z axis. As a rule, the electron rotation angle $\theta_\mathrm{s} = K/\gamma$ is small and the longitudinal electron velocity v_z is close to the velocity of light, $v_z \simeq c$.

Let us consider a circularly polarized electromagnetic wave propagating parallel to the electron beam. The field of the electromagnetic wave has only transverse components, so the energy exchange between the electron and the electromagnetic wave is due to the transverse component of the electron velocity. The rate of electron energy change is

$$\frac{d\mathcal{E}}{dt} = m_e c^2 \frac{d\gamma}{dt} = - e\boldsymbol{v}_\perp \cdot \boldsymbol{E}_\perp ,$$

where $\boldsymbol{E}_\perp$ is the vector of the electric field of the wave:

$$\boldsymbol{E}_\perp = E \{\boldsymbol{e}_x \cos[\omega(z/c - t)] + \boldsymbol{e}_y \sin[\omega(z/c - t)]\} .$$

Remembering that $dz = v_z dt$ we find

$$\begin{aligned}\frac{d\mathcal{E}}{dz} &= - \frac{e}{v_z}(v_x E_x + v_y E_y) \\ &\simeq - e\theta_s E \{\cos(k_w z)\cos[\omega(z/c-t)] - \sin(k_w z)\sin[\omega(z/c-t)]\} \\ &= - e\theta_s E \cos[k_w z + \omega(z/c-t)] \\ &= - e\theta_s E \cos\psi .\end{aligned} \qquad (1.7)$$

The phase ψ has a simple physical interpretation and is equal to the angle between the transverse velocity of the particle, $\boldsymbol{v}_\perp$, and the vector of the electric field, $\boldsymbol{E}_\perp$. For an effective energy exchange between the electron and the wave, the scalar product $(e\boldsymbol{v}_\perp \cdot \boldsymbol{E}_\perp)$ should be kept nearly constant along the whole undulator length, i.e. a synchronism should be provided. From (1.7) this resonance condition may be written as

$$d\psi = k_w dz + \frac{\omega}{c} dz - \omega dt = 0 .$$

Remembering that $dz = v_z dt$, we have

$$k_w + \frac{\omega}{c} - \frac{\omega}{v_z} = 0 . \qquad (1.8)$$

Thus, we see that synchronization takes place when the wave advances the electron beam by one wavelength at one undulator period:

$$\frac{\lambda_w}{v_z} = \frac{\lambda}{c - v_z} , \qquad (1.9)$$

where $\lambda = 2\pi c/\omega$ is the radiation wavelength. Since $v_z \simeq c$, this resonance condition may be written as

$$\lambda \simeq \frac{\lambda_w}{2\gamma_z^2} = \lambda_w \frac{1 + K^2}{2\gamma^2} . \qquad (1.10)$$

When the resonance condition takes place, the electrons with different relative phases with respect to the wave acquire different values of the energy increments (positive or negative), which results in the modulation of the longitudinal velocity of the electrons v_z within the radiation wavelength λ. This velocity modulation is transformed into density modulation of the electron beam. Under some circumstances the electron bunches fall in the decelerating phase of the wave and the average energy of the electrons decreases while the field amplitude of the wave grows due to coherent radiation of the evenly spaced electron bunches, and the process of field amplification takes place.

Let us quantitatively describe the amplification process for the case of a low gain per undulator pass. This situation is typical for an FEL oscillator. The gain can be derived in several different ways, but here a straightforward elementary approach is used. We perform the study in the framework of a one-dimensional model which assumes the amplification of a plane wave by an infinitely wide electron beam. We begin with the equations of the longitudinal motion of the particles. Convenient dynamical variables for the description of this motion are the phase ψ, introduced above, and the kinetic energy of the particle, $\mathcal{E}$. The relevant value of the phase ψ is that at the location of the particle. Hence, the total derivative of ψ is given by

$$\frac{d\psi}{dz} = \frac{\partial \psi}{\partial z} + \frac{\partial \psi}{\partial t}\frac{dt}{dz} = k_w + \frac{\omega}{c} - \frac{\omega}{v_z(\mathcal{E})} \ .$$

Thus, the phase of the particle changes when the resonance condition (1.8) is not satisfied exactly. When the particle energy $\mathcal{E}$ does not differ significantly from a nominal value $\mathcal{E}_0$, the total derivative of ψ can be written as

$$\frac{d\psi}{dz} = k_w + \frac{\omega}{c} - \frac{\omega}{v_z(\mathcal{E}_0)} + \frac{\omega}{v_z^2(\mathcal{E}_0)}\frac{dv_z}{d\mathcal{E}}(\mathcal{E} - \mathcal{E}_0) \ .$$

The rate of electron energy change is given by (1.7). Taking into account that $v_z \simeq c$ and $(dv_z/d\mathcal{E})|_{\mathcal{E}=\mathcal{E}_0} \simeq c/(\gamma_z^2 \mathcal{E}_0)$, we rewrite the equations for the phase and energy in the form:

$$\frac{d\psi}{dz} = C + \frac{\omega}{c\gamma_z^2 \mathcal{E}_0}P \ , \tag{1.11}$$

$$\frac{dP}{dz} = -e\theta_s E \cos\psi \ , \tag{1.12}$$

where $P = \mathcal{E} - \mathcal{E}_0$ is the energy deviation from the nominal value $\mathcal{E}_0$ and

$$C = k_w + \frac{\omega}{c} - \frac{\omega}{v_z(\mathcal{E}_0)}$$

is the detuning from resonance of the particle with energy $\mathcal{E}_0$. Now we differentiate (1.11) with respect to z and substitute dP/dz from (1.12). As a result, we obtain the equation for the phase:

$$\frac{d^2\psi}{dz^2} + \frac{e\omega\theta_s E}{c\gamma_z^2 \mathcal{E}_0}\cos\psi = 0 \ .$$

To make the following consideration more compact we rewrite the latter equation in the normalized form:

$$\frac{d^2\psi}{d\hat{z}^2} + \hat{u}\cos\psi = 0 \ , \tag{1.13}$$

where the longitudinal coordinate is normalized to the undulator length l_w as $\hat{z} = z/l_w$ and $\hat{u} = e\omega_s E l_w^2/(c\gamma_z^2 \mathcal{E}_0)$. The normalized equation (1.13) has the same form as the pendulum equation. The normalized electric field of the wave, $\hat{u}$, can be interpreted as the amplitude of the effective potential.

Let us consider the case when the monoenergetic unmodulated electron beam is fed to the undulator entrance. In this case all the electrons have equal energy $\mathcal{E}_0$ and their phases are uniformly distributed in the interval $(0, 2\pi)$. Thus, the initial conditions at $\hat{z} = 0$ can be written as

$$(d\psi/d\hat{z})|_{\hat{z}=0} = \hat{C} , \qquad 0 \leq \psi_0 \leq 2\pi ,$$

where $\hat{C} = Cl_w$ is the detuning parameter. In the present study we introduce the following simplifying assumptions. First, we study the linear mode of FEL operation (the small-signal regime) which means that the effective potential is small, $\hat{u} \ll 1$. Second, we consider low gain so that $\hat{u}$ in (1.13) may be regarded as a constant along the undulator, $\hat{u} = \hat{u}_{\text{ext}}$. Here $\hat{u}_{\text{ext}} = e\omega_s E_{\text{ext}} l_w^2/(c\gamma_z^2 \mathcal{E}_0)$ and E_{ext} is the field amplitude at the undulator entrance. Integrating (1.13) under these assumptions, we obtain the evolution of the electron phase along the undulator:

$$\psi(\hat{z}, \psi_0) = \psi_0 + \hat{C}\hat{z} + \Delta\psi(\hat{z}, \psi_0) , \tag{1.14}$$

where $\Delta\psi$ is a small phase perturbation given by

$$\Delta\psi(\hat{z}, \psi_0) = -\hat{u}_{\text{ext}} \int_0^{\hat{z}} d\hat{z}' \int_0^{\hat{z}'} d\hat{z}'' \cos(\psi_0 + \hat{C}\hat{z}'')$$

$$= \frac{\hat{u}_{\text{ext}}}{\hat{C}^2}\left[\cos(\psi_0 + \hat{C}\hat{z}) - \cos\psi_0\right] + \frac{\hat{u}_{\text{ext}}\hat{z}}{\hat{C}}\sin\psi_0 . \tag{1.15}$$

The gain per undulator pass is defined as

$$g_s = \frac{(E_{\text{ext}} + \Delta E)^2 - E_{\text{ext}}^2}{E_{\text{ext}}^2} ,$$

where ΔE is the field increment. Since we consider the low-gain limit, the latter expression may be simplified to

$$g_s \simeq \frac{2\Delta E}{E_{\text{ext}}} .$$

To calculate the gain, we use the energy conservation law. The power flow density of the electromagnetic wave is given by the expression:

$$\Pi = \frac{cE^2}{4\pi} .$$

The increment of Π after the undulator pass is

$$\Delta\Pi = \frac{c(E_{\text{ext}} + \Delta E)^2}{4\pi} - \frac{cE_{\text{ext}}^2}{4\pi} \simeq \frac{cE_{\text{ext}}\Delta E}{2\pi} ,$$

and must be equal to

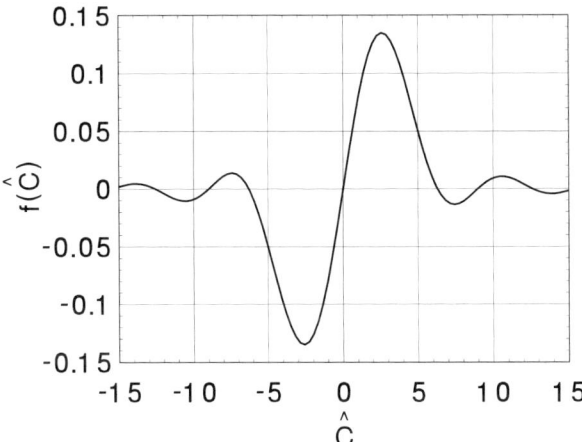

Fig. 1.2. Small-signal gain function, $f(\hat{C})$, versus the reduced detuning $\hat{C}$

$$\Delta \Pi = -\frac{j_0 \langle P \rangle}{e},$$

where $\langle P \rangle$ is the averaged loss of the energy by the electron and j_0 is the beam current density. Thus, the expression for the gain can be written in the form

$$g_s = \frac{2\Delta E}{E_{\text{ext}}} = -\frac{4\pi j_0 \langle P \rangle}{c e E_{\text{ext}}^2}. \tag{1.16}$$

Substituting (1.14) into (1.12) and integrating over z, we obtain the averaged loss of the energy by the electron:

$$\langle P \rangle = -e\theta_s E_{\text{ext}} l_w \left\langle \int_0^1 \cos\left[\psi_0 + \hat{C}\hat{z} + \Delta\psi(\hat{z}, \psi_0)\right] d\hat{z} \right\rangle$$

$$\simeq e\theta_s E_{\text{ext}} l_w \left\langle \int_0^1 \Delta\psi(\hat{z}, \psi_0) \sin(\psi_0 + \hat{C}\hat{z}) d\hat{z} \right\rangle. \tag{1.17}$$

Here the brackets $\langle ... \rangle$ denote averaging over initial phases ψ_0 homogeneously distributed on the interval $(0, 2\pi)$. Performing the integration in (1.17), averaging over the initial phases, and substituting the expression for $\langle P \rangle$ into (1.16), we obtain the final result:

$$g_s = \tau f(\hat{C}). \tag{1.18}$$

Here the gain parameter, τ, and the gain function, $f(\hat{C})$, are given by

$$\tau = \frac{2\pi j_0 \theta_s^2 \omega l_w^3}{c \gamma_z^2 \gamma I_A},$$

$$f(\hat{C}) = \frac{2}{\hat{C}^3}\left(1 - \cos\hat{C} - \frac{\hat{C}}{2}\sin\hat{C}\right) = -2\frac{\mathrm{d}}{\mathrm{d}\hat{C}}\frac{\sin^2(\hat{C}/2)}{\hat{C}^2},$$

where $I_\mathrm{A} = m_e c^3/e \simeq 17$ kA is the Alfven current.

The small-signal gain curve, $f(\hat{C}) = g_\mathrm{s}/\tau$, is shown in Fig. 1.2. The gain curve is antisymmetric and vanishes at exact resonance, $\hat{C} = 0$. It takes a maximal value of 0.135 at the detuning parameter equal to 2.6. The width of the gain curve is $\Delta\hat{C} \simeq \pi$ which corresponds to the frequency bandwidth $\Delta\omega/\omega \simeq (2N_\mathrm{w})^{-1}$ and to the "energy bandwidth" $\Delta\mathcal{E}/\mathcal{E}_0 \simeq (4N_\mathrm{w})^{-1}$. Here N_w is the number of undulator periods.

We deduced the gain formula using a classical treatment. So, the reasonable question arises of whether such a classical model describes correctly the physical processes in the FEL. Simple physical considerations show that such a description is valid for any practical FEL device where the radiation is generated by the electron beam passing a conventional undulator with static magnetic field. Indeed, the amplification process in the free electron laser displays resonance behavior. Effective amplification takes place within a narrow "energy bandwidth" $\Delta\mathcal{E} \simeq \mathcal{E}_0/(4N_\mathrm{w})$. The classical approach can be used when the energy of the radiated photon $\hbar\omega$ is much less than the "energy bandwidth", $\Delta\mathcal{E}$, of the FEL. As a rule, $\Delta\mathcal{E}/\mathcal{E}_0 \simeq 10^{-2}$, the energy range of the FEL photons is 0.01–100 eV, and the energy of the electrons is $\mathcal{E}_0 \simeq 10$–1000 MeV; parameter $\hbar\omega/\Delta\mathcal{E}$ is of the order of 10^{-5} and quantum effects are negligible. On the other hand, many of the first papers on FEL theory were based on a quantum approach. In these papers, for instance, the radiation process was considered as the scattering of virtual photons of the undulator wave in the electron frame of reference and the process of the interaction of the electron with the electromagnetic wave was described as the process of induced radiation and absorption of laser photons. In our opinion, the FEL description in terms of quantum physics is an artificial one. Indeed, the terms "spontaneous" and "induced" radiation are necessary to describe quantum lasers, but this approach is not fruitful for describing vacuum-tube devices to which the FEL belongs. Nevertheless, though giving no principally new results, investigations based on the quantum approach have influenced significantly the terminology of FEL physics and essentially quantum notions such as the Compton regime, the Raman regime and the self-amplified spontaneous emission (SASE) regime are now widely used.

It became a tradition in many popular reviews and books to derive the FEL resonance condition using a Lorentz transformation. Here the interaction of the electron with the combined electromagnetic field of the undulator and the wave is considered in the electron frame of reference. In this approach, the factor $2\gamma_z^2$ appears in the resonance condition as a consequence of the Doppler effect. Our experience has shown that such an introduction to FEL physics usually forces the reader to believe that a description of FEL operation is impossible without detailed knowledge of the special theory of relativity. We have shown above that the resonance condition can be simply derived in the laboratory frame of reference and there is no need to use the laws of relativistic kinematics. In connection with this we should note that the only relativistic formula necessary for the description of the FEL operation is

$$d\boldsymbol{p}/dt = \boldsymbol{F} = -e\boldsymbol{E} - e\frac{\boldsymbol{v}}{c} \times \boldsymbol{H} ,$$

where $\boldsymbol{p} = m_e \gamma \boldsymbol{v}$. In other words, to describe the processes in the FEL, it is necessary only to take into account the relativistic dependence of the electron momentum on the velocity. When the electron may be treated as a point particle, such an approach always gives a reliable way to describe the processes of the electron beam dynamics and radiation in the given electromagnetic fields. Such a situation, for instance, takes place in the theory of particle accelerators.

1.3 Suggested Bibliography

The first successful demonstration of FEL operation in the optical wavelength range was achieved by J.M.J. Madey and co-workers in 1976 [1.1,1.2]. A brief history of free electron lasers and practical details of particular machines may be found in the text by C. Brau [1.3]. Future industrial applications of FELs are analyzed in [1.4]. During the last few years the problem of industrial applications of a ten kW level power FEL has been intensively discussed in the FEL community. Recently an industrial FEL project has been launched by a consortium of industrial firms including DuPont, Xerox, and IBM [1.5]. Complete information about the status of X-ray and VUV free electron laser projects can be found in [1.6–1.9].

2. One-Dimensional Theory of the FEL Amplifier

The problem of electromagnetic wave amplification in the undulator refers to a class of self-consistent problems. It can be separated into two parts:

- solution of the dynamical problem, i.e. finding the motion of the particles under the action of the given electromagnetic fields;
- solution of the electrodynamic problem, i.e. finding the electromagnetic fields generated by a given distribution of charges and currents.

To close the problem, the field equations and equations of motion should be solved simultaneously. In principle, modern supercomputers allow one to perform direct simulation of the FEL process. The results of such simulations depend on a large number of problem parameters. They provide the possibility of obtaining a numerical answer for a specific set of input data, but hardly help to understand the FEL physics. A deeper insight into FEL physics can be obtained only by introducing some simplifying assumptions about the properties of the electron beam and of the electromagnetic field. Theoretical investigation of the free electron laser should be performed in two stages. In the beginning one should study the general properties of the FEL, namely the ideal mechanism of amplification. At the next stage different complications can be introduced into the FEL model allowing it to extend the number of additional effects influencing the operation of the FEL amplifier. These factors can be divided into two groups. The space charge effects and diffraction effects are fundamental. On the other hand, there are a lot of other factors such as energy spread effects, nonideality of the undulator field, etc. The principal difference of the fundamental effects and all the others is that the fundamental effects depend on the same physical parameters as the ideal FEL amplification mechanism itself. Nonfundamental effects depend on the problem parameters additional to those defining the ideal FEL mechanism. The energy spread of electrons in the beam is an example of such a nonfundamental effect.

The one-dimensional model is an important one from the methodological point of view. The following assumptions are used in the one-dimensional, steady-state model of the FEL amplifier:

- the electron beam has a uniform density distribution in the direction perpendicular to the undulator axis;

14   2. One-Dimensional Theory of the FEL Amplifier

– the electrons move along identical trajectories parallel to the undulator axis;
– the amplified wave is a monochromatic plane wave;
– the electron beam is infinitely long.

This model allows one to study the ideal mechanism of amplification. Additionally, space charge and energy spread effects can be studied in the framework of the one-dimensional model. We mentioned above that there are only two fundamental effects in the theory of the FEL amplifier, the space charge effect and the diffraction effect. The latter cannot be studied in the framework of the one-dimensional approximation. The corresponding extension of the FEL amplifier theory, described in Chap. 4, allows one to obtain the applicability region of the one-dimensional approximation. It is shown that there exists a region of physical parameters correctly described by the one-dimensional theory.

We begin our study with the linear mode of FEL amplifier operation when the output radiation power is proportional to the input power. Physically this means that the first harmonic of the beam density modulation significantly dominates the higher harmonics and a linearized Vlasov equation can be used for the description of the beam dynamics. A self-consistent solution of the Vlasov equation and Maxwell's equations leads to a unique integro-differential equation for the amplitude of the electromagnetic field. Solution of the latter equation with appropriate initial conditions for the electron beam and for the electromagnetic wave parameters at the entrance of the undulator allows one to calculate the evolution of the radiation field along the undulator.

In this chapter we solve the initial-value problem by means of the Laplace transform technique. The rigorous results obtained in reduced form furnish universal plots for calculating the output characteristics of the FEL amplifier in the linear mode of operation. These analytical solutions serve as a reliable basis for the development of numerical methods. The analysis of nonlinear processes refers to problems solvable only numerically by a computer. On the other hand, testing of the numerical simulation codes would be difficult without the use of rigorous results of FEL amplifier linear theory as a primary standard.

Similarity techniques play a dominant role in numerical simulation of processes observed in an FEL. For instance, within the scope of the one-dimensional approximation the output characteristics of the FEL amplifier are controlled by eight dimensional parameters of the beam, undulator, and input radiation. The system of self-consistent field equations describing the beam-wave interaction in the undulator may be formulated as a relation between dimensionless quantities. The equations show that a family of similar modes of operation of the FEL amplifier with untapered undulator is controlled by the values of five dimensionless parameters. For the high-gain FEL amplifier, the saturation field amplitude is independent of both undulator length and input signal amplitude. In this prominent practical case the maximum amplifier efficiency is a function of only three dimensionless parameters: detuning, space charge and energy spread.

The calculation scheme of the FEL amplifier output characteristics in saturation mode which is suitable for engineering practice is presented in this chapter. This scheme stems from similarity techniques and numerical simulation results given as design formulae and universal plots. All stages of the numerical experiment, i.e. the physical formulation of the problem, the construction of a mathematical model, the realization of an algorithm and the computation process itself, are discussed consecutively in this book.

A promising way to increase FEL amplifier efficiency up to about unity is the variation of the undulator parameters along its axis (the so-called undulator tapering). Further extensions of similarity techniques for efficiency calculations of the FEL amplifier with a tapered undulator are discussed. A procedure for the calculation of the optimal undulator parameters and FEL amplifier output characteristics with an optimal undulator is also given.

2.1 Linear Mode of Operation

2.1.1 Effective Hamiltonian

The simplicity of the one-dimensional model offers the opportunity for an almost complete analytical description of the linear mode of FEL amplifier operation. We begin the derivation process from "first principles" and write the equations of motion using the Hamiltonian formalism. Let us consider an FEL amplifier with a helical undulator. The magnetic field on the undulator axis has the form:

$$\boldsymbol{H}_\mathrm{w}(z) = \boldsymbol{e}_x H_\mathrm{w} \cos(k_\mathrm{w} z) - \boldsymbol{e}_y H_\mathrm{w} \sin(k_\mathrm{w} z) \ .$$

It is convenient to use the complex representation for the following consideration:

$$H_x + \mathrm{i} H_y = H_\mathrm{w} \exp(-\mathrm{i} k_\mathrm{w} z) \ .$$

The electron, moving in the undulator, interacts with the radiation and space charge fields. The electric field of the amplified wave may be written in the complex form:

$$E_x + \mathrm{i} E_y = \tilde{E}(z) \exp[\mathrm{i}\omega(z/c - t)] \ ,$$

where ω is the frequency of the wave. The complex field amplitude, $\tilde{E}$, does not depend on time at any space point. This corresponds to the formulation of the initial problem with time-independent initial conditions at the undulator entrance at $z = 0$. In the framework of the one-dimensional model the space charge field has only a longitudinal component E_z.

Let us consider the Hamiltonian formalism for the equations of motion. The Hamiltonian is defined as

2. One-Dimensional Theory of the FEL Amplifier

$$\mathcal{H}(p_z, z, t) = \left[(p_z c + eA_z)^2 + e^2(\boldsymbol{A}_\perp + \boldsymbol{A}_w)^2 + m_e^2 c^4\right]^{1/2} - e\phi , \quad (2.1)$$

where p_z is the longitudinal component of the generalized momentum of the particle, $\boldsymbol{A}_\perp$ is the vector potential of the wave, and ϕ and A_z are the scalar potential and the vector potential of the space charge field, respectively. The vector potential of the undulator magnetic field is given by the expression:

$$\boldsymbol{A}_w(z) = -\boldsymbol{e}_z \times \int \boldsymbol{H}_w dz .$$

In the one-dimensional approximation the transverse generalized momentum of the particle is an integral of the motion, and we let it be zero. We transform the Hamiltonian $\mathcal{H}$ from the variables p_z, z and t to variables convenient for describing the amplification process. We use the extended Hamiltonian formalism when t is considered as a canonical coordinate conjugated to the canonical momentum $p_0 = -\mathcal{H}$ (see Appendix A.1). Then we choose the coordinate z as a new time and the phase

$$\psi = k_w z + \omega(z/c - t) \quad (2.2)$$

as a new canonical coordinate. According to formulae (A.1.8) and (A.1.10), transformation (2.2) is canonical when

$$\mathcal{P} = -p_0/\omega , \qquad \mathcal{P}_0 = p_z + (p_0/\omega)(k_w + \omega/c) ,$$

where (p_0, p_z) and $(\mathcal{P}_0, \mathcal{P})$ are the old and new canonical momenta conjugated to (t, z) and (z, ψ), respectively. Hence, the new Hamiltonian $\tilde{\mathcal{H}}(\mathcal{P}, \psi, z) = -\mathcal{P}_0$ is given by

$$\begin{aligned}\tilde{\mathcal{H}}(\mathcal{P}, \psi, z) &= (k_w + \omega/c)\mathcal{P} - p_z(\mathcal{P}, z, \psi) \\ &= (k_w + \omega/c)\mathcal{P} + eA_z/c \\ &\quad - c^{-1}\left[(\mathcal{P}\omega + e\phi)^2 - e^2(\boldsymbol{A}_\perp + \boldsymbol{A}_w)^2 - m_e^2 c^4\right]^{1/2} ,\end{aligned} \quad (2.3)$$

and the canonical equations of motion have the form

$$d\psi/dz = \partial\tilde{\mathcal{H}}/\partial\mathcal{P} , \qquad d\mathcal{P}/dz = -\partial\tilde{\mathcal{H}}/\partial\psi .$$

In the framework of the one-dimensional model the vector potential of the wave, $\boldsymbol{A}_\perp(z)$, and of the undulator magnetic field, $\boldsymbol{A}_w(z)$, are gauge invariants. This is a natural consequence of the fact that they are perpendicular to the z axis and depend on the z coordinate only. The scalar potential, ϕ, and the longitudinal component of the vector potential, A_z, can be subjected to the following gauge transformation:

$$\phi \Rightarrow \phi' = \phi - c^{-1}\partial\tilde{\chi}/\partial t , \qquad A_z \Rightarrow A'_z = A_z + \partial\tilde{\chi}/\partial z ,$$

where $\tilde{\chi}$ is an arbitrary function of the z coordinate and of time t. In this case the longitudinal component of the electric field,

$$E_z = -\partial\phi/\partial z - c^{-1}\partial A_z/\partial t ,$$

remains unchanged. We choose the following function for the gauge transformation:

$$\tilde{\chi} = c \int \mathrm{d}t \phi(z,t) \ .$$

At such a choice for the gauge transformation, the scalar potential is equal to zero and the space charge field is uniquely described by the longitudinal component of the vector potential. In other words, this component of the vector potential entering the Hamiltonian can be expressed in terms of the space charge field E_z. Thus, we can rewrite the expression for the Hamiltonian (2.3) in the following form:

$$\tilde{\mathcal{H}} = (k_\mathrm{w} + \omega/c)\mathcal{E}/\omega - c^{-1}[\mathcal{E}^2 - e^2(\boldsymbol{A}_\perp + \boldsymbol{A}_\mathrm{w})^2 - m_\mathrm{e}^2 c^4]^{1/2}$$
$$+ \frac{e}{\omega}\int \mathrm{d}\psi E_z(z,\psi) \ . \tag{2.4}$$

The canonical momentum, $\mathcal{P}$, is equal to the electron kinetic energy $\mathcal{E}$ divided by the frequency ω at the chosen gauge transformation.

It is convenient to simplify the expression for the Hamiltonian. First, we expand the Hamiltonian $\tilde{\mathcal{H}}$ to the first order of the radiation field amplitude $\boldsymbol{A}_\perp$, assuming that $|\boldsymbol{A}_\perp| \ll |\boldsymbol{A}_\mathrm{w}|$:

$$\tilde{\mathcal{H}} = \frac{\mathcal{E}}{\omega}(k_\mathrm{w} + \omega/c) - \frac{1}{c}\left[\mathcal{E}^2 - e^2|\boldsymbol{A}_\mathrm{w}|^2 - m_\mathrm{e}^2 c^4\right]^{1/2}$$
$$+ \frac{e^2}{c}(\boldsymbol{A}_\perp \cdot \boldsymbol{A}_\mathrm{w})\left[\mathcal{E}^2 - e^2|\boldsymbol{A}_\mathrm{w}|^2 - m_\mathrm{e}^2 c^4\right]^{-1/2} + \frac{e}{\omega}\int \mathrm{d}\psi E_z \ . \tag{2.5}$$

Second, we expand the Hamiltonian (2.5) up to the second order over the energy deviation $\mathcal{E}$ from the nominal value $\mathcal{E}_0$:

$$\tilde{\mathcal{H}}\omega = H(P,\psi,z)$$
$$= CP + \frac{\omega}{2c\gamma_z^2 \mathcal{E}_0}P^2 - \left(U\mathrm{e}^{i\psi} + U^*\mathrm{e}^{-i\psi}\right)(1 - P/\mathcal{E}_0) + \int \mathrm{d}\psi e E_z \ , \tag{2.6}$$

where $P = \mathcal{E} - \mathcal{E}_0$, $C = k_\mathrm{w} - \omega/(2c\gamma_z^2)$ is the detuning of the electron with the nominal energy $\mathcal{E}_0$,

$$U = -\frac{e\theta_\mathrm{s}\tilde{E}(z)}{2\mathrm{i}} \tag{2.7}$$

is the complex amplitude of the effective potential[1] of the particle interaction with the electromagnetic wave, and

$$\theta_\mathrm{s} = eH_\mathrm{w}/(\mathcal{E}_0 k_\mathrm{w}) = K/\gamma \ , \qquad \gamma_z^{-2} = \gamma^{-2} + \theta_\mathrm{s}^2 \ , \qquad \gamma = \mathcal{E}_0/(m_\mathrm{e}c^2) \ .$$

When writing down (2.6) we have used the fact that in the one-dimensional approximation the transverse vector potential and the electric field of the electromagnetic wave are connected by the relation $cE_{x,y} = -\partial A_{x,y}/\partial t$. In addition, the ultrarelativistic approximation has been used, $\gamma_z^2 \gg 1$.

---

[1] The effective potential is a synonym for the ponderomotive potential. The latter definition is used frequently in the literature.

2.1.2 Self-Consistent Equations

The motion of the particles is determined not only by the external electromagnetic fields, but also by the fields produced by the electron beam itself. The simplest description of this process can be obtained under the assumption that the interaction of a single electron with collective fields produced by the whole electron beam is much stronger than its interaction with its nearest neighbors. In this case collective fields (radiation and space charge fields) can be treated in the same way as external fields. Such an approximation allows one to use the Vlasov equation for describing the evolution of the electron beam distribution function $f(P, \psi, z)$:

$$\frac{\partial f}{\partial z} + \frac{\partial H}{\partial P}\frac{\partial f}{\partial \psi} - \frac{\partial H}{\partial \psi}\frac{\partial f}{\partial P} = 0 \ . \tag{2.8}$$

Equation (2.8) is Liouville's equation in the phase space (P, ψ). Its physical sense is that the distribution function f does not change along the trajectory of the particle despite the presence of the collective interaction. Simultaneous solution of (2.8) and Maxwell's equations allows one to find the evolution of the collective fields and distribution function. We solve (2.8) using the perturbation method. For an electron beam with a small density perturbation we seek solutions for f and E_z in the form

$$f = f_0 + \tilde{f}_1 e^{i\psi} + \tilde{f}_1^* e^{-i\psi} \ , \qquad E_z = \tilde{E}_z e^{i\psi} + \tilde{E}_z^* e^{-i\psi} \ .$$

Here $f_0(P)$ is the unperturbed distribution function and $\tilde{f}_1(P, z)$ is the small perturbation ($|\tilde{f}_1| \ll |f_0|$). Using (2.6) and (2.8) we get:[2]

$$\frac{\partial \tilde{f}_1}{\partial z} + \mathrm{i}\left[C + \omega P/(c\gamma_z^2 \mathcal{E}_0)\right] \tilde{f}_1 + \left(\mathrm{i}U - e\tilde{E}_z\right) \frac{\partial f_0}{\partial P} = 0 \ . \tag{2.9}$$

We assume that the electron beam is modulated neither in velocity, nor in density at the undulator entrance, i.e.

$$\tilde{f}_1|_{z=0} = 0 \ , \qquad f_0 = n_0 F(P) \ , \qquad \int F \mathrm{d}P = 1 \ , \tag{2.10}$$

where n_0 is the beam density. The solution of (2.9) with the initial condition $\tilde{f}_1|_{z=0} = 0$ has the form

$$\tilde{f}_1 = -n_0 \frac{\mathrm{d}F}{\mathrm{d}P} \int_0^z \mathrm{d}z' (\mathrm{i}U - e\tilde{E}_z) \exp\left\{\mathrm{i}\left[C + \omega P/(c\gamma_z^2 \mathcal{E}_0)\right](z' - z)\right\} \ . \tag{2.11}$$

In the ultrarelativistic approximation, $v_z \simeq c$, the beam current density and the perturbation of the distribution function $\tilde{f}_1$ are connected by the relation:

$$j_z = -j_0 + \tilde{j}_1 e^{i\psi} + \text{C.C.} \ , \qquad \tilde{j}_1 \simeq -ec \int \tilde{f}_1 \mathrm{d}P \ ,$$

---

[2] We neglect here the terms $U\tilde{f}_1/\mathcal{E}_0$ and $(\mathrm{i}U - e\tilde{E}_z)(\mathrm{d}\tilde{f}_1/\mathrm{d}P)$ with respect to the term $(\mathrm{i}U - e\tilde{E}_z)(\mathrm{d}f_0/\mathrm{d}P)$.

2.1 Linear Mode of Operation

where $-j_0 \simeq -ecn_0$ is the longitudinal component of the beam current density at the undulator entrance.

In the framework of the one-dimensional model we can derive from Maxwell's equations the following equation for E_z:

$$\partial E_z/\partial t = -\mathrm{i}\omega \tilde{E}_z \mathrm{e}^{\mathrm{i}\psi} + \text{C.C.} = -4\pi \tilde{j}_1 \mathrm{e}^{\mathrm{i}\psi} + \text{C.C.} \tag{2.12}$$

As a result, we have

$$\tilde{E}_z = -\mathrm{i}4\pi \tilde{j}_1(z)/\omega \ . \tag{2.13}$$

Substituting (2.13) into (2.11) and integrating over P we obtain the following integral equation for $\tilde{j}_1(z)$:

$$\tilde{j}_1(z) = \mathrm{i}j_0 \int_0^z \mathrm{d}z' \left[U + \frac{4\pi e \tilde{j}_1(z')}{\omega} \right]$$

$$\times \int \mathrm{d}P \frac{\mathrm{d}F(P)}{\mathrm{d}P} \exp\left\{ \mathrm{i} \left[C + \frac{\omega P}{c\gamma_z^2 \mathcal{E}_0} \right] (z' - z) \right\} . \tag{2.14}$$

The latter equation represents the final result of the analysis of the dynamical problem. The case $C \ll k_\mathrm{w}$ will be of special interest for later work. If the complex amplitude $\tilde{E}(z)$ is a slowly varying function on the scale of the undulator period, it follows from (2.14) that the complex amplitude $\tilde{j}_1(z)$ is a slowly varying function, too.

One more relation between the amplitudes $\tilde{j}_1(z)$ and $\tilde{E}(z)$ is given by the solution of the electrodynamic problem. The vector potential of the radiation field can be found from the wave equation

$$\partial^2 \boldsymbol{A}_\perp/\partial z^2 - c^{-2}\partial^2 \boldsymbol{A}_\perp/\partial t^2 = -(4\pi/c)\boldsymbol{j}_\perp \ ,$$

where the perturbed transverse current density is given by

$$j_x + \mathrm{i}j_y = \theta_\mathrm{s} \exp(-\mathrm{i}k_\mathrm{w}z)(\tilde{j}_1 \mathrm{e}^{\mathrm{i}\psi} + \text{C.C.}) \ .$$

We seek the solution for $\boldsymbol{A}_\perp(z,t)$ in the form

$$A_{x,y} = \tilde{A}_{x,y}(z) \exp\left[\mathrm{i}\omega(z/c - t)\right] + \text{C.C.}$$

From the wave equation we have:[3]

$$\exp\left[\mathrm{i}\omega(z/c - t)\right] \left\{ \frac{2\mathrm{i}\omega}{c} \frac{\partial}{\partial z} \begin{pmatrix} \tilde{A}_x \\ \tilde{A}_y \end{pmatrix} + \frac{\partial^2}{\partial z^2} \begin{pmatrix} \tilde{A}_x \\ \tilde{A}_y \end{pmatrix} \right\} + \text{C.C.}$$

$$= -\frac{4\pi\theta_\mathrm{s}}{c} \begin{pmatrix} \cos(k_\mathrm{w}z) \\ -\sin(k_\mathrm{w}z) \end{pmatrix} (\tilde{j}_1 \mathrm{e}^{\mathrm{i}\psi} + \text{C.C.}) \ . \tag{2.15}$$

To simplify this equation, we make the following assumptions:

– The complex amplitude $\tilde{j}_1(z)$ is a slowly varying function on the scale of the undulator period.

---

[3] One can see from (2.15) that the radiation of the modulated electron beam in the undulator has resonance behaviour and takes its maximum at $\tilde{j}_1(z) = \text{const}$.

2. One-Dimensional Theory of the FEL Amplifier

– The distance, z, is much larger than the undulator period.

Under these assumptions we can neglect rapidly oscillating terms in the right-hand side, and the second derivatives of $\tilde{A}_{x,y}$ in the left-hand side of (2.15). Finally, the wave equation can be written in the form:

$$\mathrm{d}\tilde{E}/\mathrm{d}z = -2\pi\theta_s c^{-1}\tilde{j}_1(z) \ . \tag{2.16}$$

Here we have used the relation $cE_{x,y} = -\partial(A_{x,y})/\partial t$ connecting the vector potential and the electric field of the electromagnetic wave in the one-dimensional approximation.

Thus, we have obtained two equations, (2.14) and (2.16), for the slowly varying amplitudes, $\tilde{E}(z)$ and $\tilde{j}_1(z)$, which should be solved simultaneously. Substituting (2.14) into the right-hand side of (2.16), we obtain a single integro-differential equation for the field amplitude $\tilde{E}$:

$$\frac{\mathrm{d}\tilde{E}}{\mathrm{d}z} = \frac{\pi e j_0 \theta_s^2}{c} \int_0^z \mathrm{d}z' \left\{ \tilde{E}(z') + \frac{4\mathrm{i}c}{\omega\theta_s^2}\frac{\mathrm{d}\tilde{E}(z')}{\mathrm{d}z'} \right\}$$

$$\times \int_{-\infty}^{\infty} \mathrm{d}P \frac{\mathrm{d}F(P)}{\mathrm{d}P} \exp\left\{ \mathrm{i}\left(C + \frac{\omega P}{c\gamma_z^2 \mathcal{E}_0}\right)(z'-z) \right\} \ . \tag{2.17}$$

The next important step is to write down a self-consistent equation in dimensionless form. We introduce the gain parameter Γ:

$$\Gamma = \left[\frac{\pi j_0 \theta_s^2 \omega}{c\gamma_z^2 \gamma I_A}\right]^{1/3} , \tag{2.18a}$$

where $I_A = m_e c^3/e \simeq 17$ kA is the Alfven current. The longitudinal coordinate z is normalized via the gain parameter as

$$\hat{z} = \Gamma z \ . \tag{2.18b}$$

The detuning parameter $\hat{C}$ and the space charge parameter $\hat{\Lambda}_p^2$ are defined as

$$\hat{C} = C/\Gamma \ , \qquad \hat{\Lambda}_p^2 = \Lambda_p^2/\Gamma^2 \ , \tag{2.18c}$$

where

$$\Lambda_p = \left[(4\pi j_0)/(\gamma_z^2 \gamma I_A)\right]^{1/2}$$

is the wavenumber of the longitudinal plasma oscillations. The energy deviation $P = \mathcal{E} - \mathcal{E}_0$ is normalized as

$$\hat{P} = (\mathcal{E} - \mathcal{E}_0)/(\rho \mathcal{E}_0) \ , \tag{2.18d}$$

and the efficiency parameter ρ is given by the expression

$$\rho = \gamma_z^2 \Gamma c/\omega \ . \tag{2.18e}$$

Using this normalization procedure we rewrite (2.17) in the reduced form:

$$\frac{d\tilde{E}}{d\hat{z}} = \int_0^{\hat{z}} d\hat{z}' \left\{ \tilde{E}(\hat{z}') + i\hat{\Lambda}_p^2 \frac{d\tilde{E}(\hat{z}')}{d\hat{z}'} \right\}$$

$$\times \int_{-\infty}^{\infty} d\hat{P} \frac{d\hat{F}}{d\hat{P}} \exp\left\{ i\left(\hat{P} + \hat{C}\right)(\hat{z}' - \hat{z}) \right\}, \quad (2.19)$$

where $\hat{F}(\hat{P})$ is the distribution function in the reduced momentum $\hat{P}$ satisfying the normalization condition $\int \hat{F}(\hat{P}) d\hat{P} = 1$. When the energy spread in the electron beam is Gaussian,

$$F(\mathcal{E} - \mathcal{E}_0) = \left(2\pi \langle (\Delta\mathcal{E})^2 \rangle\right)^{-1/2} \exp\left(-\frac{(\mathcal{E} - \mathcal{E}_0)^2}{2\langle (\Delta\mathcal{E})^2 \rangle}\right), \quad (2.20a)$$

the corresponding distribution in the reduced canonical momentum $\hat{P}$ has the form:

$$\hat{F}(\hat{P}) = \left(2\pi \hat{\Lambda}_T^2\right)^{-1/2} \exp\left(-\frac{\hat{P}^2}{2\hat{\Lambda}_T^2}\right). \quad (2.20b)$$

The energy spread parameter, $\hat{\Lambda}_T^2$, is related to the rms energy spread, $\langle (\Delta\mathcal{E})^2 \rangle$, as

$$\hat{\Lambda}_T^2 = \langle (\Delta\mathcal{E})^2 \rangle / (\rho^2 \mathcal{E}_0^2). \quad (2.20c)$$

The self-consistent field equation (2.19) has been derived using specific initial conditions (2.10) at the undulator entrance. Namely, it describes the amplification process when an electromagnetic wave and an unmodulated electron beam are fed to the undulator entrance. A more general study of the initial conditions will be given below.

It is relevant to comment on the region of validity of the self-consistent equation (2.19). This equation has been derived using the assumption that the complex amplitudes, $\tilde{E}(z)$ and $\tilde{j}_1(z)$, are slowly varying functions on the scale of the undulator period, i.e.

$$\Gamma |d\tilde{E}/d\hat{z}| \ll k_w |\tilde{E}|, \quad \Gamma |d\tilde{j}_1/d\hat{z}| \ll k_w |\tilde{j}_1|. \quad (2.21)$$

Remembering the definition of the efficiency parameter, ρ, we rewrite this condition as

$$\rho |d\tilde{E}/d\hat{z}| \ll |\tilde{E}|, \quad \rho |d\tilde{j}_1/d\hat{z}| \ll |\tilde{j}_1|. \quad (2.22)$$

It can be shown that the latter conditions lead to the following constraints:

$$\left(\rho, \quad \rho\hat{\Lambda}_T, \quad \rho\hat{\Lambda}_p, \quad \rho\hat{C}\right) \ll 1. \quad (2.23)$$

It is seen that the value of the efficiency parameter, ρ, must be much less than unity in the applicability region of the FEL theory described in our book.[4]

---

[4] When deriving equation of motion, we expanded the Hamiltonian (2.4) to the first order of the radiation field amplitude. It can be shown that this expansion is within the constraints given by (2.23).

2.1.3 Solution of the Initial-Value Problem by the Laplace Technique

Equation (2.19) is an integro-differential equation for the field amplitude $\tilde{E}(\hat{z})$ and can be solved by the Laplace transform technique. The Laplace transformation of the field amplitude $\tilde{E}$ is given by the integral ($\operatorname{Re} p > 0$)

$$\bar{E}(p) = \int_0^\infty \mathrm{d}\hat{z}\, \exp(-p\hat{z})\tilde{E}(\hat{z}) \,.$$

Multiplying (2.19) by $\exp(-p\hat{z})$ with $\operatorname{Re} p > 0$ and integrating over $\hat{z}$ from 0 to ∞, we obtain the equation:

$$p\bar{E}(p) - \tilde{E}_{\text{ext}} = \left[\bar{E}(p) + \mathrm{i}\hat{\Lambda}_{\text{p}}^2\left(p\bar{E}(p) - \tilde{E}_{\text{ext}}\right)\right] \int_{-\infty}^{\infty} \mathrm{d}\hat{P}\, \frac{\hat{F}'(\hat{P})}{p + \mathrm{i}(\hat{P} + \hat{C})}, \quad (2.24)$$

where $\hat{F}'(\hat{P}) = \mathrm{d}\hat{F}/\mathrm{d}\hat{P}$. Here we have used the relations:

$$\int_0^\infty \mathrm{d}\hat{z}\, \exp(-p\hat{z})\mathrm{d}\tilde{E}(\hat{z})/\mathrm{d}\hat{z} = p\bar{E}(p) - \tilde{E}_{\text{ext}} \,,$$

$$\int_0^\infty \mathrm{d}\hat{z}\, \exp(-p\hat{z}) \left\{ \int_0^{\hat{z}} \mathrm{d}\hat{z}'\, \tilde{E}(\hat{z}') \exp\left[\mathrm{i}\left(\hat{P} + \hat{C}\right)(\hat{z}' - \hat{z})\right] \right\}$$

$$= \int_0^\infty \mathrm{d}\hat{z}'\, \tilde{E}(\hat{z}') \exp\left[\mathrm{i}\left(\hat{P} + \hat{C}\right)\hat{z}'\right] \int_{\hat{z}'}^{\infty} \mathrm{d}\hat{z}\, \exp\left[-\left(p + \mathrm{i}\hat{P} + \mathrm{i}\hat{C}\right)\hat{z}\right]$$

$$= \frac{\bar{E}(p)}{p + \mathrm{i}(\hat{P} + \hat{C})}, \qquad (2.25)$$

where $\tilde{E}_{\text{ext}}$ is the amplitude of the radiation field at the undulator entrance at $\hat{z} = 0$. To be specific, we assume $\tilde{E}_{\text{ext}}$ to be real and positive, $\tilde{E}_{\text{ext}} = E_{\text{ext}} > 0$.

The solution of (2.24) is given by the expression

$$\bar{E}(p) = E_{\text{ext}} \left[p - \frac{\hat{D}}{1 - \mathrm{i}\hat{\Lambda}_{\text{p}}^2\hat{D}}\right]^{-1}, \qquad (2.26)$$

where

$$\hat{D} = \int_{-\infty}^{\infty} \mathrm{d}\hat{P}\, \frac{\hat{F}'(\hat{P})}{p + \mathrm{i}(\hat{P} + \hat{C})} \,. \qquad (2.27)$$

The field amplitude $\tilde{E}(\hat{z})$ is calculated by means of the inverse Laplace transformation of (2.26):

$$\tilde{E}(\hat{z}) = \frac{1}{2\pi i} \int_{\gamma'-i\infty}^{\gamma'+i\infty} d\lambda \bar{E}(\lambda) \exp(\lambda\hat{z})$$

$$= \frac{E_{\text{ext}}}{2\pi i} \int_{\gamma'-i\infty}^{\gamma'+i\infty} d\lambda \exp(\lambda\hat{z}) \left[\lambda - \frac{\hat{D}}{1 - i\hat{\Lambda}_p^2 \hat{D}}\right]^{-1}. \quad (2.28)$$

The integration in the complex λ plane runs parallel to the imaginary axis. The constant γ' is a real positive number, larger than the real parts of all the singularities of the integrand. The linear integral (2.28) in some cases can be calculated by closing the integration contour by a semicircle at infinity in the left half-plane with subsequent application of Cauchy's residue theorem. For this trick to be applicable to integral (2.28), its integrand must have an analytic continuation into the left half-plane, and the coefficient of $\exp(\lambda\hat{z})$ must satisfy the conditions of Jordan's lemma.

According to (2.27), the function $\hat{D}$ is given as a complex integral and has a discontinuity on the imaginary λ axis. This integral becomes an analytic function if, following the Landau method, the function $\hat{D}$ is defined as

$$\hat{D} = \begin{cases} \displaystyle\int_{-\infty}^{\infty} d\hat{P} \frac{\hat{F}'(\hat{P})}{\lambda + i(\hat{P} + \hat{C})} & \text{for } \operatorname{Re}\lambda > 0, \\[2ex] \mathcal{P} \displaystyle\int_{-\infty}^{\infty} d\hat{P} \frac{\hat{F}'(\hat{P})}{\lambda + i(\hat{P} + \hat{C})} + \pi \hat{F}'(i\lambda - \hat{C}) & \text{at } \operatorname{Re}\lambda = 0, \\[2ex] \displaystyle\int_{-\infty}^{\infty} d\hat{P} \frac{\hat{F}'(\hat{P})}{\lambda + i(\hat{P} + \hat{C})} + 2\pi \hat{F}'(i\lambda - \hat{C}) & \text{for } \operatorname{Re}\lambda < 0. \end{cases} \quad (2.29)$$

Here $\mathcal{P}(\ldots)$ denotes the principal value. If the distribution function $\hat{F}(\hat{P})$ is such that the coefficient

$$\left[\lambda - \frac{\hat{D}}{1 - i\hat{\Lambda}_p^2 \hat{D}}\right]^{-1}$$

of $\exp(\lambda\hat{z})$ in the integrand in (2.28) satisfies the conditions of Jordan's lemma, the radiation field can be represented as a superposition of partial waves. Using Cauchy's residue theorem, we can write the following expression for $\tilde{E}(\hat{z})$:

$$\tilde{E}(\hat{z}) = E_{\text{ext}} \sum_j \exp(\lambda_j \hat{z}) \left[1 - \frac{\hat{D}'_j}{(1 - i\hat{\Lambda}_p^2 \hat{D}_j)^2}\right]^{-1}, \quad (2.30)$$

where

$$\hat{D}_j = \hat{D}|_{\lambda=\lambda_j}, \qquad \hat{D}'_j = \left.\frac{d\hat{D}}{d\lambda}\right|_{\lambda=\lambda_j},$$

the function $\hat{D}$ is given by expression (2.29), and λ_j is the jth root of the equation

$$\lambda - \frac{\hat{D}}{1 - i\hat{\Lambda}_{\mathrm{p}}^2 \hat{D}} = 0 \ . \tag{2.31}$$

Cold Electron Beam. In the limit of a small energy spread the distribution function $\hat{F}$ can be replaced by the delta function, $\hat{F}(\hat{P}) = \delta(\hat{P})$. In this case the function $\hat{D}$ is given by the expression

$$\hat{D} = i\left(\lambda + i\hat{C}\right)^{-2}$$

in the entire complex λ plane. Since the coefficient of $\exp(\lambda \hat{z})$ in (2.28),

$$\left[\lambda - \frac{\hat{D}}{1 - i\hat{\Lambda}_{\mathrm{p}}^2 \hat{D}}\right]^{-1} = \left[\lambda - \frac{i}{\left(\lambda + i\hat{C}\right)^2 + \hat{\Lambda}_{\mathrm{p}}^2}\right]^{-1} ,$$

is of the order of $O(\lambda^{-1})$ as $|\lambda| \to \infty$, the conditions of Jordan's lemma are satisfied. According to (2.30), the radiation field can be written as a superposition of partial waves:

$$\tilde{E}(\hat{z}) = E_{\mathrm{ext}} \sum_j \frac{\exp(\lambda_j \hat{z})}{1 - 2i(\lambda_j + i\hat{C})\lambda_j^2} ,$$

where λ_j are the roots of the cubic equation

$$\lambda = i\left[\left(\lambda + i\hat{C}\right)^2 + \hat{\Lambda}_{\mathrm{p}}^2\right]^{-1} . \tag{2.32}$$

Using the relations between the roots of the cubic equation (2.32),

$$\lambda_1 \lambda_2 \lambda_3 = i \ ,$$
$$\lambda_1 \lambda_2 + \lambda_2 \lambda_3 + \lambda_1 \lambda_3 = -\hat{C}^2 + \hat{\Lambda}_{\mathrm{p}}^2 \ ,$$
$$\lambda_1 + \lambda_2 + \lambda_3 = -2i\hat{C} \ ,$$

we obtain an expression for the field amplitude:

$$\tilde{E}(\hat{z}) = E_{\mathrm{ext}} \left[\frac{\lambda_2 \lambda_3 \exp(\lambda_1 \hat{z})}{(\lambda_1 - \lambda_2)(\lambda_1 - \lambda_3)} + \frac{\lambda_1 \lambda_3 \exp(\lambda_2 \hat{z})}{(\lambda_2 - \lambda_3)(\lambda_2 - \lambda_1)} \right.$$
$$\left. + \frac{\lambda_1 \lambda_2 \exp(\lambda_3 \hat{z})}{(\lambda_3 - \lambda_1)(\lambda_3 - \lambda_2)} \right] . \tag{2.33}$$

Exactly at resonance, $\hat{C} = 0$, for the case of a negligibly small space charge field, $\hat{\Lambda}_{\mathrm{p}}^2 \to 0$, the solution for $\tilde{E}(\hat{z})$ takes the form

$$\tilde{E}(\hat{z}) = \frac{E_{\mathrm{ext}}}{3} \left[\exp\left(\frac{\sqrt{3} + i}{2} \hat{z}\right) + \exp\left(\frac{-\sqrt{3} + i}{2} \hat{z}\right) + \exp(-i\hat{z}) \right] . \tag{2.34}$$

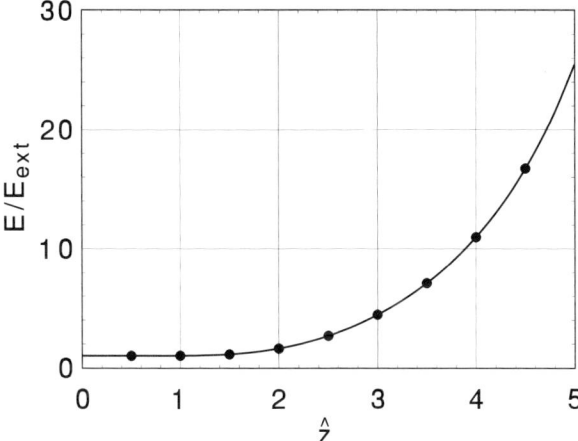

Fig. 2.1. Field gain E/E_{ext} versus the reduced undulator length. The solid curve is calculated with the analytical formula (2.34) and the circles are the results of numerical simulations with (2.93) and (2.94). Here $\hat{C} = 0$, $\hat{\Lambda}_{\text{p}}^2 \to 0$ and $\hat{\Lambda}_{\text{T}}^2 = 0$

We define the field gain as E/E_{ext}, where $E = |\tilde{E}|$. Figure 2.1 illustrates the dependence of the field gain on the reduced undulator length $\hat{z}$. In the high-gain limit the contribution of the growing partial wave dominates all the other terms in (2.34), so we can write an asymptotic expression for $\tilde{E}(\hat{z})$ ($\hat{z} \gg 1$):

$$\tilde{E}(\hat{z}) = \frac{E_{\text{ext}}}{3} \exp\left(\frac{\sqrt{3}+\mathrm{i}}{2}\hat{z}\right) . \tag{2.35}$$

The power gain, $G = |\tilde{E}|^2/E_{\text{ext}}^2$, is an important characteristic of the amplifier. An analytical expression for the power gain of the FEL amplifier operating in the linear regime can be found using (2.34). When the FEL amplifier is tuned to exact resonance, $\hat{C} = 0$, and when the space charge field can be neglected, $\hat{\Lambda}_{\text{p}}^2 \to 0$, the power gain is

$$G = \frac{1}{9}\left[1 + 4\cosh\frac{\sqrt{3}}{2}\hat{z}\left(\cosh\frac{\sqrt{3}}{2}\hat{z} + \cos\frac{3}{2}\hat{z}\right)\right] .$$

Using practical units we can write the asymptotic expression for the power gain G in the high-gain limit ($\hat{z} \gg 1$):

$$G(\text{dB}) = 10\log\left(|\tilde{E}|^2/E_{\text{ext}}^2\right) = 7.5\hat{z} - 9.5 .$$

In the limit of a negligibly small space charge field, $\hat{\Lambda}_{\text{p}}^2 \to 0$, the eigenvalue equation (2.32) takes the form:

$$\lambda\left(\lambda + \mathrm{i}\hat{C}\right)^2 = \mathrm{i} . \tag{2.36}$$

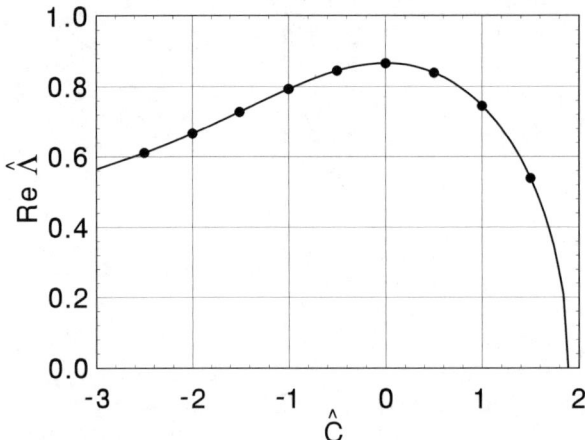

Fig. 2.2. Reduced field growth rate $\operatorname{Re}\hat{\Lambda}$ versus the detuning parameter $\hat{C}$. The solid curve is the solution of the eigenvalue equation (2.36) and the circles are the results of numerical simulations with (2.93) and (2.94). Here $\hat{\Lambda}_p^2 \to 0$ and $\hat{\Lambda}_T^2 = 0$

It is clearly seen from this equation that the propagation constants of the partial waves λ_1, λ_2, and λ_3 are universal functions of the detuning parameter $\hat{C}$ only. Investigating the roots of the eigenvalue equation (2.36) we find that one of the partial waves grows, another decays exponentially, and the third one oscillates when the detuning parameter is $\hat{C} < 3/2^{2/3} \simeq 1.89$. We denote the propagation constant of the growing partial wave as $\hat{\Lambda}$. When the FEL amplifier is tuned to $\hat{C} < 1.89$, the asymptotic expression for the power gain in the high-gain limit can be written in the form:

$$G = A\exp\left(2\operatorname{Re}\hat{\Lambda}\hat{z}\right), \qquad (2.37)$$

where the field growth rate, $\operatorname{Re}\hat{\Lambda}$, and input coupling factor, A, are universal functions of the detuning parameter $\hat{C}$. Graphs of these functions are shown in Figs. 2.2 and 2.3. The maximal value of the field growth rate is achieved at exact resonance, at $\hat{C} = 0$. At small deviations of the detuning, $\hat{C}^2 \ll 1$, the field growth rate is approximately equal to

$$\operatorname{Re}\hat{\Lambda} \simeq \frac{\sqrt{3}}{2}\left[1 - \frac{1}{9}\hat{C}^2\right]. \qquad (2.38)$$

At large negative values of the detuning parameter ($\hat{C} < 0$, $|\hat{C}| \gg 1$) the asymptotic expressions for the field growth rate and for the input coupling factor have the form:

$$\operatorname{Re}\hat{\Lambda} \simeq |\hat{C}|^{-1/2}, \qquad A \simeq (4|\hat{C}|^3)^{-1}.$$

The amplification process displays resonance behavior and the power gain depends strongly on the value of the detuning parameter $\hat{C}$. The notion of the

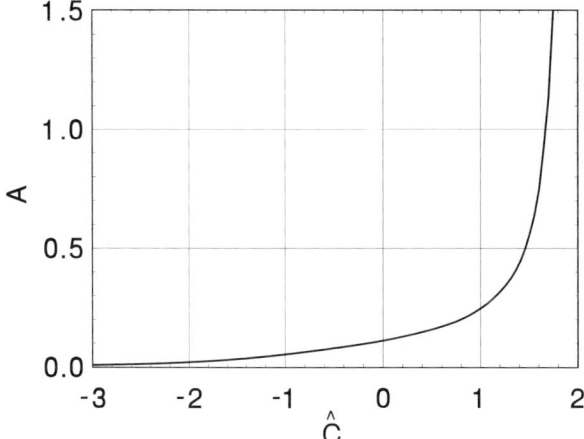

Fig. 2.3. Input coupling factor A entering (2.37) versus the detuning parameter $\hat{C}$. Here $\hat{\Lambda}_\mathrm{p}^2 \to 0$ and $\hat{\Lambda}_\mathrm{T}^2 = 0$

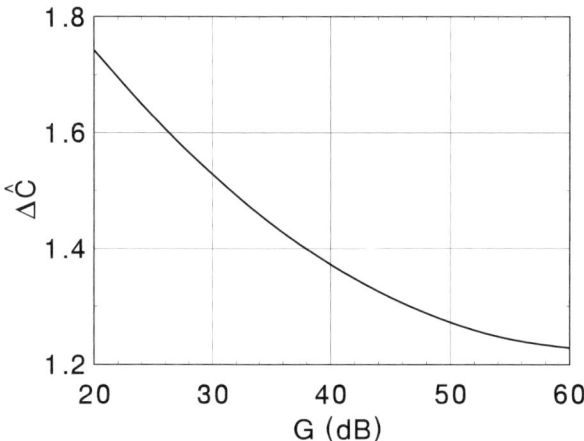

Fig. 2.4. Amplification bandwidth $\Delta\hat{C}$ versus maximal power gain. Here $\hat{\Lambda}_\mathrm{p}^2 \to 0$ and $\hat{\Lambda}_\mathrm{T}^2 = 0$

amplification bandwidth is introduced in microwave electronics and is defined as the difference between the frequencies $\Delta\omega = |\omega_1 - \omega_2|$ corresponding to the decrease of the output power of the device by a factor of two (FWHM). In FEL theory it is more convenient to introduce the notion of the amplification bandwidth in terms of the FWHM detuning parameter, $\Delta\hat{C}$. The value of $\Delta\hat{C}$ is connected by simple relations with the FWHM frequency deviation $\Delta\omega$, with the FWHM energy deviation $\Delta\mathcal{E}$, and with the FWHM undulator field deviation ΔH_w:

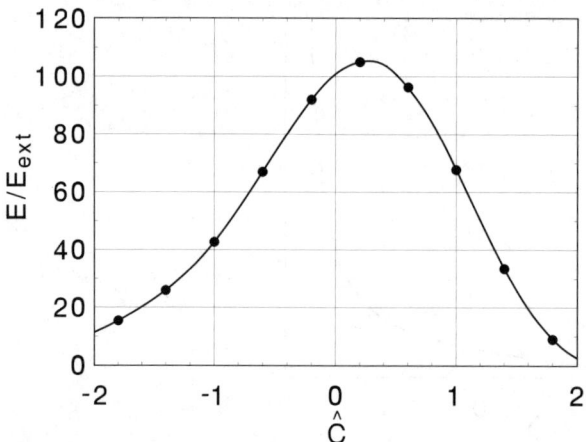

Fig. 2.5. Field gain E/E_{ext} versus the detuning parameter $\hat{C}$. The solid curve is calculated with the analytical formula (2.33) and the circles are the results of numerical simulations with (2.93) and (2.94). Here the reduced length of the undulator is $\hat{z} = 6.6$, $\hat{\Lambda}_{\text{p}}^2 \to 0$ and $\hat{\Lambda}_{\text{T}}^2 = 0$

$$\frac{\Delta\omega}{\omega_0} = 2\rho\Delta\hat{C} ,$$
$$\frac{\Delta\mathcal{E}}{\mathcal{E}_0} = \rho\Delta\hat{C} ,$$
$$\frac{\Delta H_{\text{w}}}{H_{\text{w}}} = \frac{1 + K^2}{K^2}\rho\Delta\hat{C} . \tag{2.39}$$

The amplification bandwidth can be obtained by solving the initial-value problem. In Fig. 2.4 we present the dependence of the amplification bandwidth $\Delta\hat{C}$ on the maximal power gain. This graph has been calculated using (2.33) and the roots of the eigenvalue equation (2.36). Figure 2.5 shows the dependence of the field gain on the detuning parameter. The value of the power gain is equal to $G = 40$ dB at exact resonance.

Using (2.37) we can write the expression for the amplification bandwidth in the high-gain limit:

$$\Delta\hat{C} \simeq \frac{10.4}{\sqrt{G(\text{dB}) + 9.5}} .$$

For values of the power gain G exceeding 40 dB this asymptotic expression provides an accuracy of a few per cent with respect to the rigorous solution of the initial-value problem.

Let us now study the influence of the space charge field on the amplification process. When the space charge parameter is small, $\hat{\Lambda}_{\text{p}}^2 \ll 1$, the space charge effects can be taken into account using perturbation theory. Using (2.32) we find that the maximal field growth rate

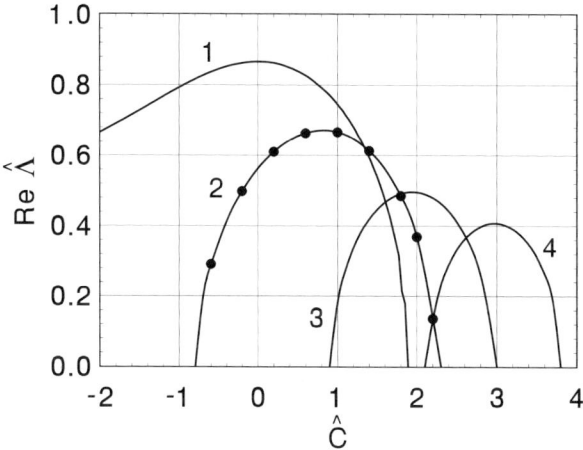

Fig. 2.6. Reduced field growth rate Re $\hat{\Lambda}$ versus the detuning parameter $\hat{C}$. Curve 1: $\hat{\Lambda}_\mathrm{p}^2 \to 0$. Curve 2: $\hat{\Lambda}_\mathrm{p}^2 = 1$. Curve 3: $\hat{\Lambda}_\mathrm{p}^2 = 4$. Curve 4: $\hat{\Lambda}_\mathrm{p}^2 = 9$. Solid curves are the solutions of the eigenvalue equation (2.32) and circles are the results of numerical simulations with (2.93) and (2.94). Here $\hat{\Lambda}_\mathrm{T}^2 = 0$

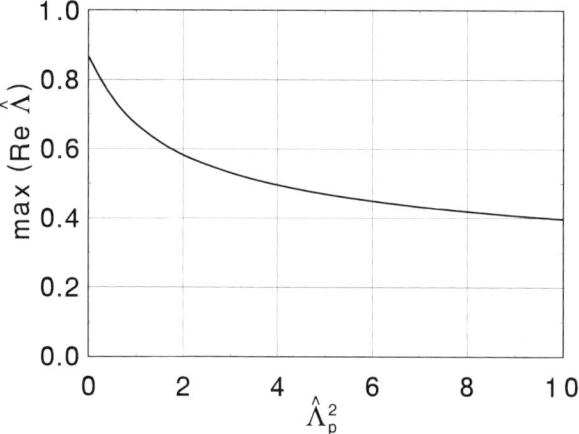

Fig. 2.7. Maximal reduced field growth rate, max(Re $\hat{\Lambda}$), versus the space charge parameter $\hat{\Lambda}_\mathrm{p}^2$. Here $\hat{\Lambda}_\mathrm{T}^2 = 0$

$$\max\left(\operatorname{Re}\hat{\Lambda}\right) \simeq \frac{\sqrt{3}}{2}\left(1 - \frac{\hat{\Lambda}_\mathrm{p}^2}{3}\right)$$

is achieved at the value of the detuning parameter $\hat{C}_\mathrm{m} \simeq \hat{\Lambda}_\mathrm{p}^2$. At large values of the space charge parameter, $\hat{\Lambda}_\mathrm{p}^2 \gg 1$, the space charge field suppresses significantly the field growth rate. The maximal value of the field growth rate is reached at $\hat{C}_\mathrm{m} \simeq \hat{\Lambda}_\mathrm{p}$:

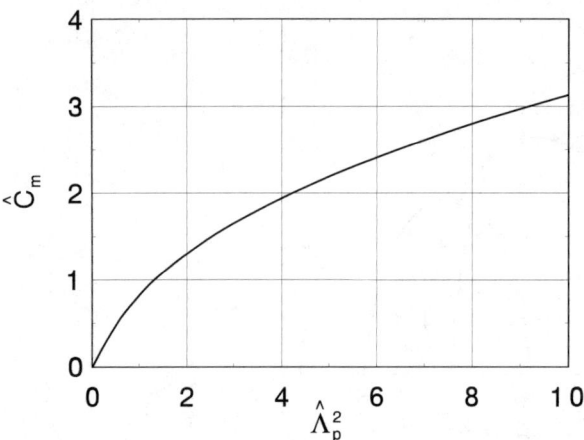

Fig. 2.8. Optimal value of the detuning parameter $\hat{C}_m$ versus the space charge parameter $\hat{\Lambda}_p^2$. Here $\hat{\Lambda}_T^2 = 0$

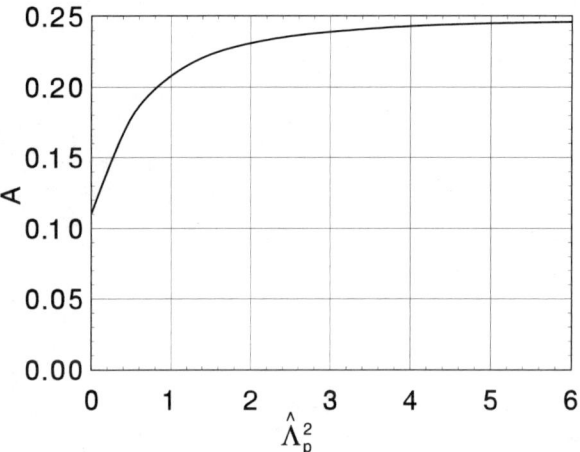

Fig. 2.9. Input coupling factor A entering (2.37) versus the space charge parameter $\hat{\Lambda}_p^2$. Here $\hat{C} = \hat{C}_m$ and $\hat{\Lambda}_T^2 = 0$

$$\max\left(\operatorname{Re}\hat{\Lambda}\right) \simeq \frac{1}{\sqrt{2\hat{\Lambda}_p}},$$

and the value of the input coupling factor A entering (2.37) tends to $1/4$.

When the space charge parameter $\hat{\Lambda}_p^2$ is of the order of unity, the field growth rate can be found by solving the cubic eigenvalue equation (2.32). In Fig. 2.6 we present the dependence of the field growth rate on the detuning parameter for several values of the space charge parameter $\hat{\Lambda}_p^2$. It is seen that for each value of the space charge parameter there is always a value of the detuning parameter $\hat{C}_m$ corresponding to the maximum of the field growth rate. The maximal value of the field growth rate and optimal detuning $\hat{C}_m$ are universal functions of the space charge parameter $\hat{\Lambda}_p^2$. The input coupling factor A entering the asymptotic expression (2.37) at tuning to $\hat{C} = \hat{C}_m$ is also a universal function of the space charge parameter $\hat{\Lambda}_p^2$. Graphs of these functions are presented in Figs. 2.7–2.9.

Lorentzian Energy Spread. Let us consider the case of an electron beam with Lorentzian energy distribution:

$$F(\mathcal{E} - \mathcal{E}_0) = \frac{1}{\pi} \frac{q}{(\mathcal{E} - \mathcal{E}_0)^2 + q^2} \ .$$

The corresponding expression for the reduced distribution function $\hat{F}$ has the form

$$\hat{F}(\hat{P}) = \frac{1}{\pi} \frac{\hat{q}}{\hat{P}^2 + \hat{q}^2} \ ,$$

where $\hat{q} = q/(\mathcal{E}_0 \rho)$. Substituting the Lorentzian distribution function into (2.29) we find that $\hat{D}$ is given by the expression:

$$\hat{D} = \mathrm{i} \left(\lambda + \hat{q} + \mathrm{i}\hat{C} \right)^{-2}$$

in the entire complex λ plane. The coefficient of $\exp(\lambda \hat{z})$ in the integrand (2.28) is equal to

$$\left[\lambda - \frac{\hat{D}}{1 - \mathrm{i}\hat{\Lambda}_p^2 \hat{D}} \right]^{-1} = \left[\lambda - \frac{\mathrm{i}}{\left(\lambda + \hat{q} + \mathrm{i}\hat{C} \right)^2 + \hat{\Lambda}_p^2} \right]^{-1} \ ,$$

and obviously satisfies the conditions of Jordan's lemma. Using (2.30) we can write:

$$\tilde{E}(\hat{z}) = E_{\mathrm{ext}} \sum_j \frac{\exp(\lambda_j \hat{z})}{1 - 2\mathrm{i} \left(\lambda_j + \hat{q} + \mathrm{i}\hat{C} \right) \lambda_j^2} \ ,$$

where λ_j are the roots of the cubic equation

$$\lambda = \mathrm{i} \left[\left(\lambda + \hat{q} + \mathrm{i}\hat{C} \right)^2 + \hat{\Lambda}_p^2 \right]^{-1} \ .$$

Therefore, in the case of a Lorentzian energy spread in the electron beam the electromagnetic wave in the undulator can be represented as the sum of three partial waves.

It should be noted that the Lorentzian energy distribution does not give a good representation of the energy distribution in actual electron beams. A more realistic distribution for the energy spread is a Gaussian one.

Gaussian Energy Spread. The distribution function of the electron beam with a Gaussian energy spread is given by (2.20). Let us derive an explicit expression for the function $\hat{D}$ at $\operatorname{Re}\lambda > 0$. Using (2.29), we obtain:

$$\hat{D} = \int_{-\infty}^{\infty} d\hat{P} \frac{\hat{F}'(\hat{P})}{\lambda + i(\hat{P} + \hat{C})} = i \int_{-\infty}^{\infty} d\hat{P} \frac{\hat{F}(\hat{P})}{\left[\lambda + i(\hat{P} + \hat{C})\right]^2} \cdot \quad (2.40)$$

Substituting

$$\frac{1}{\left[\lambda + i(\hat{P} + \hat{C})\right]^2} = \int_0^{\infty} \xi \exp\left\{-\left[\lambda + i(\hat{P} + \hat{C})\right]\xi\right\} d\xi \qquad \text{for } \operatorname{Re}\lambda > 0$$

into (2.40) and integrating over $\hat{P}$, we obtain

$$\hat{D} = i \int_0^{\infty} \xi \exp\left\{-\frac{\hat{\Lambda}_T^2 \xi^2}{2} - \left(\lambda + i\hat{C}\right)\xi\right\} d\xi \qquad \text{for } \operatorname{Re}\lambda > 0 \,. \quad (2.41)$$

Using (2.29) and the expression

$$\frac{1}{\left[\lambda + i(\hat{P} + \hat{C})\right]^2} = \int_0^{\infty} \xi \exp\left\{\left[\lambda + i(\hat{P} + \hat{C})\right]\xi\right\} d\xi \qquad \text{for } \operatorname{Re}\lambda < 0,$$

the function $\hat{D}$ can be reduced to the following form in the left half-plane:

$$\hat{D} = i \int_0^{\infty} \xi \exp\left\{-\frac{\hat{\Lambda}_T^2 \xi^2}{2} + \left(\lambda + i\hat{C}\right)\xi\right\} d\xi$$

$$- i \frac{\sqrt{2\pi}}{\hat{\Lambda}_T^3}\left(\lambda + i\hat{C}\right) \exp\left\{\frac{\left(\lambda + i\hat{C}\right)^2}{2\hat{\Lambda}_T^2}\right\} \qquad \text{for } \operatorname{Re}\lambda < 0 \,. \quad (2.42)$$

This expression contains a term proportional to $\exp(\lambda^2)$ which means that the function $\hat{D}$ has a singularity at infinity. The integrand in (2.28) has an infinite number of poles in the left half-plane located near the lines $\arg(\lambda) = \pm 3\pi/4$, and their density increases at infinity. On the other hand, Jordan's lemma states that the calculation of the linear integral (2.28) by closing the integration contour with an infinite semicircle in the left-hand plane is possible only when the function $\bar{E}(\lambda)$ tends to zero as $|\lambda| \to \infty$ uniformly in the argument of λ. So, this condition is not satisfied for the Gaussian distribution function. In this case the Laplace transform technique does not lead to complete analytical solution of the initial-value problem. However, it allows us to obtain an analytical asymptotic for the high-gain limit. This can be done using the following mathematical trick. Shifting the integration path in (2.28) to the left half-plane, we use Cauchy's residue theorem and transform the linear integral to the form

$$\int\limits_{\gamma'-i\infty}^{\gamma'+i\infty} d\lambda \bar{E}(\lambda)\exp(\lambda\hat{z}) = \int\limits_{-\alpha'-i\infty}^{-\alpha'+i\infty} d\lambda \bar{E}(\lambda)\exp(\lambda\hat{z})$$

$$+ \sum_{-\alpha'<\mathrm{Re}\,\lambda} \mathrm{Res}\,\bar{E}(\lambda_j)\exp(\lambda_j\hat{z})\,, \qquad (2.43)$$

where the summation is performed over the roots of (2.31) lying on the right-hand side of the line $(-\alpha'-i\infty, -\alpha'+i\infty)$. Here α' is a real positive number, and $\mathrm{Res}\,\bar{E}(\lambda_j)$ is the residue of the function $\bar{E}(\lambda)$ corresponding to the pole λ_j. For a Gaussian energy spread there exists only one root of (2.31) in the right half-plane and the number of roots in the interval $-\alpha' < \mathrm{Re}\,\lambda < 0$ is always finite at any finite value of α'. In the high-gain limit the contribution of the term proportional to $\exp(\lambda_j\hat{z})$ with $\mathrm{Re}\,\lambda_j > 0$ is larger than the contribution of all the other terms in (2.43). So, the asymptotic expression for the field amplitude $\tilde{E}(\hat{z})$ can be written as

$$\tilde{E}(\hat{z}) = E_{\mathrm{ext}}\exp(\hat{\Lambda}\hat{z})\left[1 - \frac{\hat{D}'}{\left(1-i\hat{\Lambda}_{\mathrm{p}}^2\hat{D}\right)^2}\right]^{-1}, \qquad (2.44)$$

where $\hat{\Lambda}$ is the growing root of (2.31). The values of $\hat{D}$ and $\hat{D}'$ at $\lambda = \hat{\Lambda}$ are calculated using (2.41).

The next step is to write down (2.44) in a form convenient for numerical calculations. Using tables of integrals we can write (for $\mathrm{Re}\,\lambda > 0$):

$$\hat{D} = i\int\limits_0^\infty \xi\exp\left\{-\frac{\hat{\Lambda}_{\mathrm{T}}^2\xi^2}{2} - \left(\lambda+i\hat{C}\right)\xi\right\}d\xi$$

$$= \frac{i}{\hat{\Lambda}_{\mathrm{T}}^2} - \frac{i\sqrt{\pi/2}}{\hat{\Lambda}_{\mathrm{T}}^3}\left(\lambda+i\hat{C}\right)\exp\left[\frac{\left(\lambda+i\hat{C}\right)^2}{2\hat{\Lambda}_{\mathrm{T}}^2}\right]$$

$$\times\left[1-\mathrm{erf}\left(\frac{\lambda+i\hat{C}}{\sqrt{2}\hat{\Lambda}_{\mathrm{T}}}\right)\right], \qquad (2.45)$$

where $\mathrm{erf}(\zeta)$ is the error function:

$$\mathrm{erf}(\zeta) = 2\pi^{-1/2}\int\limits_0^\zeta \exp(-u^2)du\,.$$

According to (2.31), the function $\hat{D}$ can be expressed in terms of the root $\hat{\Lambda}$ at $\lambda = \hat{\Lambda}$:

$$\hat{D}(\hat{\Lambda}) = \hat{\Lambda}\left[1+i\hat{\Lambda}_{\mathrm{p}}^2\hat{\Lambda}\right]^{-1}. \qquad (2.46)$$

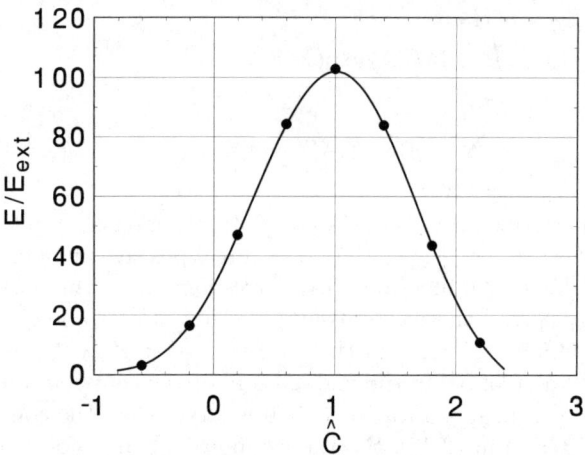

Fig. 2.10. Field gain E/E_{ext} versus the detuning parameter $\hat{C}$. The solid curve is calculated with the analytical formula (2.48) and the circles are the results of numerical simulations with (2.93) and (2.94). Here the reduced length of the undulator is $\hat{z} = 8.5$, $\hat{\Lambda}_{\text{p}}^2 = 1$ and $\hat{\Lambda}_{\text{T}}^2 = 0.1$

An explicit expression for the derivative of the function $\hat{D}$ has the form:

$$\begin{aligned}
\hat{D}' &= -\mathrm{i} \int_0^\infty \xi^2 \exp\left\{-\frac{\hat{\Lambda}_{\text{T}}^2 \xi^2}{2} - \left(\lambda + \mathrm{i}\hat{C}\right)\xi\right\} \mathrm{d}\xi \\
&= \frac{\mathrm{i}\left(\lambda + \mathrm{i}\hat{C}\right)}{\hat{\Lambda}_{\text{T}}^4} - \frac{\mathrm{i}\sqrt{2\pi}}{2\hat{\Lambda}_{\text{T}}^5}\left[(\lambda + \mathrm{i}\hat{C})^2 + \hat{\Lambda}_{\text{T}}^2\right] \\
&\quad \times \exp\left[\frac{\left(\lambda + \mathrm{i}\hat{C}\right)^2}{2\hat{\Lambda}_{\text{T}}^2}\right]\left[1 - \operatorname{erf}\left(\frac{\lambda + \mathrm{i}\hat{C}}{\sqrt{2}\hat{\Lambda}_{\text{T}}}\right)\right] \\
&= \frac{\mathrm{i}\left(\lambda + \mathrm{i}\hat{C}\right)}{\hat{\Lambda}_{\text{T}}^4} + \frac{\left(\lambda + \mathrm{i}\hat{C}\right)^2 + \hat{\Lambda}_{\text{T}}^2}{\hat{\Lambda}_{\text{T}}^2 \left(\lambda + \mathrm{i}\hat{C}\right)}\left(\hat{D} - \frac{\mathrm{i}}{\hat{\Lambda}_{\text{T}}^2}\right).
\end{aligned} \quad (2.47)$$

Using (2.46)–(2.47), we write the asymptotic expression for the field amplitude (2.44) in a form convenient for numerical calculations:

$$\tilde{E}(\hat{z}) = E_{\text{ext}} \exp\left(\hat{\Lambda}\hat{z}\right) \left\{1 + \mathrm{i}\left(\mathrm{i} - \hat{\Lambda}_{\text{p}}^2 \hat{\Lambda}\right)^2 \left[\left(\frac{\hat{\Lambda}}{\mathrm{i} - \hat{\Lambda}_{\text{p}}^2 \hat{\Lambda}} - \frac{1}{\hat{\Lambda}_{\text{T}}^2}\right)\right.\right.$$
$$\left.\left. \times \left(\frac{1}{\hat{\Lambda} + \mathrm{i}\hat{C}} + \frac{\hat{\Lambda} + \mathrm{i}\hat{C}}{\hat{\Lambda}_{\text{T}}^2}\right) + \frac{\hat{\Lambda} + \mathrm{i}\hat{C}}{\hat{\Lambda}_{\text{T}}^4}\right]^{-1}\right\}. \quad (2.48)$$

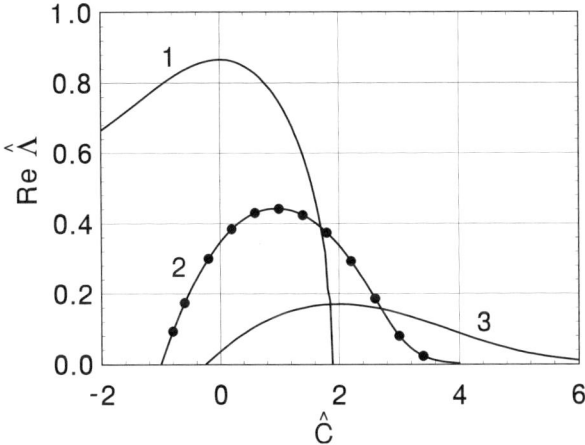

Fig. 2.11. Reduced field growth rate $\operatorname{Re}\hat{\Lambda}$ versus the detuning parameter $\hat{C}$. Curve 1: $\hat{\Lambda}_T^2 = 0$. Curve 2: $\hat{\Lambda}_T^2 = 1$. Curve 3: $\hat{\Lambda}_T^2 = 4$. Solid curves are calculated with the solution of the eigenvalue equation (2.50) and the circles are the results of numerical simulations with (2.93) and (2.94). Here $\hat{\Lambda}_p^2 \to 0$

We present in Fig. 2.10 the dependence of the field gain on the detuning parameter $\hat{C}$ under the influence of the space charge field and the energy spread in the beam ($\hat{z} = 8.5$, $\hat{\Lambda}_p^2 = 1$, $\hat{\Lambda}_T^2 = 0.1$ and the maximal field gain is equal to 40 dB). Calculations have been performed using the asymptotic formula (2.48).

In the high-gain limit the expression for the power gain, G, can be written in the form (2.37):

$$G = |\tilde{E}|^2/E_{\text{ext}}^2 = A \exp\left[2\operatorname{Re}\hat{\Lambda}\hat{z}\right],$$

where A and $\hat{\Lambda}$ are functions of the detuning parameter $\hat{C}$, of the space charge parameter $\hat{\Lambda}_p^2$ and of the energy spread parameter $\hat{\Lambda}_T^2$.

In the limit of small values of the space charge parameter, $\hat{\Lambda}_p^2 \to 0$, (2.48) transforms to

$$\tilde{E}(\hat{z}) = \frac{\hat{\Lambda}_T^2\left[\hat{\Lambda} + i\hat{C}\right]}{i\left[\hat{C}\hat{\Lambda}_T^2 + 1\right] - \hat{\Lambda}\left[\hat{\Lambda} + i\hat{C}\right]^2} E_{\text{ext}} \exp\left(\hat{\Lambda}\hat{z}\right), \tag{2.49}$$

where $\hat{\Lambda}$ is the growing root of the eigenvalue equation ($\operatorname{Re}\hat{\Lambda} > 0$):

$$\hat{\Lambda} = i\int_0^\infty \exp\left[-\hat{\Lambda}_T^2\xi^2/2 - \left(\hat{\Lambda} + i\hat{C}\right)\xi\right]\xi\,d\xi. \tag{2.50}$$

For the parameter region when space charge effects can be neglected, for $\hat{\Lambda}_p^2 \to 0$, the field growth rate can be found by solving the eigenvalue equation

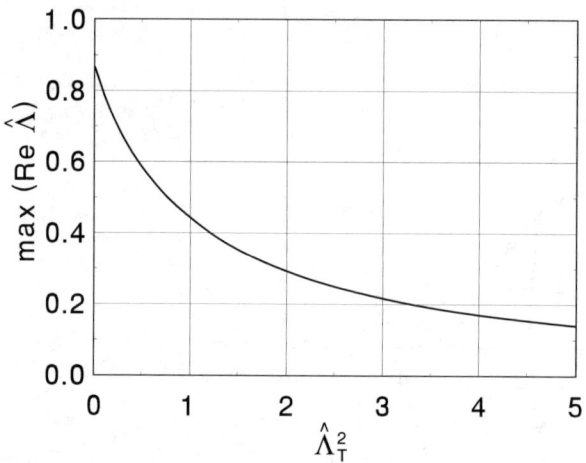

Fig. 2.12. Maximal reduced field growth rate $\max(\operatorname{Re} \hat{\Lambda})$ versus the energy spread parameter $\hat{\Lambda}_\mathrm{T}^2$. Here $\hat{\Lambda}_\mathrm{p}^2 \to 0$

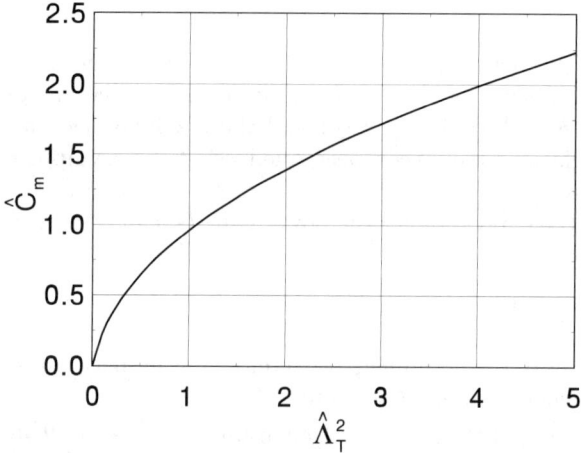

Fig. 2.13. Optimal value of the detuning parameter $\hat{C}_\mathrm{m}$ versus the energy spread parameter $\hat{\Lambda}_\mathrm{T}^2$. Here $\hat{\Lambda}_\mathrm{p}^2 \to 0$

(2.50). In Fig 2.11 we present the dependence of the field growth rate on the detuning parameter $\hat{C}$ for several values of the energy spread parameter $\hat{\Lambda}_\mathrm{T}^2$.

For small values of the energy spread parameter, $\hat{\Lambda}_\mathrm{T}^2 \ll 1$, the maximal field growth rate is equal to

$$\max(\operatorname{Re} \hat{\Lambda}) \simeq \frac{\sqrt{3}}{2} \left(1 - \hat{\Lambda}_\mathrm{T}^2\right) ,$$

and is reached at the value of the detuning parameter $\hat{C}_\mathrm{m} \simeq 3\hat{\Lambda}_\mathrm{T}^2$.

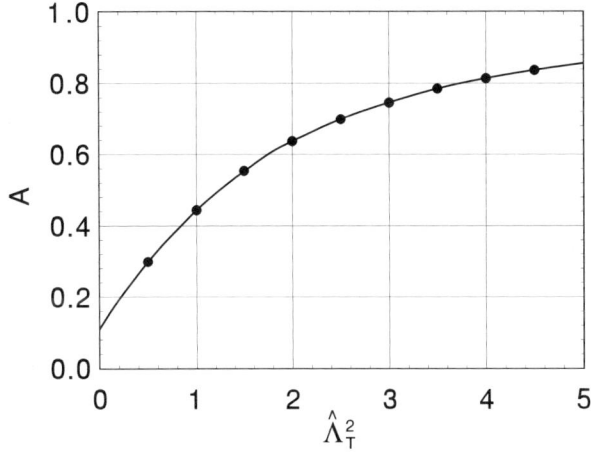

Fig. 2.14. Input coupling factor A entering (2.37) versus the energy spread parameter $\hat{\Lambda}_T^2$. Here $\hat{C} = \hat{C}_m$ and $\hat{\Lambda}_p^2 \to 0$. The solid curve is calculated with the analytical formula (2.48) and the circles are the results of the numerical solution of (2.19)

For large values of the energy spread parameter, $\hat{\Lambda}_T^2 \gg 1$, the energy spread suppresses significantly the field growth rate. The maximal value of the field growth rate is reached at $\hat{C}_m \simeq \hat{\Lambda}_T$ and is equal to

$$\max(\operatorname{Re}\hat{\Lambda}) \simeq \sqrt{\frac{\pi}{2\mathrm{e}}}\frac{1}{\hat{\Lambda}_T^2} \simeq \frac{0.76}{\hat{\Lambda}_T^2},$$

where $\mathrm{e} = 2.718...$ is the base of natural logarithms. The value of the input coupling factor A in (2.37) approaches unity asymptotically for large values of the energy spread parameter.

When the value of the energy spread parameter $\hat{\Lambda}_T^2$ is of the order of unity, the field growth rate can be found by solving the eigenvalue equation (2.50). At a fixed value of the energy spread parameter there is a value of the detuning parameter $\hat{C}_m$ corresponding to the maximal field growth rate (see Fig. 2.11). The value of the maximal field growth rate and the value of the optimal detuning $\hat{C}_m$ are universal functions of the energy spread parameter $\hat{\Lambda}_T^2$. The input coupling factor A entering (2.37), is also a universal function of $\hat{\Lambda}_T^2$ at tuning to $\hat{C} = \hat{C}_m$. Graphs of these functions are presented in Figs. 2.12–2.14.

It is interesting to compare the analytical results corresponding to the high gain limit with the results of numerical solution of the integro-differential equation (2.19). This equation has been integrated with the Runge Kutta technique and we present the results of these calculations in Fig 2.14. It is seen that in the high-gain limit there is a good agreement between the numerical and analytical results.

2.1.4 General Solution of the Initial-Value Problem

In the previous sections we performed an analysis of the FEL amplifier using the self-consistent field equation (2.17). This equation has been derived for specific initial conditions when an unmodulated electron beam and external electromagnetic wave are fed to the undulator entrance. In this section we extend our study to more general initial conditions. As in the previous sections, we begin with the Vlasov equation:

$$\frac{\partial f}{\partial z} + \frac{\partial H}{\partial P}\frac{\partial f}{\partial \psi} - \frac{\partial H}{\partial \psi}\frac{\partial f}{\partial P} = 0 \;.$$

In the linear approximation we get

$$\frac{\partial \tilde{f}_1}{\partial z} + i\left[C + \omega P/(c\gamma_z^2 \mathcal{E}_0)\right]\tilde{f}_1 + \left(iU - e\tilde{E}_z\right)\frac{\partial f_0}{\partial P} = 0 \;, \qquad (2.51)$$

where $f_0 = n_0 F(P)$. The general solution for $\tilde{f}_1$ is

$$\tilde{f}_1 = -n_0 \frac{\mathrm{d}F}{\mathrm{d}P}\int_0^z \mathrm{d}z'(iU - e\tilde{E}_z)\exp\left\{i\left[C + \omega P/(c\gamma_z^2 \mathcal{E}_0)\right](z' - z)\right\}$$

$$+ \tilde{f}_1|_{z=0}\exp\left\{-i\left[C + \omega P/(c\gamma_z^2 \mathcal{E}_0)\right]z\right\} \;. \qquad (2.52)$$

Integration of (2.51) over P with the limits $(-\infty, \infty)$ gives us the following result:

$$\frac{\mathrm{d}\tilde{j}_1}{\mathrm{d}z} + iC\tilde{j}_1 - \frac{ie\omega}{\gamma_z^2 \mathcal{E}_0}\int_{-\infty}^{\infty} P\tilde{f}_1 \mathrm{d}P = 0 \;. \qquad (2.53)$$

Then we take the derivative of (2.53) with respect to z and, using the expression for $\partial \tilde{f}_1/\partial z$ from (2.51), we obtain the relation

$$\frac{\mathrm{d}^2 \tilde{j}_1}{\mathrm{d}z^2} + iC\frac{\mathrm{d}\tilde{j}_1}{\mathrm{d}z} - \frac{e\omega C}{\gamma_z^2 \mathcal{E}_0}\int_{-\infty}^{\infty} P\tilde{f}_1 \mathrm{d}P$$

$$- \frac{e\omega^2}{c\gamma_z^4 \mathcal{E}_0^2}\int_{-\infty}^{\infty} P^2 \tilde{f}_1 \mathrm{d}P + \frac{ie\omega}{\gamma_z^2 \mathcal{E}_0}\left(iU - e\tilde{E}_z\right)\int_{-\infty}^{\infty} P\frac{\mathrm{d}f_0}{\mathrm{d}P}\mathrm{d}P = 0 \;. \qquad (2.54)$$

Now we integrate the last term in (2.54) and use (2.53). With the help of (2.7) and (2.13), we finally obtain

$$\frac{\mathrm{d}^2 \tilde{j}_1}{\mathrm{d}z^2} + 2iC\frac{\mathrm{d}\tilde{j}_1}{\mathrm{d}z} + \left[\frac{4\pi e j_0}{c\gamma_z^2 \mathcal{E}_0} - C^2\right]\tilde{j}_1$$

$$+ \frac{ie\theta_s \omega j_0}{2c\gamma_z^2 \mathcal{E}_0}\tilde{E} - \frac{e\omega^2}{c\gamma_z^4 \mathcal{E}_0^2}\int_{-\infty}^{\infty} P^2 \tilde{f}_1 \mathrm{d}P = 0 \;. \qquad (2.55)$$

Let us consider the case of a negligibly small energy spread in the electron beam. In this limit we can replace the initial distribution function by the delta function, $f_0 = n_0\delta(P)$. Analysis of (2.52) shows that in the framework of the linear approximation the last term in (2.55) is negligibly small with respect to all other terms and can be omitted. Then we obtain the equation for the first harmonic of the beam current density:

$$\frac{\mathrm{d}^2 \tilde{j}_1}{\mathrm{d}z^2} + 2\mathrm{i}C\frac{\mathrm{d}\tilde{j}_1}{\mathrm{d}z} + \left[\frac{4\pi e j_0}{c\gamma_z^2 \varepsilon_0} - C^2\right]\tilde{j}_1 + \frac{\mathrm{i}e\theta_s \omega j_0}{2c\gamma_z^2 \varepsilon_0}\tilde{E} = 0 \ . \qquad (2.56)$$

Maxwell's equations give us a relation connecting the amplitude of the electric field and the first harmonic of the beam current density

$$\mathrm{d}\tilde{E}/\mathrm{d}z = -2\pi \theta_s c^{-1} \tilde{j}_1(z) \ ,$$

which allows us to obtain the self-consistent field equation

$$\tilde{E}''' + 2\mathrm{i}\hat{C}\tilde{E}'' + \left(\hat{\Lambda}_\mathrm{p}^2 - \hat{C}^2\right)\tilde{E}' = \mathrm{i}\tilde{E} \ . \qquad (2.57)$$

This equation is valid for the general case of initial conditions for the electron beam and the radiation at the undulator entrance. The equation is written down in the reduced form using standard normalization of the parameters described in the previous section. The prime denotes differentiation with respect to $\hat{z}$. Equation (2.57) is a linear ordinary differential equation with fixed coefficients, so its general solution is given by a superposition of three linearly independent solutions:

$$\tilde{E} = \sum_{j=1}^{3} C_j \exp(\lambda_j \hat{z}) \ ,$$

where the C_j are constants. According to (2.57), the factors λ_j are the solutions of the cubic equation

$$\lambda\left[\left(\lambda + \mathrm{i}\hat{C}\right)^2 + \hat{\Lambda}_\mathrm{p}^2\right] = \mathrm{i} \ . \qquad (2.58)$$

To solve the initial-value problem, we should set the initial conditions for

$$\tilde{E}(0) \ , \quad \tilde{E}'(0) \ , \quad \tilde{E}''(0) \ ,$$

which correspond to the field amplitude and its first and second derivatives with respect to $\hat{z}$ at the undulator entrance at $\hat{z} = 0$. The field amplitude and its derivatives at any longitudinal $\hat{z}$ coordinate are calculated via the initial conditions as follows:

$$\begin{bmatrix}\tilde{E} \\ \tilde{E}' \\ \tilde{E}''\end{bmatrix}_{\hat{z}} = M(\hat{z}|0)\begin{bmatrix}\tilde{E} \\ \tilde{E}' \\ \tilde{E}''\end{bmatrix}_0 \ , \qquad (2.59)$$

where the transition matrix $M(\hat{z}|0)$ is equal to

$$M = \begin{bmatrix} \tilde{E}_1 & \tilde{E}_2 & \tilde{E}_3 \\ \tilde{E}'_1 & \tilde{E}'_2 & \tilde{E}'_3 \\ \tilde{E}''_1 & \tilde{E}''_2 & \tilde{E}''_3 \end{bmatrix}_{\hat{z}} \times \begin{bmatrix} \tilde{E}_1 & \tilde{E}_2 & \tilde{E}_3 \\ \tilde{E}'_1 & \tilde{E}'_2 & \tilde{E}'_3 \\ \tilde{E}''_1 & \tilde{E}''_2 & \tilde{E}''_3 \end{bmatrix}_{0}^{-1} .$$

The explicit expressions for the matrix elements M_{ij} are:

$$M_{11} = \lambda_2\lambda_3 B_1 + \lambda_1\lambda_3 B_2 + \lambda_1\lambda_2 B_3$$
$$M_{12} = -(\lambda_2 + \lambda_3)B_1 - (\lambda_1 + \lambda_3)B_2 - (\lambda_1 + \lambda_2)B_3$$
$$M_{13} = B_1 + B_2 + B_3$$
$$M_{21} = \lambda_1\lambda_2\lambda_3 M_{13}$$
$$M_{22} = -\lambda_1(\lambda_2 + \lambda_3)B_1 - \lambda_2(\lambda_1 + \lambda_3)B_2 - \lambda_3(\lambda_1 + \lambda_2)B_3$$
$$M_{23} = \lambda_1 B_1 + \lambda_2 B_2 + \lambda_3 B_3$$
$$M_{31} = \lambda_1\lambda_2\lambda_3 M_{23}$$
$$M_{32} = -\lambda_1^2(\lambda_2 + \lambda_3)B_1 - \lambda_2^2(\lambda_1 + \lambda_3)B_2 - \lambda_3^2(\lambda_1 + \lambda_2)B_3$$
$$M_{33} = \lambda_1^2 B_1 + \lambda_2^2 B_2 + \lambda_3^2 B_3 , \tag{2.60}$$

where

$$B_1 = \frac{\exp(\lambda_1 \hat{z})}{(\lambda_1 - \lambda_2)(\lambda_1 - \lambda_3)}$$
$$B_2 = \frac{\exp(\lambda_2 \hat{z})}{(\lambda_2 - \lambda_1)(\lambda_2 - \lambda_3)}$$
$$B_3 = \frac{\exp(\lambda_3 \hat{z})}{(\lambda_3 - \lambda_1)(\lambda_3 - \lambda_2)} . \tag{2.61}$$

The values of $\tilde{E}'$ and $\tilde{E}''$ at the undulator entrance can be expressed in terms of the complex amplitude of the first harmonic of the particle density in the phase space $\tilde{f}_1$. Using the kinetic equation (2.9) and the wave equation (2.16), we have

$$\tilde{E}'(0)/E_0 = -2\tilde{j}_1(0)/j_0 ,$$
$$\tilde{E}''(0)/E_0 = -2\tilde{j}'_1(0)/j_0 = 2\mathrm{i}\left[\hat{C}\tilde{j}_1(0)/j_0 - \int \hat{P}\hat{f}_1(0,\hat{P})\mathrm{d}\hat{P}\right] ,$$
$$\tilde{j}_1(0)/j_0 = -\int \hat{f}_1(0,\hat{P})\mathrm{d}\hat{P} , \tag{2.62}$$

where $E_0 = \rho \mathcal{E}_0 \Gamma/(e\theta_\mathrm{s})$.

Let us consider specific initial conditions when an unmodulated electron beam and an electromagnetic wave of amplitude E_ext are fed to the undulator entrance. The initial conditions at $\hat{z} = 0$ are as follows:

$$\tilde{E}(0) = E_\mathrm{ext} , \quad \tilde{E}'(0) = 0 , \quad \tilde{E}''(0) = 0 .$$

According to (2.59), we obtain that

$$\tilde{E}(\hat{z}) = M_{11}(\hat{z}|0)E_\mathrm{ext} .$$

This expression is identical to (2.33).

Another important practical kind of initial condition refers to the case when there is no electromagnetic field at the undulator entrance and the modulation of the beam density serves as the input signal for the FEL amplifier. So, the initial conditions at $\hat{z} = 0$ should be written as

$$\tilde{E}(0) = 0 , \quad \tilde{E}'(0)/E_0 = -2\tilde{j}_1(0)/j_0 , \quad \tilde{E}''(0)/E_0 = 2\mathrm{i}\hat{C}\tilde{j}_1(0)/j_0 , \quad (2.63)$$

and the evolution of the radiation field amplitude is given by the expression

$$\tilde{E}(\hat{z})/E_0 = -2M_{12}(\hat{z}|0)\tilde{j}_1(0)/j_0 + 2\mathrm{i}\hat{C}M_{13}(\hat{z}|0)\tilde{j}_1(0)/j_0 .$$

Thorough study of the initial conditions (2.63) and comparison with (2.62) indicates that the initial condition for the second derivative $\tilde{E}''(0)/E_0$ seems to be paradoxical. Namely, at non-zero detuning the derivative of the first harmonic of the current density, $\tilde{j}'_1(0)$, has a non-zero value even in the case when there is no initial energy modulation. Simple analysis shows that the derivative $\tilde{j}'_1(0) = -\mathrm{i}\hat{C}\tilde{j}_1(0)$ takes its origin from the definition of the phase ψ.

2.1.5 Linear Theory of the FEL Amplifier with a Planar Undulator

All the considerations presented above refer to the case of an FEL amplifier with helical undulator. A specific feature of the particle motion in the helical undulator is that the longitudinal velocity of the particle is constant. Another popular undulator configuration is a planar one. The constrained motion of the electron in the planar undulator differs from that in the helical one. An important feature of this motion is that the longitudinal velocity v_z of the electron oscillates along the undulator axis which creates definite problems for the description of the FEL process. Moreover, the reasonable question arises of whether the physics of the FEL amplifier with helical undulator is similar to that of the FEL amplifier with planar undulator. There could be two approaches to the analysis of the FEL amplifier with planar undulator. The first one is the derivation of the self-consistent equations, their averaging and subsequent solution. Another approach consists of the rigorous solution of nonaveraged equations with subsequent investigation of the obtained solution using the resonance approximation. Both methods should lead to identical results. In this section we demonstrate both approaches, although the method for the rigorous solution of nonaveraged equations is bulky. The reason for this is that there are papers stating that the physics of the FEL amplifier with a planar undulator differs from that of the FEL amplifier with a helical undulator. In particular, this refers to the space charge effects. In this section we demonstrate that all the solutions for the planar undulator are identical to those for the helical undulator. The only difference is the appearance of different numerical factors taking their origin from the averaging procedure. Equations written down in reduced form are identical for both undulator configurations.

Self-Consistent Field Equation. The magnetic field of the planar undulator is of the form:

$$\boldsymbol{H}_\ell(z) = \boldsymbol{e}_x H_\ell \cos(k_w z),$$

where $\boldsymbol{e}_x$ is the unit vector directed along the x axis of the Cartesian coordinate system (x, y, z). The Lorentz force is used to derive the equation of motion of an electron with energy $\mathcal{E}_0$ in the presence of a magnetic field. Integration of this equation gives

$$\boldsymbol{v}_y(z) = -\boldsymbol{e}_y c \theta_\ell \sin(k_w z),$$

where $\theta_\ell = eH_\ell/(\mathcal{E}_0 k_w)$. The electrons in the planar undulator move along sinusoidal trajectories parallel to the z axis. The amplitude of the transverse velocity of the electron is considered to be small and the longitudinal velocity of the electron v_z is close to the velocity of light, $v_z \simeq c$. Only a linearly polarized plane electromagnetic wave can be amplified in an FEL amplifier with a planar undulator. In the framework of the one-dimensional model the electric field vector of the amplified wave can be presented in the form:

$$\boldsymbol{E}_\perp = \boldsymbol{e}_y \tilde{E}_y(z) \exp\left[i\omega(z/c - t)\right] + \text{C.C.},$$

where ω is the frequency of the amplified wave. It is assumed that the complex amplitude of the electric field $\tilde{E}_y$ is a slowly varying function of the z coordinate, such that $|\mathrm{d}\tilde{E}_y/\mathrm{d}z| \ll k_w |\tilde{E}_y|$.

The Hamiltonian of the particle is given by expression (2.1). In the same way as has been done before, we choose the z coordinate as a new time and the phase

$$\psi = k_w z + \omega(z/c - t)$$

as a new generalized coordinate. The form of the Hamiltonian in the new variables coincides with that given by (2.3). In the first order of expansion in $\boldsymbol{A}_\perp$ and at small deviations of the electron energy $\mathcal{E}$ from the nominal energy $\mathcal{E}_0$, the Hamiltonian takes the form:[5]

$$H(P, \psi, z) = \left[C + \frac{\omega \theta_\ell^2}{4c} \cos(2k_w z)\right] P + \frac{\omega}{2c\gamma_\ell^2 \mathcal{E}_0} P^2 \\ - \left[U e^{i\psi} + U^* e^{-i\psi}\right](1 - P/\mathcal{E}_0) + \int \mathrm{d}\psi e E_z, \quad (2.64)$$

where $P = \mathcal{E} - \mathcal{E}_0$, $C = k_w - \omega/(2c\gamma_\ell^2)$ is the detuning of the electron with energy $\mathcal{E} = \mathcal{E}_0$ from the resonance (we assume here that $C \ll k_w$),

$$U = -\left[1 - \exp(-2ik_w z)\right] e\theta_\ell \tilde{E}_y(z)/2$$

is the complex amplitude of the effective potential of interaction, and

$$\gamma_\ell^{-2} = \gamma^{-2} + \theta_\ell^2/2, \qquad \gamma = \mathcal{E}_0/(m_e c^2).$$

---

[5] Here we omit the term proportional to $P^2 \cos(2k_w z)$. One can find that this term will result in a correction of Q by $Q(1 + P/\mathcal{E}_0)$ in (2.67) which falls outside the accuracy of the accepted approximations.

2.1 Linear Mode of Operation

The evolution of the distribution function $f(P, \psi, z)$ is governed by the kinetic equation (2.8). For the electron beam with a small density perturbation, the distribution function and the space charge field can be written in the following way:

$$f = f_0 + \tilde{f}_1 e^{i\psi} + \tilde{f}_1^* e^{-i\psi} , \qquad E_z = \tilde{E}_z e^{i\psi} + \tilde{E}_z^* e^{-i\psi} .$$

We consider the case when the beam is modulated neither in velocity nor in density at the undulator entrance, i.e.

$$\tilde{f}_1|_{z=0} = 0 , \qquad f_0 = n_0 F(P) , \qquad \int F(P) \mathrm{d}P = 1 , \qquad (2.65)$$

where n_0 is the beam density. The beam current density is connected with the distribution function $\tilde{f}_1$ by

$$j_z = -j_0 + \tilde{j}_1 e^{i\psi} + \text{C.C.} , \qquad \tilde{j}_1 \simeq -ec \int \tilde{f}_1 \mathrm{d}P ,$$

where $-j_0 \simeq -ecn_0$ is the longitudinal component of the beam current density at the undulator entrance.

In the framework of the one-dimensional model we can derive from Maxwell's equations the following relation between $\tilde{E}_z$ and $\tilde{j}_1(z)$ (see (2.12)):

$$\tilde{E}_z = -\mathrm{i} 4\pi \tilde{j}_1(z)/\omega .$$

Using (2.8) and (2.64) we find the equation for the complex amplitude $\tilde{f}_1$:

$$\frac{\partial \tilde{f}_1}{\partial z} + \mathrm{i} \left[C + \frac{\omega \theta_\ell^2}{4c} \cos(2k_\mathrm{w} z) + \frac{\omega}{c\gamma_\ell^2 \mathcal{E}_0} P \right] \tilde{f}_1$$

$$+ \mathrm{i} n_0 \left[U + \frac{4\pi e \tilde{j}_1(z)}{\omega} \right] \frac{\mathrm{d}F}{\mathrm{d}P} = 0 . \qquad (2.66)$$

The solution of this equation has the form:

$$\tilde{j}_1(z) = \mathrm{i} j_0 \int_0^z \mathrm{d}z' \left\{ -[1 - \exp(-2\mathrm{i} k_\mathrm{w} z')] e\theta_\ell \tilde{E}_y(z')/2 + 4\pi e \tilde{j}_1(z')/\omega \right\}$$

$$\times \int \mathrm{d}P \, (\mathrm{d}F(P)/\mathrm{d}P) \exp\left\{ \mathrm{i} \left[C + \omega P/(c\gamma_\ell^2 \mathcal{E}_0) \right] (z' - z) \right.$$

$$\left. + \mathrm{i}Q \left[\sin(2k_\mathrm{w} z') - \sin(2k_\mathrm{w} z) \right] \right\} , \qquad (2.67)$$

where $Q = \theta_\ell^2 \omega/(8ck_\mathrm{w}) = K^2/(4 + 2K^2)$ and $K = eH_\ell/(k_\mathrm{w} m_e c^2)$ is the undulator parameter for the planar undulator.

One more relation connecting the complex amplitudes $\tilde{j}_1(z)$ and $\tilde{E}_y(z)$ follows from the wave equation

$$\partial^2 \boldsymbol{A}_\perp / \partial z^2 - c^{-2} \partial^2 \boldsymbol{A}_\perp / \partial t^2 = -(4\pi/c) \boldsymbol{j}_\perp ,$$

where $\boldsymbol{j}_\perp$ is the perturbation of the transverse current density:

$$\boldsymbol{j}_\perp = -\boldsymbol{e}_y \theta_\ell \sin(k_\mathrm{w} z)(\tilde{j}_1 e^{i\psi} + \text{C.C.}) .$$

It follows from the relation $cE_y = -\partial A_y/\partial t$ that

$$\frac{d\tilde{E}_y}{dz} = -i\pi\theta_\ell c^{-1} \left[\exp(2ik_w z) - 1\right] \tilde{j}_1(z) . \tag{2.68}$$

It is convenient to rewrite the expression for amplitude $\tilde{j}_1(z)$ in the form:

$$\tilde{j}_1(z) = \tilde{j}_a(z) \exp\left[-iQ\sin(2k_w z)\right] . \tag{2.69}$$

where $\tilde{j}_a$ is a slowly varying function of the z coordinate. The latter statement should be proven. Let us study (2.67). Since $\tilde{j}_a$ is equal to the integral over z, the contribution of the rapidly oscillating terms can be neglected at $k_w z \gg 1$. In other words, we use the resonance approximation here.

Using (2.69) and the expansion

$$\exp\left[iQ\sin(2k_w z)\right] = \sum_{n=-\infty}^{n=+\infty} J_n(Q) \exp(2ink_w z) ,$$

where J_n is the Bessel function of nth order, we rewrite (2.67) and (2.68) as follows

$$\tilde{j}_a(z) = ij_0 \int_0^z dz' \left\{ -\frac{e\theta_\ell}{2} \tilde{E}_y(z') \sum_{n=-\infty}^{n=+\infty} J_n(Q) \{\exp\left[2ink_w z'\right] \right.$$
$$\left. - \exp\left[2i(n-1)k_w z'\right]\} + \frac{4\pi e}{\omega} \tilde{j}_a(z') \right\}$$
$$\times \int_{-\infty}^{\infty} dP (dF(P)/dP) \exp\left\{i\left[\omega P/(c\gamma_\ell^2 \mathcal{E}_0) + C\right](z'-z)\right\} , \tag{2.70}$$

$$\frac{d\tilde{E}_y}{dz} = -i\pi\theta_\ell c^{-1} \tilde{j}_a(z) \sum_{n=-\infty}^{n=+\infty} (-1)^n J_n(Q)$$
$$\times \{\exp\left[2i(n+1)k_w z\right] - \exp\left[2ink_w z\right]\} . \tag{2.71}$$

Keeping in mind that $\tilde{j}_a$ and $\tilde{E}_y$ are slowly varying amplitudes and neglecting rapidly oscillating terms in (2.70) and (2.71), we obtain the following equations

$$\tilde{j}_a(z) = ij_0 \int_0^z dz' \left\{ -\frac{A_{JJ} e\theta_\ell}{2} \tilde{E}_y(z') + \frac{4\pi e}{\omega} \tilde{j}_a(z') \right\}$$
$$\times \int_{-\infty}^{\infty} dP (dF(P)/dP) \exp\left[i\left(\omega P/(c\gamma_\ell^2 \mathcal{E}_0) + C\right)(z'-z)\right] , \tag{2.72}$$

$$\frac{d\tilde{E}_y}{dz} = i\pi\theta_\ell c^{-1} A_{JJ} \tilde{j}_a(z) , \tag{2.73}$$

where

$$A_{\mathrm{JJ}} = [J_0(Q) - J_1(Q)] \ .$$

Substituting (2.72) into the right-hand side of (2.73) and using the normalization procedure we obtain the equation for the field amplitude $\tilde{E}_y$:

$$\frac{\mathrm{d}\tilde{E}_y}{\mathrm{d}\hat{z}} = \int_0^{\hat{z}} \mathrm{d}\hat{z}' \left\{ \tilde{E}_y(\hat{z}') + \mathrm{i}\hat{\Lambda}_{\mathrm{p}}^2 \frac{\mathrm{d}\tilde{E}_y(\hat{z}')}{\mathrm{d}\hat{z}'} \right\}$$

$$\times \int_{-\infty}^{\infty} \mathrm{d}\hat{P} \frac{\mathrm{d}\hat{F}}{\mathrm{d}\hat{P}} \exp\left\{ \mathrm{i}\left(\hat{P} + \hat{C}\right)(\hat{z}' - \hat{z}) \right\} \ , \qquad (2.74)$$

where

$$\hat{z} = \Gamma z \ , \qquad \Gamma = \left[\frac{\pi \theta_\ell^2 j_0 \omega A_{\mathrm{JJ}}^2}{2 c \gamma_\ell^2 \gamma I_{\mathrm{A}}} \right]^{1/3} \ ,$$

$$\hat{C} = C/\Gamma \ , \qquad \hat{\Lambda}_{\mathrm{p}}^2 = \Lambda_{\mathrm{p}}^2/\Gamma^2 = 4\pi j_0/(\gamma_\ell^2 \gamma I_{\mathrm{A}} \Gamma^2) \ .$$

The energy deviation $P = \mathcal{E} - \mathcal{E}_0$ is normalized as

$$\hat{P} = (\mathcal{E} - \mathcal{E}_0)/(\rho \mathcal{E}_0) \ , \qquad (2.75)$$

and the efficiency parameter ρ is given by the expression:

$$\rho = \gamma_\ell^2 \Gamma c/\omega \ . \qquad (2.76)$$

One can obtain from (2.74) and (2.17) that the self-consistent field equation for the FEL amplifier with planar undulator fully agrees with the corresponding equation for the FEL amplifier with helical undulator.

Solution of Nonaveraged Equations by the Laplace Transform. The system of two coupled equations (2.70) and (2.71) can be solved using the Laplace transform technique. The Laplace transformations of (2.70) and (2.71) have the form:

$$\bar{j}_{\mathrm{a}}(p) = \frac{\mathrm{i} j_0 \omega D(p)}{\gamma_\ell^2 c \mathcal{E}_0} \left\{ -\frac{e\theta_\ell}{2} \sum_{n=-\infty}^{n=+\infty} J_n(Q) \left[\bar{E}(p - 2\mathrm{i} n k_{\mathrm{w}}) \right. \right.$$

$$\left. \left. - \bar{E}(p - 2\mathrm{i}(n-1) k_{\mathrm{w}}) \right] + \frac{4\pi e}{\omega} \bar{j}_{\mathrm{a}}(p) \right\} \ , \qquad (2.77)$$

$$p\bar{E}(p) - E_{\mathrm{ext}} = -\frac{\mathrm{i}\pi\theta_\ell}{c} \sum_{n=-\infty}^{n=+\infty} (-1)^n J_n(Q)$$

$$\times \left[\bar{j}_{\mathrm{a}}(p - 2\mathrm{i}(n+1) k_{\mathrm{w}}) - \bar{j}_{\mathrm{a}}(p - 2\mathrm{i} n k_{\mathrm{w}}) \right] \ , \qquad (2.78)$$

where the following notation has been introduced ($\operatorname{Re} p > 0$):

$$E_{\mathrm{ext}} = \tilde{E}_y(0) \ ,$$

$$\bar{j}_a(p) = \int_0^\infty \mathrm{d}z e^{-pz} \tilde{j}_a(z) ,$$

$$\bar{E}(p) = \int_0^\infty \mathrm{d}z e^{-pz} \tilde{E}_y(z) ,$$

$$D(p) = \int_{-\infty}^\infty \mathrm{d}P (\mathrm{d}F(P)/\mathrm{d}P) \left[p + iP + iC\right]^{-1} . \tag{2.79}$$

The solution of (2.77) is given by

$$\bar{j}_a(p) = -\frac{i\theta_\ell j_0 \omega}{2\gamma_\ell^2 \gamma I_A} \frac{D(p)}{[1 - i\Lambda_p^2 D(p)]}$$

$$\times \sum_{n=-\infty}^{n=+\infty} J_n(Q) \left[\bar{E}(p - 2ink_w) - \bar{E}(p - 2i(n-1)k_w)\right] , \tag{2.80}$$

where $\Lambda_p^2 = 4\pi j_0/(I_A \gamma_\ell^2 \gamma)$. Substituting the expression for $\bar{j}_a(p)$ into (2.78) we obtain:

$$p\bar{E}(p) - E_{\text{ext}} = -\frac{\pi \theta_\ell^2 j_0 \omega}{2c\gamma_\ell^2 \gamma I_A} \sum_{n=-\infty}^{n=+\infty} (-1)^n J_n(Q)$$

$$\times \left\{ \frac{D(p - 2i(n+1)k_w)}{1 - i\Lambda_p^2 D(p - 2i(n+1)k_w)} \right.$$

$$\times \sum_{m=-\infty}^{m=+\infty} J_m(Q) \left[\bar{E}(p - 2i(m+n+1)k_w) - \bar{E}(p - 2i(m+n)k_w)\right]$$

$$- \frac{D(p - 2ink_w)}{1 - i\Lambda_p^2 D(p - 2ink_w)}$$

$$\times \sum_{k=-\infty}^{k=+\infty} J_k(Q) \left[\bar{E}(p - 2i(k+n)k_w) \right.$$

$$\left. \left. - \bar{E}(p - 2i(k+n-1)k_w)\right] \right\} . \tag{2.81}$$

The inverse Laplace transformation is defined by the integral:

$$\tilde{E}_y(z) = \frac{1}{2\pi i} \int_{\gamma'-i\infty}^{\gamma'+i\infty} \mathrm{d}\lambda \bar{E}(\lambda) e^{\lambda z} . \tag{2.82}$$

The constant γ' is a real positive number larger than the real parts of all the singularities of the function $\bar{E}(\lambda)$.

2.1 Linear Mode of Operation   47

We assume the energy spread in the beam to be Gaussian with distribution function (2.20a) and the function $D(\lambda)$ is given by the expression ($\operatorname{Re}\lambda > 0$):

$$D(\lambda) = \mathrm{i} \int_0^\infty \exp\left[-\Lambda_\mathrm{T}^2 \xi^2/2 - (\lambda + \mathrm{i}C)\xi\right] \xi \mathrm{d}\xi , \tag{2.83}$$

where $\Lambda_\mathrm{T}^2 = \omega^2 \langle(\Delta\mathcal{E})^2\rangle/(c^2\gamma_\ell^4\mathcal{E}_0^2)$. As it is difficult to analyze (2.81) in the general form, it seems to be interesting to consider the important case of the high-gain limit. In this case the expression for the radiation field (2.82) reduces to the single residue of the integrand taken in the pole located at the point with the largest positive real part

$$\tilde{E}_y(z) = \operatorname{Res} \bar{E}(\Lambda) \exp(\Lambda z) , \qquad \operatorname{Re}\Lambda > 0 . \tag{2.84}$$

The radiation field changes significantly within the gain length which is much larger than the undulator period. This means that

$$(|\Lambda| , \Lambda_\mathrm{p} , \Lambda_\mathrm{T} , |C|) \ll k_\mathrm{w} . \tag{2.85}$$

Using (2.83) and (2.85) one can show that the following inequality holds for $n \neq 0$:

$$|D(\Lambda)| \gg |D(\Lambda - 2\mathrm{i}nk_\mathrm{w})| \propto (2nk_\mathrm{w})^{-2} .$$

Therefore, all terms of type $D(\lambda - 2\mathrm{i}nk_\mathrm{w})$ can be neglected with respect to the term $D(\lambda)$ in the right-hand part of (2.81) for $\lambda \to \Lambda$. Taking into account that Λ is the pole of the function $\bar{E}(\lambda)$, we can also omit all terms of type $\bar{E}(\lambda - 2\mathrm{i}nk_\mathrm{w})$ except of $\bar{E}(\lambda)$ in the right-hand part of (2.81) for $\lambda \to \Lambda$. It follows from (2.81) that the integrand in (2.82) takes the following form near the pole $\lambda \simeq \Lambda$:

$$\bar{E}(\lambda) \simeq E_\mathrm{ext} \left[\lambda - \frac{D(\lambda)\Gamma^3}{1 - \mathrm{i}\Lambda_\mathrm{p}^2 D(\lambda)}\right]^{-1} , \tag{2.86}$$

where

$$\Gamma = \left[\frac{\pi\theta_\ell^2 j_0 \omega A_\mathrm{JJ}^2}{2c\gamma_\ell^2 \gamma I_\mathrm{A}}\right]^{1/3} .$$

According to (2.86), the eigenvalue equation of the FEL amplifier with planar undulator is reduced to

$$\hat{\Lambda} - \frac{\hat{D}}{1 - \mathrm{i}\hat{\Lambda}_\mathrm{p}^2 \hat{D}} = 0 , \tag{2.87}$$

where $\hat{\Lambda} = \Lambda/\Gamma$, $\hat{\Lambda}_\mathrm{p}^2 = \Lambda_\mathrm{p}^2/\Gamma^2$, $\hat{\Lambda}_\mathrm{T}^2 = \Lambda_\mathrm{T}^2/\Gamma^2$, $\hat{C} = C/\Gamma$ and

$$\hat{D} = \mathrm{i} \int_0^\infty \exp\left[-\hat{\Lambda}_\mathrm{T}^2 \xi^2/2 - (\hat{\Lambda} + \mathrm{i}\hat{C})\xi\right] \xi \mathrm{d}\xi .$$

For small values of the energy spread parameter, $\hat{\Lambda}_T^2 \to 0$, the function $\hat{D}$ tends to $\hat{D} \to i(\hat{\Lambda} + i\hat{C})^{-2}$ and the eigenvalue equation (2.87) takes the form:

$$\left[\left(\hat{\Lambda} + i\hat{C}\right)^2 + \hat{\Lambda}_p^2\right]\hat{\Lambda} = i . \tag{2.88}$$

Using (2.84) and (2.86) we can write the high-gain asymptotic formula:

$$\tilde{E}_y = E_{\text{ext}} \exp(\hat{\Lambda}\hat{z}) \left\{1 - \frac{\hat{D}'}{(1 - i\hat{\Lambda}_p^2 \hat{D})^2}\right\}^{-1} , \tag{2.89}$$

where $\hat{z} = \Gamma z$ and $\hat{D}' = d\hat{D}(\lambda)/d\lambda|_{\lambda = \hat{\Lambda}}$.

One can obtain from (2.30), (2.31), (2.41), (2.83) and (2.87)–(2.89) that all formulae written down in reduced form are identical for both planar and helical undulator.

2.2 Saturation Effects

In the previous section we studied the linear mode of FEL amplifier operation when an increase of the input power W_{ext} leads to a proportional increase of the output power W_{out}. When the input power is increased further, the operation of the amplifier becomes nonlinear: the output power increases more slowly than the input power, and at a certain value of W_{ext} the output power reaches a maximum. To find the FEL characteristics at saturation, it is necessary to solve the equations of the nonlinear theory of the FEL amplifier. Analytical methods are of limited use in the study of the nonlinear regime, and numerical simulations must be used. Application of similarity techniques allows one to present the numerical results in such a form that they are both general and can be applicable for the calculation of specific devices.

Within the scope of the one-dimensional theory the output characteristics of the FEL amplifier are controlled by eight dimensional parameters for the electron beam, undulator and external radiation:[6]

$$l_w , \quad k_w , \quad H_w , \quad \omega , \quad \mathcal{E}_0 , \quad j_0 , \quad \langle(\Delta\mathcal{E})^2\rangle , \quad E_{\text{ext}},$$

where l_w is the length of the undulator, $k_w = 2\pi/\lambda_w$ is the undulator wavenumber, H_w is the magnetic field of the undulator, ω is the frequency of the amplified wave, $\mathcal{E}_0$ is the nominal energy of the electrons, j_0 is the electron beam current density, $\langle(\Delta\mathcal{E})^2\rangle$ is the energy spread in the electron beam, and E_{ext} is the amplitude of the master signal. To be specific, we consider here the case of a helical undulator and Gaussian energy spread in the electron beam.

---

[6] This list contains only parameters which can be changed and optimized during the process of FEL design. Formally the list of dimensional parameters should be extended by the charge, e, and the mass, m, of the particles. However, in practice only electrons are used.

2.2 Saturation Effects

The system of self-consistent field equations describing the beam-wave interaction in the undulator can be formulated as a relation between dimensionless quantities. At an appropriate normalization of the FEL equations, the operation of the FEL amplifier is described by six dimensionless parameters:

$$\hat{l}_w, \; \hat{C}, \; \hat{\Lambda}_p^2, \; \hat{\Lambda}_T^2, \; \rho, \; \hat{E}_{ext},$$

where $\hat{l}_w = \Gamma l_w$ is the dimensionless length of the undulator, $\hat{C}$ is the detuning parameter, $\hat{\Lambda}_p^2$ is the space charge parameter, $\hat{\Lambda}_T^2$ is the energy spread parameter, ρ is the efficiency parameter and

$$\hat{E}_{ext} = E_{ext}/E_0 = e\theta_s E_{ext}/(\rho \mathcal{E}_0 \Gamma)$$

is the normalized initial field amplitude. It is relevant to note that only four dimensionless parameters: $\hat{l}_w, \hat{C}, \hat{\Lambda}_p^2$ and $\hat{\Lambda}_T^2$ are sufficient for a full description of the linear mode of FEL amplifier operation.

The region of applicability of the theory presented in this book requires the efficiency parameter to have a small value, $\rho \ll 1$. As a result, when describing the low-efficiency approximation (for instance, in the nonlinear theory of the FEL amplifier with uniform undulator) we omit the efficiency parameter ρ in the self-consistent field equations. Under this approximation the normalized output field amplitude $\hat{E}_{out} = E_{out}/E_0$ is a function of five parameters:

$$\hat{E}_{out} = \mathcal{D}(\hat{l}_w, \hat{C}, \hat{\Lambda}_p^2, \hat{\Lambda}_T^2, \hat{E}_{ext}) \; .$$

When the FEL amplifier operates in the high-gain regime (at $\hat{E}_{ext} \ll 1$), the normalized field amplitude at saturation is independent of both the length of the undulator and the input signal amplitude. In this important practical case it is a function of only three dimensionless parameters: of the detuning parameter $\hat{C}$, of the space charge parameter $\hat{\Lambda}_p^2$, and of the energy spread parameter $\hat{\Lambda}_T^2$:

$$\max(\hat{E}_{out}) = \mathcal{D}(\hat{C}, \hat{\Lambda}_p^2, \hat{\Lambda}_T^2) \; .$$

In the general case the universal function $\mathcal{D}$ should be calculated numerically by solving the reduced self-consistent equations for the electromagnetic field and for the motion of the particles.

2.2.1 Self-Consistent Equations

The equations of motion can be obtained using the Hamiltonian written in the energy-phase variables $\mathcal{E}$ and the canonically conjugate phase $\psi = k_w z + \omega(z/c - t)$. At small deviations of the electron energy from the nominal value and neglecting the space charge field, the Hamiltonian (2.6) takes the form

$$H = CP + \omega P^2/(2\gamma_z^2 \mathcal{E}_0 c) - (1 - P/\mathcal{E}_0) u(z, \mathcal{E}_0) \sin(\psi + \psi_0) \; , \qquad (2.90)$$

where $P = \mathcal{E} - \mathcal{E}_0$ and C is the detuning from resonance for a particle with nominal energy $\mathcal{E}_0$:

$$C = k_w + \omega/c - \omega/v_z(\mathcal{E}_0) \simeq k_w - \omega/(2c\gamma_z^2) .$$

The amplitude and the phase of the effective potential are connected with the complex amplitude of the electric field $\tilde{E}$ by the relation ($u > 0$):

$$(u/2)\exp(i\psi_0) = -e^2 H_w \tilde{E}/(2\mathcal{E}_0 k_w) = -e\theta_s \tilde{E}/2 = iU .$$

The equations of motion corresponding to the Hamiltonian (2.90) have the form:

$$\begin{aligned} dP/dz &= u\cos(\psi + \psi_0) , \\ d\psi/dz &= \omega P/(\gamma_z^2 \mathcal{E}_0 c) + C + (u/\mathcal{E}_0)\sin(\psi + \psi_0) . \end{aligned} \quad (2.91)$$

The next step consists in writing down the equations in reduced form. The normalization procedure is similar to that used for normalization of the linear equations (see (2.18)):

$$\begin{aligned} \hat{z} &= \Gamma z , \\ \hat{P} &= P/(\rho \mathcal{E}_0) , \\ \hat{u} &= u/(\rho \mathcal{E}_0 \Gamma) = |\tilde{E}|/E_0 , \\ \hat{C} &= C/\Gamma . \end{aligned} \quad (2.92)$$

The equations of motion (2.91) written down in the reduced variables take the form:

$$\begin{aligned} d\hat{P}/d\hat{z} &= \hat{u}\cos(\psi + \psi_0) , \\ d\psi/d\hat{z} &= \hat{P} + \hat{C} + \rho\hat{u}\sin(\psi + \psi_0) . \end{aligned} \quad (2.93)$$

Using (2.16) we find that the amplitude and the phase of the effective potential are governed by the dimensionless equations:

$$\begin{aligned} d\hat{u}/d\hat{z} &= \hat{j}_1 \cos(\psi_0 - \psi_1) , \\ d\psi_0/d\hat{z} &= -(\hat{j}_1/\hat{u})\sin(\psi_0 - \psi_1) , \end{aligned} \quad (2.94)$$

where $\hat{j}_1$ and ψ_1 are, respectively, the amplitude and the phase of the first harmonic of the reduced beam current density $\hat{j}_z = j_z/j_0$:

$$\begin{aligned} \hat{j}_1 \cos\psi_1 &= \frac{1}{\pi}\int_0^{2\pi} \hat{j}_z \cos\psi d\psi , \\ \hat{j}_1 \sin\psi_1 &= -\frac{1}{\pi}\int_0^{2\pi} \hat{j}_z \sin\psi d\psi . \end{aligned} \quad (2.95)$$

The complex amplitude of the first harmonic of the beam current density $\tilde{j}_1$, introduced in Sect. 2.1, is connected with $(\hat{j}_1, \psi_1)$ by the relation:

$$\hat{j}_1 e^{i\psi_1} = 2\tilde{j}_1/j_0 \ .$$

Equations (2.93) and (2.94) form a system of self-consistent equations describing the nonlinear mode of FEL amplifier operation.

The efficiency parameter ρ is inversely proportional to number of the undulator periods per gain length $1/\Gamma$ and is always small. We begin the study of the nonlinear mode of FEL amplifier operation for the case of an untapered undulator. The term proportional to ρ should be omitted in the second equation of (2.93). Keeping this term would give result outside the accuracy of accepted physical approximation.

2.2.2 Numerical Simulation Algorithm

We simulate the electron beam with N macroparticles per interval $(0, 2\pi)$ over the phase ψ. The reduced beam current density $\hat{j}_z = j_z/j_0$ is periodic in the phase ψ and is calculated as[7]

$$\hat{j}_z = -\frac{2\pi}{N} \sum_{j=1}^{N} \delta(\psi - \psi_{(j)}) \ , \qquad 0 \leq \psi \leq 2\pi$$
$$\hat{j}_z(\psi + 2\pi n, z) = \hat{j}_z(\psi, z) \ , \qquad |n| = 0, 1, 2, \ldots \qquad (2.96)$$

where $\psi_{(j)}$ are the phases of the particles and $\delta(\psi - \psi_{(j)})$ is the delta function. It follows from (2.96) that $\hat{j}_z$ is normalized as:

$$\frac{1}{2\pi} \int_0^{2\pi} \hat{j}_z d\psi = -1 \ .$$

The amplitude $\hat{j}_1$ and the phase ψ_1 of the first harmonic of the beam current density are given by the expressions:

$$\hat{j}_1 \cos \psi_1 = \frac{1}{\pi} \int_0^{2\pi} \hat{j}_z \cos \psi d\psi = -\frac{2}{N} \sum_{j=1}^{N} \cos \psi_{(j)} \ ,$$

$$\hat{j}_1 \sin \psi_1 = -\frac{1}{\pi} \int_0^{2\pi} \hat{j}_z \sin \psi d\psi = \frac{2}{N} \sum_{j=1}^{N} \sin \psi_{(j)} \ . \qquad (2.97)$$

The equations of motion (2.93) and the field equations (2.94) form a system of $2N+2$ equations describing the amplification process in the FEL amplifier.

2.2.3 Power Balance

In the framework of the one-dimensional approximation the average power flow density transported by the electromagnetic wave is given by the expression:

---

[7] The minus sign appears in the expression for $\hat{j}_z$ because the electrons have charge $(-e)$ and move in the positive direction along the z axis.

$$\Pi = c|\tilde{E}|^2/(4\pi) \ .$$

In the high-gain limit the efficiency of the FEL amplifier is defined as the ratio of the output radiation power flow density to the flow density of the electron beam power:

$$\eta = e\Pi/(\mathcal{E}_0 j_0) = \rho \hat{u}^2/4 \ . \tag{2.98}$$

The electromagnetic power radiated by the electron beam must be equal to the electron beam power losses, $\Pi = -\langle P \rangle j_0/e$, where $\langle P \rangle$ is the mean energy loss by the electron. Therefore, the FEL efficiency can be defined via the power losses by the electron beam:

$$\eta = -\langle P \rangle / \mathcal{E}_0 \ . \tag{2.99}$$

We should prove that the self-consistent FEL equations provide an energy conservation law, i.e. that relations (2.98) and (2.99) are equivalent. Using the first equation (2.93), we can write the following expression for the rate of the mean energy loss by the electron:

$$\langle \mathrm{d}\hat{P}/\mathrm{d}\hat{z} \rangle = \hat{u} N^{-1} \sum_{j=1}^{N} \cos(\psi_{(j)} + \psi_0) \ .$$

The sum in the latter expression can be rewritten in terms of the amplitude and the phase of the first harmonic of the beam density:

$$N^{-1} \sum_{j=1}^{N} \cos(\psi_{(j)} + \psi_0) = -\frac{1}{2}\hat{j}_1 \cos(\psi_1 - \psi_0) \ .$$

Using the first equation of (2.94), we obtain

$$\left\langle \frac{\mathrm{d}\hat{P}}{\mathrm{d}\hat{z}} \right\rangle = -\frac{\hat{u}}{2}\frac{\mathrm{d}\hat{u}}{\mathrm{d}\hat{z}} \ ,$$

which means that

$$\langle \hat{P} \rangle = -\hat{u}^2/4 \ .$$

Remembering that $\hat{P} = (\mathcal{E} - \mathcal{E}_0)/(\rho \mathcal{E}_0)$, we find that expressions (2.98) and (2.99) are equivalent, i.e. the power is balanced.

In what follows it is convenient to introduce the notion of the reduced efficiency defined as

$$\hat{\eta} = \eta/\rho \ . \tag{2.100}$$

2.2.4 Saturation in the High-Gain FEL Amplifier

We consider the initial conditions when the electron beam is neither modulated in velocity nor in density and there is an electromagnetic wave of amplitude E_{ext} at the undulator entrance at $z = 0$ ($j = 1, \ldots, N$):

$$\hat{P}_{(j)}(0) = 0 \ , \qquad \hat{j}_1(0) = 0 \ , \qquad \hat{u}(0) = \hat{u}_{\mathrm{ext}} = E_{\mathrm{ext}}/E_0 \ .$$

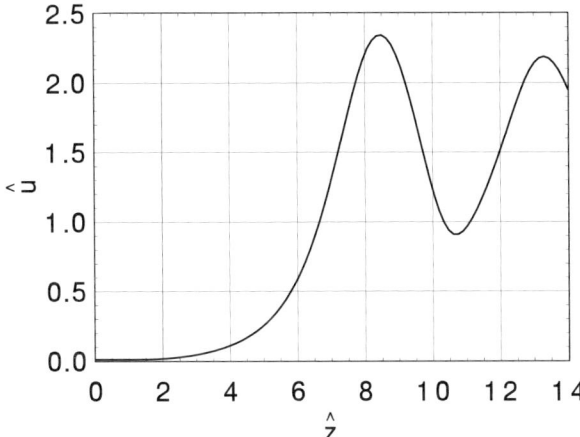

Fig. 2.15. Reduced field amplitude $\hat{u}$ versus reduced length of the undulator, $\hat{z}$. Here $\hat{C} = 0$, $\hat{\Lambda}_{\mathrm{p}}^2 \to 0$, $\hat{\Lambda}_{\mathrm{T}}^2 = 0$ and $\hat{u}_{\mathrm{ext}} = 0.01$

According to (2.92), the saturation field parameter is given by

$$E_0 = \rho \mathcal{E}_0 \Gamma / (e \theta_{\mathrm{s}}) \ . \tag{2.101}$$

We start with the simplest situation neglecting the influence of the space charge and the energy spread effects on the operation of the FEL amplifier. The plot in Fig. 2.15 presents the dependence of the reduced field amplitude $\hat{u}$ on the reduced undulator length $\hat{z}$ for $\hat{C} = 0$ and $\hat{u}_{\mathrm{ext}} = 0.01$. The field stops growing at the saturation point when the beam is overmodulated and a significant fraction of the electrons falls into the accelerating phase of the effective potential. The maximal value of the reduced field amplitude at $\hat{C} = 0$ is equal to

$$\hat{u}_{\mathrm{max}} = E_{\mathrm{max}}/E_0 = 2.34 \ , \tag{2.102}$$

which corresponds to the value of the reduced efficiency (see (2.98) and (2.100)):

$$\hat{\eta}_{\mathrm{max}} = 1.37 \ . \tag{2.103}$$

It should be noted that the saturation efficiency of the FEL amplifier does not depend on the amplitude of the master signal when the FEL amplifier operates in the high-gain regime, i.e. when $E_{\mathrm{ext}}/E_0 \ll 1$.

To analyze the dynamics of the particles in the undulator, it is convenient to study their distribution in the phase plane $(\hat{P}, \Delta\psi)$, where $\Delta\psi = \psi + \psi_0$. Figure 2.16 presents such distributions when the FEL amplifier operates in the linear regime (plot (a)), just before saturation (plot (b)), and at saturation (plot (c)).

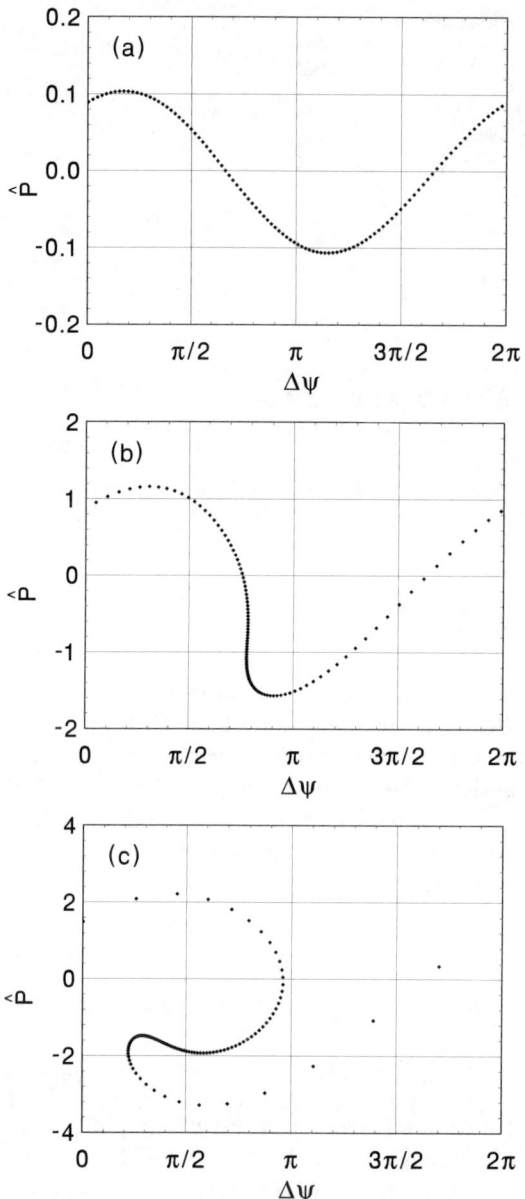

Fig. 2.16. Phase space distribution of the particles at different stages of the amplification. Graph (**a**): $\hat{z} = 4$, graph (**b**): $\hat{z} = 7$, and graph (**c**): $\hat{z} = 8.4$ (saturation point, see Fig. 2.15). Here $\hat{C} = 0$, $\hat{\Lambda}_p^2 \to 0$, $\hat{\Lambda}_T^2 = 0$ and $\hat{u}_{\text{ext}} = 0.01$

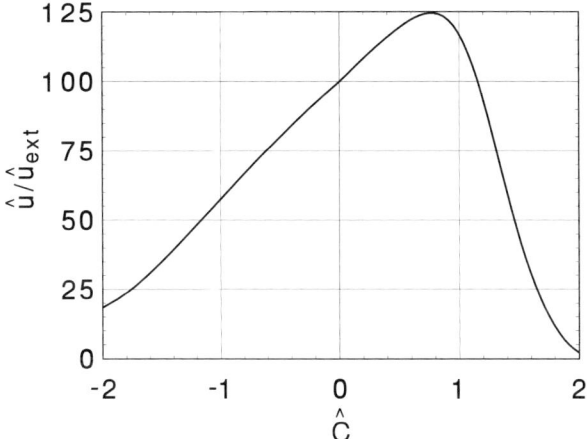

Fig. 2.17. Field gain $\hat{u}/\hat{u}_{\text{ext}}$ versus the detuning parameter $\hat{C}$. Here the reduced undulator length is $\hat{z} = 7.4$, $\hat{\Lambda}_p^2 \to 0$, $\hat{\Lambda}_T^2 = 0$ and $\hat{u}_{\text{ext}} = 0.0234$. At exact resonance, $\hat{C} = 0$, the FEL amplifier operates at the saturation point

When the FEL amplifier is tuned to exact resonance, $\hat{C} = 0$, and when the reduced length of the amplifier is $\hat{z} > 4$, the power gain at saturation can be calculated with the approximate formula

$$G_{\text{max}} = \frac{1}{38} \exp\left(\sqrt{3}\hat{z}\right), \qquad (2.104)$$

or, in decibels:

$$G_{\text{max}}(\text{dB}) = 10 \log G_{\text{max}} = 7.5\hat{z} - 15.8 \ .$$

The field amplitude at the amplifier entrance, E_{ext}, providing the saturation regime at the undulator exit, can be found using the expression:

$$E_{\text{ext}}/E_0 = \hat{u}_{\text{ext}} = 2.34/\sqrt{G_{\text{max}}}, \qquad (2.105)$$

where the value of G_{max} is given by (2.104). These simple formulae provide an accuracy of several per cent with respect to the results of numerical simulations.

The plot presented in Fig. 2.15 enables one to calculate the growth of the field amplitude for any amplifier tuned to the exact resonance $\hat{C} = 0$ and having reduced length $\hat{z} > 4$. It is seen from this plot and relations (2.104) and (2.105) that the coordinate of the saturation point, $\hat{z}_{\text{max}}$, is given by the relation:

$$\hat{z}_{\text{max}} = 3.1 + \frac{2}{\sqrt{3}} \ln\left(\hat{u}_{\text{ext}}^{-1}\right) . \qquad (2.106)$$

The field amplitude at the point $\hat{z} = \hat{z}_{\text{max}} - \Delta\hat{z}$ is equal to the field amplitude corresponding to that at $\hat{z} = 8.4 - \Delta\hat{z}$ presented in the plot in Fig. 2.15. When $\Delta\hat{z} > 4$, the linear approximation becomes applicable, and the field amplitude

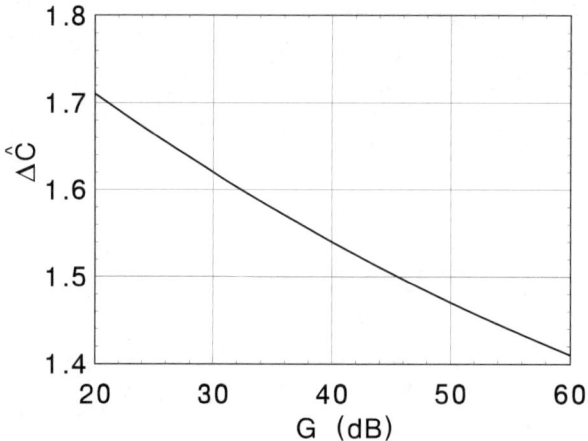

Fig. 2.18. FWHM power amplification bandwidth $\Delta \hat{C}$ versus maximal power gain. At exact resonance, $\hat{C} = 0$, the FEL amplifier operates at the saturation point. Here $\hat{\Lambda}_p^2 \to 0$ and $\hat{\Lambda}_T^2 = 0$

E can be found using (2.35) where E_{ext} should be calculated using (2.104) and (2.105).

Let us consider a specific numerical example for calculating the parameters of the FEL amplifier tuned to exact resonance. Let the reduced length of the amplifier be $\hat{z} = 7.4$. The problem is to define the value of the input signal providing saturation at the given undulator length. According to (2.104) and (2.105), saturation is reached at the value of the input field amplitude $\hat{u}_{\text{ext}} = 2.34 \times 10^{-2}$ and the power gain is equal to $G = 40$ dB.

An important characteristic of the FEL amplifier is the amplification bandwidth. Figure 2.17 presents a specific example of the dependence of the field gain on the detuning parameter. We obtain from this plot that the power gain is 40 dB at tuning to the exact resonance $\hat{C} = 0$. The FWHM power amplification bandwidth is equal to $\Delta \hat{C} = 1.54$. The plot presented in Fig. 2.18 can be used to find $\Delta \hat{C}$ for an FEL amplifier with a different value of the gain G (i.e. with different reduced length).

In the general case, the maximal reduced efficiency of the FEL amplifier $\hat{\eta}_{\text{max}}$ is a universal function of the detuning parameter $\hat{C}$. The graph of this function is presented in Fig. 2.19. One can see that the maximal efficiency is an increasing function of the detuning parameter. This phenomenon can be simply explained: the electrons interact with the wave for a longer distance when the detuning parameter is increased.

The power gain at the saturation point is also a universal function of the detuning parameter $\hat{C}$:

$$G_{\text{max}} = A_{\text{m}}(\hat{C}) \exp\left[2 \operatorname{Re} \hat{\Lambda} \hat{z}\right] . \tag{2.107}$$

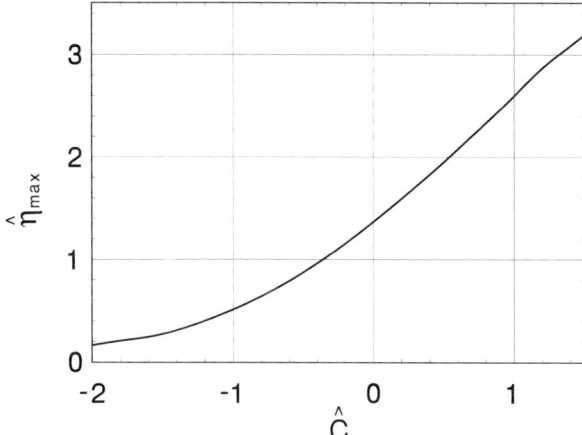

Fig. 2.19. Maximal reduced efficiency $\hat{\eta}_{\mathrm{max}}$ of the FEL amplifier versus the detuning parameter $\hat{C}$. Here $\hat{\Lambda}_{\mathrm{p}}^2 \to 0$ and $\hat{\Lambda}_{\mathrm{T}}^2 = 0$

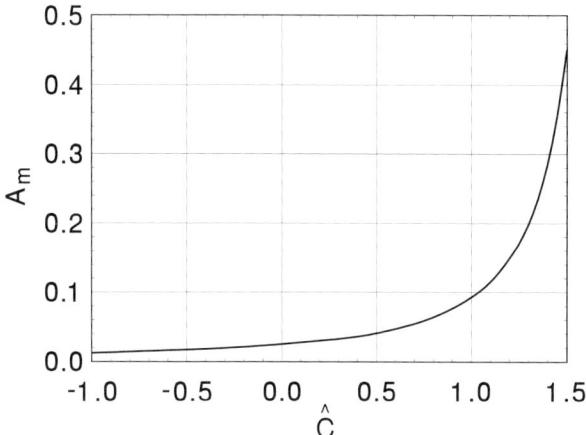

Fig. 2.20. Input coupling factor A_{m} entering (2.107) versus the detuning parameter $\hat{C}$

The graph of the function $A_{\mathrm{m}}(\hat{C})$ is presented in Fig. 2.20. The values of the reduced field growth rate, $\operatorname{Re}\hat{\Lambda}$, can be found with the help of the plot presented in Fig. 2.2. The amplitude of the input signal providing saturation at the exit of the undulator can be obtained using the graphs presented in Figs. 2.2, 2.19 and 2.20 and using the expression

$$\hat{u}_{\mathrm{ext}} = 2\left(\hat{\eta}_{\mathrm{max}}/G_{\mathrm{max}}\right)^{1/2}.$$

Another important characteristic of the amplifier is the amplitude characteristic, i.e. the dependence of the output field amplitude on the input

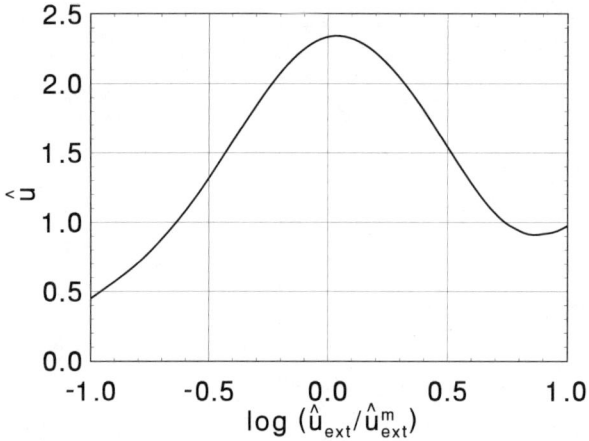

Fig. 2.21. Output field amplitude $\hat{u}$ versus the input field amplitude $\hat{u}_{\text{ext}}$. At $\hat{u}_{\text{ext}} = \hat{u}_{\text{ext}}^{\text{m}}$ the FEL amplifier operates at the saturation point. Here $\hat{C} = 0$, $\hat{\Lambda}_{\text{p}}^2 \to 0$ and $\hat{\Lambda}_{\text{T}}^2 = 0$

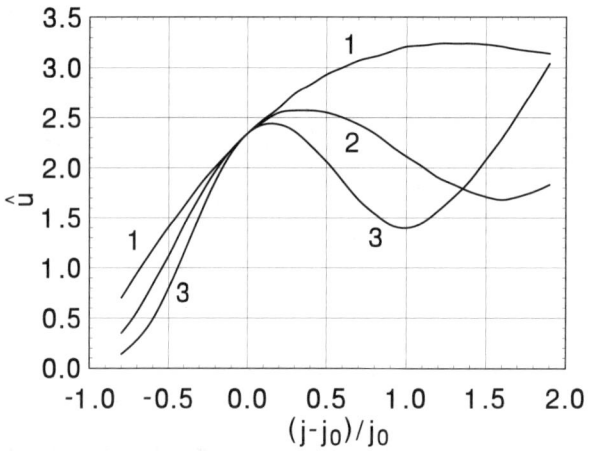

Fig. 2.22. Output field amplitude versus the beam current density. At $j = j_0$ the amplifier operates at the saturation point. The reduced parameters are calculated at $j = j_0$. Curve 1: $G = 20$ dB. Curve 2: $G = 40$ dB. Curve 3: $G = 60$ dB. Here $\hat{C} = 0$, $\hat{\Lambda}_{\text{p}}^2 \to 0$ and $\hat{\Lambda}_{\text{T}}^2 = 0$

amplitude. The dependence of $\hat{u}$ on $\hat{u}_{\text{ext}}$ for the FEL amplifier operating in the saturation regime is presented in Fig. 2.21. It should be noted that the amplitude characteristic of the amplifier is independent of the gain value when the FEL amplifier operates in the high-gain limit.

One of the problems for FEL designers is to define tolerances of the beam current deviation from a nominal value. This can be done using the plots in

Fig. 2.22 showing the dependence of the field amplitude at the amplifier exit on the value of the beam current for different values of the power gain G.

As a rule, the number of macroparticles for numerical simulations of the FEL amplifier with a cold electron beam has to be chosen in the range $N = 100$–200. In this case the results of simulations are independent of the value of N and the accuracy of simulations is better than 0.1 %. The simulation code can be carefully tested by simulating the linear mode of FEL amplifier operation. In Fig. 2.1 we compare the results of the calculations of the linear stage of FEL amplifier operation obtained with the analytical formulae and with the numerical simulation algorithm. In Fig. 2.2 we compare analytical results (2.36) for the field growth rate with those obtained from numerical simulations. In Fig. 2.5 we compare the amplitude-frequency characteristics obtained by numerical simulations and by analytic solution of the initial-value problem using (2.33). One can show that there is good agreement between the numerical and analytical results.

2.2.5 Space Charge Effects

All the results presented in the previous section refer to the case when there is no influence of the space charge field on the amplification process. In this section we study the influence of the space charge effects on FEL amplifier operation.

In the framework of the one-dimensional model the space charge field can be calculated using a method equivalent to the Green's function method. In the nonlinear regime the beam current j_z is a periodic function of the phase ψ and can be expanded in a Fourier series:

$$j_z = \sum_{n=1}^{\infty} j_n \cos(n\psi + \psi_n) ,$$

with j_n and ψ_n are given by the expression

$$j_n \left\{ \begin{array}{c} \cos(\psi_n) \\ \sin(\psi_n) \end{array} \right\} = \frac{1}{\pi} \int_0^{2\pi} j_z \left\{ \begin{array}{c} \cos(n\psi) \\ -\sin(n\psi) \end{array} \right\} d\psi .$$

In the framework of the one-dimensional approximation we obtain from Maxwell's equations that

$$E_z = 4\pi\omega^{-1} \sum_{n=1}^{\infty} n^{-1} j_n \sin(n\psi + \psi_n) . \tag{2.108}$$

When performing simulations, we represent the electron beam with N layers per interval $(0, 2\pi)$ over the phase ψ, and represent the beam current j_z in the form

$$j_z = -\frac{2\pi}{N} j_0 \sum_{j=1}^{N} \delta(\psi - \psi_{(j)}) ,$$

2. One-Dimensional Theory of the FEL Amplifier

where $\delta(\psi - \psi_{(j)})$ is the delta function and $\psi_{(j)}$ is the phase of the jth layer. The amplitude j_n and phase ψ_n of the beam current Fourier harmonic are given by the following expression:

$$j_n \begin{Bmatrix} \cos(\psi_n) \\ \sin(\psi_n) \end{Bmatrix} = \frac{2}{N} j_0 \sum_{j=1}^{N} \begin{Bmatrix} -\cos(n\psi_{(j)}) \\ \sin(n\psi_{(j)}) \end{Bmatrix}.$$

Substituting this expression into eq. (2.108) for E_z and using the relation

$$\sum_{n=1}^{\infty} n^{-1} \sin(n\xi) = (\pi - \xi)/2, \qquad (0 < \xi < 2\pi),$$

we get an expression for the longitudinal electric field at the position of the ith layer:

$$E_z^{(i)} = -\frac{4\pi j_0}{N\omega} \sum_{j \neq i} \left[\pi \, \text{sgn}\left(\psi_{(i)} - \psi_{(j)}\right) - \left(\psi_{(i)} - \psi_{(j)}\right) \right], \qquad (2.109)$$

where

$$\begin{aligned} \text{sgn}\left(\psi_{(i)} - \psi_{(j)}\right) &= 1 & \text{for } \left(\psi_{(i)} - \psi_{(j)}\right) > 0, \\ \text{sgn}\left(\psi_{(i)} - \psi_{(j)}\right) &= -1 & \text{for } \left(\psi_{(i)} - \psi_{(j)}\right) < 0. \end{aligned}$$

Using the Hamiltonian (2.6) and the expression (2.109) we can write the equations of motion for the ith layer:

$$\frac{d\hat{P}_{(i)}}{d\hat{z}} = \hat{u} \cos\left(\psi_{(i)} + \psi_0\right)$$

$$+ \hat{\Lambda}_P^2 \left\{ \frac{1}{N} \sum_{j \neq i} \left[\pi \, \text{sgn}\left(\psi_{(i)} - \psi_{(j)}\right) - \left(\psi_{(i)} - \psi_{(j)}\right) \right] \right\}$$

$$\frac{d\psi_{(i)}}{d\hat{z}} = \hat{P}_{(i)} + \hat{C}. \qquad (2.110)$$

It should be noted that there is a significant difference between the linear and nonlinear modes of FEL operation concerning the calculation of the space charge field. In the linear mode of FEL operation, both the radiation fields and space charge fields are calculated using only the first harmonic of the electron beam density modulation. In the nonlinear mode the radiation fields are calculated in the same way, but the space charge fields must be calculated using all higher harmonics of the beam density modulation. We have shown above that the Fourier series of the space charge harmonics is reduced to the summing up of the trigonometric series, and the latter results in a simple algebraic function. Therefore, the approach developed here gives a radical solution of the problem providing a strict result for the space charge field, so that all the space charge harmonics are taken into account in the algorithm.

When the space charge field is taken into account, the maximal reduced efficiency of the FEL amplifier operating in the high-gain limit is a universal function of two parameters, the detuning parameter $\hat{C}$ and the space charge

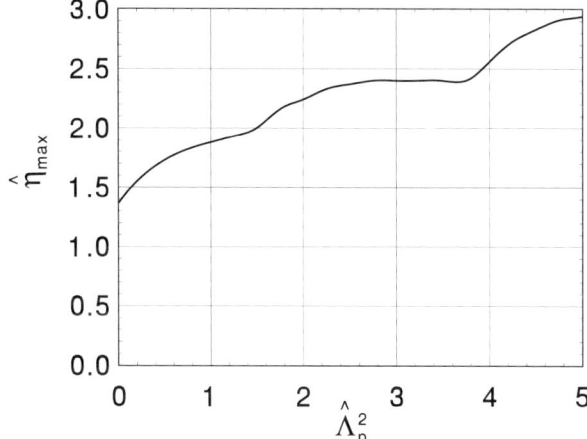

Fig. 2.23. Maximal reduced efficiency $\hat{\eta}_{\mathrm{max}}$ of the FEL amplifier versus the space charge parameter $\hat{\Lambda}_{\mathrm{p}}^2$. Here $\hat{\Lambda}_{\mathrm{T}}^2 = 0$ and the value of the detuning parameter $\hat{C} = \hat{C}_{\mathrm{m}}(\hat{\Lambda}_{\mathrm{p}}^2)$ corresponds to the maximal gain in the linear mode of operation (see Fig. 2.8)

parameter $\hat{\Lambda}_{\mathrm{p}}^2$ (we assume here that there is no energy spread in the electron beam, $\hat{\Lambda}_{\mathrm{T}}^2 = 0$). When the amplifier is tuned to the maximal field growth rate in the linear mode of operation, the FEL amplifier efficiency at saturation is a universal function of the space charge parameter $\hat{\Lambda}_{\mathrm{p}}^2$ only. This dependence is presented in Fig. 2.23. One can see that the maximal efficiency of the FEL amplifier is an increasing function of the space charge parameter. The reason for this is that the space charge fields prevent overmodulation of the beam near the saturation point and the interaction of the modulated electron beam with the wave is prolonged. It should be noted, however, that the undulator length is also increased in this case.

As we mentioned above, the numerical simulation code can be thoroughly tested using the results of the linear theory. Figure 2.6 presents comparative results of numerical simulations and analytical calculations. The difference between the analytical and simulation results is less than 0.1% when the number of macroparticles is $N = 200$.

2.2.6 Energy Spread Effects

To be specific, we assume the energy spread in the electron to be Gaussian. The maximal reduced efficiency of the amplifier operating in the high-gain limit is a universal function of three parameters: the detuning parameter $\hat{C}$, the space charge parameter $\hat{\Lambda}_{\mathrm{p}}^2$ and the energy spread parameter $\hat{\Lambda}_{\mathrm{T}}^2$. When the space charge field can be neglected, $\hat{\Lambda}_{\mathrm{p}}^2 \to 0$, and when the amplifier is tuned to the maximal field growth rate in the linear mode of operation, the

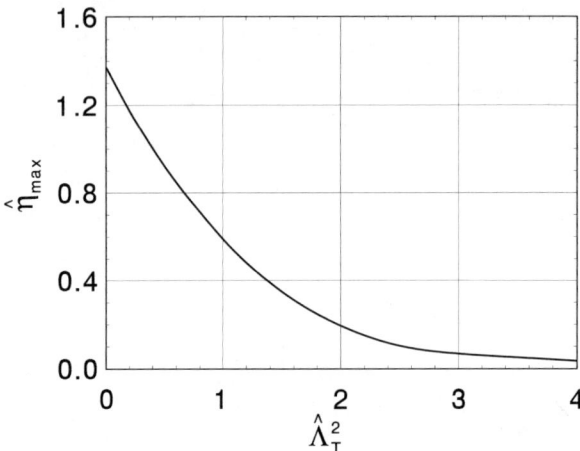

Fig. 2.24. Maximal reduced efficiency $\hat{\eta}_{\mathrm{max}}$ of the FEL amplifier versus the energy spread parameter $\hat{\Lambda}_{\mathrm{T}}^2$. Here $\hat{\Lambda}_{\mathrm{p}}^2 \to 0$ and the value of the detuning parameter $\hat{C} = \hat{C}_{\mathrm{m}}(\hat{\Lambda}_{\mathrm{T}}^2)$ corresponds to the maximal gain in the linear mode of operation (see Fig. 2.13)

reduced efficiency at saturation is a universal function of the energy spread parameter $\hat{\Lambda}_{\mathrm{T}}^2$ only. The plot of this function is presented in Fig. 2.24. One can obtain from this plot that the energy spread drastically decreases the FEL amplifier efficiency.

Simulation of the energy spread effect does not require modification of the self-consistent equations presented in the previous sections. The only difference consists in the preparation of an initial ensemble of macroparticles. The simplest way is to distribute the macroparticles in the phase space in accordance with the initial distribution function. Nevertheless, this method is noisy and requires too big a number of macroparticles in order to achieve the required accuracy. Also, there is a danger of simulating non-physical effects connected with uncontrolled values of the harmonics appearing due to the limited number of macroparticles. We recommend the following method which allows us to achieve higher accuracy of the simulations with a limited number of macroparticles. When simulating cold electron beam we have distributed macroparticles evenly over the phase interval $(0, 2\pi)$ putting one particle at each point. For simulation of the energy spread we put at each phase slot an ensemble of the macroparticles corresponding to the desired energy spread model. An effective method of particle distribution is the so-called method of prescribed positioning of the particles. The idea of the method consists in the exact calculation of the initial energy deviations of the particles in accordance with the initial distribution function. The value of the first harmonic is always equal to zero in this case. After preparing an initial ensemble, the drift space of length $\hat{l}_{\mathrm{p}}$ should be introduced to let the particles distribute more homogeneously in the phase ψ. This results in a strong suppression of the unwilling higher harmonics of the density modulation, while the energy distribution remains the same as the initial one. When simulating the Gaussian energy distribution the optimal value of the length of such a drift space is given by the expression:

$$\hat{l}_p = \frac{0.18}{\hat{A}_T} \frac{N_p}{N_\psi},$$

where N_ψ is the number of divisions in phase ψ and N_p is the number of macroparticles at each phase point used for simulation of the energy spread. Such a technique for the preparation of the initial macroparticle ensemble enables one to simulate rather well an unmodulated electron beam with the number of divisions $N_\psi = 4$ and $N_p = 200$.

The simulation code described above can be tested at the linear stage. Figures 2.6 and 2.13 present comparative results obtained by means of numerical simulations (circles) and analytical results (solid curves calculated with (2.48) and (2.49)). It is seen from these plots that there is good agreement between the numerical and analytical results.

2.2.7 FEL Amplifier with a Planar Undulator

In the previous section we studied saturation effects in the FEL amplifier with the helical undulator amplifying a circularly polarized electromagnetic wave. All the results and plots obtained above refer also to the case of the linear polarized radiation

$$\boldsymbol{E} = \boldsymbol{e}_y \tilde{E}_y(z) \exp[i\omega(z/c - t)] + \text{C.C.}$$

and the planar undulator with the field

$$\boldsymbol{H} = \boldsymbol{e}_x H_\ell \cos(k_w z)$$

after the renormalization procedure considered below.

Let us demonstrate the calculation of the characteristics of the FEL amplifier with planar undulator. The equations of motion of the electrons in the field of the planar undulator can be obtained using the Hamiltonian (2.64) written down in the energy-phase variables $P = \mathcal{E} - \mathcal{E}_0$ and the canonically conjugate phase $\psi = k_w z + \omega(z/c - t)$. In the limit of small influence of the space charge field the equations of motion have the form:

$$dP/dz = iU \exp(i\psi) + \text{C.C.},$$

$$d\psi/dz = C + \frac{\omega \theta_\ell^2}{4c} \cos(2k_w z) + \frac{\omega P}{c\gamma_\ell^2 \mathcal{E}_0},$$

where $C = k_w - \omega/(2\gamma_\ell^2 c)$ is the detuning of the electron with the nominal energy $\mathcal{E}_0$,

$$U = -[1 - \exp(-2ik_w z)] e \theta_\ell \tilde{E}_y(z)/2$$

is the complex amplitude of the effective potential, and

$$\theta_\ell = eH_\ell/(\mathcal{E}_0 k_w) , \qquad \gamma_\ell^{-2} = \gamma^{-2} + \theta_\ell^2/2 .$$

The complex amplitude of the amplified wave can be found from (2.68):

$$d\tilde{E}_y/dz = -i\pi\theta_\ell c^{-1}[\exp(2ik_w z) - 1]\tilde{j}_1(z) ,$$

where $\tilde{j}_1(z)$ is the complex amplitude of the first harmonic of the beam current density. Then we choose the new definition of the phase

$$\psi' = \psi - Q\sin(2k_w z) ,$$

where $Q = \omega\theta_\ell^2/(8k_w c)$. Rewriting the equations of motion and the field equation in terms of the new phase ψ' and averaging over the undulator period, we get

$$dP/dz = -(i/2)A_{\mathrm{JJ}}e\theta_\ell \tilde{E}_y e^{i\psi'} + \mathrm{C.C.} ,$$
$$d\psi'/dz = C + \omega P/(c\gamma_\ell^2 \mathcal{E}_0) ,$$
$$d\tilde{E}_y/dz = i\pi A_{\mathrm{JJ}}\theta_\ell c^{-1}\tilde{j}_{\mathrm{a}} ,$$

where

$$A_{\mathrm{JJ}} = [J_0(Q) - J_1(Q)] ,$$
$$\tilde{j}_{\mathrm{a}}(z) = \tilde{j}_1(z)\exp[iQ\sin(2k_w z)] .$$

In the same way as was done in Sect. 2.2.1 we perform the normalization procedure

$$\hat{z} = \Gamma z ,$$
$$\hat{P} = P/(\rho\mathcal{E}_0) ,$$
$$\hat{C} = C/\Gamma ,$$
$$\hat{u} = |\tilde{E}_y|/E_0 ,$$

where

$$\Gamma = \left[\pi j_0 \theta_\ell^2 \omega A_{\mathrm{JJ}}^2 \gamma^{-1}\gamma_\ell^{-2} I_{\mathrm{A}}^{-1}(2c)^{-1}\right]^{1/3} ,$$
$$E_0 = c\mathcal{E}_0 \gamma_\ell^2 \Gamma^2/(e\theta_\ell \omega A_{\mathrm{JJ}}) ,$$
$$\rho = c\gamma_\ell^2 \Gamma/\omega .$$

As a result, we can write the self-consistent equations in the following reduced form:

$$d\hat{P}/d\hat{z} = \hat{u}\cos(\psi' + \psi_0) ,$$
$$d\psi'/d\hat{z} = \hat{P} + \hat{C} , \qquad (2.111)$$

$$d\hat{u}/d\hat{z} = \hat{j}_{\mathrm{a}}\cos(\psi_0 - \psi_{\mathrm{a}}) ,$$
$$d\psi_0/d\hat{z} = -(\hat{j}_{\mathrm{a}}/\hat{u})\sin(\psi_0 - \psi_{\mathrm{a}}) , \qquad (2.112)$$

where $(\hat{u}, \psi_0)$ and $(\hat{j}_a, \psi_a)$ are connected with the complex amplitudes $\tilde{E}_y$ and $\tilde{j}_a$ by the relations

$$\hat{u}e^{i\psi_0} = -i\tilde{E}_y/E_0, \qquad \hat{j}_a e^{i\psi_a} = 2\tilde{j}_a/j_0.$$

One can obtain from (2.93), (2.94) and (2.111), (2.112) that the reduced self-consistent equations are identical for both the cases of the helical and the planar undulator.

The next problem is the calculation of the space charge fields. The expression for the space charge field E_z may be obtained in the same way as in Sect. 2.2.5. Indeed, independently of the type of undulator, the beam current density j_z is a periodic function of time and may be expanded in a Fourier series. Using Maxwell's equation, in the one-dimensional approximation we get expression (2.108) for E_z. After summation over n we find that formula (2.109) is valid also for the case of the planar undulator. It is clearly seen from (2.109) that the field E_z depends only on the phase difference between the macroparticles, $(\psi_{(i)} - \psi_{(j)})$, so it is not a rapidly oscillating function of z (while the phases ψ of the macroparticles are rapidly oscillating). Using the definition of the new phase ψ' we can rewrite (2.110) in the form:

$$\frac{d\hat{P}_{(i)}}{d\hat{z}} = \hat{u}\cos\left(\psi'_{(i)} + \psi_0\right),$$

$$+\hat{\Lambda}_p^2 \left\{ \frac{1}{N} \sum_{j \neq i} \left[\pi \operatorname{sgn}\left(\psi'_{(i)} - \psi'_{(j)}\right) - \left(\psi'_{(i)} - \psi'_{(j)}\right)\right] \right\}$$

$$\frac{d\psi'_{(i)}}{d\hat{z}} = \hat{P}_{(i)} + \hat{C}, \qquad (2.113)$$

where the space charge parameter for the FEL amplifier with planar undulator is defined as

$$\hat{\Lambda}_p^2 = \Lambda_p^2/\Gamma^2, \qquad \Lambda_p^2 = 4\pi j_0 \gamma^{-1} \gamma_\ell^{-2} I_A^{-1}.$$

In the case of a Gaussian energy spread in the electron beam (2.20a), the following definition of the energy spread parameter for the FEL amplifier with planar undulator should be used:

$$\hat{\Lambda}_T^2 = \Lambda_T^2/\Gamma^2, \qquad \Lambda_T^2 = c^{-2}\omega^2 \gamma_\ell^{-4}\langle(\Delta\mathcal{E})^2\rangle\mathcal{E}_0^{-2}.$$

2.3 FEL Amplifier with Tapered Undulator

The operation of the FEL amplifier is based on the prolonged (resonance) interaction of the electron beam with the electromagnetic wave in the undulator. The amplification process can be divided into two stages, linear and nonlinear. During the linear stage of amplification, exponential growth of the electromagnetic field amplitude and of the beam modulation amplitude takes place. Nevertheless, the beam modulation is much less than unity in the linear regime, and the largest fraction of the radiation power is produced at the nonlinear stage of the operation, when the beam modulation becomes about unity. In the case of an untapered undulator, the bunched beam effectively interacts with the electromagnetic wave along a length which is of the order of the gain length $l_\mathrm{g} \simeq \Gamma^{-1}$. At this stage of the amplification electrons lose the visual fraction of their energy which results in the violation of the resonance condition. As a result, the beam is overmodulated, most electrons fall into the accelerating phase of the effective potential and the electron beam starts to absorb power from the electromagnetic wave. Remembering that the field amplitude at saturation (see (2.101) and (2.102)) is of the order of $E_0 = \rho \mathcal{E}_0 \Gamma/(e\theta_\mathrm{s})$, we estimate the saturation efficiency of the FEL amplifier to be of the order of $\eta \simeq eE_0\theta_\mathrm{s}/(\mathcal{E}_0\Gamma) = \rho$. So, we see that the efficiency of the FEL amplifier with an untapered undulator is limited by the value of the efficiency parameter ρ which is always much less than unity.

The problem to be solved is how to prolong the interaction of the bunched electron beam with the electromagnetic wave. A reliable method to increase the FEL amplifier efficiency consists in an adiabatic change of the undulator parameters (or, in other words, by the use of so-called undulator tapering). Analytical techniques do not provide any reliable basis for a description of such a complicated problem as trapping of the particles in the regime of coherent deceleration, and we perform the corresponding analysis using the results of numerical simulations. Nevertheless, application of similarity techniques makes the results of these investigations physically transparent and the entire process can be simply characterized with a few dimensionless parameters.

To make our discussion clearer, we begin our study with the case of a low efficiency approximation, when the FEL amplifier efficiency increases significantly with respect to the case of an untapered undulator but still remains much less than unity. Then we study the more complicated case of a high-efficiency FEL amplifier. All the results are obtained by means of similarity techniques and possess a high degree of generality.

2.3.1 Low-Efficiency Approximation

So far only the motion in a uniform undulator has been considered. The situation with a tapered undulator can be seen as a simple extension of what has been treated already. First, we transform the Hamiltonian (2.1) from variables p_z and t to variables convenient for describing the amplification process in the tapered undulator. We introduce the phase of a particle with respect to the wave by

$$\psi = \int k_\mathrm{w}(z)\mathrm{d}z + \omega(z/c - t) \ .$$

According to formulae (A.1.8) and (A.1.10) the transformation $(t, z) \to (z, \psi)$ is canonical when

$$\mathcal{P} = -p_0/\omega , \qquad \mathcal{P}_0 = p_z + (p_0/\omega)(k_w(z) + \omega/c) ,$$

where (p_0, p_z) and $(\mathcal{P}_0, \mathcal{P})$ are the old and new canonical momenta conjugated to (t, z) and (z, ψ), respectively. Hence, the new Hamiltonian $\tilde{H}(\mathcal{P}, \psi, z)$ is given by expression (2.3), where k_w and $\boldsymbol{A}_w$ are now functions of z. For small deviations of the electron energy from the nominal value and neglecting the space charge field, the Hamiltonian (2.5) takes the form (2.90). In the same way as was done in Sect. 2.2.1 we find that the equations of motion have the form (2.93). According to the system of equations (2.93), the phase motion of the particles is determined by the detuning

$$C = k_w - \frac{\omega}{2c\gamma_z^2} = k_w - \frac{\omega(1 + K^2)}{2c\gamma^2} ,$$

which is a function of the undulator period $\lambda_w = 2\pi/k_w$ and of the undulator parameter $K = eH_w/(m_e c^2 k_w)$. When the undulator parameters change along the undulator length, the detuning is a function of the z coordinate. In this Section we study the undulator tapering at a fixed value of the undulator parameter $K = \text{const}$. In this case the magnetic field of the undulator should be changed inversely proportionally to the undulator period, $H_w \propto \lambda_w^{-1}$.

The magnetic field on the axis of a helical undulator with bifilar winding has the form:

$$H_w \propto I_w R_w^{-1} \left[(k_w R_w)^2 K_0(k_w R_w) + (k_w R_w) K_1(k_w R_w) \right] ,$$

where K_0 and K_1 are modified Bessel functions, and I_w and R_w are the current and radius of the winding, respectively. The undulator parameter K for such an undulator is a universal function of the parameter $k_w R_w$ for fixed current through the winding. Therefore, the undulator tapering in this case can be organized by shaping the radius of the winding proportionally to the undulator period, $R_w \propto \lambda_w$.

In this section we study the case of a low-efficiency approximation, when the FEL amplifier efficiency increases significantly with respect to the case of an untapered undulator, but still remains much less than unity. The condition of the small value of the efficiency η enables one to use the system of equations (2.93) and (2.94) which has been used in the previous Sections for simulations of the FEL amplifier with an untapered undulator. Here we also neglect the term proportional to ρ in (2.93). The only difference is that the detuning parameter $\hat{C}$ in the second equation of (2.93) becomes a function of the z coordinate, $\hat{C} = \hat{C}(0) + \Delta k_w(z)/\Gamma$ at $K = \text{const}$.

Optimization of Undulator Parameters. The undulator consists of two sections. Let us consider the case when the detuning is fixed in the initial section and grows as a quadratic polynomial after the beginning of the tapering:

$$C(z) = C(0) + \Delta k_w(z) = b_0 + b_1(z - z_i) + b_2(z - z_i)^2 ,$$

where z_i is the coordinate where the tapering begins. The detuning parameter $\hat{C}$ entering (2.93) can be written in reduced form as

$$\hat{C} = \hat{b}_0 + \hat{b}_1(\hat{z} - \hat{z}_i) + \hat{b}_2(\hat{z} - \hat{z}_i)^2 ,$$

$$\hat{z}_i = \Gamma z_i, \qquad \hat{b}_0 = b_0/\Gamma , \qquad \hat{b}_1 = b_1/\Gamma^2 , \qquad \hat{b}_2 = b_2/\Gamma^3 . \qquad (2.114)$$

The choice of the quadratic law of tapering can be easily understood from analysis of a simple model situation. Let us consider the case of a completely bunched electron beam. It follows from the first equation of (2.94) that the field amplitude grows proportionally to the undulator length. Then, it follows from the first equation of (2.93) that the change of the particle energy is proportional to the squared length of the undulator. Finally, from the second equation of (2.93) we find that the electrons remain resonant when the detuning parameter $\hat{C}$ is changed quadratically, too. This qualitative consideration allows one to find asymptotic behavior of the detuning parameter $\hat{C}(\hat{z})$.

To find optimal values of the tapering parameters $\hat{b}_0$, $\hat{b}_1$, $\hat{b}_2$ and $\hat{z}_i$, we performed a set of calculations to maximize the output amplitude at

$$(\hat{z} - \hat{z}_i) \gg 1 ,$$

and obtained the following values:

$$\hat{b}_0 = 0 , \qquad \hat{b}_1 = 1.44 , \qquad \hat{b}_2 = 0.36 . \qquad (2.115)$$

The coordinate of the beginning of the tapering is given by the relation:

$$\hat{z}_i = 1.7 + \frac{2}{\sqrt{3}} \ln\left(\hat{u}_{\text{ext}}^{-1}\right) . \qquad (2.116)$$

Comparison of relations (2.106) and (2.116) shows that the undulator tapering must start at a distance of $\Delta\hat{z} = 1.4$ before the saturation point of the untapered undulator.

Phase analysis shows that 65% of the particles are trapped in the regime of coherent deceleration at optimal parameters of the tapering (2.115) and (2.116). The trapping factor depends strongly on the value of the coefficient $\hat{b}_2$. Particles do not trap in the regime of coherent deceleration for $\hat{b}_2 > 0.4$. The trapping process is stable and the fraction of the trapped particles is even increased for $\hat{b}_2 < 0.36$. However, the equilibrium decelerating phase shifts closer to 90^0 in this case which results in a slower increase of the field amplitude than in the case of $\hat{b}_2 = 0.36$.

Calculation of the Optimized FEL Amplifier Characteristics. In Fig. 2.25 we present the dependence of the field amplitude on the undulator length for the FEL amplifier with tapered parameters. The input field amplitude is equal to $\hat{u}_{\text{ext}} = 0.01$. Undulator tapering is performed in accordance with the law:

$$\hat{C} = \begin{cases} 0 & \text{for } \hat{z} < 7 \\ 1.44(\hat{z} - 7) + 0.36(\hat{z} - 7)^2 & \text{for } \hat{z} > 7 . \end{cases} \qquad (2.117)$$

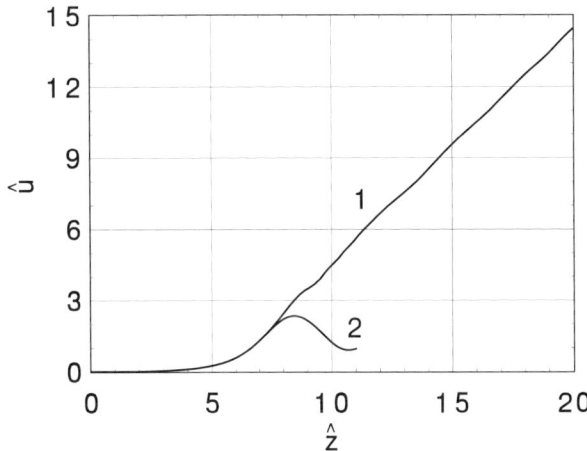

Fig. 2.25. Reduced field amplitude $\hat{u}$ versus the reduced undulator length $\hat{z}$. Here $\hat{\Lambda}_{\rm p}^2 \to 0$, $\hat{\Lambda}_{\rm T}^2 = 0$ and $\hat{u}_{\rm ext} = 0.01$. Curve 1 corresponds to the case of undulator tapering according to (2.117). Curve 2 corresponds to the case of an untapered undulator at $\hat{C} = 0$

In the same figure we present the dependence of $\hat{u}$ on $\hat{z}$ for an untapered undulator. Figure 2.26 presents the phase distributions of the macroparticles at different coordinates of the tapered section.

The plot in Fig. 2.25 can be used for the calculation of the output field amplitude on the undulator length for an arbitrary value of the input signal $\hat{u}_{\rm ext}$. Indeed, the field amplitude $\hat{u}(\hat{z})$ in the untapered section, for $\hat{z} < \hat{z}_{\rm i}$, can be calculated using the technique described in Sect. 3.4. The field amplitude in the tapered section, at $\hat{z} = \hat{z}_{\rm i} + \Delta\hat{z}$, is equal to the value of $\hat{u}(7 + \Delta\hat{z})$ given by the plot in Fig. 2.25.

When the undulator parameters are tapered according to (2.114)–(2.116), the field amplitude for $(\hat{z} - \hat{z}_{\rm i}) > 3$ can be calculated with the following approximate formula:

$$\hat{u}(\hat{z}) \simeq \hat{z} - \frac{2}{\sqrt{3}} \ln\left(\hat{u}_{\rm ext}^{-1}\right) . \tag{2.118}$$

It should be noted once more that this formula is valid in the low efficiency approximation only, i.e. when the output efficiency at the amplifier exit at $\hat{z} = \hat{z}_{\rm f}$ is small:

$$\eta = [\hat{u}(\hat{z}_{\rm f})]^2 \frac{\rho}{4} \ll 1 .$$

In Fig. 2.27 we present the plots of the dependence of the output field amplitude on the initial detuning parameter $\hat{C}_0 = \hat{b}_0$ for different lengths of the undulator. The initial value of the detuning is equal to

$$C_0 = k_{\rm w}(0) - \omega(1 + K^2)/(2\gamma^2 c) ,$$

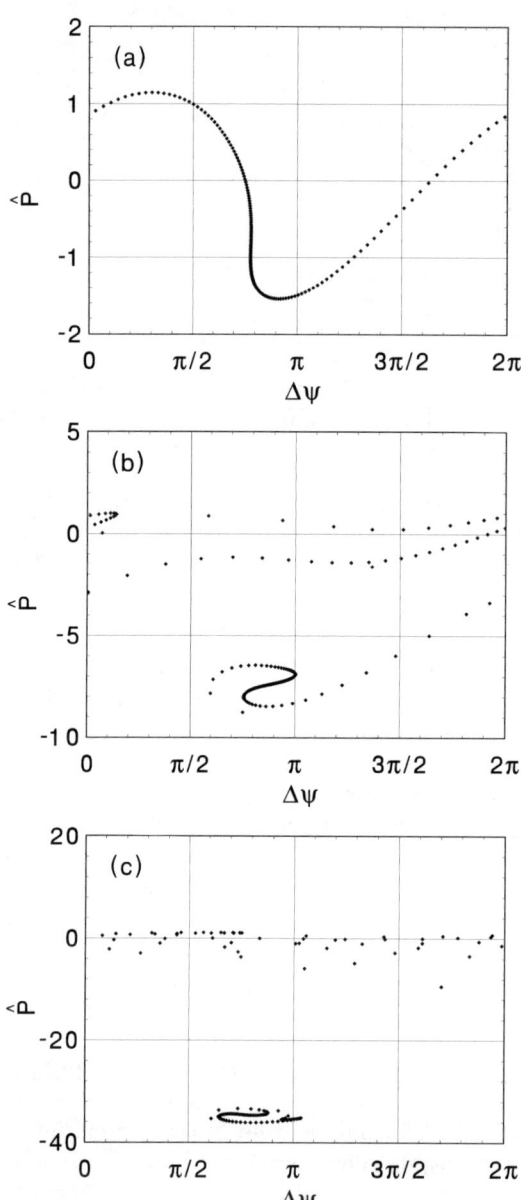

Fig. 2.26. Phase space distribution of the particles in the case of undulator tapering according to (2.117). Graph (**a**): $\hat{z} = 7$, graph (**b**): $\hat{z} = 10$, and graph (**c**): $\hat{z} = 15$. Here $\hat{A}_p^2 \to 0$, $\hat{A}_T^2 = 0$ and $\hat{u}_{\text{ext}} = 0.01$

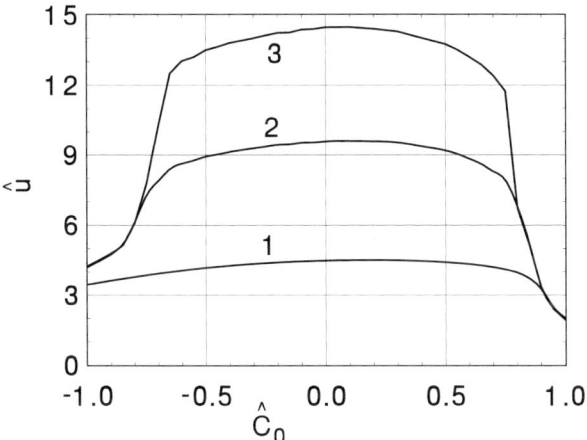

Fig. 2.27. Output field amplitude versus the detuning parameter. Here $\hat{\Lambda}_{\rm p}^2 \to 0$, $\hat{\Lambda}_{\rm T}^2 = 0$ and $\hat{u}_{\rm ext} = 0.01$. The detuning parameter is equal to $\hat{C} = \hat{C}_0$ at $\hat{z} < \hat{z}_{\rm i}$ and $\hat{C} = \hat{C}_0 + 1.44 \times (\hat{z} - \hat{z}_{\rm i}) + 0.36 \times (\hat{z} - \hat{z}_{\rm i})^2$ at $\hat{z} > \hat{z}_{\rm i}$. The values of $\hat{z}_{\rm i}$ are given by (2.116). Curve 1: $\hat{z} - \hat{z}_{\rm i} = 3$. Curve 2: $\hat{z} - \hat{z}_{\rm i} = 8$. Curve 3: $\hat{z} - \hat{z}_{\rm i} = 13$

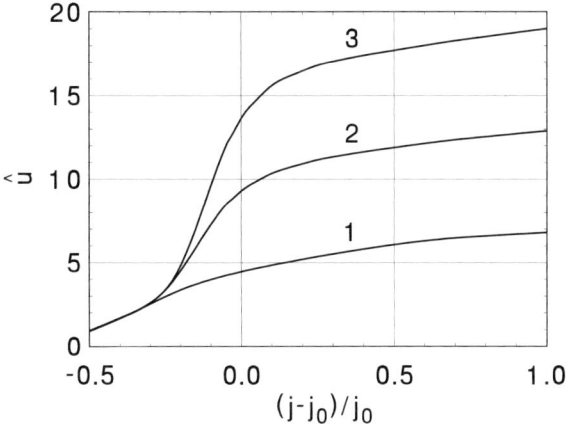

Fig. 2.28. Output field amplitude versus the beam current density. The reduced parameters are calculated at $j = j_0$. Here $\hat{\Lambda}_{\rm p}^2 \to 0$, $\hat{\Lambda}_{\rm T}^2 = 0$, and $\hat{u}_{\rm ext} = 0.01$. The detuning parameter $\hat{C}$ changes according to (2.114) with coefficients given by (2.115) and (2.116). Curve 1: $\hat{z} - \hat{z}_{\rm i} = 3$. Curve 2: $\hat{z} - \hat{z}_{\rm i} = 8$. Curve 3: $\hat{z} - \hat{z}_{\rm i} = 13$

where $k_{\rm w}(0)$ is the wavenumber of the undulator at its entrance. One can see from Fig. 2.27 that the bandwidth of the amplifier with the tapered undulator is close to that of the amplifier with untapered undulator. This means that the tolerances to the electron energy deviations remain the same.

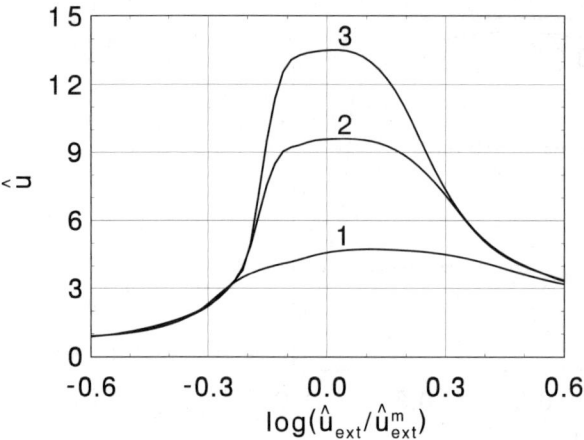

Fig. 2.29. Output field amplitude $\hat{u}$ versus the input field amplitude $\hat{u}_{\text{ext}}$. Here $\hat{\Lambda}_p^2 \to 0$, $\hat{\Lambda}_T^2 = 0$ and $\hat{u}_{\text{ext}} = 0.01$. The detuning parameter $\hat{C}$ changes according to (2.114) with coefficients given by (2.115). The values of $\hat{z}_i$ are given by (2.116). Curve 1: $\hat{z} - \hat{z}_i = 3$. Curve 2: $\hat{z} - \hat{z}_i = 8$. Curve 3: $\hat{z} - \hat{z}_i = 13$

Limitations on the deviation of the beam current density from a nominal value (at which the tapering is optimal) can be obtained with the help of Fig. 2.28 which shows the plots of the dependencies of the output field amplitude on the deviation of the beam current density, $\Delta j = j - j_0$. When calculating these plots, we have performed the normalization procedure at the nominal value of the beam current density, j_0. It is seen from these plots that the regime of coherent deceleration remains stable with increase of the beam current density with respect to the nominal value. This can be explained with the help of (2.114). It follows from this relation that for an increase of the beam current density with all the other parameters fixed, the actual value of the tapering coefficient $\hat{b}_2$ decreases, which, as we discussed above, does not destroy the process of coherent deceleration.

Amplitude characteristics of the FEL amplifier with a tapered undulator are presented in Fig. 2.29.

It is interesting to analyze the energy distribution of the electrons at the undulator exit. One can see from Fig. 2.26 that the electrons in the beam are separated into two fractions when the FEL amplifier operates in a tapered regime. When the value of the FEL efficiency η increases significantly with respect to the untapered case, the average energy of the trapped particles is equal to:

$$\langle \mathcal{E} \rangle = \mathcal{E}_0 (1 - 0.38 \rho \hat{u}^2) \, .$$

The rms energy spread of the trapped particles oscillates with the undulator length. It is connected with the fact that the phase density of the particles is inhomogeneous inside the separatrix. As the particles perform slow energy oscillations with respect to the equilibrium energy, these result in slow oscillations of the rms energy spread. Numerical simulations show that the maximal value of the reduced rms energy spread $\langle(\Delta\hat{P})^2\rangle$ is of the order of unity. Remembering that

$$\langle(\Delta\mathcal{E}/\mathcal{E}_0)^2\rangle = \rho^2\langle(\Delta\hat{P})^2\rangle,$$

we write the following expression for the rms energy spread of the trapped particles:

$$\langle(\Delta\mathcal{E}/\mathcal{E}_0)^2\rangle^{1/2} \lesssim \rho.$$

The rms energy spread and average energy of untrapped particles are almost independent of the length of the undulator and are equal to

$$\langle(\Delta\mathcal{E}/\mathcal{E}_0)^2\rangle^{1/2} \simeq 2\rho, \qquad \langle\mathcal{E}\rangle \simeq \mathcal{E}_0(1 - 0.6\rho).$$

Space Charge and Energy Spread Effects. Now we complicate our study by taking into account the space charge field. In this case the detuning parameter is

$$\hat{C} = \begin{cases} \hat{C}_m & \text{for } \hat{z} < \hat{z}_i, \\ \hat{C} = \hat{C}_m + \hat{b}_1(\hat{z} - \hat{z}_i) + \hat{b}_2(\hat{z} - \hat{z}_i)^2 & \text{for } \hat{z} > \hat{z}_i, \end{cases} \qquad (2.119)$$

where $\hat{C}_m$ is the value of the detuning parameter corresponding to the maximal field growth rate in the linear mode of operation. At each value of the space charge parameter, optimization of the tapering coefficients $\hat{b}_1$, $\hat{b}_2$, and $\hat{z}_i$ on the field maximum at $\hat{z} - \hat{z}_i \gg 1$ has been performed using the system of self-consistent equations (2.94) and (2.110). Figure 2.30 presents the dependence of the trapping factor on the space charge parameter.

Using equations (2.93) and (2.94) we have also performed a similar study of the influence of the energy spread on the trapping efficiency. These results

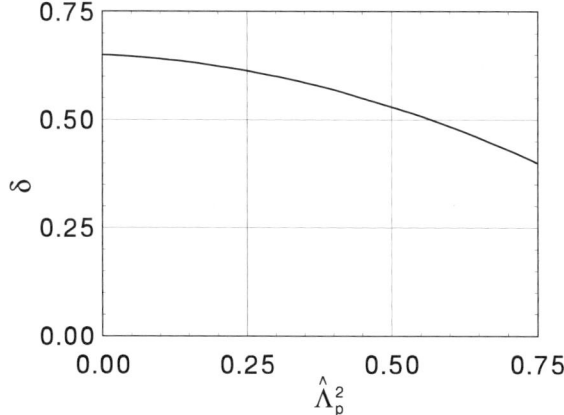

Fig. 2.30. Trapping factor versus the space charge parameter $\hat{\Lambda}_p^2$. The undulator tapering is performed according to (2.119). Here $\hat{\Lambda}_T^2 = 0$

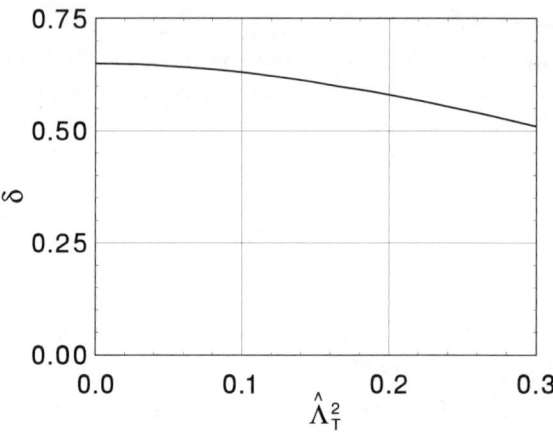

Fig. 2.31. Trapping factor versus the energy spread parameter $\hat{\Lambda}_T^2$. The undulator tapering is performed according to (2.119). Here $\hat{\Lambda}_p^2 \to 0$

are presented in Fig. 2.31. It is seen from Figs. 2.30 and 2.31 that the space charge field and the energy spread significantly limit the possibility of increasing the FEL amplifier efficiency by means of the undulator tapering.

Region of Applicability of the Low-Efficiency Approximation. Numerical simulation of an undulator tapering in the low-efficiency approximation has been performed using the equations of motion (2.93) obtained from the Hamiltonian (2.90). These equations correctly describe the motion of the particles only for small deviations of the energy $\mathcal{E}$ from the nominal value of $\mathcal{E}_0$. This approximation is not valid for large values of the beam energy loss, so when calculating a high-efficiency FEL amplifier, the original Hamiltonian (2.3) should be used. In the first order in the expansion of the vector potential of the wave $\boldsymbol{A}_\perp$, and neglecting the space charge field, the Hamiltonian can be written as

$$H = \int^{\mathcal{E}} \left[k_w(z) - \omega \left(v_z^{-1}(\mathcal{E}, z) - c^{-1} \right) \right] d\mathcal{E} - u(\mathcal{E}, z) \sin(\psi + \psi_0) , \quad (2.120)$$

where $v_z(\mathcal{E}, z)$ is the longitudinal component of the electron velocity and $\psi = \int k_w(z) dz + \omega(z/c - t)$. Expression (2.120) is a generalization of the Hamiltonian (2.5) to the case of a tapered undulator. The equations of motion corresponding to the Hamiltonian (2.120) have the form:

$$d\mathcal{E}/dz = -\partial H/\partial \psi = u \cos(\psi + \psi_0) ,$$
$$d\psi/dz = \partial H/\partial \mathcal{E} = k_w - \omega \left(v_z^{-1} - c^{-1} \right) - \partial u/\partial \mathcal{E} \sin(\psi + \psi_0) . \quad (2.121)$$

Equations (2.121) can be written in normalized form. It is convenient to calculate normalized variables using the physical parameters of the beam and of the undulator at the entrance to the FEL amplifier. We consider the case of the undulator tapering for a fixed value of the undulator parameter, $K =$ const. The expression for the electron rotation angle has the form:

$$\theta_{\rm s}(\mathcal{E}) = \theta_{\rm s}(\mathcal{E}_0)/(1 + \Delta\mathcal{E}/\mathcal{E}_0) = \theta_{\rm s}(\mathcal{E}_0)/(1 + \rho\hat{P}) \,, \tag{2.122}$$

and the equations of motion are as follows:

$$\begin{aligned}\frac{\mathrm{d}\hat{P}}{\mathrm{d}\hat{z}} &= \frac{\hat{u}}{1+\rho\hat{P}}\cos(\psi+\psi_0) \,, \\ \frac{\mathrm{d}\psi}{\mathrm{d}\hat{z}} &= \frac{\hat{P}(1+\rho\hat{P}/2)}{(1+\rho\hat{P})^2} + \hat{C} + \rho\hat{u}\frac{\sin(\psi+\psi_0)}{(1+\rho\hat{P})^2} \,,\end{aligned} \tag{2.123}$$

where the detuning parameter is a function of the $\hat{z}$ coordinate ($\gamma = \mathcal{E}_0/(m_{\rm e}c^2)$):

$$\hat{C} = k_{\rm w}(\hat{z})/\Gamma - \omega(1 + K^2)/(2\gamma^2 c\Gamma) \,.$$

The equations for the field amplitude $\hat{u}$ and for the field phase ψ_0 have the form:

$$\begin{aligned}\frac{\mathrm{d}\hat{u}}{\mathrm{d}\hat{z}} &= -\frac{2}{N}\sum_{j=1}^{N}\frac{\cos\left(\psi_{(j)} + \psi_0\right)}{1 + \rho\hat{P}_{(j)}} \,, \\ \frac{\mathrm{d}\psi_0}{\mathrm{d}\hat{z}} &= \frac{2}{\hat{u}N}\sum_{j=1}^{N}\frac{\sin\left(\psi_{(j)} + \psi_0\right)}{1 + \rho\hat{P}_{(j)}} \,.\end{aligned} \tag{2.124}$$

Here the dependence of the rotation angle (2.122) on energy was taken into account.

We begin the study using optimal tapering parameters for a low-efficiency approximation given by relations (2.114) and (2.115). For $\hat{z} > \hat{z}_{\rm i}$ the detuning parameter changes in accordance with:

$$\hat{C} = 1.44(\hat{z} - \hat{z}_{\rm i}) + 0.36(\hat{z} - \hat{z}_{\rm i})^2 \,.$$

The results of numerical simulations for several values of the efficiency parameter ρ are presented in Fig. 2.32. For a sufficiently short undulator length the energy losses of the trapped particles are small, $\Delta\mathcal{E}/\mathcal{E}_0 = \rho\hat{P} \ll 1$. In this case the system of equations (2.123) and (2.124) transforms to the system of equations (2.93) and (2.94). At the initial stage of tapering the average energy of trapped particles decreases quadratically with the undulator length. Synchronism is maintained by the quadratic increase of the value of the detuning parameter $\hat{C}$ (see the second equation (2.123)). The average change of the phase of the trapped particles is zero, $\langle \mathrm{d}\psi/\mathrm{d}\hat{z}\rangle = 0$, and their motion corresponds to phase oscillations about the equilibrium decelerating phase $\psi_{\rm e} + \psi_0 =$ const. (according to the second equation (2.124), the change of the phase of the effective potential ψ_0 can be neglected when $\hat{u} \gg 1$). As the length of the undulator is increased, the difference between the approximate system of equations (2.93) and (2.94) and the original system (2.123) and (2.124) becomes significant. It is seen from the equation for the phase

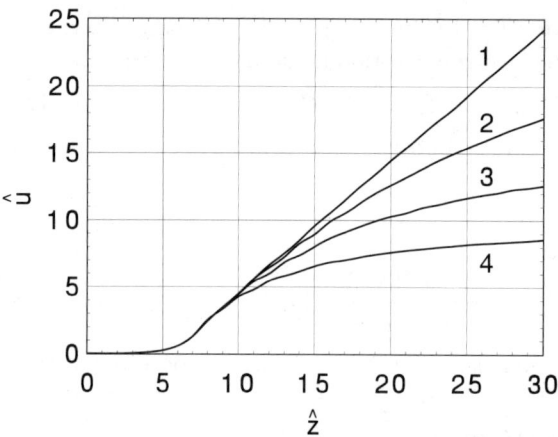

Fig. 2.32. Reduced field amplitude $\hat{u}$ versus the reduced undulator length $\hat{z}$ at different values of the efficiency parameter ρ. Here $\hat{\Lambda}_p^2 \to 0$, $\hat{\Lambda}_T^2 = 0$ and $\hat{u}_{\text{ext}} = 0.01$. The detuning parameter changes according to (2.117). Curve 1: $\rho \to 0$. Curve 2: $\rho = 0.003$. Curve 3: $\rho = 0.01$. Curve 4: $\rho = 0.03$

$$\frac{d\psi}{d\hat{z}} = \frac{\hat{P}(1+\rho\hat{P}/2)}{(1+\rho\hat{P})^2} + \hat{C},$$

that the compensation of the quadratic growth of the detuning parameter $\hat{C}$ will take place at a smaller decelerating rate of the trapped particles. Numerical simulations show that at the final stage of deceleration, the number of trapped particles does not change and they perform phase oscillations about the equilibrium decelerating phase $\psi_e + \psi_0$, which decreases and approaches adiabatically the value of $90°$. As a result, the growth of the field amplitude is slowed down and further increase of the undulator length becomes ineffective. The plots in Fig. 2.32 give an idea of the region of applicability of the low-efficiency approximation considered in the previous Section.

2.3.2 The High-Efficiency FEL Amplifier

We have found above that the quadratic law of undulator tapering, which is optimal in the low-efficiency case, becomes nonoptimal when one needs to achieve high efficiency of the amplifier comparable to unity. To achieve this goal, the undulator parameters must be changed faster than the quadratic polynomial (2.114). Analyzing (2.123) for the phase, one can suppose that the linear law of the change in the field amplitude will take place also in the high-efficiency case when, for $\hat{z} > \hat{z}_i = \hat{z}_{\max} - 1.4$ (see (2.106) and (2.116)), the detuning parameter will change as

$$\hat{C} = \mathcal{F}(\hat{z}) = T(\hat{z})\left[1 - \rho T(\hat{z})/2\right]\left[1 - \rho T(\hat{z})\right]^{-2},$$
$$T(\hat{z}) = 1.44(\hat{z} - \hat{z}_i) + 0.36(\hat{z} - \hat{z}_i)^2. \tag{2.125}$$

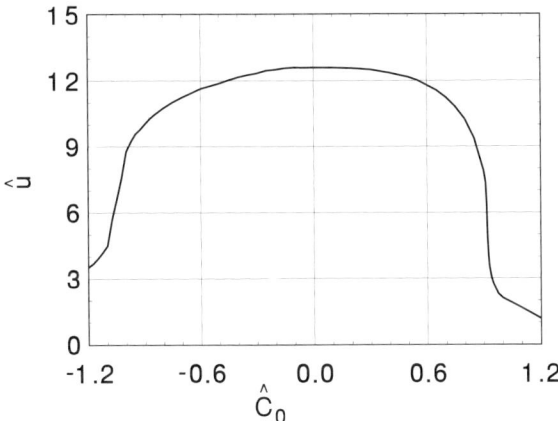

Fig. 2.33. Output field amplitude versus the detuning parameter in the high-efficiency case. Here $\hat{\Lambda}_p^2 \to 0$ and $\hat{\Lambda}_T^2 = 0$. The detuning parameter changes according to the law $\hat{C} = \hat{C}_0$ for $\hat{z} < \hat{z}_i$ and $\hat{C} = \hat{C}_0 + \mathcal{F}(\hat{z})$ for $\hat{z} > \hat{z}_i$ (see (2.125)). The value of $\hat{z}_i$ is given by (2.116). At $\hat{C}_0 = 0$ the efficiency is equal to 40% and the power gain at $\hat{z} = \hat{z}_i$ is equal to 40 dB

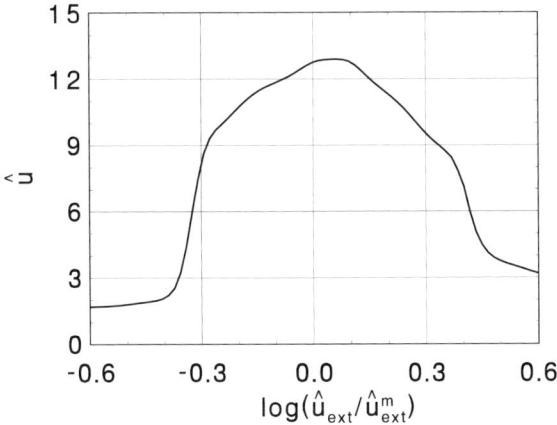

Fig. 2.34. Output field amplitude $\hat{u}$ versus the input field amplitude $\hat{u}_{\text{ext}}$ in the high-efficiency case. Here $\hat{\Lambda}_p^2 \to 0$ and $\hat{\Lambda}_T^2 = 0$. The detuning parameter $\hat{C}$ changes according to (2.125) where $\hat{z}_i$ corresponds to the nominal value of the input signal $\hat{u}_{\text{ext}}^m$. At $\hat{u}_{\text{ext}} = \hat{u}_{\text{ext}}^m$ the efficiency is equal to 40% and the power gain at $\hat{z} = \hat{z}_i$ is equal to 40 dB

Results of numerical simulations confirm this assumption. Indeed, when the detuning parameter is changed as in (2.125), the field amplitude changes according to the linear law (2.118):

$$\hat{u}(\hat{z}) \simeq \hat{z} - \hat{z}_{\max} + 3 \, ,$$

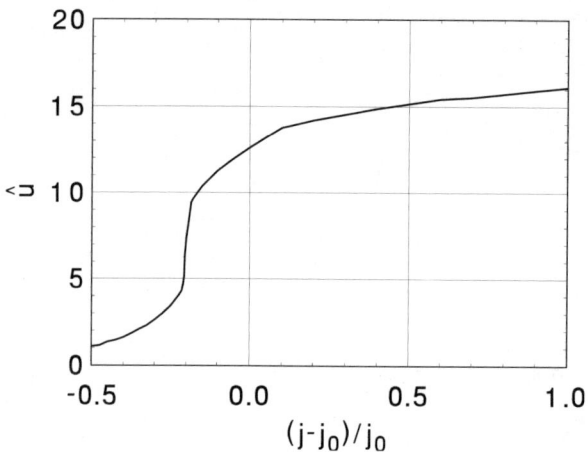

Fig. 2.35. Output field amplitude versus the beam current density in the high-efficiency case. Here $\hat{\Lambda}_p^2 \to 0$ and $\hat{\Lambda}_T^2 = 0$. The detuning parameter $\hat{C}$ changes according to (2.125). The reduced parameters are calculated at $j = j_0$. When $j = j_0$, the efficiency is equal to 40% and the power gain at $\hat{z} = \hat{z}_i$ is equal to 40 dB

even when the amplifier efficiency becomes comparable with unity. It is amazing that in this regime the output field is a linear function of $\hat{z}$ and does not depend on ρ. Indeed, the rotation angle of the trapped electrons is a function of ρ and increases while the trapped particles lose their energy. On the other hand, the equilibrium decelerating phase $\psi_e + \psi_0$ decreases, approaching 90°. These two effects compensate each other and effective interaction of the beam with the electromagnetic wave takes place up to an amplifier efficiency of about unity.

Using relations (2.98) and (2.118) we can calculate the total length $\hat{z}_f$ of the undulator required to achieve a reduced efficiency of $\hat{\eta}$:

$$\hat{z}_f = 2\sqrt{\hat{\eta}} + \frac{2}{\sqrt{3}} \ln(\hat{u}_{\text{ext}}^{-1}) \ . \tag{2.126}$$

In accordance with (2.125), the total change of the undulator wavenumber is equal to

$$\begin{aligned} C &= \Delta k_w(z_f) \\ &= 2\pi \left[\lambda_w^{-1}(z_f) - \lambda_w^{-1}(z_i) \right] \\ &= \Gamma T(\hat{z}_f) \left[1 - \rho T(\hat{z}_f)/2 \right] \left[1 - \rho T(\hat{z}_f) \right]^{-2} \ , \end{aligned} \tag{2.127}$$

where $T(\hat{z}_f)$ and $\hat{z}_f$ are calculated using (2.125) and (2.126).

The output characteristics of the amplifier with an efficiency of 40% are presented in Figs. 2.33 and 2.34. One can see that the characteristics of the high-efficiency FEL amplifier do not differ significantly from those of the low-efficiency amplifier. In Fig. 2.35 we present the dependence of the output field amplitude on the beam current. One can see that the output field amplitude is almost constant with increasing beam current, and the amplifier efficiency decreases only due to the increase of the beam power. For instance, the efficiency decreases from 40% down to 33% when the beam current is increased by a factor of 1.5.

2.3.3 Some Generalizations

Tapering for a Fixed Undulator Period. We have studied above undulator tapering for a fixed undulator parameter, $K = \text{const}$. Another way of undulator tapering is tapering for a fixed undulator period $\lambda_w = \text{const}$. From the theoretical point of view this method seems to be more complicated, and the results obtained are not so general as those obtained above for tapering at $K = \text{const}$.

It follows from the definition of the detuning C that the undulator field H_w must decrease when we deal with tapering for the fixed undulator period $\lambda_w = \text{const}$. The simplest law of magnetic field change is quadratic:

$$[H_w(z_i) - H_w(z)]/H_w(z_i) = b_1(z - z_i) + b_2(z - z_i)^2 . \tag{2.128}$$

The normalization procedure is performed in usual way. We calculate normalization factors using the parameters of the beam and of the undulator at the amplifier entrance. In accordance with (2.121) and (2.16) the system of reduced self-consistent equations has the form

$$\frac{d\hat{P}}{d\hat{z}} = \frac{1}{1+\rho\hat{P}}\left[1 - \frac{1+s}{s}\rho T(\hat{z})\right]\hat{u}\cos(\psi+\psi_0) ,$$

$$\frac{d\psi}{d\hat{z}} = \frac{1}{(1+\rho\hat{P})^2}\left\{\hat{P}\left[1+\frac{\rho\hat{P}}{2}\right] + T(\hat{z})\left[1 - \frac{1+s}{2s}\rho T(\hat{z})\right]\right.$$

$$\left. + \rho\hat{u}\left[1 - \frac{1+s}{s}\rho T(\hat{z})\right]\sin(\psi+\psi_0)\right\} , \tag{2.129}$$

$$\frac{d\hat{u}}{d\hat{z}} = -\left[1 - \frac{1+s}{s}\rho T(\hat{z})\right]\left[\frac{2}{N}\sum_{j=1}^{N}\frac{\cos(\psi_{(j)}+\psi_0)}{1+\rho\hat{P}_{(j)}}\right] ,$$

$$\frac{d\psi_0}{d\hat{z}} = \frac{1}{\hat{u}}\left[1 - \frac{1+s}{s}\rho T(\hat{z})\right]\left[\frac{2}{N}\sum_{j=1}^{N}\frac{\sin(\psi_{(j)}+\psi_0)}{1+\rho\hat{P}_{(j)}}\right] , \tag{2.130}$$

where $s = K^2$. The value of the undulator parameter, K, is calculated for an untapered section.[8] The function $T(\hat{z})$ is given by the expression:

$$T(\hat{z}) = \hat{b}_1(\hat{z} - \hat{z}_i) + \hat{b}_2(\hat{z} - \hat{z}_i)^2 , \tag{2.131}$$

---

[8] One can find that $\rho(1+s)/s = \hat{\Lambda}_p^2/4$. The system of equations (2.129) is written for the case of negligibly small space charge effects. This requires the value of $\rho(1+s)/s$ to be much less than the unity.

where

$$\hat{b}_1 = \frac{s}{(1+s)} \frac{b_1}{\rho \Gamma}, \qquad \hat{b}_2 = \frac{s}{(1+s)} \frac{b_2}{\rho \Gamma^2}.$$

In the initial part of the tapered section, when the change of the energy of the trapped particles is small, we can expand (2.129) and (2.130) in the small parameter $\rho \hat{P}$ and obtain the system of equations describing the low-efficiency approximation:

$$\frac{d\hat{P}}{d\hat{z}} = \hat{u} \cos(\psi + \psi_0), \qquad \frac{d\psi}{d\hat{z}} = \hat{P} + T(\hat{z}),$$

$$\frac{d\hat{u}}{d\hat{z}} = \hat{j}_1 \cos(\psi_0 - \psi_1), \qquad \frac{d\psi_0}{d\hat{z}} = -\frac{\hat{j}_1}{\hat{u}} \sin(\psi_0 - \psi_1).$$

This system is identical to the system of equations (2.93) and (2.94) describing the low-efficiency approximation of the undulator tapering for a fixed value of the undulator parameter, $K = $ const. The detuning parameter $\hat{C}(\hat{z})$ is equal to $T(\hat{z})$ in this case. Therefore, we can conclude that in the low-efficiency approximation, when $\eta \ll 1$, both tapering methods are equivalent.

Now let us consider the case of a high-efficiency FEL with tapering with a fixed period. Using the system of equations (2.129) and (2.130) we performed an optimization of the coefficients $\hat{b}_1$, $\hat{b}_2$, and $\hat{z}_i$ on the maximum of the output field at $(\hat{z} - \hat{z}_i) \gg 1$. It was assumed that $s \simeq 1$ and $\rho \simeq 10^{-2}$. The following values were obtained:

$$\hat{b}_1 = 1.44, \qquad \hat{b}_2 = 0.3, \qquad \hat{z}_i = 1.7 + \frac{2}{\sqrt{3}} \ln(\hat{u}_{\text{ext}}^{-1}).$$

Phase analysis has shown that 68% of the particles are trapped in the regime of coherent deceleration.

The fact that the value of the coefficient $\hat{b}_2$ is slightly less than that of the undulator tapering in the low-efficiency approximation requires some explanation. Numerical simulations performed with the complete equations (2.123), (2.124), (2.129) and (2.130) show that as the undulator field is decreased, the phase motion of the particles becomes less stable with respect to that calculated with (2.93) and (2.94). The main losses of the trapped particles occur within the first period of phase oscillation, at

$$(\hat{z} - \hat{z}_i) \simeq 3,$$

when the particles come quite close to the boundary of the stability region. At the values $s \simeq 1$ and $\rho \simeq 10^{-2}$, the difference in the simulation results obtained with the complete and approximate equations becomes visible, and for $\hat{b}_2 = 0.36$ it causes the number of trapped particles to decrease visibly when passing through the critical point. Therefore, the more stable regime becomes possible at $\hat{b}_2 = 0.3$. We should recall here that in the case of tapering at a fixed undulator parameter the optimal value of $\hat{b}_2$ is 0.36 also for a high-efficiency FEL amplifier.

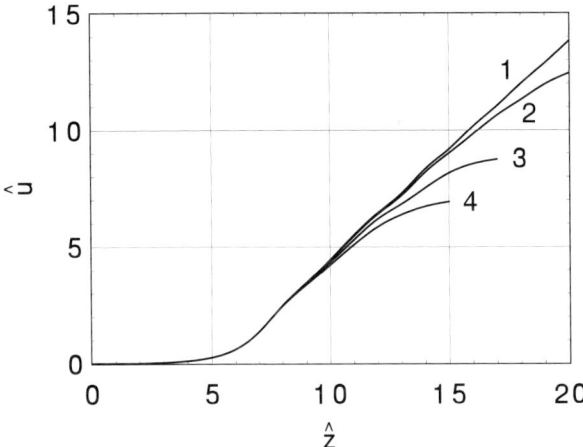

Fig. 2.36. Reduced field amplitude $\hat{u}$ versus $\hat{z}$. Here $\hat{\Lambda}_{\rm p}^2 \to 0$, $\hat{\Lambda}_{\rm T}^2 = 0$, $\rho = 0.01$, and $\hat{u}_{\rm ext} = 0.01$. The undulator field changes according to (2.128), (2.131), and (2.134). The length of the tapered section is given by (2.133). Curve 1: $s = \infty$. Curve 2: $s = 5$. Curve 3: $s = 1$. Curve 4: $s = 0.5$

The length of the tapering section $(z_{\rm f} - z_{\rm i})$ obviously cannot exceed the distance over which the undulator field $H_{\rm w}$ decreases to zero as in (2.128) and can be found from the equation:

$$0.3\rho \frac{1+s}{s}(\hat{z}_{\rm f} - \hat{z}_{\rm i})^2 + 1.44\rho \frac{1+s}{s}(\hat{z}_{\rm f} - \hat{z}_{\rm i}) - 1 = 0 \ . \tag{2.132}$$

To be exact, the minimal value of the undulator field is limited from below not by the value zero, but by the condition $H_{\rm w} \gg E/(2\gamma_z^2)$. Using (2.128) and (2.131), this condition can be written in the following reduced form:

$$\rho^2 \hat{u}(\hat{z}) \ll s \left[1 - \frac{1+s}{s}\rho T(\hat{z})\right] \ . \tag{2.133}$$

However, at $s > 0.1$ and $\rho < 0.03$, the length of the tapered section obtained with (2.133) does not differ significantly from that obtained with (2.132).

In Fig. 2.36 we present plots of the field amplitude for the undulator tapering with a fixed undulator period. The tapering function $T(\hat{z})$ is given by the expression:

$$T(\hat{z}) = 1.44(\hat{z} - \hat{z}_{\rm i}) + 0.3(\hat{z} - \hat{z}_{\rm i})^2 \ , \tag{2.134}$$

and the beginning of the undulator tapering is connected with the amplitude of the external field by the relation:

$$\hat{z}_{\rm i} = 1.7 + \frac{2}{\sqrt{3}} \ln\left(\hat{u}_{\rm ext}^{-1}\right) \ . \tag{2.135}$$

We present in the same figure the plot for the field amplitude $\hat{u}(\hat{z})$ calculated with the system of approximate equations (2.93) and (2.94). One can obtain from these plots that the larger parameter s, the less the field amplitude

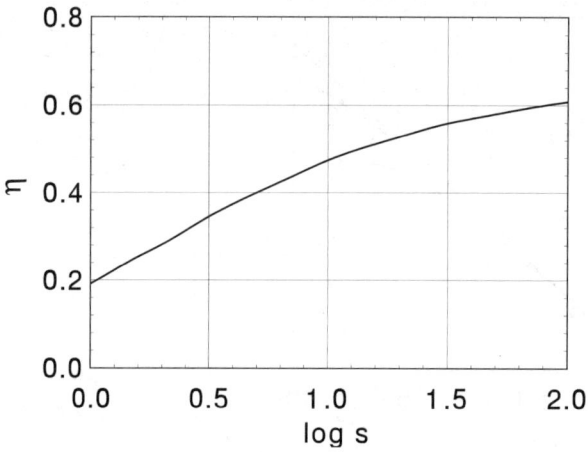

Fig. 2.37. Efficiency of the amplifier versus the parameter s. Here $\hat{\Lambda}_{\rm p}^2 \to 0$, $\hat{\Lambda}_{\rm T}^2 = 0$, $\rho = 0.01$, and $\hat{u}_{\rm ext} = 0.01$. The undulator field changes according to (2.128), (2.131), and (2.134). The length of the tapered section is given by (2.133)

$\hat{u}(\hat{z}, s, \rho)$ deviates from the linear behavior in the final part of the tapered section. The dependence of the maximal efficiency of the amplifier on the value of the parameter s is presented in Fig. 2.37. It is seen that the efficiency increases with an increase in the value of s. When $s \gg 1$, undulator tapering with a fixed undulator period provides nearly the same results as tapering with a fixed undulator parameter.

The results of simulations presented in Fig. 2.37 have been obtained at a specific value of the efficiency parameter $\rho = 0.01$. Simulations performed at different values of ρ show that the maximal reduced efficiency does not differ significantly from that presented in Fig. 2.37 when the efficiency parameter is within the limits $0.003 \lesssim \rho \lesssim 0.03$.

FEL Amplifier with a Planar Undulator. When performing a study of the FEL amplifier with a tapered undulator, we assumed the undulator to be helical. The case of a planar undulator is more complicated and should be studied separately. The only exception is the low-efficiency approximation and the tapering at a fixed value of the undulator parameter. All the results and universal plots for this particular case are applicable to the case of the planar undulator. The only difference is that the normalization procedure should be performed taking into account the corresponding numerical factors appearing due to the averaging over the undulator period (see Sect. 2.2.7).

2.4 Concluding Remarks

It is relevant to make some remarks on the region of applicability of the one-dimensional theory of FEL amplifier. One of the basic assumptions of the theory is that the amplitudes of the electromagnetic wave and of the first harmonic of the current density change slowly on the scale of the undulator period. This takes place when the efficiency parameter is small, $\rho \ll 1$. Let us express this condition in terms of physical parameters. Let the beam current and the beam radius be I_0 and r_b, respectively. The efficiency parameter, ρ, and the gain parameter,

$$\Gamma = \left[\frac{I_0 \theta_\mathrm{s}^2 \omega}{c\gamma_z^2 \gamma I_A r_\mathrm{b}^2}\right]^{1/3},$$

are connected by the relation $\rho = c\gamma_z^2 \Gamma/\omega$. When writing down the expression for the gain parameter, we have taken into account that the beam current density is $j_0 = I_0/(\pi r_\mathrm{b}^2)$. Thus, the requirement for the parameter ρ to be small can be written as

$$\left[\frac{c\gamma_z}{\omega r_\mathrm{b}}\right]^{2/3} \left[\frac{I_0}{\gamma I_A}\right]^{1/3} \left[\frac{K^2}{1+K^2}\right]^{1/3} \ll 1.$$

As a rule, this condition is well satisfied in all practical problems.

Formally the one-dimensional theory of the FEL amplifier deals with the amplification of a plane electromagnetic wave by an infinitely wide electron beam. In practice, the electron beam and the electromagnetic wave have finite transverse dimensions, and diffraction effects always take place. For the one-dimensional model to be applicable, the diffraction losses must be small. Let us perform a qualitative study of this problem assuming that the walls of the vacuum chamber are placed far from the electron beam. In practice such an assumption is valid for FEL amplifiers operating in the near-infrared or shorter-wavelength range. This approximation can be referred to as the approximation of an open electron beam (see Chap. 4 for more details). Consider an electron beam of radius r_b moving in a helical undulator. According to diffraction theory, a parallel beam of radiation with frequency ω and radius r_b expands into an angle of about $c/(\omega r_\mathrm{b})$. In the linear regime the typical length of the change in the radiation field amplitude is about the gain length l_g:

$$l_\mathrm{g} \simeq \Gamma^{-1},$$

A rough estimate for the diffraction losses to be small is

$$r_\mathrm{b} \gg c l_\mathrm{g}/(\omega r_\mathrm{b}),$$

which simply means that the diffraction expansion of the radiation at one gain length must be much less than the size of the beam. The parameter

$$N_\mathrm{F} = r_\mathrm{b}^2 \omega/(c l_\mathrm{g})$$

can be referred to as the Fresnel number. The requirement for the diffraction losses to be small holds at large values of the Fresnel number, $N_\mathrm{F} \gg 1$. It is convenient to rewrite this condition in a different way:

$$B = N_\mathrm{F}^{3/2} = r_\mathrm{b}^2/r_\mathrm{diff}^2 \gg 1,$$

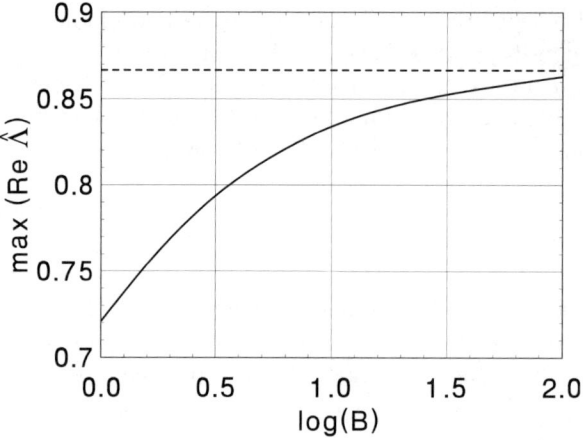

Fig. 2.38. Normalized field growth rate versus diffraction parameter. Calculations have been performed within the framework of the three-dimensional theory for an axisymmetric electron beam with stepped profile. The dotted line represents a one-dimensional growth rate, $3^{1/2}/2$

where

$$r_{\text{diff}}^2 = \frac{c^2 \gamma_z}{\omega^2 \theta_s} \left(\frac{I_A \gamma}{I_0} \right)^{1/2} .$$

The ratio r_b^2/r_{diff}^2 is known as the diffraction parameter B, which is used in Chaps. 4 and 5 as a measure of diffraction effects. At the value of the diffraction parameter $B \gtrsim 10$ the one-dimensional approximation gives a rather good estimate for the field growth rate with an accuracy better than 10% (see Fig. 2.38).[9] The saturation length is also calculated with sufficient accuracy. It is relevant to note that the value of the diffraction parameter for X-ray FELs is typically much larger than unity and the one-dimensional approximation is frequently used for a rapid estimation of their parameters. On the other hand, we should warn the reader against directly transferring the tapering technique, optimal for the one-dimensional model, to the case of an open electron beam. This problem is discussed in Chap. 4.

The one-dimensional model can also be used for the calculation of FEL amplifiers operating in the millimeter wavelength range. It is typical for these devices that the waveguide walls significantly influence the process of radiation amplification. When the structure of the radiation field is close to that of an empty waveguide mode, one can use the equations of the one-dimensional

---

[9] Note that the plot in Fig. 2.38 represents the actual field growth rate, normalized to the value of the one-dimensional gain parameter, Γ. Since the value of Γ scales with the beam radius as $r_b^{-2/3}$, the actual growth rate will decrease with an increase of the beam size (or, of the diffraction parameter B) and fixed value of the beam current.

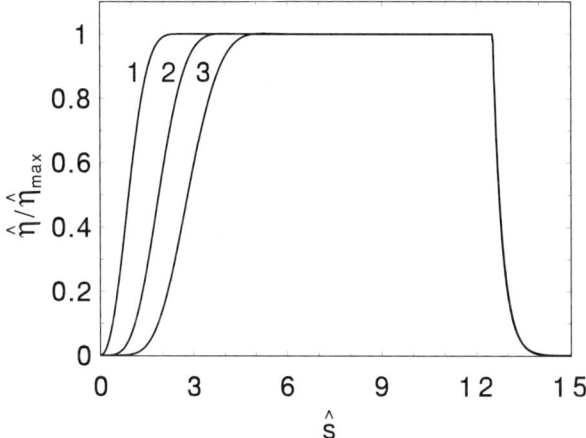

Fig. 2.39. Radiation power normalized to maximal value versus $\hat{s} = \rho\omega(z/v_z - t)$ for different undulator lengths: $\hat{z} = 5$ (curve 1), $\hat{z} = 10$ (curve 2) and $\hat{z} = 15$ (curve 3). Calculations have been performed with the one-dimensional, time-dependent simulation code for a rectangular electron bunch with $\rho\omega T = 12.5$

theory with an appropriate redetermination of the problem parameters. In particular, it is necessary to take into account that the phase velocity of the wave in the waveguide differs from the velocity of light c. Also, the effective density of the beam current should be calculated taking into account the field distribution of the waveguide mode. Rigorous study of waveguide effects (see Chap. 5) shows that the eigenvalue equations for FEL amplifiers operating in a single-mode regime are always similar to those of the one-dimensional model. Nonlinear simulations can be performed in the same way as we have done in this chapter. This also refers to the case of undulator tapering, including the high-efficiency case. All the reduced formulae and universal plots presented in this chapter can be used in an appropriate calculation of the dimensionless parameters.

In this chapter we have restricted our attention to the steady-state theory of the FEL amplifier. To concentrate on the ideal mechanism of amplification, we assumed that a continuous electron beam with current density constant in time is fed to the undulator entrance. In practical situations the electron beam has a finite pulse duration, and the question arises of when one can use the results of this chapter. Let us consider first a rectangular bunch of duration T. If the slippage of the radiation with respect to electrons per unit gain length is much less than the bunch length,

$$\rho\omega T \gg 1 , \tag{2.136}$$

then one can neglect the edge effects and use the steady-state approach. Using the plots presented in Fig. 2.39, one can give a quantitative answer to the question about the region of applicability of the steady-state model. Now let us consider the electron pulse with a gradient axial profile of the current $I_0(t)$. Let the typical duration of such a pulse T satisfy the condition (2.136). Then this situation can be described in the framework of the steady-state theory in the following way. As an approximation, the smooth profile $I_0(t)$ may be replaced on the interval $(0, T)$ by a "boxcar" function. The interval $(0, T)$ is divided into N_s subintervals of equal length. Within each subinterval, the approximation to $I_0(t)$ is constant. At the end of each subinterval, the approximate profile jumps to a new constant value. When

$$\rho \omega T / N_s \gg 1 \, ,$$

we can simulate the FEL process separately within each subinterval.

2.5 Suggested Bibliography

To provide the reader with an opportunity to form his own opinion of the topics discussed in this chapter, we present here selected references to the original papers. The eigenvalue equation of the FEL amplifier taking into account space charge and energy spread effects was derived in [2.1]. Analytical study of the initial-value problem was performed in [2.2, 2.3]. The most comprehensive study of the initial-value problem is presented in [2.4]. Analytical solution of the initial-value problem for the case of a Gaussian energy spread is reported in [2.5]. The influence of the space charge field on the operation of the FEL amplifier with a planar undulator has been studied in [2.5, 2.6]. A high-efficiency FEL amplifier with a tapered undulator was investigated theoretically in [2.7–2.9]. Experimental results on the high-efficiency FEL amplifier can be found in [2.10]. An essential feature of our book consists in the application of similarity techniques. Such an approach has been used in [2.5, 2.11–2.13]. When describing particle motion, we used the Hamiltonian formalism. For a general discussion of the extended Hamiltonian formalism we suggest reading the book [2.14]. A treatment about Landau method for solution of Vlasov equation may be found in the text by G. Ecker [2.15].

3. One-Dimensional Theory of the FEL Oscillator

The FEL oscillator consists of a resonator with an active medium – an electron beam in an undulator. A schematic illustration of the FEL oscillator is presented in Fig. 1.1. The problem of the interaction between the radiation and the active medium in the resonator refers to a class of self-consistent problems. To describe the FEL oscillator, the equations for the particle motion and Maxwell's equations should be solved simultaneously taking into account the initial and boundary conditions for the electron beam and for the radiation. When analyzing the lasing process in the FEL oscillator, one should take into account that it starts from the shot noise in the electron beam. During the lasing process many longitudinal modes with different frequencies can be excited in the resonator. Their amplitudes change with time. These factors make the study much more complicated with respect to that performed in Chap. 2 for the steady-state theory of the FEL amplifier.

A deeper insight into FEL physics can be obtained by introducing some simplified assumptions about the properties of the electron beam and of the electromagnetic field. All the investigations presented in this chapter are performed in the framework of the one-dimensional model of the FEL oscillator. The following assumptions are used in this model:

- the undulator is placed between two plane parallel mirrors;
- the electron beam has a uniform density distribution in the direction perpendicular to the undulator axis;
- the electrons move along identical trajectories parallel to the undulator axis;
- the amplified wave is a plane wave;
- the electron beam is infinitely long.

We begin the study of the FEL oscillator by considering it as an FEL amplifier with feedback. The electromagnetic wave travels in the resonator forward and back reflecting from the mirrors. When the wave travels in the same direction as the electron beam, it is amplified due to the beam-wave interaction in the undulator. The latter process is treated in the same way as for the FEL amplifier. This approach enables one to study the FEL process in a single-mode model as well as in a multimode model. To simulate the latter case, several waves with different frequencies (corresponding to the longitudinal modes of the resonator) are fed to the undulator entrance. The basic principles of FEL oscillator operation can be described most clearly in the framework of the single-mode model. Additionally, we assume the field gain per resonator pass to be small and do not take into account finite-pulse effects. This relatively simple model enables one to take into account almost all the main physical effects influencing FEL oscillator operation and in many cases provides a correct description of the processes in the FEL oscillator.

We study in detail the most popular FEL oscillator configuration with a uniform undulator. The linear mode of FEL oscillator operation is investigated by means of analytical techniques. We derive analytic solutions for the small-signal gain taking into account the space charge fields and the energy spread of electrons in the electron beam. These rigorous solutions are written down in reduced form and are used to obtain the universal dependencies for the linear mode of FEL oscillator operation. The nonlinear mode is investigated by means of numerical simulations. Rigorous results of the linear theory serve as a primary standard for testing the numerical simulation code. The main emphasis is put on finding optimal conditions for FEL operation to reach maximal FEL efficiency at saturation. The results of numerical simulations are generalized by means of similarity techniques. In particular, we show that the maximal FEL efficiency at saturation and the maximal amplitude of the radiation field in the resonator are universal functions of only three reduced parameters: the reduced damping factor of the resonator, the space charge and the energy spread parameter.

Then we extend our study to an FEL oscillator equipped with a nonuniform undulator. We consider the prominent practical case when the undulator consists of two sections separated by a drift (or dispersion) section. The first undulator section is uniform, and the parameters of the second section may be linearly tapered. Two popular configurations can be described in this way. The first one, an optical klystron, is used for increasing the small-signal gain. The second configuration is the FEL oscillator with tapered main undulator. Sometimes a prebuncher is installed in front of the main undulator. The latter scheme allows one to increase the efficiency at saturation. The linear mode of the FEL oscillator with nonuniform undulator is studied analytically. The nonlinear mode of operation is investigated analytically and numerically with subsequent application of similarity techniques. We present universal reduced dependencies allowing one to calculate the characteristics of the FEL oscillator with a nonuniform undulator.

Most of the results presented in this chapter were obtained in the framework of the single-mode model. This is a very fruitful approach allowing one to study many features of the FEL oscillator by means of relatively simple tools. Nevertheless, some basic problems cannot be studied in the framework of this approach. The first problem is which value of the input intracavity power should be used in the simulations. The second one is the region of applicability of the single-mode model itself. To answer these questions, at the end of this chapter we perform an investigation of the operation of a multimode FEL oscillator. Rigorous solutions are obtained describing the linear mode of FEL oscillator operation starting from the shot noise in the electron beam. The study is based on the description of the electromagnetic field in terms of high-Q modes of an optical resonator. The active medium (the electron beam in the undulator) is described in terms of the susceptibility. The electromagnetic field is coupled to the active medium via the transverse current density entering Maxwell's equations. The field damping in the resonator is calculated using Leontovich's boundary conditions on a mirror surface. The results obtained give the value of the effective initial intracavity power in the resonator. Also, we discuss the region of applicability for the single-mode model of the FEL oscillator and the problem of sideband instability.

3.1 Small-Signal Gain

3.1.1 Basic Relations

We consider a plane Fabry Perot resonator equipped with two plane parallel mirrors. The distance between the mirrors is equal to L. An undulator of length l_w is placed inside the resonator and its axis coincides with the resonator axis. To be specific, we write all the formulae for the case of a helical undulator. The coordinate of the undulator entrance is $z = 0$. We neglect the transverse variations of the undulator field and assume that the electrons move along the constrained helical trajectories in parallel with the z-axis (on average over the undulator period). The electron rotation angle θ_s is considered to be small and the longitudinal electron velocity v_z is close to the velocity of light ($v_z \simeq c$).

We suppose the electromagnetic field in the resonator to be circularly polarized because of the helical magnetic field of the undulator. Using the complex representation, we can write the expression for the electric field of the wave synchronous with the electron beam:

$$E_x + \mathrm{i}E_y = \tilde{E}(z)\exp[\mathrm{i}\omega(z/c - t)] \ .$$

To calculate the evolution of the radiation in the FEL oscillator, we should take into account two processes: amplification of the electromagnetic wave by the electron beam and losses in the mirrors. The amplitude of the electromagnetic wave changes along the undulator axis due to the interaction with the electron beam. We consider the case of a low-gain approximation which means that the field increment per undulator pass is small, i.e.

$$|\tilde{E}(l_\mathrm{w}) - \tilde{E}(0)|/|\tilde{E}(0)| \ll 1 \ . \tag{3.1}$$

Such a situation is typical for many practical devices. The gain per pass of the undulator is defined as

$$g = |\tilde{E}(l_\mathrm{w})|^2/|\tilde{E}(0)|^2 - 1 \ . \tag{3.2}$$

Introducing notation

$$Z = \tilde{E}(l_w)/\tilde{E}(0) - 1 ,$$

and using condition (3.1) we rewrite (3.2) in the following form:

$$g \simeq 2\,\mathrm{Re}\,Z .$$

The phase of the electromagnetic wave changes in the amplification process, too. Introducing the notation $\arg\{\tilde{E}(0)\} = \psi_0$ and $\arg\{\tilde{E}(l_w)\} = \psi_0 + \Delta\psi_0$, we may write:

$$\Delta\psi_0 \simeq \mathrm{Im}\,Z .$$

In this section we study the linear mode of FEL oscillator operation (or, in other words, the small-signal regime). This mode of FEL oscillator operation can be described by means of the methods developed in Chap. 2. The change of the field amplitude along the undulator axis is given by (2.16):

$$d\tilde{E}/dz = -2\pi\theta_s \tilde{j}_1(z)/c , \qquad (3.3)$$

where the perturbed transverse current density is given by

$$(j_x + ij_y) = \theta_s \exp(-ik_w z)(\tilde{j}_1 e^{i\psi} + \mathrm{C.C.}) . \qquad (3.4)$$

Evolution of the first harmonic of the beam current density can be found from the solution of the linearized Vlasov equation and is given by (2.14):

$$\tilde{j}_1(z) = ij_0 \int_0^z dz' \left[U + 4\pi e \tilde{j}_1(z')/\omega \right]$$

$$\times \int dP(dF(P)/dP) \exp\left\{ i\left[C + \omega P/(c\gamma_z^2 \mathcal{E}_0) \right] (z' - z) \right\} . \qquad (3.5)$$

Here we have used the same notation as in Chap. 2: $P = \mathcal{E} - \mathcal{E}_0$ is the deviation of the electron energy $\mathcal{E}$ from the nominal value $\mathcal{E}_0$, $C = k_w - \omega/(2c\gamma_z^2)$ is the detuning of the electron with the nominal energy $\mathcal{E}_0$,

$$U = -e\theta_s \tilde{E}(z)/(2i)$$

is the complex amplitude of the effective potential, $\theta_s = eH_w/(\mathcal{E}_0 k_w)$, $\gamma_z^{-2} = \gamma^{-2} + \theta_s^2$, and $\gamma = \mathcal{E}_0/(m_e c^2)$. Using (3.3) and (3.5) we can calculate the gain and the field phase increment in the small-signal regime.

Radiation losses in the resonator are usually described by the parameter α equal to the relative power losses per resonator round-trip. This parameter is also referred to as the damping factor. In the framework of the one-dimensional model $\alpha = T_1 + T_2 + \Gamma_1 + \Gamma_2$, where (T_1, Γ_1) and (T_2, Γ_2) are the transmission and absorption coefficients of the first and second resonator mirror, respectively.

In the linear mode of FEL oscillator operation the values of g and $\Delta\psi_0$ do not depend on the value of the field stored in the resonator. We denote the small-signal gain as g_s. It is a function of the frequency and the parameters of the electron beam and of the undulator. When the maximal value of the gain, $\max(g_s)$, exceeds the relative radiation losses in the resonator,

$$\max(g_s) > \alpha \,, \tag{3.6}$$

the radiation power in the resonator begins to grow, i.e. lasing takes place. The intracavity radiation power during the nth radiation round-trip in the resonator can be expressed in terms of the initial intracavity power $W_{\text{in}}(0)$:

$$W_{\text{in}}(n) = (1 + \max(g_s) - \alpha)^n W_{\text{in}}(0) \simeq W_{\text{in}}(0) \exp\left[(\max(g_s) - \alpha) n\right] \,.$$

The initial intracavity power is usually estimated as the spontaneous undulator radiation power within the FEL bandwidth and within the solid angle corresponding to the angular divergence of the ground transverse mode of the resonator. We present in Sect. 3.4 a rigorous derivation of the expression for the initial intracavity power. The radiation power loss in the resonator (due to absorption and transmission) during the nth round-trip is

$$W(n) = \alpha W_{\text{in}}(0) \exp\left[(\max(g_s) - \alpha) n\right] \,. \tag{3.7}$$

3.1.2 Cold Electron Beam

Let us find the gain for the simplest case of FEL oscillator operation, when we can neglect the effects of the space charge field and of the energy spread. The distribution function of a monoenergetic electron beam is given by the delta function, $F(P) = \delta(P)$. It follows from (3.5) that

$$\tilde{j}_1(z) = \mathrm{i} j_0 \theta_s e\omega (2\mathcal{E}_0 \gamma_z^2 c)^{-1} \int_0^z \mathrm{d}z'(z'-z)\tilde{E}(z')\exp[\mathrm{i}C(z'-z)] \,.$$

Integrating (3.3) over z under condition (3.1), we obtain the expression for the function Z:

$$Z = -\mathrm{i}\pi j_0 \theta_s^2 e\omega (\mathcal{E}_0 \gamma_z^2 c^2)^{-1} \int_0^{l_w} \mathrm{d}z \int_0^z \mathrm{d}z'(z'-z)\exp[\mathrm{i}C(z'-z)] \,.$$

It is convenient to rewrite this expression in the following form:

$$Z = \mathrm{i}\frac{\tau}{2} \int_0^1 \mathrm{d}\xi \int_0^\xi \mathrm{d}\xi' \xi' \exp(-\mathrm{i}\hat{C}\xi') \,, \tag{3.8}$$

where $\tau = 2\pi \theta_s^2 \omega j_0 l_w^3 (c\gamma_z^2 \gamma I_A)^{-1}$ is the FEL oscillator gain parameter,[1] $I_A = m_e c^3/e \simeq 17$ kA is the Alfven current and $\hat{C} = Cl_w$ is the detuning

---

[1] In the theory of the FEL amplifier the gain parameter, Γ, is a dimensional value and is used for normalization of the longitudinal coordinate. The value of Γ^{-1} is of the order of the field gain length. In the theory of the FEL oscillator the normalization of the longitudinal coordinate is performed with respect to the total length of the undulator. The gain parameter is a dimensionless value, and τ^{-1} is a measure of time in terms of the resonator round-trips.

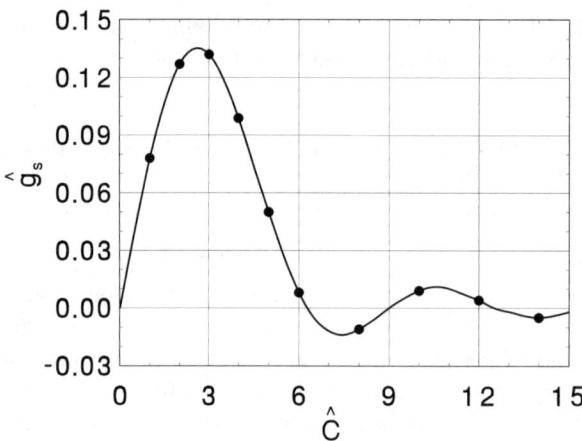

Fig. 3.1. The reduced small-signal gain $\hat{g}_s$ versus the reduced detuning $\hat{C}$. Here $\hat{\Lambda}_T^2 = 0$ and $\hat{\Lambda}_p^2 \to 0$. The full curve is calculated with the analytical formula (3.9) and the circles are the results of numerical simulations with (3.13) and (3.14)

parameter. Introducing the notation $\hat{g}_s = g_s/\tau$ and $\hat{Z} = 2Z/\tau$ and integrating (3.8), we reduce the function $\hat{Z}$ to the simpler expression:

$$\hat{Z} = 2\mathrm{i}\hat{C}^{-2}\left[2(\hat{C})^{-1}\sin(\hat{C}/2) - \cos(\hat{C}/2)\right]\exp(-\mathrm{i}\hat{C}/2) \ . \tag{3.9a}$$

The gain and the field phase increment per undulator pass are expressed in terms of the function $\hat{Z}$ as

$$\hat{g}_s = \operatorname{Re}\hat{Z} \ , \qquad \Delta\psi_0/\tau = (1/2)\operatorname{Im}\hat{Z} \ . \tag{3.9b}$$

In the case under study the maximal gain is a function only of the detuning parameter, $\hat{C}$. The gain takes its maximum value at $\hat{C}_m = 2.6$:

$$\max(\hat{g}_s) = 0.135 \ .$$

The lasing condition can be written in the normalized form:

$$\max(\hat{g}_s) > \hat{\alpha} \ ,$$

where $\hat{\alpha} = \alpha/\tau$ is the reduced damping factor.

The dependence of the reduced small-signal gain, $\hat{g}_s$, on the detuning parameter $\hat{C}$ is presented in Fig. 3.1 (note that the gain curve is antisymmetric with respect to $\hat{C} = 0$).

3.1.3 Gaussian Energy Spread

Let us consider an electron beam with a Gaussian energy spread. The initial distribution function is given by (2.20a). In the case of a negligibly small space charge field, we obtain:

3.1 Small-Signal Gain   93

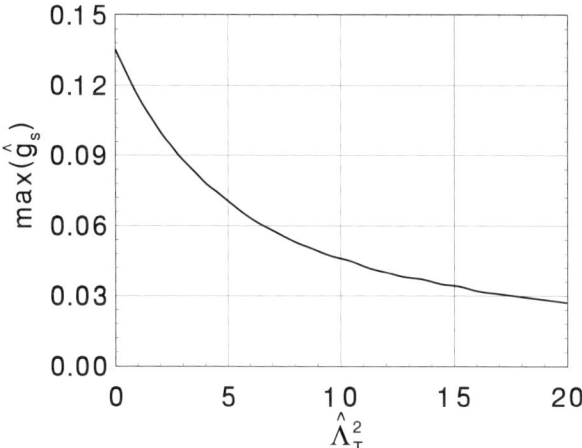

Fig. 3.2. The maximal reduced small-signal gain $\hat{g}_s$ versus the energy spread parameter $\hat{\Lambda}_T^2$. Here $\hat{\Lambda}_p^2 \to 0$. The calculations have been performed with (3.9b) and (3.10)

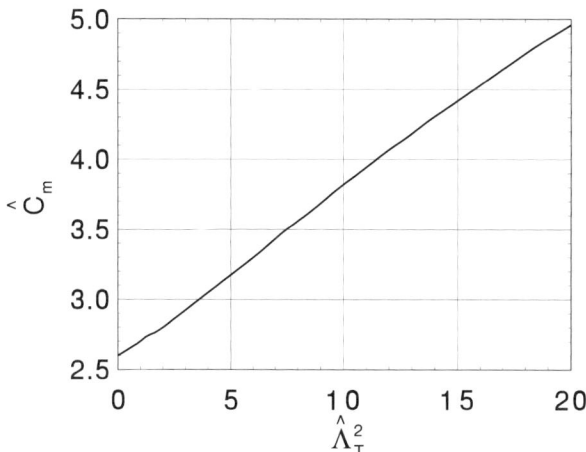

Fig. 3.3. The optimal reduced detuning $\hat{C}_m$ versus the energy spread parameter $\hat{\Lambda}_T^2$. Here $\hat{\Lambda}_p^2 \to 0$. The calculations have been performed with (3.9b) and (3.10)

$$\hat{Z} = i \int_0^1 d\xi \int_0^\xi d\xi' \xi' \exp\left[-i\hat{C}\xi' - \hat{\Lambda}_T^2 (\xi')^2/2\right] , \qquad (3.10)$$

where $\hat{\Lambda}_T^2 = \langle (\Delta \mathcal{E})^2 \rangle / (\beta^2 \mathcal{E}_0^2)$ is the FEL oscillator energy spread parameter, $\beta = (4\pi N_w)^{-1}$ is the efficiency parameter and N_w is the number of undulator periods.

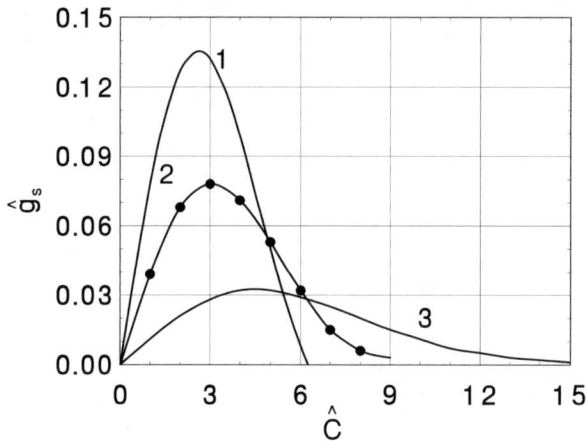

Fig. 3.4. The reduced small-signal gain $\hat{g}_s$ versus the reduced detuning $\hat{C}$. Here $\hat{\Lambda}_p^2 \to 0$. The full curves are calculated with the analytic formulae (3.9b) and (3.10). The circles are the results of numerical simulations with (3.13) and (3.14). Curve 1: $\hat{\Lambda}_T^2 = 0$. Curve 2: $\hat{\Lambda}_T^2 = 4$. Curve 3: $\hat{\Lambda}_T^2 = 16$.

According to (3.10), the maximal gain, $\max(\hat{g}_s) = \max(\operatorname{Re}\hat{Z})$, and the detuning parameter, $\hat{C}_m$, corresponding to this maximal gain, are universal functions of the energy spread parameter $\hat{\Lambda}_T^2$. The plots of these functions are presented in Figs. 3.2 and 3.3. The gain curves for several values of $\hat{\Lambda}_T^2$ are presented in Fig. 3.4. At large values of the energy spread parameter, $\hat{\Lambda}_T^2 \gg 1$, we have asymptotically:

$$\hat{C}_m \simeq \hat{\Lambda}_T , \qquad \max(\hat{g}_s) \simeq \sqrt{\pi/(2e)}\,\hat{\Lambda}_T^{-2} \simeq 0.76\hat{\Lambda}_T^{-2} ,$$

where $e = 2.718...$ is the base of natural logarithms.

3.1.4 Space Charge Effects

Let us study the influence of the space charge field on the gain of the FEL oscillator. We consider a monoenergetic electron beam with initial distribution function $F(P) = \delta(P)$. The complex amplitude $\tilde{j}_1(z)$ of the first harmonic of the beam current density can be calculated using the Laplace transform. Multiplying (3.5) by $\exp(-pz)$ and integrating over z from 0 to ∞ under condition (3.1), we obtain:

$$\bar{j}_1(p) = \int_0^\infty e^{-pz}\tilde{j}_1(z)\mathrm{d}z = -\frac{\omega j_0 U}{\gamma_z^2 \varepsilon_0 c}\frac{1}{p\left[(p+iC)^2 + \Lambda_p^2\right]} ,$$

where $\Lambda_p^2 = 4\pi j_0/(\gamma_z^2\gamma I_A)$. To obtain $\tilde{j}_1$, we must perform the inverse Laplace transformation of the function $\bar{j}_1(p)$. The complex function $\bar{j}_1(p)$ satisfies the Jordan's lemma. So, using Cauchy's residue theorem we obtain:

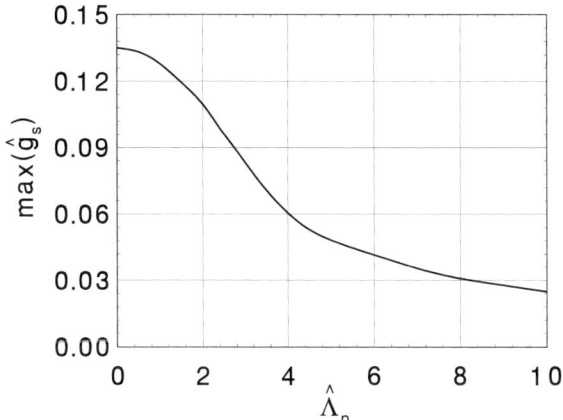

Fig. 3.5. The maximal reduced small-signal gain $\hat{g}_s$ versus $\hat{\Lambda}_p$. Here $\hat{\Lambda}_T^2 = 0$. The calculations have been performed with (3.9b) and (3.12)

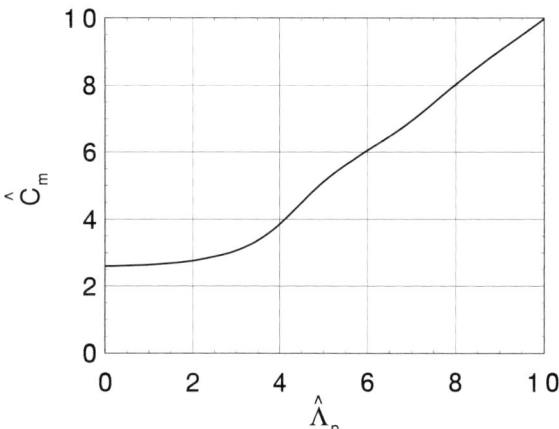

Fig. 3.6. The optimal reduced detuning $\hat{C}_m$ versus $\hat{\Lambda}_p$. Here $\hat{\Lambda}_T^2 = 0$. The calculations have been performed with (3.9b) and (3.12)

$$\tilde{j}_1(z) = \frac{i\tilde{E}\omega e \theta_s j_0}{2c\gamma_z^2 \mathcal{E}_0}$$
$$\times \left\{ \frac{1}{C^2 - \Lambda_p^2} + \frac{\exp\left[-i(C + \Lambda_p)z\right]}{2\Lambda_p(C + \Lambda_p)} - \frac{\exp\left[-i(C - \Lambda_p)z\right]}{2\Lambda_p(C - \Lambda_p)} \right\}. \quad (3.11)$$

Substituting this expression into (3.3) and integrating over z, we find

$$\hat{Z} = i\left(\hat{C}^2 - \hat{\Lambda}_p^2\right)^{-2} \left\{ \hat{\Lambda}_p^2 - \hat{C}^2 - 2i\hat{C}\left[1 - \cos(\hat{\Lambda}_p)\exp(-i\hat{C})\right] \right.$$
$$\left. - \hat{\Lambda}_p^{-1}\left(\hat{C}^2 + \hat{\Lambda}_p^2\right)\sin(\hat{\Lambda}_p)\exp(-i\hat{C}) \right\}, \quad (3.12)$$

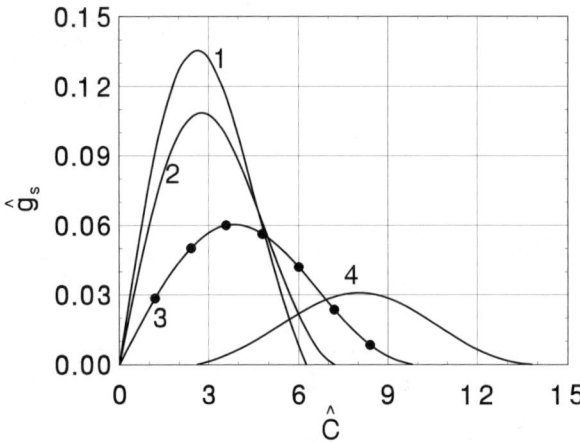

Fig. 3.7. The reduced small-signal gain $\hat{g}_s$ versus $\hat{C}$. Here $\hat{\Lambda}_T^2 = 0$. The curves are calculated with the analytical formulae (3.9b) and (3.12) and the circles are the results of numerical simulations with (3.14) and (3.22) Curve 1: $\hat{\Lambda}_p \to 0$. Curve 2: $\hat{\Lambda}_p = 2$. Curve 3: $\hat{\Lambda}_p = 4$. Curve 4: $\hat{\Lambda}_p = 8$

where $\hat{\Lambda}_p^2 = \Lambda_p^2 l_w^2$ is the FEL oscillator space charge parameter. It follows from (3.12) that the maximal gain, $\max(\hat{g}_s) = \max(\operatorname{Re} \hat{Z})$, and the detuning parameter, $\hat{C}_m$, corresponding to this maximum, are universal functions of the space charge parameter $\hat{\Lambda}_p^2$. The plots of these functions are presented in Figs. 3.5 and 3.6. The gain curves for several values of $\hat{\Lambda}_p$ are presented in Fig. 3.7. For large values of the space charge parameter, $\hat{\Lambda}_p^2 \gg 1$, the maximal gain is achieved at $\hat{C}_m \simeq \hat{\Lambda}_p$ and is equal to

$$\max(\hat{g}_s) \simeq (4\hat{\Lambda}_p)^{-1} \, .$$

3.2 Saturation Effects in the FEL Oscillator

All the results obtained in the previous section refer to the initial stage of FEL oscillator operation. In the linear mode of operation the radiation output power grows exponentially in time, while the radiation field phase changes linearly in time. Near the saturation point the electron motion becomes nonlinear and the radiation output power achieves asymptotically its saturation level. To find the saturation power of the FEL oscillator one should solve the equations of nonlinear FEL oscillator theory.

In this section the analysis of nonlinear processes in the FEL oscillator is performed using the single-mode model. The lasing frequency ω corresponds to the maximum of the gain in the linear stage, i.e. the value of the detuning parameter $\hat{C}$ is assumed to be equal to $\hat{C}_m$, when the gain achieves its maximum at the linear stage. The electron beam pulse duration is assumed to be infinitely long, i.e. we do not study effects connected with finite pulse duration.

In the framework of the accepted model, the output characteristics of the FEL oscillator are functions of eight dimensional parameters:

$$\mathcal{E}_0\,,\quad \hat{j}_0\,,\quad \langle(\Delta\mathcal{E})^2\rangle\,,\quad k_{\mathrm{w}}\,,\quad H_{\mathrm{w}}\,,\quad l_{\mathrm{w}}\,,\quad L\,,\quad \alpha\,.$$

The resonator base L defines only the time scale of the development of the FEL process, but not the characteristics of radiation at saturation. Application of similarity techniques allows one to describe the characteristics of the FEL oscillator at saturation with only five dimensionless parameters:

$$\hat{\alpha}\,,\quad \tau\,,\quad \beta\,,\quad \hat{\Lambda}_{\mathrm{p}}^2\,,\quad \hat{\Lambda}_{\mathrm{T}}^2\,,$$

where $\hat{\alpha} = \alpha/\tau$ is the reduced damping factor, τ is the gain parameter, $\beta = (4\pi N_{\mathrm{w}})^{-1}$ is the efficiency parameter, N_{w} is number of undulator periods, $\hat{\Lambda}_{\mathrm{p}}^2$ is the space charge parameter, and $\hat{\Lambda}_{\mathrm{T}}^2$ is the energy spread parameter. In this book we limit our consideration to the case of a low-efficiency FEL oscillator. In this approximation the efficiency parameter β is excluded from the self-consistent field equations and the number of parameters of the problem is reduced to four. In many practical situations the gain per pass is low, $g \ll 1$, which corresponds to small values of the gain parameter, $\tau \lesssim 1$. In this case the FEL characteristics at saturation are determined by three reduced parameters: $\hat{\alpha}$, $\hat{\Lambda}_{\mathrm{p}}^2$ and $\hat{\Lambda}_{\mathrm{T}}^2$.

3.2.1 Self-Consistent Equations

The self-consistent equations describing the amplification process of the wave in the undulator are similar to (2.93) and (2.94) describing the amplification process in the FEL amplifier (the only difference consists in another normalization procedure):

$$\frac{\mathrm{d}\hat{P}}{\mathrm{d}\hat{z}} = \hat{u}\cos(\psi + \psi_0)\,,$$
$$\frac{\mathrm{d}\psi}{\mathrm{d}\hat{z}} = \hat{P} + \hat{C} + \beta\hat{u}\sin(\psi + \psi_0)\,, \tag{3.13}$$

$$\frac{\mathrm{d}\hat{u}}{\mathrm{d}\hat{z}} = \frac{\tau\hat{j}_1}{2}\cos(\psi_0 - \psi_1)\,,$$
$$\frac{\mathrm{d}\psi_0}{\mathrm{d}\hat{z}} = -\frac{\tau\hat{j}_1}{2\hat{u}}\sin(\psi_0 - \psi_1)\,, \tag{3.14}$$

where $\hat{z} = z/l_{\mathrm{w}}$, $\hat{u} = ul_{\mathrm{w}}/(\beta\mathcal{E}_0)$ is the normalized effective potential of the interaction of the particle with the electromagnetic wave, $(u/2)\exp(i\psi_0) = iU = -e\theta_{\mathrm{s}}\tilde{E}/2$, $\hat{P} = P/(\beta\mathcal{E}_0)$, $P = \mathcal{E} - \mathcal{E}_0$ and $\beta = c\gamma_z^2/(\omega l_{\mathrm{w}}) = (4\pi N_{\mathrm{w}})^{-1}$ is the efficiency parameter. The definition of the detuning parameter $\hat{C}$ and the gain parameter τ were introduced in Sect. 3.1. The parameter β is inversely proportional to the number of undulator periods N_{w} and is always small.

Hence, we neglect the summand proportional to β in the second equation (3.13).

The value of the detuning parameter $\hat{C}$ should be set to $\hat{C}_m$ corresponding to the maximum of the small-signal gain calculated in Sect. 3.1. The amplitude and phase of the first harmonic of the beam current density, $\hat{j}_1 = j_1/j_0$ and ψ_1, are calculated as

$$\hat{j}_1 \cos\psi_1 = \frac{1}{\pi} \int_0^{2\pi} \hat{j}_z \cos\psi \, d\psi \;,$$

$$\hat{j}_1 \sin\psi_1 = -\frac{1}{\pi} \int_0^{2\pi} \hat{j}_z \sin\psi \, d\psi \;, \quad (3.15)$$

where $\hat{j}_z = j_z/j_0$.

3.2.2 Nonlinear Simulation Algorithm

We simulate the electron beam with N macroparticles per interval $(0, 2\pi)$ over the phase ψ. The reduced beam current density $\hat{j}_z$ is calculated as:

$$\hat{j}_z = -\frac{2\pi}{N} \sum_{k=1}^{N} \delta(\psi - \psi_{(k)}) \;,$$

where $\psi_{(k)}$ are the phases of the particles and $\delta(\psi - \psi_{(k)})$ is the delta function. The function $\hat{j}_z$ has the following normalization:

$$\frac{1}{2\pi} \int_0^{2\pi} \hat{j}_z \, d\psi = -1 \;.$$

The amplitude and phase of the first harmonic of the beam current density are given by the expressions:

$$\hat{j}_1 \cos\psi_1 = \frac{1}{\pi} \int_0^{2\pi} \hat{j}_z \cos\psi \, d\psi = -\frac{2}{N} \sum_{k=1}^{N} \cos\psi_{(k)} \;,$$

$$\hat{j}_1 \sin\psi_1 = -\frac{1}{\pi} \int_0^{2\pi} \hat{j}_z \sin\psi \, d\psi = \frac{2}{N} \sum_{k=1}^{N} \sin\psi_{(k)} \;. \quad (3.16)$$

The procedure of numerical simulation is organized as follows. At the moment of time t_j we have at the undulator entrance the unmodulated electron beam and the electromagnetic field with amplitude $E^{(j)} = |\tilde{E}(t_j)|$ and phase $\psi_0^{(j)}$, i.e. at $z=0$ we have ($k=1,\ldots,N$):

$$\hat{P}_{(k)} = 0 \;, \qquad \hat{j}_1 = 0 \;, \qquad \hat{u} = \hat{u}^{(j)} = E^{(j)}/E_0 \;, \qquad \psi_0 = \psi_0^{(j)} \;,$$

where $E_0 = \beta \mathcal{E}_0/(e\theta_s l_w)$ is the saturation field parameter. The equation of motion (3.13) and the field equations (3.14) are solved numerically with the Runge Kutta technique. After one undulator pass we calculate the increase of the field amplitude $\Delta \hat{u}^{(j)}$ and its phase $\Delta \psi_0^{(j)}$. Then, after the radiation round-trip in the resonator, i.e. at the moment of time $t_{j+1} = t_j + 2L/c$, we obtain the following initial conditions at the undulator entrance:

$$\hat{P}_{(k)} = 0, \quad \hat{j}_1 = 0, \quad \hat{u} = \hat{u}^{(j+1)} = [1 - \alpha/2]\left[\Delta \hat{u}^{(j)} + \hat{u}^{(j)}\right],$$

$$\psi_0 = \psi_0^{(j)} + \Delta \psi_0^{(j)}.$$

We do not take into account the phase change after reflection from the resonator mirrors because this effect does not influence FEL oscillator operation. The multiple use of this procedure under the given initial conditions for the radiation field at the moment of time $t_0 = 0$ enables one to calculate the field evolution in time. The initial field amplitude $\hat{u}^{(0)} = \hat{u}_{\text{in}}$ can be estimated using (3.110) obtained in Sect. 3.4.

3.2.3 Resonator Losses and Efficiency Optimization

In the linear mode of operation, at $\hat{u} \ll 1$, the field amplitude stored in the resonator grows exponentially in time. Near saturation, at $\hat{u} \simeq 1$, the electron beam is overmodulated, which leads to a slowing down of the growth of the field amplitude. When the power gain becomes equal to the resonator losses, the field amplitude $\hat{u}$ achieves asymptotically its maximal value $\hat{u}^{(\infty)}$. So, the field amplitude at saturation depends on the value of the resonator losses.

We start the investigation of efficiency optimization assuming that the space charge and energy spread effects are negligibly small. When the field amplification per one resonator pass is small, the saturation condition may be written as:

$$\Delta \hat{u}^{(\infty)} = \alpha \hat{u}^{(\infty)}/2 . \tag{3.17}$$

Let us perform a qualitative analysis of (3.13) and (3.14). The value of $\Delta \hat{u}^{(j)}$ is proportional to the gain parameter τ. The amplitude $\hat{j}_1$ and the phase $(\psi_1 - \psi_0)$ depend only on the field amplitude, $\hat{u}$. So, the increment of the field amplitude near saturation, $\Delta \hat{u}^{(\infty)}$, may be represented as:

$$\Delta \hat{u}^{(\infty)} = \tau f(\hat{u}^{(\infty)}) . \tag{3.18}$$

It follows from (3.17) and (3.18) that the field amplitude at saturation is a function only of the reduced damping factor, $\hat{\alpha} = \alpha/\tau$:

$$\hat{u}^{(\infty)} = \hat{u}^{(\infty)}(\hat{\alpha}) .$$

The FEL efficiency at saturation is defined as the ratio of the radiation power losses in the resonator to the electron beam power. In the case under study the density of the radiation power losses is given by

$$\Pi = cE^{(\infty)} \Delta E^{(\infty)}/(2\pi) ,$$

so we have for the FEL efficiency

$$\eta = e\Pi/(\mathcal{E}_0 j_0) = \beta \hat{u}^{(\infty)} \Delta \hat{u}^{(\infty)}/\tau \,,$$

where β is the efficiency parameter. It is convenient to introduce the reduced efficiency $\hat{\eta} = \eta/\beta$. Using relation (3.17), we get

$$\hat{\eta} = \eta/\beta = \hat{\alpha}(\hat{u}^{(\infty)})^2/2 \,. \tag{3.19}$$

Now let us show that the energy conservation law takes holds in the FEL oscillator. We introduced above the FEL efficiency in terms of the radiation power. The FEL efficiency may also be expressed in terms of the electron energy losses: $\eta = -\langle P \rangle/\mathcal{E}_0$, where $P = (\mathcal{E} - \mathcal{E}_0)$, $\mathcal{E}_0$ is the nominal electron energy at the undulator entrance, and the symbol $\langle \ldots \rangle$ means averaging over the beam electrons. Using the definition of the reduced energy deviation $\hat{P}$, we find that the reduced efficiency can be expressed as $\hat{\eta} = -\langle \hat{P} \rangle$. One can find from the system of canonical equations (3.13) that the rate of the mean energy losses of an electron is equal to

$$\langle \mathrm{d}\hat{P}/\mathrm{d}\hat{z} \rangle = \hat{u} N^{-1} \sum_{k=1}^{N} \cos(\psi_{(k)} + \psi_0) \,.$$

Remembering that

$$N^{-1} \sum_{k=1}^{N} \cos(\psi_{(k)} + \psi_0) = -\frac{1}{2}\hat{j}_1 \cos(\psi_1 - \psi_0) \,,$$

and using the first equation of (3.14), we get

$$\langle \mathrm{d}\hat{P}/\mathrm{d}\hat{z} \rangle = -(\hat{u}/\tau)\mathrm{d}\hat{u}/\mathrm{d}\hat{z} \,.$$

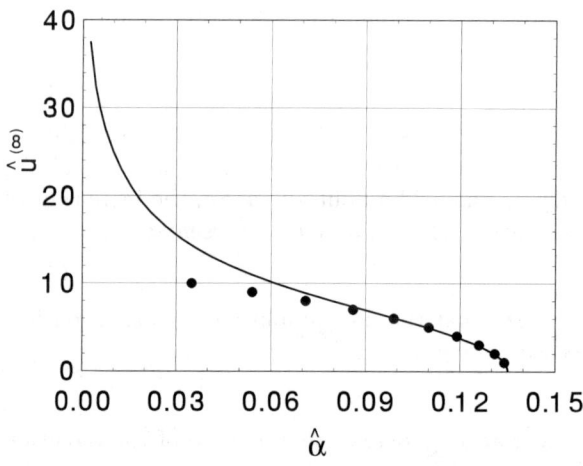

Fig. 3.8. The reduced field amplitude versus the reduced damping factor. The FEL oscillator operates at saturation. The circles are calculated with the approximate formula (3.20). Here $\hat{\Lambda}_T^2 = 0$ and $\hat{\Lambda}_p^2 \to 0$

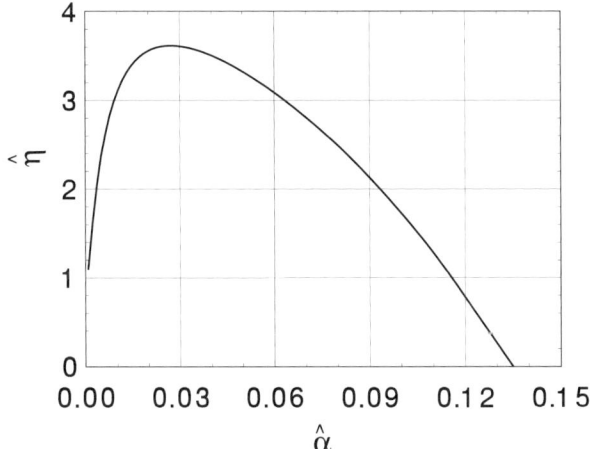

Fig. 3.9. The reduced efficiency versus the reduced damping factor. The FEL oscillator operates at saturation. Here $\hat{\Lambda}_T^2 = 0$ and $\hat{\Lambda}_p^2 \to 0$

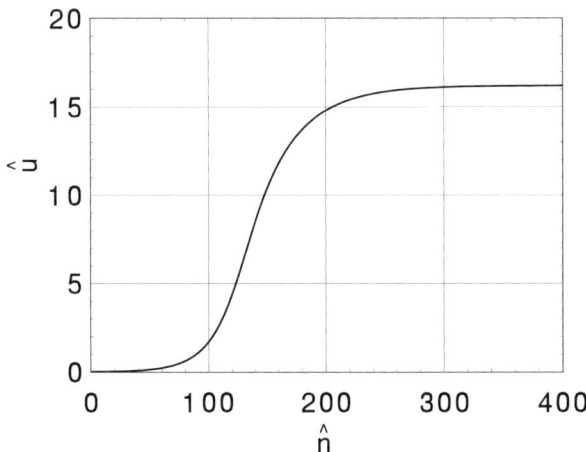

Fig. 3.10. The reduced field amplitude in the resonator versus the reduced number of resonator round-trips at the optimal value of the reduced damping factor $\hat{\alpha} = 0.028$. Here $\hat{\Lambda}_T^2 = 0$, $\hat{\Lambda}_p^2 \to 0$ and $\hat{u}^{(0)} = 0.01$

It follows from this relation that in the low-gain limit the reduced efficiency at saturation is given by

$$\hat{\eta} = -\langle \hat{P} \rangle = \hat{u}^{(\infty)} \Delta \hat{u}^{(\infty)}/\tau = \hat{\alpha}(\hat{u}^{(\infty)})^2/2.$$

Comparing this expression with (3.19), we find that the total power of the electron beam and radiation is conserved.

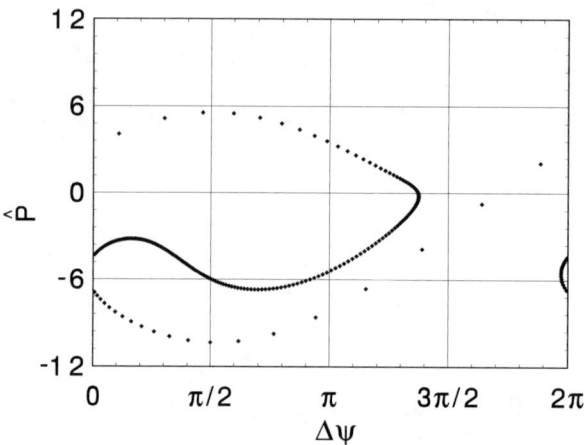

Fig. 3.11. Phase space distribution of the particles at the undulator exit. The FEL oscillator operates at saturation. Here $\hat{\Lambda}_T^2 = 0$, $\hat{\Lambda}_p \to 0$, and $\hat{\alpha}_{\rm opt} = 0.028$

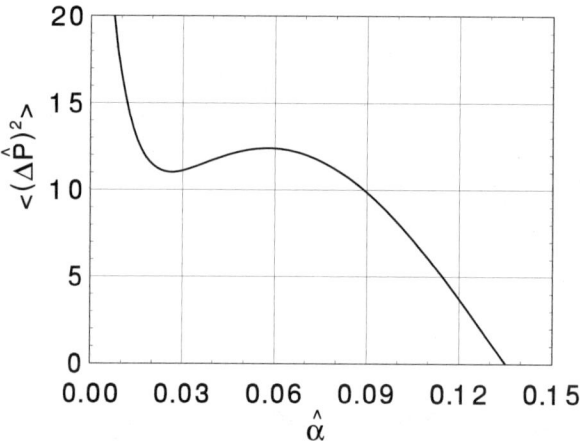

Fig. 3.12. The reduced rms energy spread of the particles at the undulator exit versus the reduced damping factor. The FEL oscillator operates at saturation. Here $\hat{\Lambda}_T^2 = 0$ and $\hat{\Lambda}_p^2 \to 0$

Using the approach presented above we have calculated universal characteristics of the FEL oscillator at saturation. Figure 3.8 shows the reduced field amplitude at saturation as a function of the reduced damping factor. Near the lasing threshold, $\hat{\alpha} = 0.135$, there is a simple relation between $\hat{u}^{(\infty)}$ and $\hat{\alpha}$ which can be approximated as

$$\hat{\alpha} \simeq 0.135 - 0.001(\hat{u}^{(\infty)})^2 \; . \tag{3.20}$$

It is relevant to note that this quadratic dependence is a general feature of a weakly saturated medium. For instance, it is widely used in the theory of conventional lasers when describing the saturation mechanism just above the lasing threshold. When the FEL oscillator operates well above threshold, the field amplitude at saturation cannot be described by simple laws similar to those applicable to conventional lasers. The reason for this is connected with the more complicated saturation mechanism in the FEL oscillator. For instance, when the damping factor tends to zero, the saturation field in the FEL oscillator tends to infinity more slowly than in the case of a single-mode quantum laser with a homogeneously broadened line.

The plot of the reduced FEL efficiency versus the reduced damping factor $\hat{\alpha}$ is presented in Fig. 3.9. It is clearly seen that there is an optimum value of $\hat{\alpha}$ when the FEL efficiency achieves its maximum. Using the plot in Fig. 3.9, one may find the values of the maximum reduced efficiency and the optimum value of the reduced damping factor:

$$\hat{\eta}_{\max} = 3.62, \qquad \hat{\alpha}_{\mathrm{opt}} = 0.028 .$$

The value of the FEL efficiency is given by $\eta_{\max} = 0.29/N_{\mathrm{w}}$. The plot in Fig. 3.10 illustrates the time evolution of the field stored in the resonator at the optimal value of the resonator losses α_{opt}. The value of the initial normalized field is $\hat{u}^{(0)} = 0.01$. The reduced number of round-trips $\hat{n} = \tau n$ is chosen as a measure of time. At such a normalization the plot in Fig. 3.10 can be simply scaled to an arbitrary value of the initial field, $\hat{u}^{(0)} \ll 1$, by means of shifting the horizontal axis by

$$\Delta \hat{n} = [\max(\hat{g}_{\mathrm{s}}) - \hat{\alpha}_{\mathrm{opt}}]^{-1} \ln \frac{0.01}{\hat{u}^{(0)}} = 9.3 \left[\ln \left(\frac{1}{\hat{u}^{(0)}} \right) - 4.6 \right] .$$

Figure 3.11 presents the distributions of electrons in the phase space $(\hat{P}, \Delta\psi)$ at the undulator exit when the FEL oscillator operates at saturation. Here $\Delta\psi = \psi + \psi_0$ and ψ_0 is the phase of the effective potential. Figure 3.12 illustrates the reduced energy spread of the electrons at the undulator exit as a function of the reduced damping factor $\hat{\alpha}$. It is seen from this plot that the energy spread diverges as $\hat{\alpha} \to 0$. This effect is connected with the specific behavior of the stored field as a function of the parameter $\hat{\alpha}$ (see Fig. 3.8). When $\hat{\alpha}$ increases, the energy spread decreases. At the value of $\hat{\alpha} = \hat{\alpha}_{\mathrm{opt}} = 0.028$, the reduced energy spread $\langle(\Delta\hat{P})^2\rangle = f(\hat{\alpha})$ achieves its local minimum. The energy spread $\langle(\Delta\mathcal{E})^2\rangle/\mathcal{E}_0^2$ and the reduced energy spread $\langle(\Delta\hat{P})^2\rangle$ are connected by the following relation:

$$\begin{aligned}
\langle(\Delta\mathcal{E})^2\rangle/\mathcal{E}_0^2 &= N^{-1}\mathcal{E}_0^{-2} \sum_{k=1}^{N} (\mathcal{E}_{(k)} - \langle\mathcal{E}\rangle)^2 \\
&= \beta^2 N^{-1} \sum_{k=1}^{N} (\hat{P}_{(k)} - \langle\hat{P}\rangle)^2 \\
&= \beta^2 \langle(\Delta\hat{P})^2\rangle ,
\end{aligned} \qquad (3.21)$$

where the value of $\langle(\Delta\hat{P})^2\rangle$ may be found using the plot in Fig. 3.12. At the optimal value of the reduced damping factor, $\hat{\alpha} = \hat{\alpha}_{\mathrm{opt}} = 0.028$, the energy spread of the electrons at the undulator exit is given by the simple formula:

$$\sqrt{\langle(\Delta\mathcal{E}/\mathcal{E}_0)^2\rangle} \simeq 0.26/N_{\mathrm{w}} \ .$$

The number of macroparticles for the simulations should be $N = 100\text{–}200$. In this case the simulation results do not depend on the value of N to an accuracy better than 10^{-3}. The testing runs can be performed in the small-signal regime. In Fig. 3.1 we compare the results of the gain calculations obtained with the analytical formula (3.9) and with the numerical simulations. There is good agreement between the numerical and analytical results.

3.2.4 Space Charge and FEL Efficiency

The investigation of space charge effects requires the same modification of (3.13) as was made for the FEL amplifier equations (2.110):

$$\frac{\mathrm{d}\hat{P}_{(k)}}{\mathrm{d}\hat{z}} = \hat{u}\cos\left(\psi_{(k)} + \psi_0\right)$$

$$+ \hat{\Lambda}_{\mathrm{p}}^2\left\{\frac{1}{N}\sum_{i\neq k}\left[\pi\,\mathrm{sgn}\left(\psi_{(k)} - \psi_{(i)}\right) - \left(\psi_{(k)} - \psi_{(i)}\right)\right]\right\}\ ,$$

$$\frac{\mathrm{d}\psi_{(k)}}{\mathrm{d}\hat{z}} = \hat{P}_{(k)} + \hat{C}\ . \tag{3.22}$$

The normalization procedure is identical to that used in Sect. 3.1: $\hat{z} = z/l_{\mathrm{w}}$, $\hat{C} = Cl_{\mathrm{w}}$, $\hat{P} = P/(\beta\mathcal{E}_0)$, $\hat{\Lambda}_{\mathrm{p}}^2 = \Lambda_{\mathrm{p}}^2 l_{\mathrm{w}}^2 = 4\pi j_0 l_{\mathrm{w}}^2/(\gamma_z^2 \gamma I_{\mathrm{A}})$, and $\hat{u} = u l_{\mathrm{w}}/(\beta\mathcal{E}_0)$.

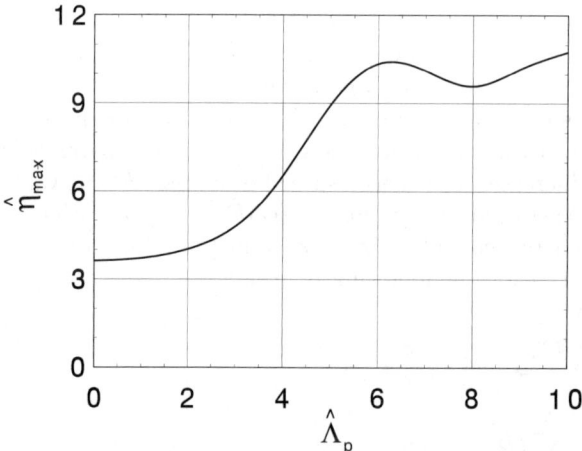

Fig. 3.13. The maximal reduced efficiency versus $\hat{\Lambda}_{\mathrm{p}}$. The FEL oscillator operates at saturation. Here $\hat{\Lambda}_{\mathrm{T}}^2 = 0$

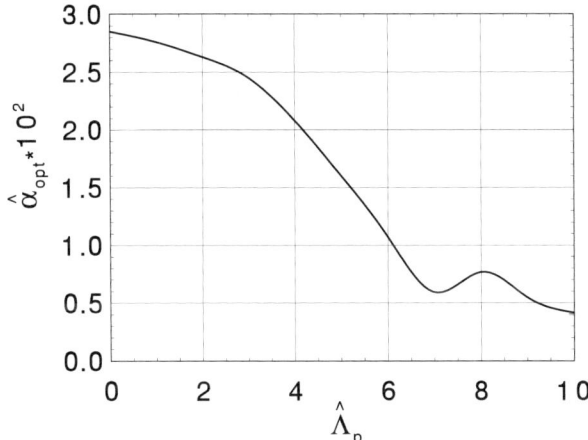

Fig. 3.14. The optimal reduced damping factor versus $\hat{\Lambda}_{\mathrm{p}}$. The FEL oscillator operates at saturation. Here $\hat{\Lambda}_{\mathrm{T}}^2 = 0$

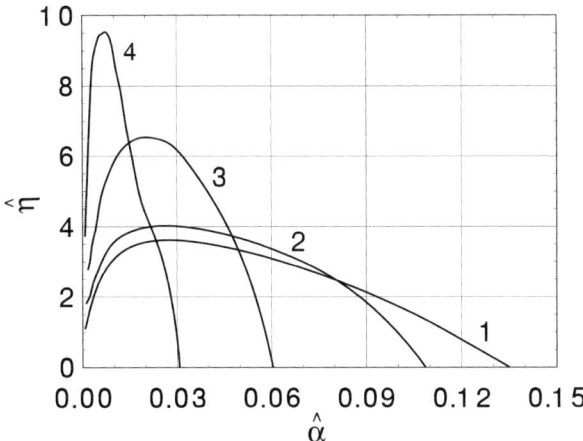

Fig. 3.15. The reduced efficiency versus the reduced damping factor. The FEL oscillator operates at saturation. Here $\hat{\Lambda}_{\mathrm{T}}^2 = 0$. Curve 1: $\hat{\Lambda}_{\mathrm{p}} \to 0$. Curve 2: $\hat{\Lambda}_{\mathrm{p}} = 2$. Curve 3: $\hat{\Lambda}_{\mathrm{p}} = 4$. Curve 4: $\hat{\Lambda}_{\mathrm{p}} = 8$

When the space charge fields are taken into account, the reduced efficiency is a universal function of two parameters: the reduced damping factor $\hat{\alpha}$ and the space charge parameter $\hat{\Lambda}_{\mathrm{p}}^2$. At each value of the space charge parameter $\hat{\Lambda}_{\mathrm{p}}^2$ there is an optimal value of the reduced damping factor, $\hat{\alpha}_{\mathrm{opt}}$, when the FEL efficiency achieves its maximum. This maximal efficiency and $\hat{\alpha}_{\mathrm{opt}}$ are universal functions of the space charge parameter, $\hat{\Lambda}_{\mathrm{p}}^2$. The dependence of the maximal FEL efficiency on $\hat{\Lambda}_{\mathrm{p}}$ is presented in Fig. 3.13. It is seen that the efficiency of the FEL oscillator grows with $\hat{\Lambda}_\mathrm{p}^2$. This can be explained by the fact that the space charge tends to prevent beam overmodulation in the nonlinear regime. Figure 3.14 illustrates the dependence of the optimal value of the reduced damping factor $\hat{\alpha}_\mathrm{opt}$ on $\hat{\Lambda}_\mathrm{p}$. The plots in Fig. 3.15 show the dependence of the reduced efficiency on the reduced damping factor $\hat{\alpha}$ for several values of $\hat{\Lambda}_\mathrm{p}$.

The simulation code can be tested at the linear stage of FEL oscillator operation (see Fig. 3.7). When the number of macroparticles is $N = 200$ the relative difference between the analytical and simulation results is less than 10^{-3}.

3.2.5 Energy Spread and FEL Efficiency

In the presence of an initial energy spread in the electron beam the FEL reduced efficiency is a universal function of three parameters: $\hat{\alpha}$, $\hat{\Lambda}_\mathrm{p}^2$, and $\hat{\Lambda}_\mathrm{T}^2$ (we assume the energy spread to be Gaussian). Let us consider the case of a negligibly small space charge field ($\hat{\Lambda}_\mathrm{p}^2 \to 0$). In this case the maximal reduced FEL efficiency and the optimal value of the reduced damping factor $\hat{\alpha}_\mathrm{opt}$ are universal functions of the energy spread parameter $\hat{\Lambda}_\mathrm{T}^2$. The plots of these functions are presented in Figs. 3.16 and 3.17. The dependencies of the FEL reduced efficiency on the reduced damping factor are plotted in Fig. 3.18.

The organization of the code for the simulation of the energy spread effects is similar to that described in Chap. 2. The macroparticle ensemble for these simulation runs has been prepared as follows. First, we prepared a micro-ensemble of 200 particles corresponding to a Gaussian energy distribution. Then we distributed four micro-ensembles evenly over the phase ψ from 0 to

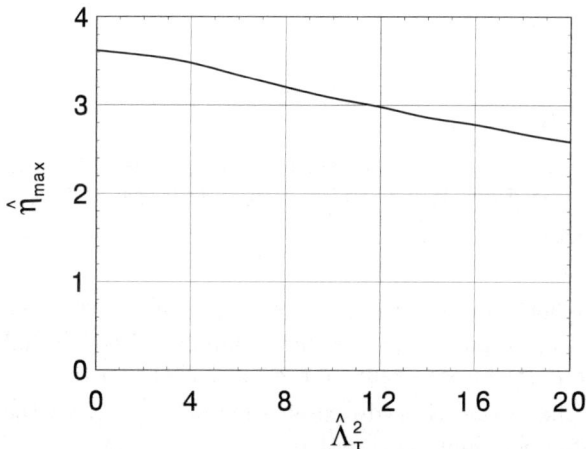

Fig. 3.16. The maximal reduced efficiency versus the energy spread parameter $\hat{\Lambda}_\mathrm{T}^2$. The FEL oscillator operates at saturation. Here $\hat{\Lambda}_\mathrm{p}^2 \to 0$

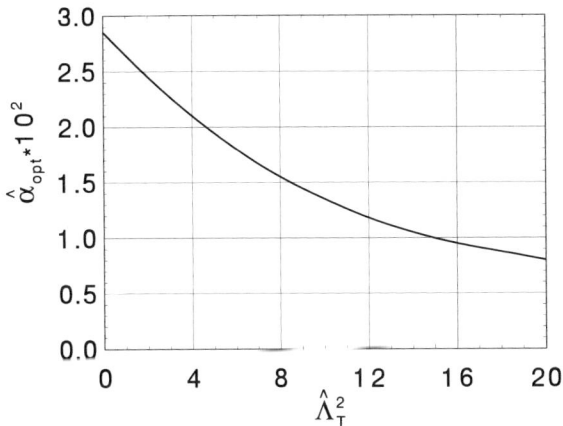

Fig. 3.17. The optimal reduced damping factor versus the energy spread parameter $\hat{\Lambda}_T^2$. The FEL oscillator operates at saturation. Here $\hat{\Lambda}_P^2 \to 0$

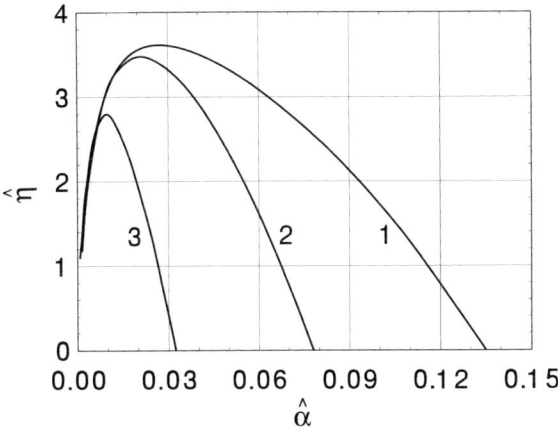

Fig. 3.18. The reduced efficiency versus the reduced damping factor. The FEL oscillator operates at saturation. Here $\hat{\Lambda}_P^2 \to 0$. Curve 1: $\hat{\Lambda}_T^2 = 0$. Curve 2: $\hat{\Lambda}_T^2 = 4$. Curve 3: $\hat{\Lambda}_T^2 = 16$

2π. After this we suppressed higher harmonics with the help of the drift space (see Sect. 2.2.6). The testing of the code was performed in the linear stage (see Fig. 3.4). It is seen that there is good agreement between the simulation and analytical results.

3.2.6 Some Generalizations

Optimization of the Output Power. The results of the efficiency optimization presented in the previous section refer to the efficiency of transfor-

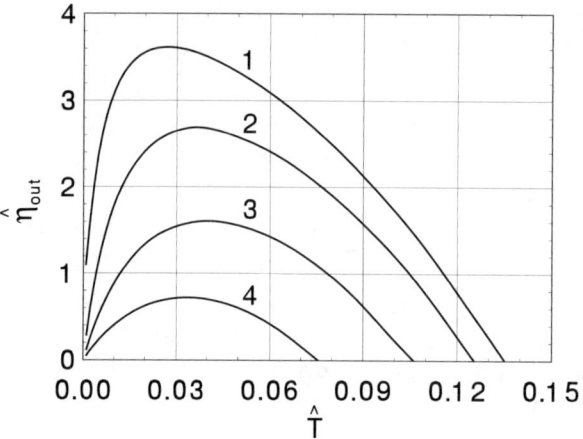

Fig. 3.19. The reduced practical efficiency versus the reduced transmission coefficient $\hat{T}$. The FEL oscillator operates at saturation. Here $\hat{\Lambda}_T^2 = 0$ and $\hat{\Lambda}_p^2 \to 0$. Curve 1: $\hat{\Gamma} = 0$. Curve 2: $\hat{\Gamma} = 0.01$. Curve 3: $\hat{\Gamma} = 0.03$. Curve 4: $\hat{\Gamma} = 0.06$

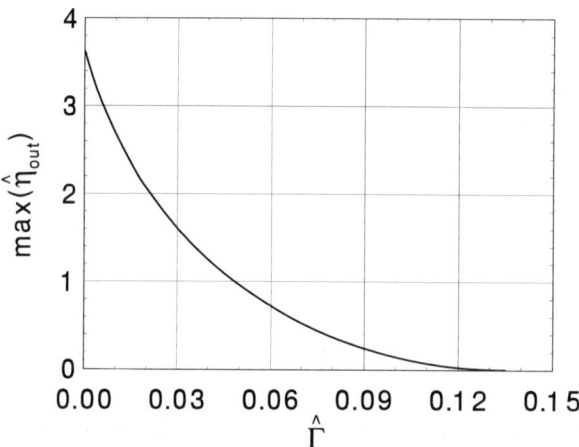

Fig. 3.20. The maximal reduced practical efficiency versus the reduced absorption coefficient $\hat{\Gamma}$. The FEL oscillator operates at saturation. Here $\hat{\Lambda}_T^2 = 0$ and $\hat{\Lambda}_p^2 \to 0$

mation of the electron beam power into the radiation power. Only a fraction of the radiation power goes out of the resonator and is of use for further application. Let us introduce the notion of the practical FEL efficiency, η_{out}, equal to the ratio of the output FEL radiation power to the electron beam power. The corresponding value of the reduced practical efficiency $\hat{\eta}_{\text{out}}$ is

$$\hat{\eta}_{\text{out}} = \eta_{\text{out}}/\beta \ .$$

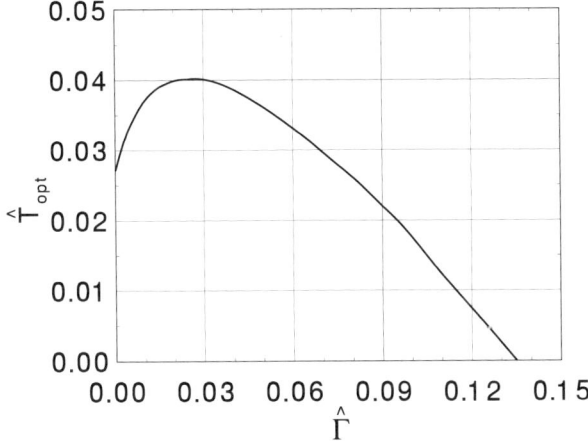

Fig. 3.21. The optimal reduced transmission coefficient $\hat{T}_{\text{opt}}$ versus the reduced absorption coefficient $\hat{\Gamma}$. The FEL oscillator operates at saturation. Here $\hat{\Lambda}_T^2 = 0$ and $\hat{\Lambda}_p^2 \to 0$

The damping factor of the resonator, α, is defined by the values of the transmission and absorption coefficients of the mirrors:

$$\alpha = T_1 + T_2 + \Gamma_1 + \Gamma_2 \;,$$

where (T_1, Γ_1) and (T_2, Γ_2) are the transmission and absorption coefficients of the first and second resonator mirror, respectively. The reduced practical efficiency, $\hat{\eta}_{\text{out}}$, may be written in the form:

$$\hat{\eta}_{\text{out}} = \eta_{\text{out}}/\beta = \hat{T}(\hat{T} + \hat{\Gamma})^{-1}\hat{\eta} \;,$$

where $\hat{T} = (T_1 + T_2)/\tau$ and $\hat{\Gamma} = (\Gamma_1 + \Gamma_2)/\tau$. Taking into account that $\hat{\alpha} = \hat{T} + \hat{\Gamma}$ and using the plot presented in Fig. 3.9, we can calculate the value of the reduced efficiency $\hat{\eta}$. The practical efficiency $\hat{\eta}_{\text{out}}$ can be optimized by an appropriate choice of the transmission coefficient $\hat{T}$ of the mirror at a fixed value of the absorption coefficient $\hat{\Gamma}$. It is seen from Fig. 3.19 that for any fixed value of the absorption coefficient $\hat{\Gamma}$ there exists an optimal value of the transmission coefficient $\hat{T}$ when the practical reduced efficiency $\hat{\eta}_{\text{out}}$ takes its maximal value. The plots of the maximal reduced efficiency and of the optimal reduced transmission coefficient are presented in Figs. 3.20 and 3.21.

The FEL Oscillator with Gain of about Unity. All the results presented above refer to an FEL oscillator with a low gain per undulator pass, $g \ll 1$. Let us consider the case of a gain of about unity. To simplify the study, we consider a resonator equipped with two mirrors: one perfectly reflecting and another with transmission coefficient T and absorption coefficient $\Gamma \simeq 0$. The space charge and energy spread effects are assumed to be negligibly small.

In the general case calculations of FEL oscillator can be performed with the self-consistent equations (3.13) and (3.14) which are valid for any value

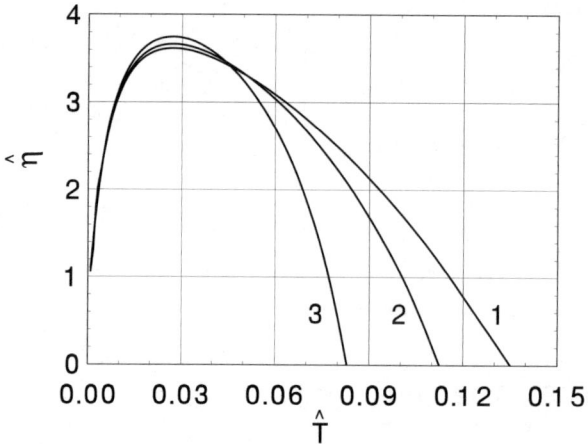

Fig. 3.22. The reduced efficiency versus the reduced transmission coefficient $\hat{T}$. The FEL oscillator operates at saturation. Here $\hat{\Lambda}_T^2 = 0$, $\hat{\Lambda}_p^2 \to 0$ and $\hat{\Gamma} = 0$. Curve 1: $\tau \to 0$. Curve 2: $\tau = 2$. Curve 3: $\tau = 6$

of the gain. Prior presenting numerical results, we should like to make a few remarks. The saturation condition written down for the case of an arbitrary gain is as follows:

$$(1 - T)(\hat{u}^{(\infty)} + \Delta\hat{u}^{(\infty)})^2 = (\hat{u}^{(\infty)})^2 \ . \tag{3.23}$$

For an arbitrary value of the gain parameter, τ, the FEL reduced efficiency is given by the following expression:

$$\hat{\eta} = \eta/\beta = T(\hat{u}^{(\infty)} + \Delta\hat{u}^{(\infty)})^2/(2\tau) \ . \tag{3.24}$$

It follows from (3.13) and (3.14) that formula (3.18) is no longer valid for an arbitrary value of the gain, but it can be generalized as:

$$\Delta\hat{u}^{(\infty)} = f(\tau, \hat{u}^{(\infty)}) \ . \tag{3.25}$$

Substituting (3.25) into (3.23), we find that the field amplitude $\hat{u}^{(\infty)}$ at saturation and the reduced efficiency (3.24) are functions of two parameters, $\hat{T} = T/\tau$ and τ:

$$\hat{u}^{(\infty)} = f_1(\hat{T}, \tau) \ , \qquad \hat{\eta} = f_2(\hat{T}, \tau) \ .$$

The results of numerical simulations are presented in Fig. 3.22 for different values of the gain. The dependence of the optimal value of the detuning parameter $\hat{C}_m$ on the parameter τ has been taken into account in these calculations. Curve 1 is calculated in the low-gain approximation (see Fig. 3.9). It is seen that in the region of practical interest, $\hat{T} \simeq 0.03$, the value of the reduced efficiency $\hat{\eta}$ is close to the asymptotic value in a wide region of the gain parameter, up to $\tau \simeq 6$. Therefore, all the plots presented in this section for the case of the low gain can be used in practice, even when the gain has an order of several tens of a per cent.

Planar Undulator. All the results presented above refer to the case of a circularly polarized radiation and a helical undulator. They may be simply transferred to the case of linearly polarized radiation

$$\boldsymbol{E} = \boldsymbol{e}_y \tilde{E}_y(z) \exp[i\omega(z/c - t)] + \text{C.C.} ,$$

and a planar undulator with magnetic field

$$H_y = 0 , \qquad H_x = H_\ell \cos(k_\text{w} z)$$

by the following normalization procedure:

$$\tau = \pi \omega \theta_\ell^2 j_0 l_\text{w}^3 A_{\text{JJ}}^2 (c \gamma_\ell^2 \gamma I_\text{A})^{-1} ,$$

$$\mathcal{E}_\text{U} = \mathcal{E}_0 \beta (e \theta_\ell l_\text{w} A_{\text{JJ}})^{-1} ,$$

$$\hat{C} = (k_\text{w} - \omega(2c)^{-1} \gamma_\ell^{-2}) l_\text{w} ,$$

$$\beta = c \gamma_\ell^2 (\omega l_\text{w})^{-1} = (4\pi N_\text{w})^{-1} ,$$

$$\hat{\Lambda}_\text{p} = l_\text{w} \left[4\pi j_0 \gamma^{-1} \gamma_\ell^{-2} I_\text{A}^{-1} \right]^{1/2} ,$$

$$\hat{\Lambda}_\text{T} = l_\text{w} \omega \gamma_\ell^{-2} \mathcal{E}_0^{-1} c^{-1} \sqrt{\langle (\Delta \mathcal{E})^2 \rangle} = 4\pi N_\text{w} \sqrt{\langle (\Delta \mathcal{E})^2 \rangle / \mathcal{E}_0^2} ,$$

$$\hat{\alpha} = \alpha / \tau ,$$

$$\hat{\eta} = \eta / \beta ,$$

$$\hat{P} = P / (\beta \mathcal{E}_0) ,$$

$$\hat{u} = |\tilde{E}_y| / E_0 .$$

Here the following notation has been used: $\theta_\ell = eH_\ell/(\mathcal{E}_0 k_\text{w})$, $\gamma_\ell^{-2} = \gamma^{-2} + \theta_\ell^2/2$, $A_{\text{JJ}} = [J_0(Q) - J_1(Q)]$, $Q = \theta_\ell^2 \omega/(8ck_\text{w})$, and J_0 and J_1 are Bessel functions.

3.3 FEL Oscillator with Nonuniform Undulator

In the previous sections of this chapter we have considered the FEL oscillator with a uniform undulator. In some experimental situations this simplest scheme is not optimal. To improve the FEL performance, different modifications are introduced into the undulator design. The most popular modification of the FEL oscillator is known as an optical klystron. It allows one to achieve a significantly larger small-signal gain with respect to a conventional FEL oscillator. On the other hand, the saturation efficiency of the optical klystron is less than that of a conventional FEL oscillator. Another popular FEL oscillator configuration uses a tapered main undulator. Sometimes a prebuncher (a short undulator) is additionally installed in front of the main undulator. Such a scheme allows one to increase the FEL efficiency. However, the small-signal gain in this case is less with respect to FEL oscillator with a uniform undulator.

112   3. One-Dimensional Theory of the FEL Oscillator

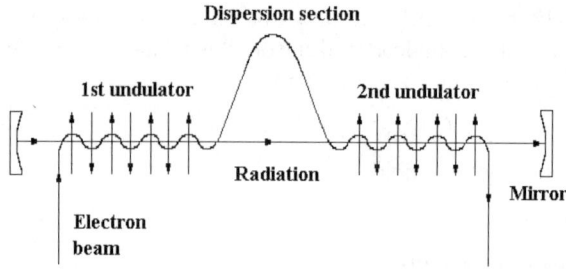

Fig. 3.23. Conceptual scheme of an optical klystron

The optical klystron consists of two undulator sections separated by a drift space or dispersion section (see Fig. 3.23). The first undulator modulates the electron beam in the energy. After passing the dispersion section, the energy modulation transforms into density modulation and a bunched electron beam amplifies the electromagnetic radiation in the second undulator. The small-signal gain for a cold electron beam can be increased significantly with respect to a uniform undulator. On the other hand, this scheme is more sensitive to the energy spread. Saturation in the optical klystron is defined by nonlinear density modulation in the dispersion section and is achieved at a relatively low value of the optical field in the resonator. As a result, the efficiency of the optical klystron is lower, too. Typically, the efficiency is decreased inversely proportionally to the gain increase. Induced energy spread in the electron beam after leaving the optical klystron is significantly less than in the case of a conventional FEL oscillator.

The optical klystron has an advantage over a conventional FEL oscillator at least in two practical cases. One of them is a kind of MOPA (master oscillator–power amplifier) scheme when the FEL oscillator is used to provide a tunable input signal for further amplification in the FEL amplifier. The same driving electron beam is subsequently used in the oscillator and then in the amplifier. For effective operation of this scheme it is necessary to have a high-quality electron beam at the amplifier entrance. As we mentioned above, the small energy perturbation of the electron beam is one of the features of the optical klystron. In addition, an experimenter can easily control the small-signal gain and the saturation power of the oscillator by tuning the magnetic field in the dispersion section.

It is a frequent situation when the optical klystron is installed at a storage ring. The reason for this is that the value the gain parameter is usually small in this case, and it becomes problematic to exceed the lasing threshold with conventional oscillator scheme. The saturation mechanism in a storage ring FEL (in particular, in the optical klystron) is more complicated than that described above. The radiation in the resonator interacts with the electron bunch circulating in the storage ring which complicates considerations, because the beam dynamics in the storage ring should be taken into account.

In this book we do not study storage ring FELs and refer the reader to the special literature (see the suggested bibliography at the end of this chapter).

In this book we also study the operation of an FEL oscillator with a tapered main undulator and a prebuncher. When analyzing undulator tapering in an FEL oscillator, one should take into account that the lasing frequency depends on the tapering depth and that the optical field changes in time. These factors make the study much more complicated with respect to that performed in Chap. 2 for the FEL amplifier. Thorough analysis shows that the potential for an efficiency increase in the FEL oscillator is rather limited. So far, in practice undulator tapering allows one to increase the efficiency by a factor of about 3 only. Further increase of the efficiency can be achieved by means of a prebuncher installed in front of the main undulator. The prebuncher is usually much shorter than the main undulator. The length of the drift space is usually chosen in such a way that it does not change significantly the beam bunching in the linear mode of operation, but enhances optimally the beam bunching in the nonlinear mode of operation. As a result, the bunched beam is fed to the input of the main undulator with tapered parameters, and the FEL oscillator efficiency can be increased. Another useful role of the prebuncher is the possibility of controlling the lasing frequency within a certain range by changing the length of the drift space. Typically, the prebuncher increases the FEL efficiency additionally by a factor of 2. Finally, we can conclude that traditional schemes of the FEL oscillator cannot provide high efficiency. The problem can be solved with some novel approaches which are discussed at the end of this chapter.

3.3.1 Basic Equations

Small-Signal Gain. We consider a helical undulator which consists of two sections of length l_1 and l_2 separated by a dispersion section (or drift space). To find the small-signal gain, we solve the Vlasov equation and Maxwell's equations. We begin with the case of untapered undulators and a cold electron beam. We also neglect the influence of the space charge field. Then the evolution of the complex amplitude $\tilde{f}_1$ of the first harmonic of the distribution function is described by (2.9):

$$\partial \tilde{f}_1/\partial z + i\left[C + \omega P/(c\gamma_z^2 \mathcal{E}_0)\right] \tilde{f}_1 + iU \partial f_0/\partial P = 0 , \qquad (3.26)$$

where $f_0 = n_0 F(P)$ is the initial distribution function. The general form of the solution for $\tilde{f}_1$ in the uniform undulator is

$$\tilde{f}_1 = -in_0 \frac{dF}{dP} \int_0^z dz' U \exp\left\{i\left[C + \omega P/(c\gamma_z^2 \mathcal{E}_0)\right](z' - z)\right\}$$
$$+ \tilde{f}_1|_{z=0} \exp\left\{-i\left[C + \omega P/(c\gamma_z^2 \mathcal{E}_0)\right](z - z_i)\right\} . \qquad (3.27)$$

We study the case when an unmodulated electron beam is fed to the entrance of the first undulator, i.e. $\tilde{f}_1|_{z=0} = 0$. The solution for $\tilde{f}_1$ in the first undulator, at $0 < z < l_1$, has the form:

$$\tilde{f}_1 = -i n_0 \frac{dF}{dP} \int_0^z dz' U \exp\left\{i\left[C + \omega P/(c\gamma_z^2 \mathcal{E}_0)\right](z'-z)\right\} . \tag{3.28}$$

The next problem consists in the calculation of the evolution of the distribution function in the dispersion section. Let us consider the case when the dispersion section of the length d consists of one undulator period, i.e. the undulator wavenumber is $k_{\mathrm{wd}} = 2\pi/d$. The undulator parameter in the dispersion section is $K_{\mathrm{d}} = eH_{\mathrm{wd}}/(k_{\mathrm{wd}} m_e c^2)$. To describe the evolution of the distribution function in the dispersion section (drift space) we use the formalism developed in Chap. 2 for the case of a tapered undulator. When developing this formalism, we did not assume that the parameters of the undulator must be continuous functions of z. So, the undulator field and period may have breaks. It should be noted, however, that the phase $\psi = \int k_{\mathrm{w}}(z) dz + \omega(z/c - t)$ is always a continuous function of z, since it is given by the integral of $k_{\mathrm{w}}(z)$.

Let us consider two regions of the parameters of the dispersion section. The first one is $k_{\mathrm{wd}} \ll k_{\mathrm{w}}$ and $K_{\mathrm{d}} \gg K$ (k_{w} and K refer to the regular undulator sections). The second region corresponds to the case of a simple drift space without magnetic field ($K_{\mathrm{d}} = 0$). In both cases there is no beam–wave interaction, and (3.26) can be written as follows:

$$\partial \tilde{f}_1 / \partial z + i\left[C_{\mathrm{d}} + \omega P/(c\gamma_{\mathrm{zd}}^2 \mathcal{E}_0)\right] \tilde{f}_1 = 0 . \tag{3.29}$$

The detuning parameter in the dispersion section is given by

$$C_{\mathrm{d}} = k_{\mathrm{wd}} - \frac{\omega}{2c\gamma_{\mathrm{zd}}^2} = \frac{\gamma_z^2}{\gamma_{\mathrm{zd}}^2} C - \frac{\gamma_z^2}{\gamma_{\mathrm{zd}}^2} k_{\mathrm{w}} + \frac{2\pi}{d} ,$$

where C, γ_z^2, and k_{w} refer to the regular undulator sections, and $\gamma_{\mathrm{zd}} = \gamma/\sqrt{1 + K_{\mathrm{d}}^2}$.

We denote the values of $\tilde{f}_1$ at the exit of the first undulator, at $z = l_1$, and at the end of the dispersion sections, by the symbols $\tilde{f}_1^{(\Leftarrow)}$ and $\tilde{f}_1^{(\Rightarrow)}$, respectively. According to (3.29), these values are connected by the relation

$$\tilde{f}_1^{(\Rightarrow)} = \tilde{f}_1^{(\Leftarrow)} \exp\left\{-i\left[C_{\mathrm{d}} + \omega P/(c\gamma_{\mathrm{zd}}^2 \mathcal{E}_0)\right] d\right\} .$$

Thus, the expression for $\tilde{f}_1^{(\Rightarrow)}$ becomes

$$\tilde{f}_1^{(\Rightarrow)} = -i n_0 \frac{dF}{dP} \exp(i\delta\psi) \int_0^{l_1} dz' U$$

$$\times \exp\left\{i\left[C + \omega P/(c\gamma_z^2 \mathcal{E}_0)\right]\left[z' - l_1 - (\gamma_z^2/\gamma_{\mathrm{zd}}^2) d\right]\right\} , \tag{3.30}$$

where $\delta\psi = (\gamma_z^2/\gamma_{\mathrm{zd}}^2) k_{\mathrm{w}} d = k_{\mathrm{w}} d (1 + K_{\mathrm{d}}^2)/(1 + K^2)$.

We formally let the total length of the system be equal to $l_{\mathrm{w}} = l_1 + l_2$. The change of the distribution function $\tilde{f}_1$ in the dispersion section is introduced

3.3 FEL Oscillator with Nonuniform Undulator

as a leap at $z = l_1$. Since there is no beam-wave interaction in the dispersion section, this formal procedure will not influence the final result for the gain in the system consisting of two undulators separated by a dispersion section. According to (3.27), the solution of (3.26) in the interval $l_1 < z < l_w$ satisfying the initial condition (3.30) is

$$\tilde{f}_1 = -in_0 \frac{dF}{dP} \left\{ \exp(i\delta\psi) \int_0^{l_1} dz'U \right.$$
$$\times \exp\left\{i\left[C + \omega P/(c\gamma_z^2 \mathcal{E}_0)\right]\left[z' - z - (\gamma_z^2/\gamma_{zd}^2)d\right]\right\}$$
$$\left. + \int_{l_1}^z dz'U \exp\left\{i\left[C + \omega P/(c\gamma_z^2 \mathcal{E}_0)\right](z' - z)\right\} \right\}. \tag{3.31}$$

We consider the case of a cold electron beam with initial distribution function $F(P) = \delta(P)$. Using the relation

$$\tilde{j}_1 \simeq -ec \int \tilde{f}_1 dP,$$

we find the expressions for the first harmonic of the beam current density

$$\tilde{j}_1 = j_0\omega/(\mathcal{E}_0\gamma_z^2 c) \int_0^z dz'(z' - z)U \exp\left[iC(z' - z)\right] \quad \text{for } 0 < z < l_1,$$

$$\tilde{j}_1 = j_0\omega/(\mathcal{E}_0\gamma_z^2 c) \left\{ \exp(i\delta\psi) \int_0^{l_1} dz' \left[z' - z - (\gamma_z^2/\gamma_{zd}^2)d\right] U \right.$$
$$\times \exp\left[iC[z' - z - (\gamma_z^2/\gamma_{zd}^2)d]\right]$$
$$\left. + \int_{l_1}^z dz'(z' - z)U \exp\left[iC(z' - z)\right] \right\} \quad \text{for } l_1 < z < l_w. \tag{3.32}$$

According to our notation, introduced in Sect. 3.1, the small-signal gain is described by the function $Z = \tilde{E}(l_w)/\tilde{E}(0) - 1$. In the same way as was done above, we perform the normalization of the equations and calculate the normalized value of $\hat{Z} = 2Z/\tau$, where τ is the gain parameter. Substituting (3.32) into (3.3) and integrating (3.3) in the limits from 0 to l_w under the condition (3.1), we find the following expression for $\hat{Z}$:

$$\hat{Z} = i\int_0^{\hat{l}_1} d\xi \int_0^\xi d\xi'\xi' \exp(-i\hat{C}\xi') + i \int_0^{1-\hat{l}_1} d\xi \int_0^\xi d\xi'\xi' \exp(-i\hat{C}\xi')$$
$$+ i\exp(i\delta\psi) \int_0^{1-\hat{l}_1} d\xi \int_{\xi+\hat{d}}^{\xi+\hat{l}_1+\hat{d}} d\xi'\xi' \exp(-i\hat{C}\xi'), \tag{3.33}$$

116   3. One-Dimensional Theory of the FEL Oscillator

where $\hat{d} = (\gamma_z^2/\gamma_{zd}^2)(d/l_w) = (1 + K_d^2)d/\left[(1 + K^2)l_w\right]$, $\hat{C} = Cl_w$, and $\hat{l}_1 = l_1/l_w$. The phase shift in the dispersion section is $\delta\psi = 2\pi N_w \hat{d}$, where N_w is the total number of undulator periods in both undulator sections. It is convenient to introduce a nonmultiple beam-wave phase shift as

$$\overline{\delta\psi} = 2\pi \left\{ N_w \hat{d} - [N_w \hat{d}] \right\},$$

where $[\ldots]$ denotes the integer part.

Let us complicate the study by introducing a Gaussian energy spread in the electron beam. The initial distribution function $F(P)$ is given by (2.20a) in this case. In addition, we assume that the second undulator is tapered according to a linear law:

$$\hat{C}(z) = \begin{cases} \hat{C}_0 & \text{for} \quad 0 < \hat{z} < \hat{l}_1 \\ \hat{C}_0 + \hat{b}_1(\hat{z} - \hat{l}_1) & \text{for} \quad \hat{l}_1 < \hat{z} < \hat{l}_w \end{cases} \quad (3.34)$$

Let us express the tapering parameter $\hat{b}_1$ in terms of the undulator parameters. Under a linear change in the undulator period $\lambda_w = 2\pi/k_w$ and fixed undulator parameter K, we get:

$$\hat{b}_1 \simeq -2\pi N_w^{(2)} \left[\lambda_w(l_w) - \lambda_w(0)\right]/\lambda_w(0),$$

where $N_w^{(2)}$ is the number of undulator periods in the second section. Under a linear change in the undulator field and fixed period we get:

$$\hat{b}_1 \simeq -\left\{4\pi N_w^{(2)} K^2(0)/\left[1 + K^2(0)\right]\right\} [H_w(l_w) - H_w(0)]/H_w(0).$$

The expressions for $\hat{b}_1$ are valid for both positive and negative signs of $\hat{b}_1$. The relative change of the undulator period (or the undulator field) is assumed to be small in the present treatment.

Omitting the details of the calculations, we present the final result for $\hat{Z}$ for the case of linear tapering of the second undulator and a Gaussian energy spread in the electron beam:

$$\hat{Z} = i \int_0^{\hat{l}_1} d\xi \int_0^{\xi} d\xi' \xi' \exp\left[-i\hat{C}_0 \xi' - \hat{\Lambda}_T^2 (\xi')^2/2\right]$$

$$+ i \int_0^{1-\hat{l}_1} d\xi \int_0^{\xi} d\xi' \xi' \exp\left[-i\hat{C}_0 \xi' + i\hat{b}_1 \xi'(\xi'/2 - \xi) - \hat{\Lambda}_T^2 (\xi')^2/2\right]$$

$$+ i \exp(i\delta\psi) \int_0^{1-\hat{l}_1} d\xi \exp(-i\hat{b}_1 \xi^2/2)$$

$$\times \int_{\xi+\hat{d}}^{\xi+\hat{l}_1+\hat{d}} d\xi' \xi' \exp\left[-i\hat{C}_0 \xi' - \hat{\Lambda}_T^2 (\xi')^2/2\right], \quad (3.35)$$

where $\hat{C}_0 = C_0 l_{\mathrm{w}}$ and $\hat{\Lambda}_{\mathrm{T}}^2 = (4\pi N_{\mathrm{w}})^2 \langle (\Delta \mathcal{E}/\mathcal{E}_0)^2 \rangle$ is the energy spread parameter.

It is relevant to make some remarks in conclusion to this section. To be specific, we assumed that the first and second undulator sections are helical and the dispersion section consists of one period of a helical undulator. It can be shown that (3.33) and (3.35) are valid also for the case of an arbitrary field profile in the dispersion section. Let the field in the dispersion section be planar and have the form $\boldsymbol{H} = \boldsymbol{e}_x H(z)$. Then the parameter K_{d}^2 entering $\hat{d}$ should be replaced by

$$K_{\mathrm{d}}^2 \to K_{\mathrm{eff}}^2 = \frac{\gamma^2}{d} \int_0^d \theta^2(z) \mathrm{d}z ,$$

where

$$\theta(z) = \frac{1}{\gamma} \int_0^z \frac{eH(\xi)}{m_{\mathrm{e}} c^2} \mathrm{d}\xi$$

is the angle of the particle velocity with respect to the z coordinate. Despite (3.33) and (3.35) have been derived for a helical geometry of the first and second undulator, they are also valid for a planar geometry at an appropriate redetermination of the problem parameters.

Nonlinear Simulation Algorithm. The self-consistent equations describing the nonlinear mode are similar to those used in Sect. 3.2:

$$\begin{aligned}
\frac{\mathrm{d}\hat{P}}{\mathrm{d}\hat{z}} &= \hat{u} \cos(\psi + \psi_0) , \\
\frac{\mathrm{d}\psi}{\mathrm{d}\hat{z}} &= \hat{P} + \hat{C}(\hat{z}) ,
\end{aligned} \quad (3.36)$$

$$\begin{aligned}
\frac{\mathrm{d}\hat{u}}{\mathrm{d}\hat{z}} &= \frac{\tau \hat{j}_1}{2} \cos(\psi_0 - \psi_1) , \\
\frac{\mathrm{d}\psi_0}{\mathrm{d}\hat{z}} &= -\frac{\tau \hat{j}_1}{2\hat{u}} \sin(\psi_0 - \psi_1) .
\end{aligned} \quad (3.37)$$

Integration of (3.36) and (3.37) should be performed in the limits from $\hat{z} = 0$ to $\hat{z} = 1$. All the definitions for the reduced variables are the same as in Sect. 3.1. It should be taken into account that the detuning parameter is a function of $\hat{z}$ given by (3.34) and that the initial detuning $\hat{C}_0$ corresponds to the maximum of $\mathrm{Re}\,\hat{Z}$ in the small-signal regime (see (3.35)). To take into account the action of the drift space, we change the phases of the particles by a leap at $\hat{z} = \hat{l}_1$:

$$\psi^{(\Rightarrow)} = \psi^{(\Leftarrow)} + \left(\hat{P} + \hat{C}_0 \right) \hat{d} - \overline{\delta \psi} , \quad (3.38)$$

where $\hat{P}$ is the reduced energy deviation at the exit of the first section, $\overline{\delta \psi} = 2\pi \left\{ N_{\mathrm{w}} \hat{d} - [N_{\mathrm{w}} \hat{d}] \right\}$, and $[\ldots]$ denotes the integer part. Due to the slippage of the electromagnetic wave in the drift space with respect to the particle, the phase shift of the particle with respect to the effective potential,

118   3. One-Dimensional Theory of the FEL Oscillator

$-\delta\psi$, is always negative. One should not wonder that the phase of the first harmonic of the beam current density, $\tilde{j}_1$, acquires a positive phase increment $\delta\psi$ after the dispersion section (see (3.32)). This is connected with its definition: $\tilde{j}_1 \propto \langle \exp\{-i\psi\}\rangle$, where the brackets denote averaging over the phases of the particles.

3.3.2 Optical Klystron

In the previous sections we derived general equations describing the linear and nonlinear modes of operation of the FEL oscillator with a nonuniform undulator. In this section we study in detail the operation of the optical klystron. The magnetic structure of the optical klystron consists of two untapered undulators separated by the dispersion section.

Small-Signal Mode of Operation. We begin our study by neglecting the energy spread in the electron beam and assuming the undulators to have equal lengths, i.e. $\hat{l}_1 = 1/2$. Integrating (3.33) for the case of $\delta\psi = 2\pi k$ ($k = 1, 2, \ldots$), we obtain

$$\hat{Z} = \frac{1}{\hat{C}^3}\left[2V(\hat{C}/2) + V(\hat{C}\hat{d}) + V(\hat{C}(\hat{d}+1)) - 2V(\hat{C}(\hat{d}+1/2)) - \hat{C} - 4i\right], \qquad (3.39)$$

where $V(\zeta) = (2i - \zeta)\exp(-i\zeta)$. First, we note that the limiting transition to the case of uniform undulator (see (3.9)) takes place as $\hat{d} \to 0$. The reduced small-signal gain $\hat{g}_s = g_s/\tau$ of the optical klystron is

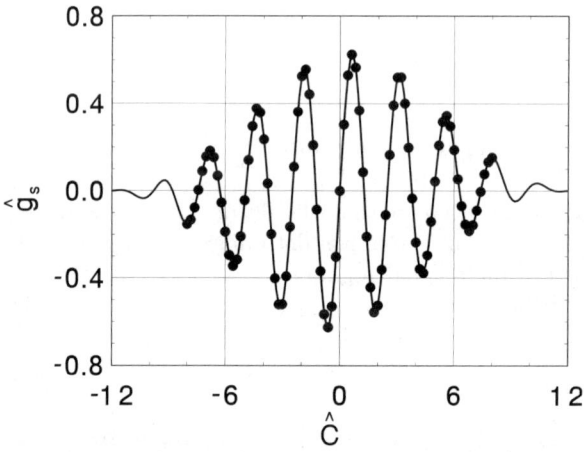

Fig. 3.24. The reduced small-signal gain $\hat{g}_s$ versus the reduced detuning $\hat{C}$. Here $\overline{\delta\psi} = 0$, $\hat{d} = 2$ and $\hat{l}_1 = 1/2$. The solid curve is calculated the with analytical formula (3.40) and the circles are the results of calculations with nonlinear simulation code

3.3 FEL Oscillator with Nonuniform Undulator 119

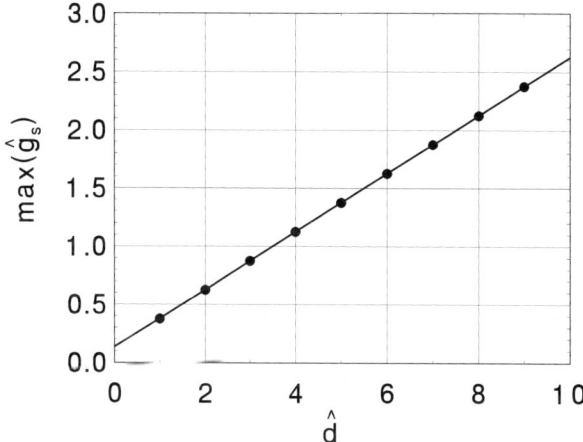

Fig. 3.25. The maximal reduced small-signal gain $\hat{g}_s$ versus the reduced length of the dispersion section. Here $\overline{\delta\psi} = 0$ and $\hat{l}_1 = 1/2$. The full curve is calculated with formula (3.40) and the circles are the results of calculations with the asymptotic formula (3.41)

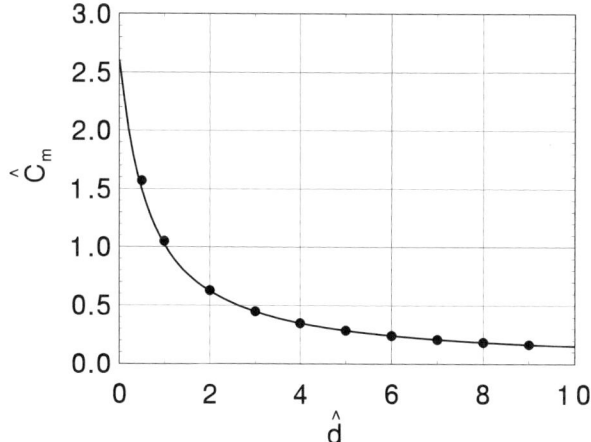

Fig. 3.26. The optimal reduced detuning $\hat{C}_m$ versus the reduced length of the dispersion section. Here $\overline{\delta\psi} = 0$ and $\hat{l}_1 = 1/2$. The full curve is calculated with formula (3.40). The circles are the results of calculations with the asymptotic formula (3.41)

$$\hat{g}_s = \operatorname{Re} \hat{Z} = -\frac{1}{\hat{C}^3}\left[2f(\zeta)\Big|_0^{\hat{C}/2} + f(\zeta)\Big|_{\hat{C}(\hat{d}+1/2)}^{\hat{C}(\hat{d}+1)} - f(\zeta)\Big|_{\hat{C}\hat{d}}^{\hat{C}(\hat{d}+1/2)}\right], \quad (3.40)$$

where $f(\zeta) = 2\cos(\zeta) + \zeta\sin(\zeta)$ and

$$f(\zeta)\Big|_a^b = f(b) - f(a).$$

Figure 3.24 presents the gain curve for a specific value of the parameter $\hat{d} = 2$ and $\overline{\delta\psi} = 0$. It is seen from this plot that the gain curve is antisymmetric with respect to $\hat{C} = 0$ in this case. We also obtain that the maxima of the gain curve are separated almost equidistantly over the detuning. It can be shown that this separation should be equal to $2\pi/\hat{d}$ at large values of the parameter $\hat{d}$. At exact resonance, $\hat{C} = 0$, the beam-wave phase shift in the drift space is equal to zero, since $\delta\psi = 2\pi k$. At a finite value of the detuning parameter, a phase shift appears equal to $\hat{C}\hat{d}$ (see (3.30)). The first maximum of the gain curve corresponds to the optimal value of the phase shift at the entrance of the second section. An additional beam-wave phase shift equal to 2π caused by the change of the detuning by the value $2\pi/\hat{d}$ corresponds to the next maximum. The gain curve is antisymmetric with respect to $\hat{C} = 0$ at $\delta\psi = 2\pi k$. When the value of $\delta\psi$ changes smoothly, the maxima are shifted inside the envelope of the gain curve and the picture is repeated each time when $\delta\psi$ changes by 2π.

When the value of the parameter $\hat{d} \gtrsim 1$, the phase shift $\delta\psi$ changes by a value of about unity for a small relative change of the parameter $\hat{d}$. Indeed, according to its definition, $\delta\psi = 2\pi N_w \hat{d}$. The number of undulator periods is always large, $N_w \gg 1$. Thus, $\delta\psi$ and $\hat{d}$ can be considered as independent parameters. For convenience we use the nonmultiple beam-wave phase shift defined as $\overline{\delta\psi} = 2\pi \left\{ N_w \hat{d} - [N_w \hat{d}] \right\}$.

In the case under study (for $\overline{\delta\psi} = 0$ and $\hat{l}_1 = 1/2$), the maximum of the small-signal gain $\max(\hat{g}_s)$, and the detuning parameter $\hat{C}_m$ corresponding to this maximum are universal functions of the parameter $\hat{d}$. In Figs. 3.25

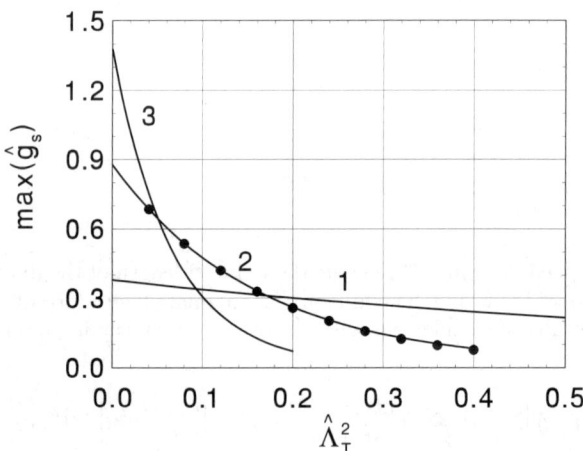

Fig. 3.27. The maximal reduced small-signal gain $\hat{g}_s$ versus the energy spread parameter $\hat{\Lambda}_T^2$. Here $\overline{\delta\psi} = 0$, $\hat{l}_1 = 1/2$, and $\hat{b}_1 = 0$. The full curves are calculated with (3.35). The circles are the results of calculations with the asymptotic formula (3.42). Curve 1: $\hat{d} = 1$. Curve 2: $\hat{d} = 3$. Curve 3: $\hat{d} = 5$

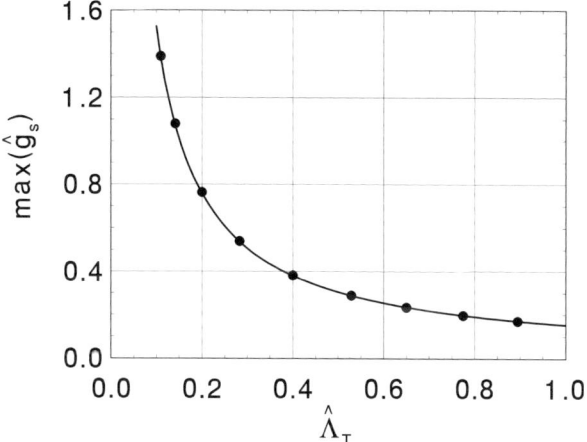

Fig. 3.28. The maximal reduced small-signal gain $\hat{g}_s$ versus $\hat{\Lambda}_T$. Here $\overline{\delta\psi} = 0$, $\hat{l}_1 = 1/2$ and $\hat{b}_1 = 0$, and $\hat{d} = \hat{d}_{\text{opt}}$. The full curves are calculated with (3.35). The circles are the results of calculations with the asymptotic formula (3.43b)

and 3.26 we present plots of these functions calculated with (3.40). The circles are calculated using the asymptotic formulae

$$\hat{C}_m = \frac{\pi}{2(\hat{d}+1/2)}, \quad \max(\hat{g}_s) = \frac{\hat{d}+1/2}{4}, \tag{3.41}$$

which are valid in the limit of $\hat{d} \gg 1$ and can be derived from (3.40). It is seen that even at $\hat{d} \simeq 1$, the asymptotic formulae (3.41) provide good accuracy.

Let us study the influence of the energy spread on the operation of the optical klystron. We use (3.35) at $\hat{b}_1 = 0$ and $\overline{\delta\psi} = 0$ for the calculation of the gain. The maximum of the small-signal gain, $\max(\hat{g}_s)$, is a function of two parameters, $\hat{d}$ and $\hat{\Lambda}_T^2$. In Fig. 3.27 we present plots of this function for several values of $\hat{d}$. The circles in these plots are calculated with the asymptotic formula

$$\max(\hat{g}_s) = \frac{\hat{d}+1/2}{4} \exp\left[-\frac{1}{2}\hat{\Lambda}_T^2(\hat{d}+1/2)^2\right]. \tag{3.42}$$

The latter formula has been derived from (3.35) assuming that $\hat{d} \gg 1$ and $\hat{\Lambda}_T^2 \hat{d}^2 \lesssim 1$. It follows from this formula that at each value of the energy spread parameter there is an optimal value of the parameter $\hat{d}$

$$\hat{d}_{\text{opt}} = \frac{1}{\hat{\Lambda}_T} - \frac{1}{2}, \tag{3.43a}$$

when the maximal value of the gain

$$\max(\hat{g}_s) = \frac{0.152}{\hat{\Lambda}_T} \tag{3.43b}$$

is achieved.

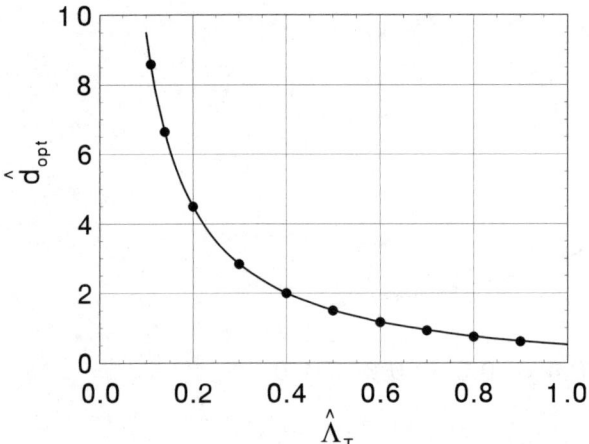

Fig. 3.29. The optimal reduced length of the dispersion section $\hat{d}_{\text{opt}}$ versus $\hat{\Lambda}_T$. Here $\overline{\delta\psi} = 0$, $\hat{l}_1 = 1/2$ and $\hat{b}_1 = 0$. The full curves are calculated with (3.35). The circles are the results of calculations with the asymptotic formula (3.43a)

Figures 3.28 and 3.29 present the corresponding dependencies. The solid curves show the rigorous solution given by (3.35) and the circles are calculated with the asymptotic formulae (3.43). Analyzing these relations and plots we can conclude that the optical klystron has a significant benefit in the small-signal gain with respect to the conventional FEL oscillator scheme (see Sect. 3.1) only in the case of a small energy spread parameter, $\hat{\Lambda}_T^2 \ll 1$.

Saturation Effects. Numerical simulations show that the nonlinear mode of optical klystron operation differs significantly from that of the conventional FEL oscillator. In the latter case nonlinear processes become significant when the value of the reduced field amplitude in the resonator becomes $\hat{u} \simeq 1$. An optimal choice of the FEL oscillator parameters allows one to reach a high value of the field stored in the resonator, $\hat{u} \gg 1$. In contrast, in the optical klystron with a large value of the parameter $\hat{d}$, the nonlinear behavior occurs at small values of the radiation field, at $\hat{u} \ll 1$. There is almost no phase motion of the particles in the undulators and the saturation effects are defined mainly by the overmodulation of the electron beam in the dispersion section. This process can be well described using analytical techniques developed for the calculations of conventional rf klystrons.

Let us start with the case when we can neglect beam bunching in the undulators. Under this assumption, the beam is modulated in the energy in the first undulator. Then the energy modulation causes modulation of the beam density in the dispersion section. Finally, the bunched beam amplifies the wave in the second undulator. According to (3.41), the detuning parameter $\hat{C}_m$ is close to zero at large values of $\hat{d}$. The amplitude of the energy

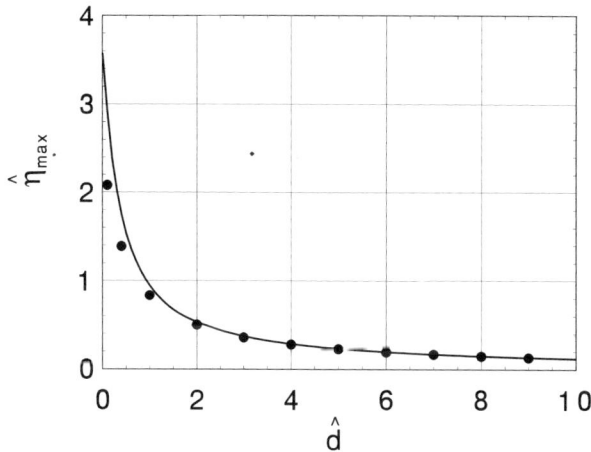

Fig. 3.30. Maximal reduced efficiency at saturation versus the reduced length of the dispersion section $\hat{d}$. Here $\overline{\delta\psi} = 0$, $\hat{l}_1 = 1/2$, $\hat{b} = 0$ and $\hat{\Lambda}_T^2 = 0$. The full curve is calculated with (3.36) and (3.37). The circles are calculated with the asymptotic formula (3.49)

modulation after the first undulator is $\Delta \hat{P} = \hat{u}\hat{l}_1$. Following the theory of a conventional klystron, we introduce the bunching parameter X:

$$X = \hat{u}\hat{l}_1\hat{d} \;.$$

It follows from the theory of the klystron that the first harmonic of the beam current, appearing in (3.37), is

$$\hat{j}_1 = 2J_1(X) \;,$$

where J_1 is a Bessel function. According to (3.37), the increment of the field in the second undulator is

$$\Delta\hat{u} = \hat{l}_2 \tau J_1(X) \;,$$

where $\hat{l}_2 = 1 - \hat{l}_1$. The gain is given by the expression:

$$g = 2\Delta\hat{u}/\hat{u} = 2(1 - \hat{l}_1)\tau J_1(X)/\hat{u} \;. \tag{3.44}$$

In the saturation regime the gain must be equal to the resonator losses, $g = \alpha$. The value of the field at saturation, $\hat{u}^{(\infty)}$, is given by the solution of the transcendental equation:

$$2(1 - \hat{l}_1)J_1(\hat{u}^{(\infty)}\hat{l}_1\hat{d}) = \hat{\alpha}\hat{u}^{(\infty)} \;. \tag{3.45}$$

The value of the reduced efficiency at saturation is given by the expression:

$$\hat{\eta} = \hat{\alpha}(\hat{u}^{(\infty)})^2/2 = (1 - \hat{l}_1)\hat{u}^{(\infty)}J_1(\hat{u}^{(\infty)}\hat{l}_1\hat{d}) \;. \tag{3.46}$$

The maximum saturation efficiency and the optimal value of the reduced damping factor $\hat{\alpha}$ can be obtained without solving the transcendental equation (3.45). Indeed, we can find the maximum of the saturation efficiency using the condition for its derivative with respect to $\hat{u}^{(\infty)}$ to be equal to zero:

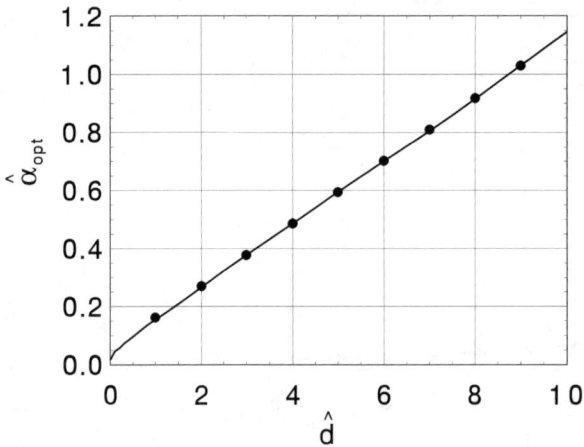

Fig. 3.31. Optimal value of the reduced damping factor versus the reduced length of the dispersion section $\hat{d}$. Here $\overline{\delta\psi} = 0$, $\hat{l}_1 = 1/2$, $\hat{b} = 0$ and $\hat{\Lambda}_T^2 = 0$. The full curve is calculated with (3.36) and (3.37). The circles are the results of calculations with the asymptotic formula (3.49)

$$\mathrm{d}\hat{\eta}/\mathrm{d}\hat{u}^{(\infty)} = (1 - \hat{l}_1)\left[J_1(X) + XJ_1'(X)\right] = 0 \ .$$

Remembering the relation

$$J_1(X) + XJ_1'(X) = J_0(X) \ ,$$

we find that the optimal value of the bunching parameter X is equal to the root of the Bessel function J_0:

$$X_{\mathrm{opt}} = 2.405 \ . \tag{3.47}$$

Using (3.45)–(3.47) and the relation $J_1(X_{\mathrm{opt}}) = 0.519$, we find:

$$\hat{\eta}_{\mathrm{max}} = 1.25(1 - \hat{l}_1)/(\hat{l}_1 \hat{d}) \ ,$$
$$\hat{\alpha}_{\mathrm{opt}} = 0.432 \hat{l}_1 \hat{d}(1 - \hat{l}_1) \ . \tag{3.48}$$

We see that the dependence of the maximal reduced efficiency $\hat{\eta}_{\mathrm{max}}$ on the length of the first undulator $\hat{l}_1$ is stronger than the corresponding dependence of $\hat{\alpha}_{\mathrm{opt}}$. For instance, the efficiency increases by a factor of 4 when the value of $\hat{l}_1$ changes from 0.5 to 0.2. On the other hand, the parameter $\hat{\alpha}_{\mathrm{opt}}$ decreases only by 30%. So, we may conclude that if there is a possibility of decreasing the reduced damping factor, the first undulator should be shorter than the second undulator and a significant benefit in the efficiency can be achieved at the same total length of undulators.

More precise formulae, valid even when $\hat{d} \simeq 1$, can be obtained taking into account the beam bunching not only in the dispersion section, but also in the undulators. Let us consider the special case of $\hat{l}_1 = 1/2$. One can obtain that in this case the bunching parameter is

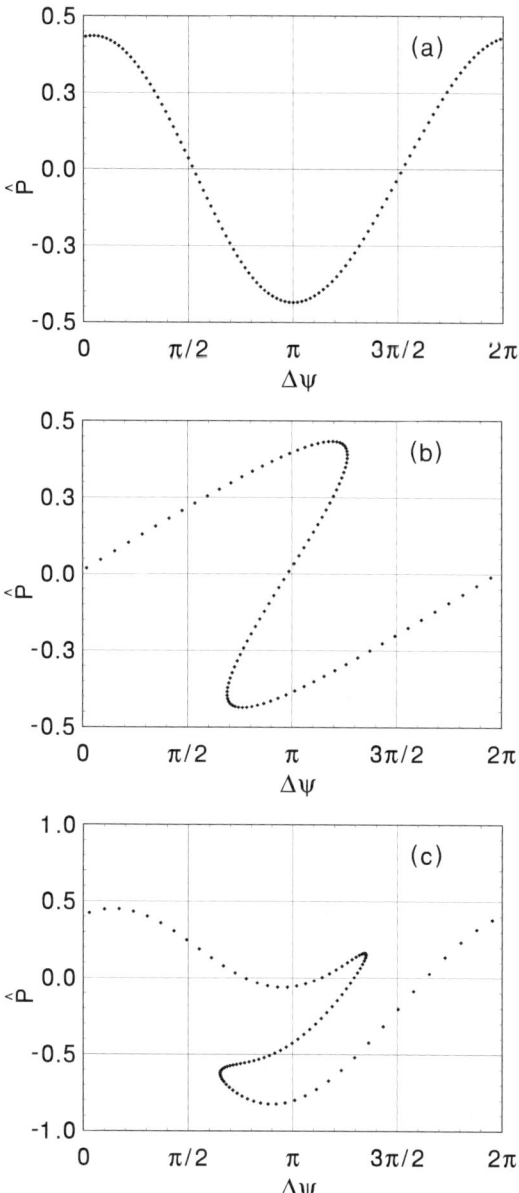

Fig. 3.32. Phase space distribution of the particles in the optical klystron operating at saturation. Here $\hat{\alpha} = 0.594$, $\overline{\delta\hat{\psi}} = 0$, $\hat{d} = 5$, $\hat{l}_1 = 1/2$, $\hat{b} = 0$ and $\hat{\Lambda}_T^2 = 0$. Plot (**a**): distribution after the first undulator section, plot (**b**): distribution after the dispersion section, and plot (**c**): distribution after the second undulator section

$$X = \begin{cases} \hat{u}\hat{z}^2/2 & \text{for} \quad \hat{z} < 1/2 \,, \\ \hat{u}/8 + \hat{u}\hat{d}/2 & \text{at the entrance of the second section}\,, \\ (\hat{u}/2)(\hat{d} + \hat{z} - 1/4) & \text{for} \quad \hat{z} > 1/2 \,. \end{cases}$$

The first harmonic of the current $\hat{j}_1$ and the bunching parameter are now functions of $\hat{z}$. Assuming that the field gain takes place only in the second section, we can write:

$$\Delta\hat{u} = \tau \int_{1/2}^{1} d\hat{z} J_1(X) \,.$$

Integration of this expression can be performed analytically which allows us to write the formula for the reduced efficiency as:

$$\hat{\eta} = 2\left[J_0\left((\hat{u}/2)(\hat{d}+1/4)\right) - J_0\left((\hat{u}/2)(\hat{d}+3/4)\right)\right]\,.$$

After performing the optimization procedure, we obtain

$$\hat{\eta}_{\max} = 1.25/(\hat{d}+1/2)\,, \qquad \hat{\alpha}_{\mathrm{opt}} = 0.108(\hat{d}+1/2)\,. \tag{3.49}$$

In Figs. 3.30 and 3.31 we present the corresponding dependencies calculated with the approximate formulae (3.49) and with numerical simulations with (3.36)–(3.38). We see that the approximate formulae (3.49) provide sufficient accuracy of calculations even at $\hat{d} \simeq 1$. For illustration, in Fig. 3.32 we present the phase distribution of the particles when the optical klystron operates at saturation.

3.3.3 FEL Oscillator with a Prebuncher and a Tapered Main Undulator

Let us now study another popular configuration of the FEL oscillator with nonuniform undulator, namely the FEL oscillator with a prebuncher and a tapered main undulator. In this case the first undulator (prebuncher) is short, $\hat{l}_1 \ll 1$, and the second undulator (main undulator) is tapered by a linear law ($\hat{b}_1 \neq 0$).

Small-Signal Mode of Operation. We begin our study with the case when there is no prebuncher and the main undulator is tapered by a linear law. When the energy spread effects can be neglected, the expression (3.35) is reduced to

$$\hat{Z} = \mathrm{i} \int_0^1 d\xi \int_0^\xi d\xi'\xi' \exp\left\{-\mathrm{i}\left[\hat{C}_0 \xi' + \hat{b}_1 \xi \xi' - \hat{b}_1(\xi')^2/2\right]\right\}\,. \tag{3.50}$$

The reduced gain can be written as

$$\hat{g}_{\mathrm{s}} = \int_0^1 d\xi \int_0^\xi d\xi'\xi' \sin\left[\hat{C}_0 \xi' + \hat{b}_1 \xi \xi' - \hat{b}_1(\xi')^2/2\right]\,. \tag{3.51}$$

3.3 FEL Oscillator with Nonuniform Undulator

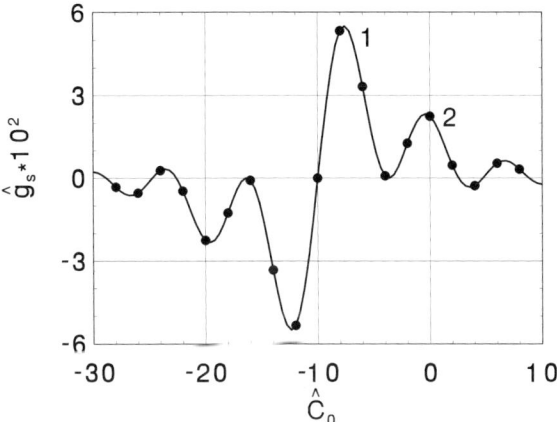

Fig. 3.33. The reduced small-signal gain $\hat{g}_s$ versus the reduced detuning $\hat{C}_0$ at the undulator entrance. Here $\hat{b}_1 = 20$, (1) is the first maximum and (2) is the second maximum. The full curve is calculated with (3.51) and the circles are the results of calculations with nonlinear simulation code

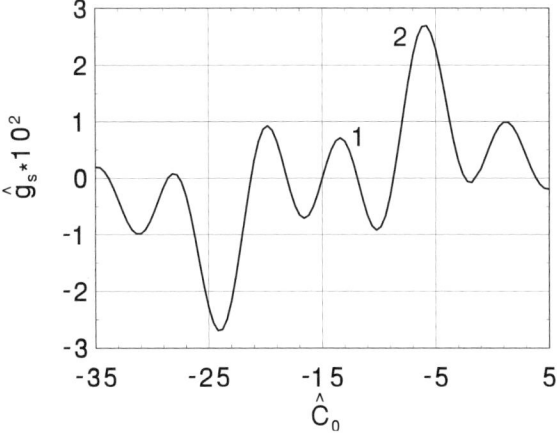

Fig. 3.34. The reduced small-signal gain $\hat{g}_s$ versus the reduced detuning $\hat{C}_0$ at the undulator entrance. Here $\hat{b}_1 = 30$, (1) is the first maximum and (2) is the second maximum

Figures 3.33 and 3.34 show the gain profiles for two different positive values of the tapering parameter $\hat{b}_1$. It is clearly seen that the curves are antisymmetric with respect to the line $\hat{C}_0 = -\hat{b}_1/2$. The value of $\hat{g}_s$ at the first maximum (dominating for the untapered case, $\hat{b}_1 = 0$) decreases rapidly with increase of the tapering depth. The first maximum becomes less than the second one at $\hat{b}_1 \simeq 26$. Further, the second maximum dominates till $\hat{b}_1 \simeq 38$, and so on. We limit our consideration to within these two maxima,

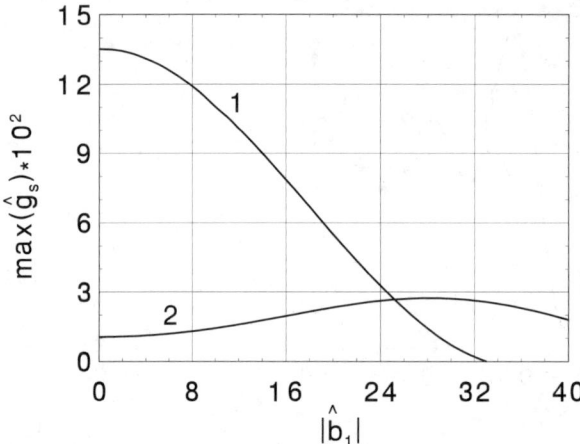

Fig. 3.35. The maximal reduced small-signal gain $\hat{g}_s$ as a function of the tapering parameter. Here (1) is the first maximum and (2) is the second maximum

i.e. with the region of the tapering coefficient $0 < |\hat{b}_1| < 38$. The position of the maxima is given approximately by the formula:

$$\hat{C}_0^m \simeq \hat{C}_0^m|_{\hat{b}_1=0} - \hat{b}_1/2 , \tag{3.52}$$

which is valid for both positive and negative values of the parameter $\hat{b}_1$. For instance, the maxima of the gain are shifted in the direction of smaller values of the detuning parameter $\hat{C}_0$ for the positive sign of $\hat{b}_1$. (i.e. the lasing frequency increases with respect to the untapered case). The values of the maximal gain $\max(\hat{g}_s)$ do not depend on the sign of $\hat{b}_1$ and are universal functions of the absolute value of $\hat{b}_1$ (see Fig. 3.35). On the other hand, the position of the maxima depends strongly on the sign and on the absolute value of the tapering parameter $\hat{b}_1$ (see (3.52)). When the FEL oscillator operates without external monochromatization, the lasing frequency is defined by the condition of maximal small-signal gain. According to Fig. 3.35 and (3.52), we can write an approximate expression for the detuning parameter $\hat{C}_0^m$ corresponding to the position of the main maximum:

$$\begin{aligned}\hat{C}_0^m &\simeq 2.6 - \hat{b}_1/2 &\text{for} \quad |\hat{b}_1| < 26 , \\ \hat{C}_0^m &\simeq 10.6 - \hat{b}_1/2 &\text{for} \quad 26 < |\hat{b}_1| < 38 .\end{aligned} \tag{3.53}$$

We will show below that such complicated behavior of the lasing frequency results in nontrivial consequences for efficiency optimization. In particular, there is a principal difference in efficiency optimization between the FEL amplifier and the FEL oscillator: undulator tapering does not provide a dramatic increase in the efficiency in the latter case.

Now we consider the case when the short prebuncher is installed in front of the main tapered undulator. In Figs. 3.36 and 3.37 we present the curves

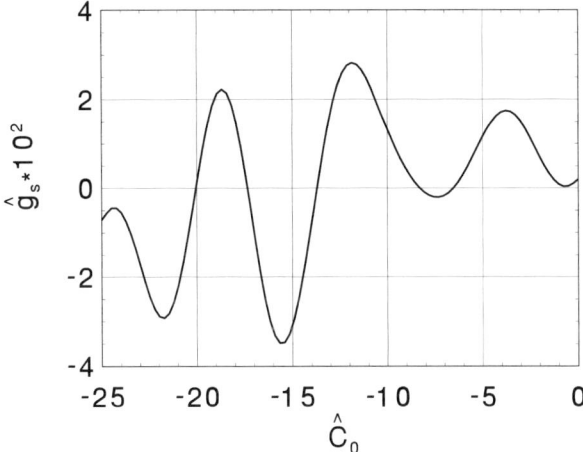

Fig. 3.36. The reduced small-signal gain $\hat{g}_s$ versus the reduced detuning $\hat{C}_0$ at the undulator entrance. Here $\overline{\delta\psi} = 0$, $\hat{d} = 0.3$, $\hat{l}_1 = 0.05$, $\hat{b}_1 = 30$ and $\hat{\Lambda}_T^2 = 0$. The calculations have been performed with (3.35)

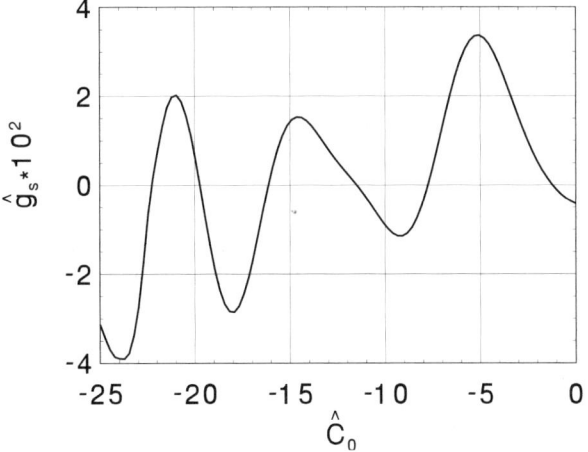

Fig. 3.37. The reduced small-signal gain $\hat{g}_s$ versus the reduced detuning $\hat{C}_0$ at the undulator entrance. Here $\overline{\delta\psi} = \pi$, $\hat{d} = 0.3$, $\hat{l}_1 = 0.05$, $\hat{b}_1 = 30$ and $\hat{\Lambda}_T^2 = 0$. The calculations have been performed with (3.35)

of the small-signal gain for $\hat{l}_1 = 0.05$, $\hat{d} = 0.3$, $\hat{b}_1 = 30$, and $\hat{\Lambda}_T^2 = 0$ calculated with (3.35). It is seen from these plots that the shape of the gain curves and the positions of the maxima depend strongly on the value of the beam-wave phase shift parameter $\overline{\delta\psi}$ (in these plots its values are equal to 0 and π). This reveals an opportunity to optimize the position of the main maximum (i.e., the lasing frequency) in order to achieve maximal efficiency at saturation.

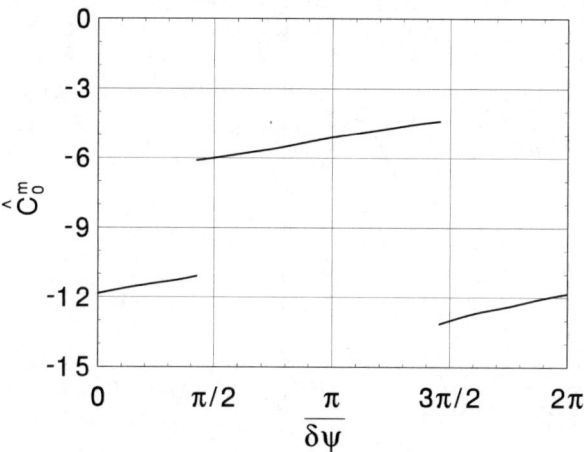

Fig. 3.38. The detuning $\hat{C}_0^m$, corresponding to the position of the main maximum, versus the $\overline{\delta\psi}$ parameter. Here $\hat{d} = 0.3$, $\hat{l}_1 = 0.05$, $\hat{b}_1 = 30$ and $\hat{A}_T^2 = 0$. The calculations have been performed with (3.35)

Figure 3.38 presents the dependence on the beam-wave phase shift $\overline{\delta\psi}$ of the detuning parameter $\hat{C}_0^m$ corresponding to the position of the main maximum of the small-signal gain. Strictly speaking, the parameters $\overline{\delta\psi}$ and $\hat{d}$ are not independent for values of $\hat{d} \lesssim 1$. Nevertheless, when the number of undulator periods is about several tens (which usually takes place in practice), the relative change of the parameter $\hat{d}$ is small when $\overline{\delta\psi}$ changes from 0 to 2π.

Efficiency Optimization. In this section we study the problem of efficiency optimization for an FEL oscillator with tapered undulator. As in the case of the small-signal gain we start with the case when there is no prebuncher and the main undulator is tapered by a linear law. We mentioned above that the lasing frequency is defined by the condition of maximal small-signal gain for the FEL oscillator without external monochromatization. In the case of a tapered undulator the lasing frequency is a complicated function of the tapering depth (see (3.53)) which results in nontrivial behavior of the efficiency at saturation. Figures. 3.39 and 3.40 present the dependence of the saturation efficiency on the tapering parameter at the value $\hat{\alpha} = 5 \times 10^{-3}$ of the normalized cavity losses.

The breaks of the curves at $\hat{b}_1 \simeq 26$ correspond to the transition between lasing at the first and the second maximum (see Fig. 3.35 and (3.53)). The breaks in Fig. 3.39 at $\hat{b}_1 \simeq 13$ and $\hat{b}_1 \simeq 32$ have their origin in the non-monotonic dependence of the gain on the value of the optical field stored in the cavity. It is seen from Fig. 3.39 that a visible increase of the efficiency can be achieved only in a narrow range of positive values of the tapering parameter $\hat{b}_1$. To explain this fact, let us analyze the behavior of the detuning parameter $\hat{C}_0^m$ corresponding to the position of the main maximum (see

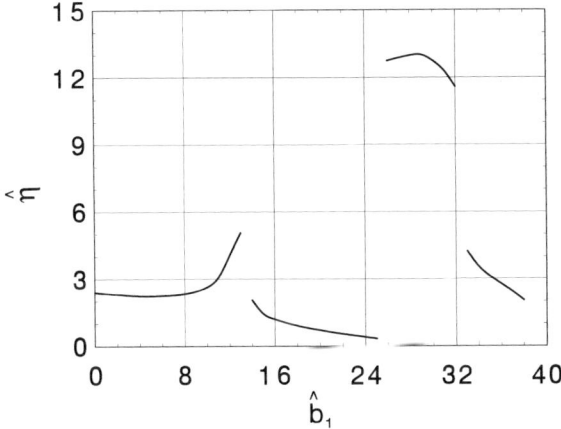

Fig. 3.39. The reduced efficiency at saturation as a function of the tapering parameter. Here $\hat{\alpha} = 0.005$ and $\hat{b}_1 > 0$

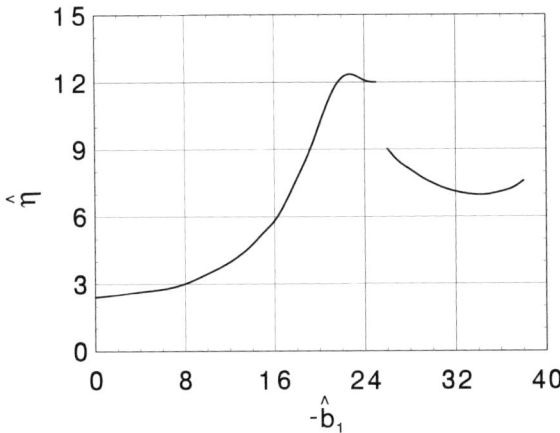

Fig. 3.40. The reduced efficiency at saturation as a function of the tapering parameter. Here $\hat{\alpha} = 0.005$ and $\hat{b}_1 < 0$

eq. (3.53)). When the tapering parameter is positive and large, $\hat{b}_1 \gg 1$, the detuning parameter $\hat{C}_0^m$ takes relatively large negative values in the region $\hat{b}_1 < 26$. Simulations show that in this case the particles are bunched at the accelerating phase of the effective potential and take away the energy from the radiation field when passing the initial part of the undulator. The process of field amplification occurs only in the final part of the undulator. When $\hat{b}_1$ is slightly larger than 26, the detuning parameter $\hat{C}_0^m$ is only slightly shifted from the resonance value and particles are trapped and decelerated almost from the very beginning of the undulator.

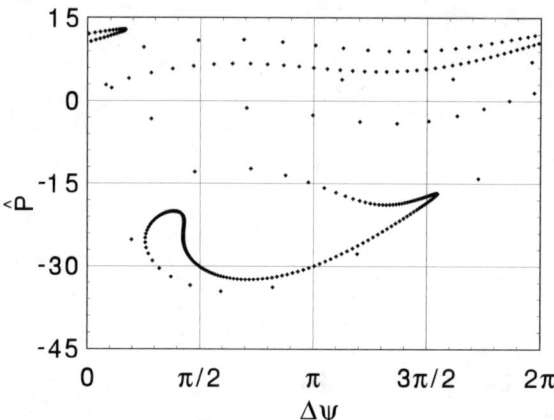

Fig. 3.41. Phase space distribution of the particles at the undulator exit. The FEL oscillator operates at saturation. Here $\hat{\alpha} = 0.005$ and $\hat{b}_1 = 29$

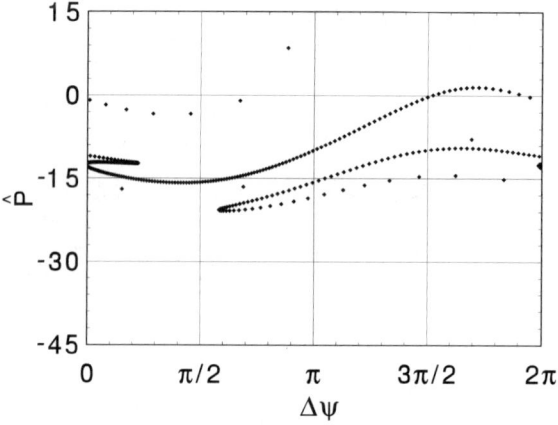

Fig. 3.42. Phase space distribution of the particles at the undulator exit. The FEL oscillator operates at saturation. Here $\hat{\alpha} = 0.005$ and $\hat{b}_1 = -23$

On the other hand, the initial detuning $\hat{C}_0$ is positive at negative values of the tapering parameter $\hat{b}_1$. The field amplification process starts from the very beginning of the undulator, but at the end of the undulator most particles fall into the accelerating phase of the effective potential and the field decreases. Analysis of Fig. 3.40 shows that the negative tapering, $\hat{b}_1 < 0$, has an advantage with respect to the positive tapering, $\hat{b}_1 > 0$, when the lasing frequency corresponds to the first maximum of the gain curve. In Figs. 3.41 and 3.42 we present the phase distribution of the particles at the undulator exit when the FEL oscillator operates in the saturation regime.

3.3 FEL Oscillator with Nonuniform Undulator   133

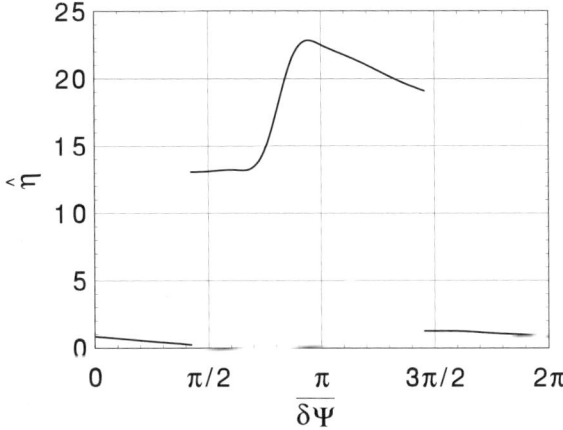

Fig. 3.43. The reduced efficiency at saturation versus the $\overline{\delta\psi}$ parameter. Here $\hat{d} = 0.3$, $\hat{l}_1 = 0.05$, $\hat{b}_1 = 30$, $\hat{A}_T^2 = 0$ and $\hat{\alpha} = 0.005$

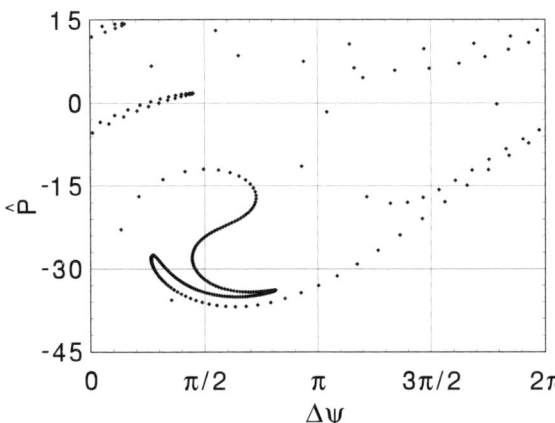

Fig. 3.44. Phase space distribution of the particles at the undulator exit. The FEL oscillator operates at saturation. Here $\hat{d} = 0.3$, $\hat{l}_1 = 0.05$, $\hat{b}_1 = 30$, $\hat{A}_T^2 = 0$, $\hat{\alpha} = 0.005$, and $\overline{\delta\psi} = \pi$

The example presented above refers to a specific value of the reduced damping factor $\hat{\alpha} = 5 \times 10^{-3}$. More general study of the problem can be found in the bibliography to this chapter. Here we present only a summary of these investigations. The maximal reduced saturation efficiency of the FEL oscillator with a tapered undulator is usually about $\hat{\eta} \simeq 12\text{--}13$ in the region of the parameters used in practical devices. On the other hand, the maximal reduced efficiency of the FEL oscillator with uniform undulator is equal to $\hat{\eta}_{max} = 3.62$ (see Sect. 3.2). So, the undulator tapering allows one to get an increase of the FEL oscillator efficiency only by a factor of about 3.5.

Let us present some results of numerical simulations of the FEL oscillator with a prebuncher and tapered main undulator. The parameters of the numerical example are the same as those used for the illustration of the linear mode of operation (see Figs. 3.36–3.38). The parameter of the resonator losses is $\hat{\alpha} = 5 \times 10^{-3}$ and the initial detuning parameter corresponds to the maximum of the small-signal gain in Fig. 3.38.

The dependence of the reduced efficiency on the beam-wave phase shift parameter $\overline{\delta\psi}$ is presented in Fig. 3.43. In Fig. 3.44 we present the phase space distribution of the particles for the value of the phase shift in the drift space equal to π. It is seen that the efficiency depends significantly on the value of the phase shift in the drift space. This is mainly connected with the dependence of the initial detuning parameter on the phase shift (see Fig. 3.38). So, there exists the possibility of optimizing the efficiency by means of an appropriate choice of the drift space. First, this allows one to fix the lasing frequency within certain limits. Also, the prebuncher improves the trapping of the particles in the strong field regime. A more thorough analysis shows that the installation of the prebuncher in front of a tapered undulator allows one to increase the efficiency additionally by a factor of 2. Finally, we can conclude that the undulator tapering provides rather limited possibilities for increasing the efficiency of the FEL oscillator. In any case, it cannot be make as high as in the FEL amplifier. This is a consequence of the fact that the lasing frequency (or, the detuning parameter) is not an independent parameter as in the case of the FEL amplifier, but is a complicated function of the tapering depth. As a result, we cannot control the detuning parameter, the main tool providing the possibility of reaching a high efficiency in the case of the FEL amplifier.

3.4 Start-Up from Shot Noise in the FEL Oscillator

In this chapter we have performed an analysis of FEL oscillator operation using the phenomenological approach which consists in considering the FEL oscillator as a low-gain, monochromatic FEL amplifier with feedback. The nonlinear mode of FEL oscillator operation has been studied assuming that the lasing frequency corresponds to the maximum of the small-signal gain. This approach provides wide possibilities for the description of different properties of the FEL oscillator, but it does not give an answer to the important practical question about the start-up of the FEL oscillator. In this section we present a detailed study of this problem.

Fluctuations of the electron beam current density is the factor triggering the lasing of the FEL oscillator. These fluctuations always exist due to the discrete nature of the electron current produced by a large number of moving particles. In other words, the shot noise is an intrinsic feature of the electron beam. Sometimes the start-up process in the FEL oscillator is explained as amplification of the spontaneously emitted radiation. Nevertheless, the spontaneously emitted radiation itself (or, the incoherent undulator radiation) has its origin in the shot noise in the electron beam.

In this section we present a rigorous analysis of the optical resonator excitation by the electron beam moving in the undulator. All the results are derived from "first principles". The investigation is performed in the framework of the one-dimensional model, using the assumption of a long electron beam. The electromagnetic field in the optical resonator is treated as a set of longitudinal modes. Leontovich's boundary conditions for the electromagnetic field are used on the mirror surfaces.

3.4.1 Basic Equations

We consider a plane Fabry Perot resonator of base L equipped with two plane parallel mirrors. A helical undulator of length l_w is placed between the mirrors and its axis coincides with the resonator axis. We suppose the electromagnetic field in the resonator to be circularly polarized because of the helical magnetic system of the undulator. Using the complex representation, in the one-dimensional approximation the radiation field in the resonator may be presented as a superposition of oscillations with different longitudinal wavenumbers. In the case of ideal mirrors, we have:

$$E_x + iE_y = \sum_m \tilde{E}_m(t) \exp(-i\omega_m t) \sin(k_m z), \tag{3.54}$$

where $k_m = m\pi/L$, $\omega_m = cm\pi/L$ and m is an integer number, $m \gg 1$. The physical sense of (3.54) is that an integer number of half-waves must fit into the resonator base. This expression may be generalized to the case when the resonator mirrors are made of material with refractive index n'. We suppose the value of n' to be a large complex number, i.e. $|n'| \gg 1$. For example, the refractive index of the metallic mirrors with conductivity σ is given by the expression $n'(\omega) = \sqrt{4\pi i\sigma/\omega}$, where ω is the frequency of the electromagnetic wave. The electromagnetic field must satisfy Leontovich's boundary conditions on the mirror surface (see Sect. 5.5 for more details). The monochromatic field may be represented as

$$\boldsymbol{E} = \boldsymbol{E}_\omega e^{-i\omega t} + \text{C.C.}, \qquad \boldsymbol{H} = \boldsymbol{H}_\omega e^{-i\omega t} + \text{C.C.}$$

Leontovich's boundary condition for this field may be written in the form:

$$(\boldsymbol{n} \times \boldsymbol{E}_\omega)\Big|_S = \frac{1}{n'} (\boldsymbol{n} \times (\boldsymbol{n} \times \boldsymbol{H}_\omega))\Big|_S ,$$

where $\boldsymbol{n}$ is the unit vector of the outward normal to the mirror surface. Using Maxwell's equation $c\boldsymbol{\nabla} \times \boldsymbol{E} = -\partial \boldsymbol{H}/\partial t$, we rewrite the boundary condition as follows:

$$\left[[E_x + iE_y] \mp \frac{ic}{\omega n'} \frac{\partial}{\partial z} [E_x + iE_y] \right]\Big|_{z=\binom{0}{L}} = 0 . \tag{3.55}$$

Assuming the refractive index to be constant within the FEL bandwidth, we may generalize (3.54) for the radiation field in the resonator for a finite value of the refraction index n':

$$E_x + iE_y = \sum_m \tilde{E}_m(t) \exp(-i\omega_m t) \sin(k_m z + \delta'), \qquad (3.56)$$

where $k_m = m\pi/L - 2i/(n'L)$ and $\delta' = i/n'$. One can show that this field satisfies Leontovich's boundary conditions (3.55).

Under accepted assumptions, the longitudinal modes of the resonator are independent of each other and the bandwidth of each mode is much narrower than the frequency difference between the modes:

$$\left| \frac{1}{\tilde{E}_m} \frac{d\tilde{E}_m}{dt} \right| \ll \frac{\pi c}{L}. \qquad (3.57)$$

In other words, this condition means that relative change of $\tilde{E}_m$ during a resonator round-trip is small.

The power flow density going out of the resonator volume is expressed in terms of Poynting's vector at the surfaces of both mirrors and can be written as follows:

$$\Pi = \sum_m \Pi_m = \frac{c}{4\pi} \operatorname{Re}\left(\frac{2}{n'}\right) \sum_m |\tilde{E}_m|^2. \qquad (3.58)$$

To prove this statement, we write an explicit expression for Poynting's vector

$$\boldsymbol{\Pi} = \frac{c}{4\pi} \boldsymbol{E} \times \boldsymbol{H}$$

at the surface of the first mirror, at $z = 0$. According to Maxwell's equation, the electric and the magnetic field are connected by the relation:

$$c\boldsymbol{\nabla} \times \boldsymbol{E} = -\partial \boldsymbol{H}/\partial t.$$

Since $|n'|^{-1}$ is a small value, the terms $|n'|^{-2}$ can be neglected when calculating Poynting's vector, and we get

$$H_x + iH_y = \sum_m \tilde{E}_m(t) \exp(-i\omega_m t) \cos(m\pi z/L).$$

The vectors $\boldsymbol{E}$ and $\boldsymbol{H}$ are perpendicular to each other and their vector product (constituting Poynting's vector) is directed towards the mirror surface

$$\begin{aligned}
\Pi_z|_{z=0} &= \frac{c}{4\pi}(E_x H_y - H_x E_y)|_{z=0} \\
&= -\frac{c}{4\pi} \operatorname{Im}\left[(E_x + iE_y)(H_x - iH_y)\right]|_{z=0} \\
&= -\frac{c}{4\pi} \operatorname{Re}\left(\frac{1}{n'}\right) \sum_n |\tilde{E}_m(t)|^2.
\end{aligned}$$

So, there is a flow of energy onto the mirror. The same calculations can be performed for the second mirror placed at $z = L$ and we finally get the total

3.4 Start-Up from Shot Noise in the FEL Oscillator

power loss per unit area given by (3.58). When the mirrors are made of materials with different refractive indexes n'_1 and n'_2, the following substitution should be made in the equations:

$$\frac{2}{n'} = \frac{1}{n'_1} + \frac{1}{n'_2}. \tag{3.59}$$

Let us consider the problem of resonator excitation by a continuous electron beam moving with velocity v_z along the undulator. We assume that there is no optical field in the resonator at some moment of time $t = 0$. Our goal is to calculate the evolution of the electromagnetic field in the resonator and to calculate the power going out of the resonator volume.

In the framework of the one-dimensional model the electric field $\boldsymbol{E}_\perp(z,t)$ of the electromagnetic wave in the resonator is subjected to the wave equation

$$c^2 \frac{\partial^2 \boldsymbol{E}_\perp}{\partial z^2} - \frac{\partial^2 \boldsymbol{E}_\perp}{\partial t^2} = 4\pi \frac{\partial \boldsymbol{j}_\perp}{\partial t}, \tag{3.60}$$

which may be obtained from Maxwell's equations. Analysis of this equation shows that the optical field in the resonator might be excited only when the derivative of the current density takes a nonzero value. Therefore, to describe correctly the resonator excitation we have to take into account the discrete nature of the electron beam current (or, the shot noise) which leads to a white spectrum of such an input signal.

The transverse component of the beam current density is connected with the longitudinal component by the relation:

$$(j_x + ij_y) = \theta_s(z) \exp(-ik_w z) j_z, \tag{3.61}$$

where

$$\theta_s(z) = \theta_s = \text{const.} \quad \text{for} \quad z_i < z < z_i + l_w$$
$$\theta_s(z) = 0 \quad \text{for} \quad 0 < z < z_i \quad \text{and} \quad z_i + l_w < z < L.$$

Here z_i is the coordinate of the undulator entrance. The total transverse current density is the sum of the external and induced components:

$$(j_x + ij_y) = (j_x + ij_y)^{\text{ext}} + (j_x + ij_y)^{\text{ind}}. \tag{3.62}$$

The longitudinal component of the external beam current density can be represented as[2]

---

[2] In this book we use the following form for the Fourier transformations:

$$\bar{F}(\omega) = \int_{-\infty}^{\infty} dt\, F(t) \exp(i\omega t),$$

$$F(t) = \frac{1}{2\pi} \int_{-\infty}^{\infty} d\omega\, \bar{F}(\omega) \exp(-i\omega t).$$

$$j_z^{\text{ext}}\left(t - \frac{z}{v_z}\right) = \frac{1}{2\pi} \int_{-\infty}^{\infty} d\omega \bar{j}(\omega) \exp\left[i\omega\left(\frac{z}{v_z} - t\right)\right]$$

$$= \frac{1}{2\pi} \int_0^{\infty} d\omega \bar{j}(\omega) \exp\left[i\omega\left(\frac{z}{v_z} - t\right)\right] + \text{C.C.} \quad (3.63)$$

Only the spectral components of the input current density (3.63), close to the eigenfrequencies ω_m, will excite the resonator. Neglecting the nonresonant background, we can write:

$$j_z^{\text{ext}}\left(t - \frac{z}{v_z}\right) \simeq \sum_m \tilde{j}_m^{\text{ext}}(t) \exp\left[i\omega_m\left(\frac{z}{v_z} - t\right)\right] + \text{C.C.}, \quad (3.64)$$

where

$$\tilde{j}_m^{\text{ext}}(t) = \frac{1}{2\pi} \int_{-\epsilon}^{\epsilon} d(\Delta\omega_m) \bar{j}(\Delta\omega_m) \exp(-i\Delta\omega_m t), \quad (3.65)$$

$\Delta\omega_m = \omega - \omega_m$ and ϵ satisfies the condition

$$\left|\frac{1}{\tilde{E}_m} \frac{d\tilde{E}_m}{dt}\right| \ll \epsilon \ll \frac{\pi c}{L}. \quad (3.66)$$

The external current excites the field in the resonator and this field induces additional density modulation. Since we consider the linear problem, the induced transverse current density is proportional to the electric field of the synchronous wave:

$$(j_x + ij_y)^{\text{ind}} \simeq \sum_m \chi_m \frac{\tilde{E}_m(t)}{2i} \exp\left[i\omega_m(z/c - t)\right]. \quad (3.67)$$

To find the susceptibility χ_m, one should solve the equations of electron motion in the given electromagnetic field. This problem has been solved in the previous sections. Here we consider the case of a cold electron beam and neglect space charge effects. Using (3.4), (3.5) and the condition of a low field gain per undulator pass, we can write:

$$\chi_m = ij_0\omega_0 e\theta_s^2(2c\gamma_z^2\mathcal{E}_0)^{-1} \int_{z_i}^{z} dz'(z' - z) \exp\left[iC_m(z' - z)\right].$$

where j_0 is the beam current density, $C_m = k_w - \omega_m/(2c\gamma_z^2)$ is the detuning of the mth mode with the particle having nominal energy $\mathcal{E}_0$, $\theta_s = eH_w/(\mathcal{E}_0 k_w)$, $\gamma_z^{-2} = \gamma^{-2} + \theta_s^2$, $\gamma = \mathcal{E}_0/(m_e c^2)$, and $\omega_0 = 2ck_w\gamma_z^2$.

Substituting (3.56) and (3.62) into (3.60), we obtain

3.4 Start-Up from Shot Noise in the FEL Oscillator

$$\sum_m \omega_m \left(\frac{2c}{n'L}\tilde{E}_m + \frac{d\tilde{E}_m}{dt}\right) \exp(-i\omega_m t) \sin(k_m z + \delta')$$

$$= -2\pi\theta_s(z) \exp(-ik_w z) \left(\sum_m \omega_m \tilde{j}_m^{\text{ext}} \exp\left[i\omega_m(z/v_z - t)\right] + \text{C.C.}\right)$$

$$+ i\pi \sum_m \omega_m \chi_m \tilde{E}_m \exp\left[i\omega_m(z/c - t)\right]. \tag{3.68}$$

Then we multiply the obtained equation by $\sin(k_m z)$ and integrate over the resonator length from 0 to L. Neglecting small nonresonant terms, we get the equation for the slowly varying complex amplitudes

$$\frac{d\tilde{E}_m}{dt} - a_m \tilde{E}_m + b_m \tilde{j}_m^{\text{ext}} = 0, \tag{3.69}$$

where

$$a_m = \frac{c\tau}{4L}\hat{Z}(\hat{C}_m) - \frac{2c}{n'L}, \tag{3.70}$$

$$b_m = i\frac{2\pi\theta_s l_w}{L}\frac{\sin(\hat{C}_m/2)}{(\hat{C}_m/2)} \exp(-i\hat{C}_m/2), \tag{3.71}$$

$\tau = 2\pi\omega_0\theta_s^2 l_w^3 j_0(c\gamma_z^2\gamma I_A)^{-1}$, and $\hat{C}_m = C_m l_w$.
The function $\hat{Z}(\hat{C}_m)$ has the form:

$$\hat{Z}(\hat{C}_m) = i\int_0^1 d\xi \int_0^\xi d\xi' \xi' \exp\left(-i\hat{C}_m \xi'\right). \tag{3.72}$$

The integral (3.72) can be calculated analytically:

$$\hat{Z}(\hat{C}_m) = 2i\hat{C}_m^{-2}\left[2(\hat{C}_m)^{-1}\sin(\hat{C}_m/2) - \cos(\hat{C}_m/2)\right] \exp(-i\hat{C}_m/2). \tag{3.73}$$

3.4.2 General Results

The next problem is to solve (3.69) with the initial condition $\tilde{E}_m(0) = 0$. This can be done by means of the Laplace technique. The Laplace transforms of the amplitudes of the radiation field and of the current density are ($\operatorname{Re} p > 0$):

$$E_m(p) = \int_0^\infty dt \exp(-pt)\tilde{E}_m(t),$$

$$j_m(p) = \int_0^\infty dt \exp(-pt)\tilde{j}_m^{\text{ext}}(t) = \frac{1}{2\pi}\int_{-\epsilon}^{\epsilon} d(\Delta\omega_m)\frac{\tilde{j}(\Delta\omega_m)}{p + i\Delta\omega_m}.$$

The solution of the equation for the Laplace transforms is

$$E_m(p) = \frac{b_m}{a_m - p} j_m(p) = \frac{b_m}{2\pi} \int_{-\epsilon}^{\epsilon} \mathrm{d}(\Delta\omega_m) \frac{\bar{j}(\Delta\omega_m)}{(a_m - p)(p + \mathrm{i}\Delta\omega_m)} . \qquad (3.74)$$

We perform an inverse Laplace transformation using Cauchy's residue theorem. Finally, we get

$$\tilde{E}_m(t) = \frac{b_m}{2\pi} \int_{-\epsilon}^{\epsilon} \mathrm{d}(\Delta\omega_m) \frac{\bar{j}(\Delta\omega_m)}{a_m + \mathrm{i}\Delta\omega_m} \left[\exp(-\mathrm{i}\Delta\omega_m t) - \exp(a_m t)\right] . \qquad (3.75)$$

It is assumed here that $\operatorname{Re} a_m \ne 0$. We should also mention that $\epsilon \gg |a_m|$ due to the condition (3.66). Thus, the value of ϵ in the integration limits can be replaced by infinity.

Let us now calculate the ensemble averaged power density of the mth mode going out of the resonator volume. According to (3.58), this can be expressed as

$$\langle \Pi_m(t) \rangle = \frac{c}{4\pi} \operatorname{Re}\left(\frac{2}{n'}\right) \langle |\tilde{E}_m(t)|^2 \rangle , \qquad (3.76)$$

where $\langle \ldots \rangle$ means averaging over the ensemble. Using (3.75), we can write

$$\langle |\tilde{E}_m(t)|^2 \rangle$$
$$= \frac{|b_m|^2}{4\pi^2} \int_{-\infty}^{\infty} \mathrm{d}(\Delta\omega_m) \int_{-\infty}^{\infty} \mathrm{d}(\Delta\omega'_m) \langle \bar{j}(\Delta\omega_m) \bar{j}^*(\Delta\omega'_m) \rangle$$
$$\times \frac{[\exp(-\mathrm{i}\Delta\omega_m t) - \exp(a_m t)][\exp(\mathrm{i}\Delta\omega'_m t) - \exp(a_m^* t)]}{(a_m + \mathrm{i}\Delta\omega_m)(a_m^* - \mathrm{i}\Delta\omega'_m)} . \qquad (3.77)$$

To calculate the correlation of the spectral components of the beam current density, we should consider a microscopic picture of the electron beam current at the undulator entrance. We start the analysis for the case of a rectangular electron pulse of finite duration T and then we go over to the limit of an infinitely long pulse. The electron beam current is made up of moving electrons randomly arriving at the entrance of the undulator:

$$I(t) = (-e) \sum_{k=1}^{N} \delta(t - t_k) ,$$

where $\delta(\ldots)$ is the delta function, $(-e)$ is the charge of the electron, N is the number of electrons in a bunch and t_k is the random arrival time of the electron at the undulator entrance. The beam current averaged over an ensemble of the bunches can be written in the form:

$$\langle I(t) \rangle = \frac{(-e)N}{T} = -I_0 \qquad \text{for} \qquad -\frac{T}{2} < t < \frac{T}{2} . \qquad (3.78)$$

3.4 Start-Up from Shot Noise in the FEL Oscillator

The probability of the arrival of an electron during a time interval $t, t + \mathrm{d}t$ is equal to $\mathrm{d}t/T$. The electron beam current $I(t)$ and its Fourier transform $\bar{I}(\omega)$ are connected by:

$$\bar{I}(\omega) = \int_{-\infty}^{\infty} e^{i\omega t} I(t) \mathrm{d}t = (-e) \sum_{k=1}^{N} e^{i\omega t_k} ,$$

$$I(t) = \frac{1}{2\pi} \int_{-\infty}^{\infty} \bar{I}(\omega) e^{-i\omega t} \mathrm{d}\omega = (-e) \sum_{k=1}^{N} \delta(t - t_k) . \quad (3.79)$$

Now we can calculate the first-order correlation of the complex Fourier harmonics $\bar{I}(\omega)$ and $\bar{I}(\omega')$:

$$\langle \bar{I}(\omega) \bar{I}^*(\omega') \rangle = e^2 \left\langle \sum_{k=1}^{N} \sum_{n=1}^{N} \exp(i\omega t_k - i\omega' t_n) \right\rangle .$$

Expanding this relation, we can write:

$$\langle \bar{I}(\omega) \bar{I}^*(\omega') \rangle$$
$$= e^2 \left\langle \sum_{k=1}^{N} \exp\left[i(\omega - \omega') t_k\right] \right\rangle + e^2 \left\langle \sum_{k \neq n} \exp(i\omega t_k - i\omega' t_n) \right\rangle$$
$$= e^2 \sum_{k=1}^{N} \langle \exp\left[i(\omega - \omega') t_k\right] \rangle + e^2 \sum_{k \neq n} \langle \exp(i\omega t_k) \rangle \langle \exp(-i\omega' t_n) \rangle . \quad (3.80)$$

It is easy to find that

$$\langle \exp(i\omega t_k) \rangle = \frac{1}{T} \int_{-T/2}^{T/2} e^{i\omega t_k} \mathrm{d}t_k = \bar{F}(\omega) , \quad (3.81)$$

where

$$\bar{F}(\omega) = \left(\frac{\omega T}{2}\right)^{-1} \sin\left(\frac{\omega T}{2}\right) .$$

Substituting (3.81) into (3.80), we obtain:

$$\langle \bar{I}(\omega) \bar{I}^*(\omega') \rangle = e^2 N \bar{F}(\omega - \omega') + e^2 N(N-1) \bar{F}(\omega) \bar{F}^*(\omega') . \quad (3.82)$$

Now we go over to the long-pulse limit, $T \to \infty$. In this limit the second summand in (3.82) is neglected and the first one can be written as

$$\langle \bar{I}(\omega) \bar{I}^*(\omega') \rangle = e I_0 \lim_{T \to \infty} \left[T \bar{F}(\omega - \omega') \right] = 2\pi e I_0 \delta(\omega - \omega') . \quad (3.83)$$

The following representation of the delta function has been used here:

$$\delta(x) = \frac{\xi}{\pi} \frac{\sin(\xi x)}{\xi x} \qquad \text{as} \qquad \xi \to \infty .$$

3. One-Dimensional Theory of the FEL Oscillator

These considerations refer to the total beam current. In the framework of the one-dimensional model we deal with the current density. To find the correlation of the spectral components of the current density, we assume that only one transverse mode of the resonator is excited and that the transversely coherent fraction of the shot noise signal is defined by the total beam current. Thus, we can write

$$\frac{\bar{j}(\omega)}{j_0} = \frac{\bar{I}(\omega)}{I_0} . \tag{3.84}$$

Finally, the result (3.83) can be rewritten for the current density:

$$\langle \bar{j}(\Delta\omega_m)\bar{j}^*(\Delta\omega'_m)\rangle = \frac{ej_0^2}{I_0} 2\pi\delta(\Delta\omega_m - \Delta\omega'_m) . \tag{3.85}$$

Substituting (3.85) in (3.77), we obtain[3]

$$\langle |\tilde{E}_m(t)|^2\rangle$$

$$= \frac{ej_0^2}{I_0} \frac{|b_m|^2}{2\pi} \int_{-\infty}^{\infty} \mathrm{d}(\Delta\omega_m)$$

$$\times \frac{1 + \exp(2\operatorname{Re} a_m t) - 2\exp(\operatorname{Re} a_m t)\cos\left[(\operatorname{Im} a_m + \Delta\omega_m)t\right]}{(\operatorname{Re} a_m)^2 + (\operatorname{Im} a_m + \Delta\omega_m)^2} . \tag{3.86}$$

Performing the integration, we finally get

$$\langle |\tilde{E}_m(t)|^2\rangle = \frac{ej_0^2|b_m|^2}{2I_0} \frac{\exp(2\operatorname{Re} a_m t) - 1}{\operatorname{Re} a_m} . \tag{3.87}$$

For further consideration it is convenient to go over to the number of resonator round-trips instead of time: $n = ct/(2L)$. Using (3.87), (3.70), and (3.71) we express (3.76) in the following form

$$\langle \Pi_m(n)\rangle = \frac{\pi e j_0^2 \theta_s^2 l_w^2}{2I_0 L} \frac{\sin^2(\hat{C}_m/2)}{(\hat{C}_m/2)^2} \frac{\alpha}{\tau\operatorname{Re}\hat{Z}(\hat{C}_m) - \alpha}$$

$$\times \left\{\exp\left\{\left[\tau\operatorname{Re}\hat{Z}(\hat{C}_m) - \alpha\right]n\right\} - 1\right\} , \tag{3.88}$$

where $\alpha = \operatorname{Re}(8/n')$. The parameter with the same notation has been used in the previous sections. The damping factor α has been defined as the relative power losses after reflections of the travelling wave from the two mirrors. It follows from the Fresnel formula that the relative power loss in the mirror is equal to $\operatorname{Re}(4/n')$ at normal incidence, so there is complete agreement of the notation used in this section and above.

---

[3] To simplify the derivation of (3.86), we have assumed the temporal profile of the electron beam to be symmetric with respect to $t = 0$. The spectral correlation (3.85) is a real function in this case. It can be shown that (3.86) is also valid when there is no symmetry with respect to $t = 0$.

3.4 Start-Up from Shot Noise in the FEL Oscillator

The total power flow density going out of the resonator is equal to

$$\langle \Pi(n) \rangle = \sum_m \langle \Pi_m(n) \rangle . \tag{3.89}$$

To calculate $\langle \Pi(n) \rangle$, one should take into account the following. The frequency range of the resonator excitation is of the order of unity in terms of the detuning parameter. On the other hand, the frequency difference between two adjacent modes is small:

$$\Delta \hat{C}_{\text{mode}} = \hat{C}_{m-1} - \hat{C}_m = \frac{\pi l_{\text{w}}}{2\gamma_z^2 L} \ll 1 .$$

This means that many longitudinal modes are excited and we can replace the summation in (3.89) by integration:

$$\langle \Pi(n) \rangle = \frac{e j_0^2 \theta_s^2 l_{\text{w}}^2 \gamma_z^2}{I_0} \int_{-\infty}^{\infty} d\hat{C} \, \frac{\sin^2(\hat{C}/2)}{(\hat{C}/2)^2}$$

$$\times \frac{\alpha}{\tau \operatorname{Re} \hat{Z}(\hat{C}) - \alpha} \left\{ \exp\left\{ \left[\tau \operatorname{Re} \hat{Z}(\hat{C}) - \alpha \right] n \right\} - 1 \right\} . \tag{3.90}$$

The normalized density of the power losses is defined as:

$$\hat{\eta}(n) = \frac{\eta(n)}{\beta} = \frac{e \langle \Pi(n) \rangle}{j_0 \mathcal{E}_0 \beta} , \tag{3.91}$$

where $\beta = (4\pi N_{\text{w}})^{-1}$ is the efficiency parameter introduced in Sect. 3.2 and $N_{\text{w}} = k_{\text{w}} l_{\text{w}}/(2\pi)$ is the number of undulator periods. Using (3.90), we write down an explicit expression for the normalized density of the power losses:

$$\hat{\eta}(\hat{n}) = \frac{1}{2\pi N_{\text{c}}} \int_{-\infty}^{\infty} d\hat{C} \, \frac{\sin^2(\hat{C}/2)}{(\hat{C}/2)^2}$$

$$\times \frac{\hat{\alpha}}{\operatorname{Re} \hat{Z}(\hat{C}) - \hat{\alpha}} \left\{ \exp\left\{ \left[\operatorname{Re} \hat{Z}(\hat{C}) - \hat{\alpha} \right] \hat{n} \right\} - 1 \right\} . \tag{3.92}$$

where $\hat{\alpha} = \alpha/\tau$ is the reduced damping factor, $\hat{n} = \tau n$ is the reduced number of resonator round-trips, $N_{\text{c}} = N_\lambda/(2\pi\tau\beta)$ is the number of cooperating electrons, and $N_\lambda = 2\pi I_0/(e\omega_0)$ is the number of electrons per radiation wavelength.

Lasing in the FEL oscillator takes place when the maximal value of the small-signal gain, $\max(g_{\text{s}}) = \tau \max\left[\operatorname{Re}(\hat{Z}(\hat{C}))\right]$, is larger than the relative power losses per round-trip, α. A large number of longitudinal modes begins to grow exponentially in the linear regime, thus increasing the total radiation field stored in the resonator. When the small-signal gain is less than the relative power losses in the resonator (the FEL oscillator operates under the lasing threshold), the averaged radiation power stored in the resonator tends to some equilibrium value after many round-trips. It is convenient for further consideration to introduce the notion of the threshold coefficient

$$K_{\text{th}} = \frac{\max(\hat{g}_s)}{\hat{\alpha}},$$

where $\max(\hat{g}_s) = \max(g_s)/\tau$ is the maximal normalized gain. For the FEL oscillator with a cold electron beam the maximal value of the small-signal gain is equal to $\max(\hat{g}_s) \simeq 0.135$ and is achieved at the value of the detuning parameter $\hat{C}_{\max} = 2.606$. The value of $K_{\text{th}} = 1$ separates two different regimes of FEL oscillator operation. Lasing takes place for $K_{\text{th}} > 1$, and the FEL oscillator operates below the threshold for $K_{\text{th}} < 1$.

3.4.3 Operation Below Threshold

Let us study the operation of the FEL oscillator below the lasing threshold, $K_{\text{th}} < 1$. For a large number of resonator round-trips, $\hat{n} \gg (\hat{\alpha} - \max(\hat{g}_s))^{-1}$, the value of the exponent in (3.92) tends to zero, and we can write

$$\hat{\eta} = (N_c)^{-1} f(K_{\text{th}}), \qquad (3.93)$$

where

$$f(K_{\text{th}}) = \frac{1}{2\pi} \int_{-\infty}^{\infty} d\hat{C} \, \frac{\sin^2(\hat{C}/2)}{(\hat{C}/2)^2} \left[1 - K_{\text{th}} \frac{\operatorname{Re} \hat{Z}(\hat{C})}{\max(\hat{g}_s)} \right]^{-1}. \qquad (3.94)$$

Analysis of this expression shows that $f(K_{\text{th}}) \to 1$ for operation far below the threshold, for $K_{\text{th}} \to 0$. Therefore, (3.93) is simplified to

$$\hat{\eta} = (N_c)^{-1} \qquad \text{at} \quad K_{\text{th}} \to 0. \qquad (3.95)$$

Now let us consider the operation of the FEL oscillator just below threshold, at $1 - K_{\text{th}} \ll 1$. In this case there is a narrow peak in the integrand of (3.94) near $\hat{C}_{\max}$, and we can use the following expansion:

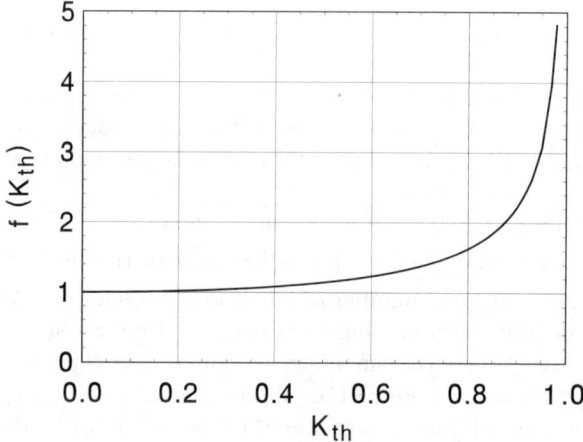

Fig. 3.45. The function $f(K_{\text{th}})$ given by (3.94)

3.4 Start-Up from Shot Noise in the FEL Oscillator   145

$$\operatorname{Re} \hat{Z}(\hat{C}) \simeq \max(\hat{g}_s) - h(\hat{C} - \hat{C}_{\max})^2 , \quad (3.96)$$

where

$$h = -\frac{1}{2}\left[\frac{d^2}{d\hat{C}^2}(\operatorname{Re}\hat{Z})\right]\bigg|_{\hat{C}=\hat{C}_{\max}} \simeq 0.0221 . \quad (3.97)$$

Substituting (3.96) into (3.94) and performing the integration, we get

$$f(K_{\mathrm{th}}) \simeq \frac{0.677}{\sqrt{1 - K_{\mathrm{th}}}} \qquad \text{at} \quad 1 - K_{\mathrm{th}} \ll 1 . \quad (3.98)$$

In the general case the function $f(K_{\mathrm{th}})$ can be calculated by numerical integration of (3.94). The plot of this function is presented in Fig. 3.45.

The results obtained for the operation of the FEL oscillator under the lasing threshold have a simple physical interpretation. Let us perform a qualitative consideration of a three-dimensional resonator assuming that the fundamental transverse TEM$_{00}$ Gaussian mode of the resonator is excited by the shot noise signal. When the transverse size of the electron beam is much less than that of the mode, the beam current density, appearing in the one-dimensional FEL equations, should be replaced by the effective current density:

$$(j_0)_{\mathrm{eff}} = 2I_0/(\pi w^2) , \quad (3.99)$$

where w is the waist size of the Gaussian mode.

Let us consider operation well below the lasing threshold. Using (3.95), (3.91), and (3.99) we calculate the contribution of the fundamental transverse resonator mode to the radiation power going out of the resonator volume:

$$W_{\mathrm{out}} = \frac{\tau W_{\mathrm{b}}}{8\pi N_{\mathrm{w}}^2 N_\lambda} = 4\frac{eI_0 l_{\mathrm{w}}}{w^2}\frac{K^2}{1+K^2} , \quad (3.100)$$

where $K = eH_{\mathrm{w}}/(k_{\mathrm{w}} m_e c^2)$ is the undulator parameter and $W_{\mathrm{b}} = I_0 \mathcal{E}_0/e$ is the electron beam power.

Now let us calculate the spontaneous undulator radiation. The spectral and angular density of the radiation energy emitted by a single electron during the undulator pass is given by the expression (at zero angle):

$$\frac{d^2\mathcal{E}}{d\omega d\Omega} = \frac{2N_{\mathrm{w}}^2 e^2 \gamma^2}{c}\left(\frac{K}{1+K^2}\right)^2 \frac{\sin^2\left[\pi N_{\mathrm{w}}(\omega-\omega_0)/(2\omega_0)\right]}{\left[\pi N_{\mathrm{w}}(\omega-\omega_0)/(2\omega_0)\right]^2} . \quad (3.101)$$

In the small-angle approximation the solid angle is equal to $d\Omega = \theta d\theta d\varphi$. Integration of (3.101) over ω and over φ gives us factors $2\omega_0/N_{\mathrm{w}}$ and 2π, respectively. We also have to integrate over θ from 0 to some effective maximal angle which is chosen to be the angle of divergence for the Gaussian mode, $\lambda/(\pi w)$, where $\lambda = 2\pi c/\omega_0$. Then we multiply the result by the value of I_0/e, and obtain an estimate for the transversely coherent fraction of the undulator radiation emitted by the electron beam:

$$W_{\mathrm{sp}} \simeq 4\frac{eI_0 l_{\mathrm{w}}}{w^2}\frac{K^2}{1+K^2} . \quad (3.102)$$

It is seen that this expression is identical to (3.100).[4] So, we can conclude that the result (3.95), obtained in the framework of the one-dimensional model, can be interpreted as the excitation of the fundamental transverse mode of the resonator by the spontaneous undulator radiation. This radiation power is distributed among a large number of longitudinal modes of the resonator.

In conclusion to this section we should like to comment the specific behavior of the function $f(K_{\text{th}})$ when K_{th} tends to unity. This means that there is a significant increase of the radiation power when approaching the lasing threshold. This effect is usually referred to as the coherent enhancement of the spontaneous emission and can be explained in a simple way. When K_{th} approaches unity, the optical field stored in the resonator is not negligibly weak and induces a density modulation at the frequencies of the longitudinal modes. This additional modulation can be comparable to (or, even much larger than) the corresponding spectral components of the shot noise. Therefore, the radiation power is increased with respect to the level of the spontaneous emission.

3.4.4 Operation Above Threshold

Let us consider the operation of the FEL oscillator above the lasing threshold. In this case the longitudinal modes of the resonator grow exponentially for which the condition $\operatorname{Re} \hat{Z}(\hat{C}_m) > \hat{\alpha}$ is satisfied. We study the operation of the FEL oscillator after a large number of round-trips:

$$\hat{n} \gg (\max(\hat{g}_{\text{s}}) - \hat{\alpha})^{-1} \ .$$

Under this condition the contribution to the sum (3.89) of the longitudinal modes with $\hat{C}_m$ being close to $\hat{C}_{\max}$ will dominate over the others. Assuming the number of modes to be large, we can go over to integration when calculating the normalized density of the power loss given by (3.92). Using the expansion for the function $\hat{Z}$ in the form of (3.96), we obtain:

$$\hat{\eta}(\hat{n}) = \frac{1}{2\pi N_{\text{c}}} \frac{\sin^2(\hat{C}_{\max}/2)}{(\hat{C}_{\max}/2)^2} \frac{\hat{\alpha}}{\max(\hat{g}_{\text{s}}) - \hat{\alpha}}$$

$$\times \exp\left[(\max(\hat{g}_{\text{s}}) - \hat{\alpha})\hat{n}\right] \int_{-\infty}^{\infty} \mathrm{d}(\Delta\hat{C}) \ H(\Delta\hat{C}) \ , \tag{3.103}$$

where $\Delta\hat{C} = \hat{C} - \hat{C}_{\max}$ and

$$H(\Delta\hat{C}) = \exp\left[-h\hat{n}(\Delta\hat{C})^2\right] \ . \tag{3.104}$$

Integration of (3.103) gives the following result:

---

[4] Despite the complete identity of the results, we should stress that they are both only estimates.

$$\hat{\eta}(\hat{n}) = \frac{1.04}{N_c \sqrt{\hat{n}}} \frac{\hat{\alpha}}{\max(\hat{g}_s) - \hat{\alpha}} \exp\left[(\max(\hat{g}_s) - \hat{\alpha})\hat{n}\right] . \tag{3.105}$$

Here we have taken into account that $\hat{C}_{\max} = 2.606$ and $h = 0.0221$. Expression (3.105) is a rigorous result of the one-dimensional theory obtained in the limit of a large number of resonator round-trips. It can be interpreted as the excitation of the fundamental transverse mode of the resonator. Indeed, many transverse modes are excited in the resonator. When the FEL oscillator operates above the lasing threshold, the amplitudes of the modes grow exponentially. Each mode is characterized by the value of the field growth rate. As a rule, the fundamental mode has the highest field growth rate. When we wait for a long time period since start-up of the FEL oscillator, we find that the total radiation power is determined by the fundamental mode. Therefore, (3.105) can be generalized to the three-dimensional case by an appropriate substitution of the value of the effective current density (3.99). Such an approximation provides good accuracy for many practical situations.

When calculating the parameters of the FEL oscillator in Sects. 3.1 and 3.2, we assumed the starting value of the input power to be a given value. In this section we performed a rigorous analysis of the start-up from shot noise. The results obtained can be used for the calculation of the initial effective power of the shot noise in the FEL oscillator. The radiation power loss in the resonator (see (3.7)) changes in the linear regime as follows

$$W_{\text{out}}(n) = \alpha W_{\text{in}}(0) \exp\left[(\max(g_s) - \alpha) n\right] , \tag{3.106}$$

where $W_{\text{in}}(0)$ is the initial intracavity power. Using (3.106), (3.105), and (3.91), we can write down the following approximate expression for the effective initial intracavity power:

$$W_{\text{in}}^{\text{eff}}(0) \simeq \frac{\beta W_b}{\tau N_c \left(\max(\hat{g}_s) - \hat{\alpha}\right) \sqrt{\hat{n}}} . \tag{3.107}$$

The value obtained for the effective initial power depends on the number of round-trips, $\hat{n}$. Exponential growth of the radiation power occurs only in the linear regime, and when $\hat{\eta}(\hat{n})$ becomes comparable with unity, saturation effects start to play a significant role. Using (3.105), we can estimate the number of round-trips when saturation in the FEL oscillator occurs:

$$\hat{n} \simeq \frac{\ln N_c}{\max(\hat{g}_s) - \hat{\alpha}} . \tag{3.108}$$

Then the approximate expression for the effective initial intracavity power becomes

$$W_{\text{in}}^{\text{eff}}(0) \simeq \frac{\beta W_b}{\tau N_c \sqrt{[\max(\hat{g}_s) - \hat{\alpha}] \ln N_c}} . \tag{3.109}$$

This value should be used for the numerical simulation of the nonlinear mode of FEL oscillator operation based on the approach developed in Sect. 3.2. It is relevant to present an expression for the normalized initial field in the resonator $\hat{u}_{\text{in}} = E_{\text{in}}/E_0$ which should be used in the simulation code. Using the definition of E_0 (see Sect. 3.2) and the relation

$$\frac{W_{\text{in}}^{\text{eff}}(0)}{W_{\text{b}}} = \frac{ceE_{\text{in}}^2}{4\pi j_0 \mathcal{E}_0} ,$$

we get

$$\hat{u}_{\text{in}}^2 \simeq \frac{2}{N_{\text{c}} \sqrt{[\max(\hat{g}_{\text{s}}) - \hat{\alpha}] \ln N_{\text{c}}}} . \qquad (3.110)$$

Let us study the coherent properties of the radiation from the FEL oscillator operating in the linear regime. We have assumed above that the number of longitudinal modes of the resonator is large. Now we obtain an explicit expression for this number. The averaged power spectrum at a large number of round-trips is the Gaussian line (3.104). The rms width of the spectrum is equal to $\hat{\sigma} = (2h\hat{n})^{-1/2}$. The number of modes in the case of the Gaussian line can be calculated as follows (compare with Chap. 6):

$$M = \frac{\hat{\sigma}\sqrt{\pi}}{\Delta \hat{C}_{\text{mode}}} = \frac{2L}{N_{\text{w}} \lambda \sqrt{2\pi h \hat{n}}} . \qquad (3.111)$$

Now we consider field correlations in the time domain. The field at a given space point can be presented in the following form:

$$E_x + iE_y = \tilde{E}(t) \exp(-i\omega_{\max} t) ,$$

where $\omega_{\max}$ corresponds to the maximum of the gain curve (in other words, it corresponds to the normalized detuning $\hat{C}_{\max}$). The first-order time correlation function, $g_1(t - t')$, is an important characteristic of the radiation field:

$$g_1(t - t') = \frac{\langle \tilde{E}(t)\tilde{E}^*(t') \rangle}{\left[\langle |\tilde{E}(t)|^2 \rangle \langle |\tilde{E}(t')|^2 \rangle \right]^{1/2}} .$$

In the framework of the accepted limitations the characteristics of the random process of $\tilde{E}(t)$ (spectrum width, correlation functions, etc.) change slowly with the value of n. In other words, this process can be considered as a quasi-stationary one. Thus, the Fourier transform of (3.104) gives us the value of the first-order time correlation function (see Chap. 6 for more details):

$$g_1(t - t') = \frac{\int\limits_{-\infty}^{\infty} d(\Delta\omega) H(\Delta\omega) \exp\left[-i\Delta\omega(t - t')\right]}{\int\limits_{-\infty}^{\infty} d(\Delta\omega) H(\Delta\omega)} ,$$

where $\Delta\omega = \omega - \omega_{\max}$. The latter expression can be rewritten as

$$g_1(t - t') = \frac{\int\limits_{-\infty}^{\infty} d(\Delta\hat{C}) \, H(\Delta\hat{C}) \, \exp\left[-i\Delta\hat{C}(t - t')c/(\lambda N_{\text{w}})\right]}{\int\limits_{-\infty}^{\infty} d(\Delta\hat{C}) \, H(\Delta\hat{C})} . \qquad (3.112)$$

The integration yields

$$g_1(t-t') = \exp\left[-\frac{c^2(t-t')^2}{4h\hat{n}\lambda^2 N_w^2}\right]. \tag{3.113}$$

The coherence time is defined by the expression

$$\tau_c = \int_{-\infty}^{\infty} \mathrm{d}(t-t')|g_1(t-t')|^2$$

and is equal to

$$\tau_c = \frac{N_w\lambda\sqrt{2\pi h\hat{n}}}{c}. \tag{3.114}$$

Comparing (3.111) and (3.114), we find that

$$M = \frac{2L}{c\tau_c},$$

i.e. the number of longitudinal modes is equal to the number of independent wavepackets, simultaneously existing inside the resonator.

The number of longitudinal modes decreases in time due to narrowing of the spectrum envelope. Nevertheless, it always remains large in the linear regime for any reasonable set of FEL oscillator parameters. This can be seen if one takes into account the estimate (3.108) for the number of round-trips corresponding to the total duration of the linear mode of operation.

3.5 Concluding Remarks

The results of the nonlinear theory, presented in this chapter, are obtained in the framework of the single-mode model. However, one should keep in mind that this model provides valid results only for a certain region of parameters of the FEL oscillator. The region of applicability of the results obtained can be found only in the framework of the multi-mode model. Detailed study of the problem is beyond the scope of our book, and we present here only a summary for the case of the FEL oscillator with uniform undulator.

It has been shown in the previous section that the number of longitudinal modes decreases in time when the FEL oscillator operates in the linear regime. This tendency also occurs in the nonlinear regime for values of the threshold coefficient $1 < K_{\mathrm{th}} < 3.6$ (corresponding to the reduced damping factor $0.135 > \hat{\alpha} > 0.037$). In this region of the parameters the spectrum width is much less than the FEL gain curve width, thus the single-mode model is applicable.

For larger values of the threshold coefficient, $K_{\mathrm{th}} > 3.6$ (or $\hat{\alpha} < 0.037$), spectrum broadening occurs in the deep nonlinear regime. This effect is usually referred to as sideband instability (or trapped-particle instability), and is due to synchrotron oscillations of the electrons in the effective potential of

150   3. One-Dimensional Theory of the FEL Oscillator

interaction in the strong electromagnetic field stored in the resonator. These oscillations lead to modulation of the radiation field at the frequencies of the synchrotron oscillations and, therefore, to the growth of the spectral components (sidebands) shifted by these frequencies with respect to the frequencies of the stored radiation field. The increase of K_{th} leads to an increase of the radiation field stored in the resonator. For the values of $K_{th} > 3.6$, the amplification of the shifted spectral components becomes larger than the losses in the resonator and they start to grow.

Summarizing this discussion, we can state that the single-mode model always provides valid results for the FEL oscillator with a uniform undulator when the value of the threshold coefficient is within the limits $1 < K_{th} < 3.6$. Sideband instability takes place for larger values of this parameter. In principle, this unwanted nonlinear phenomenon can be suppressed by introducing filtering frequency elements into the resonator, for instance, a Littrow-mounted diffraction grating. As a result, it becomes possible to apply the single-mode model for the calculation of such an FEL oscillator.

It is worth mentioning that the one-dimensional model of the FEL oscillator is widely used for the calculation of practical devices. The reason for this is that it can be simply generalized for taking into account diffraction effects. As a rule, the optical resonator of an FEL oscillator is formed by two spherical mirrors. At a small value of the field gain, the electron beam does not significantly affect the field eigenmode of the resonator and it remains to be close to that of the empty resonator. As a result, the FEL process can be calculated in the approximation of the given field. To take into account the field distribution in the resonator, the notion of a filling factor is introduced. For instance, when the field eigenmode in the resonator is a Gaussian TEM_{00} mode, the field distribution in the waist is of the form:

$$|\tilde{E}| \propto \exp(-r^2/w^2) ,$$

where r is the transverse coordinate and w is the size of the waist. When the transverse size of the electron beam is much less than w and when we can neglect the change of the Gaussian mode along the undulator, we replace in the definition of the gain parameter the value of the beam current density, j_0, by an effective value

$$(j_0)_{\text{eff}} = 2I_0/(\pi w^2) ,$$

where I_0 is the total beam current. One should also keep in mind that the space charge parameter should be calculated using the actual beam current density. In many practical situations this approach provides sufficient accuracy of the calculations.

In this chapter we have assumed that the driving electron beam is continuous in time. A frequent practical situation is when the driving electron accelerator produces a train of periodically spaced electron bunches. Let the electron bunch duration be equal to T_b and the bunch spacing be equal to $T_e \simeq 2L/c$, where L is the resonator base. Let us consider a rectangular bunch.

3.5 Concluding Remarks

The results, presented in this chapter, are applicable when the electron pulse duration is much larger than the coherence time τ_c (3.114),

$$T_b/\tau_c \gg 1 \ . \tag{3.115}$$

Now let us consider the electron pulse with a gradient axial profile of the current $I_0(t)$. Let the typical duration of such a pulse T_b satisfy the condition (3.115). Then this situation can be described in the following way. As an approximation, the smooth profile $I_0(t)$ may be replaced on the interval $(0, T_b)$ by a "boxcar" function. The interval $(0, T_b)$ is divided into N_s subintervals of equal length. Within each subinterval, the approximation to $I_0(t)$ is constant. At the end of each subinterval, the approximate profile jumps to a new constant value. When

$$T_b/(\tau_c N_s) \gg 1 \ ,$$

we can simulate the FEL process separately within each subinterval.

In Sect. 3.3 we analyzed the problem of the efficiency increase by means of undulator tapering. The study has shown that the problem of the efficiency increase is not a trivial one, and the application of undulator tapering provides only limited possibilities for an increase in efficiency. At least, for the simplest FEL oscillator without an external monochromatization of the field, the undulator tapering allows one to increase the efficiency by a factor of about 3 only. Installation of the prebuncher increases the efficiency additionally by a factor of 2. In any case, the maximal FEL oscillator efficiency does not exceed the value of a few per cent (the efficiency of the FEL oscillator with uniform undulator is usually below 1%).

On the other hand, the efficiency of the FEL amplifier with tapered undulator can be made about unity (see Chap. 2). Such a significant difference in the efficiency between the FEL amplifier and FEL oscillator is connected with the principal difference between these two FEL configurations when undulator tapering is used. In the case of the FEL amplifier, the frequency of the amplified wave is determined by a master oscillator, the initial conditions at the undulator entrance are fixed, the process of field amplification develops in space, and, as a result, spatial tapering of the undulator parameters enables one to trap a significant fraction of the electrons in the regime of coherent deceleration. In contrast to this, the lasing frequency of the FEL oscillator depends on the value of the undulator tapering depth and is defined by the condition of the maximum of the small-signal gain in the linear mode of operation. The physically obvious and technically possible solution of this problem might be to fix the desired frequency using the external monochromatization (for instance, by means of a Littrow-mounted diffraction grating). Using a narrow-band external monochromatization might also be useful for suppression of the spectrum broadening (or, sideband instability) in the FEL oscillator operating in the nonlinear regime.

We note that the initial conditions at the undulator entrance depend on the time due to the dependence on time of the field stored in the resonator of the FEL oscillator. If one optimizes the parameters in order to obtain higher gain in the strong optical field, the gain may be too small in the weak field, and vice versa. The natural solution of this problem is to change in time the depth of the undulator tapering by changing in time the magnetic field of individual undulator periods independently. At the beginning of the lasing process the undulator is uniform, thus providing the maximal small-signal gain. When the radiation field in the resonator achieves its maximum, the undulator field begins to change in time along the undulator axis. The increase of the tapering depth leads to an increase of the radiation field amplitude. At some value of the tapering depth, the maximum value of the radiation field is achieved. At this moment the process of undulator tapering is stopped and the FEL oscillator operates in the saturation regime with high efficiency. The time interval required to reach the high efficiency stationary regime depends on the rate of change of the undulator magnetic field and can be large in practice. Thus, the driving electron beam should be continuous or quasi-continuous and may be generated, for instance, by a superconducting accelerator.

There is another method for time-dependent variation of the FEL oscillator parameters. It consists of introducing time-dependent RF accelerating fields into the interaction region. From a physical point of view this method is equivalent to undulator tapering. For a correct choice of the FEL parameters, a significant fraction of the electrons is trapped in the effective potential well, thus increasing the FEL efficiency. Recent studies have shown that both options of a high-efficiency FEL oscillator can be realized technically. Numerical simulations show that an efficiency of about several tens of a per cent can be achieved. To suppress sideband instability one can use external monochromatization as mentioned above.

3.6 Suggested Bibliography

The bibliography devoted to the theory and experiments with FEL oscillators is extremely large. Here we present some selected references to the original papers which can help the reader to study the subject in more detail.

The small-signal gain formula was derived using a quantum approach in [3.1]. Later it was obtained in the framework of a classical approach [3.2, 3.3]. More detailed investigations of the small-signal gain has been performed in [3.4–3.8].

Similarity techniques in the theory of the FEL oscillator were used in [3.9–3.11].

The optical klystron was proposed in [3.12].

The analysis of undulator tapering in the FEL oscillator, presented in this book, is based on the results published in the paper [3.13], where the reader can also find additional information.

Investigations devoted to the analysis of start-up from noise are described in [3.14, 3.15]

There are several papers [3.16–3.19] dealing with the ideas of constructing a high-efficiency FEL oscillator. We have only briefly mentioned in this book that a high-efficiency FEL oscillator can be constructed by application of time-dependent tapering or time-dependent RF fields. The reader can find additional information in the above-mentioned papers.

Also, we only briefly touched on the problem of sideband instabilities which might occur in the nonlinear mode of FEL oscillator operation. More information on this subject can be found in [3.20, 3.21]. A proof-of-principle experiment using grating in an FEL for sideband suppression was reported in [3.22].

In this chapter we did not touch on finite-pulse effects in the FEL oscillator. We refer the reader to [3.23–3.28] which contain a detailed study of this problem. We should also note that the time-dependent approach presented in Chap. 6 can be simply upgraded for time-dependent simulations of the FEL oscillator.

To be exact, the results presented in this chapter cannot be applied to FEL oscillator installed at a storage ring. The present model assumes that a fresh electron beam interacts with the radiation at each resonator round-trip. This is not the case for a storage-ring FEL where the radiation interacts with the same electron beam during each round-trip. For a correct description of the storage-ring FEL, one should take into account the dynamics of the electron beam in the storage ring, which significantly complicates the problem. Relevant studies can be found in the original papers [3.29–3.32].

4. Diffraction Effects in the FEL Amplifier

In this chapter we extend the analysis of the FEL amplifier taking into account diffraction effects. The analyzed models are based on Maxwell's wave equation taken in the paraxial approximation and the description of the electron beam with the Vlasov equation. It is anticipated that electrons move (on the average over constrained motion) only along trajectories parallel to the undulator axis. It is assumed that the waveguide, or vacuum chamber walls, are placed far enough from the electron beam, formally at infinity. Such a model of an open electron beam allows one to simplify significantly the description of the amplification process taking into account important physical effects, such as diffraction of radiation, space charge fields, and energy spread of electrons in the beam. The approximation of the open beam describes rather well FEL amplifiers operating in the wavelength range from infrared down to X-ray.

The physics of the FEL amplifier with an open electron beam is much richer than that of the simplified one-dimensional model considered in Chap. 2. In particular, an interesting physical effect takes place in the linear mode of FEL amplifier operation. The radiation, produced by the FEL amplifier in the high-gain linear regime, can be presented as a superposition of the beam radiation modes. The parameters of these modes are defined by the electron beam parameters (peak current, details of the transverse distribution of the beam current, etc.) and by the parameters of the undulator. The transverse distribution of the field in each beam radiation mode is fixed, while the amplitude grows exponentially with undulator length. At a relatively large undulator length only one mode survives, having a maximal field growth rate. This effect, predicted at the beginning of 1980s, is now called optical guiding.

An effect similar to the optical guiding effect occurs in optical fibers. When the optical beam propagates in the vacuum, it diverges due to diffraction effects. Typical angular divergence is about of $\Delta\theta \simeq \lambda/D$, where D is the transverse size of the optical beam. The influence of the diffraction effects is reduced significantly when the optical beam propagates in a dielectric fiber. As a rule, this is connected with the finite value of the reflection coefficient on the boundary of the waveguide. The guiding effect can also take place in an active medium having maximum amplification on the axis. In the latter case the guiding effect is due to the following reasons. The first is the permanent radiation of the energy in the core of the active medium. The second reason is the partial reflection of the radiation due to the inhomogeneous distribution of the complex reflection index.

When the FEL amplifier operates in the linear regime, the driving electron beam can be considered as an active medium whose properties do not depend on the longitudinal z coordinate. Let us analyze the nature of the self-consistent solution of Maxwell's equations and the Vlasov equation at a fixed frequency ω. The electric field of the wave amplified in the helical undulator may be represented in the complex form:

$$E_x + \mathrm{i}E_y = \tilde{E}(z, \boldsymbol{r}_\perp) \exp[\mathrm{i}\omega(z/c - t)] \ . \tag{4.1}$$

In the high-gain limit the complex amplitude $\tilde{E}(z, \boldsymbol{r}_\perp)$ can be reduced to

$$\tilde{E}(z, \boldsymbol{r}_\perp) = \Phi(\boldsymbol{r}_\perp) \exp(\Lambda z) \ . \tag{4.2}$$

The radiation field must have finite energy, which leads to the requirement of quadratic integrability of the function $\Phi(\boldsymbol{r}_\perp)$. An important consequence of the latter requirement is that the set of propagation constants, Λ, becomes discrete. Also, a specific solution for $\Phi(\boldsymbol{r}_\perp)$ corresponds to each value of Λ. In other words, the radiation of the electron beam in the undulator can be represented as a set of partial waves. In the following we call these waves the beam radiation (or guided) modes. During the amplification process the transverse distribution of the field of the beam radiation mode remains fixed, while its amplitude grows exponentially with undulator length.

Let us consider the case when the electromagnetic wave with fixed frequency ω is fed to the undulator entrance:

$$(E_x + \mathrm{i}E_y)|_{z=0} = \tilde{E}_{\mathrm{ext}}(\boldsymbol{r}_\perp) \exp(-\mathrm{i}\omega t) \ ,$$

At a sufficient distance from the undulator entrance the radiation can be presented as a superposition of the exponentially growing beam radiation modes

$$\tilde{E}(z, \boldsymbol{r}_\perp) = \sum_j A_j \Phi_j(\boldsymbol{r}_\perp) \exp(\Lambda_j z) \ ,$$

where Λ_j and $\Phi_j(\boldsymbol{r}_\perp)$ are the eigenvalues and the eigenfunctions of the beam radiation modes, respectively, and $\mathrm{Re}\,\Lambda_j > 0$. In the following sections we illustrate the methods of finding the eigenvalues and the eigenfunctions by means of the solution of the eigenvalue problem. The coefficients A_j are calculated by means of the solution of the initial-value problem.

We mentioned above that there is an analogy between the beam radiation modes and the modes of an optical waveguide. Nevertheless, such an analogy is not complete. In particular, the modes of the optical waveguide are orthogonal, while the beam radiation modes are not. This difference comes from the different physical origin of these effects. Indeed, the eigenmodes of the optical waveguide are the solution of Maxwell's equations, while the beam radiation

4. Diffraction Effects in the FEL Amplifier

modes generated by the electron beam in the undulator are the self-consistent solution of Maxwell's equations and the Vlasov equation. For instance, the modes of the optical waveguide are the solution of the equation

$$\nabla^2 \boldsymbol{E} + \left(\frac{\omega}{c} n'(\boldsymbol{r}_\perp)\right)^2 \boldsymbol{E} = 0 \ ,$$

where $n'(\boldsymbol{r}_\perp)$ is the refractive index of the dielectric waveguide. In order to obtain an analogy with the FEL amplifier, we consider the case of an overmoded waveguide, and use the paraxial approximation. Using (4.1) for the circularly polarized radiation, we rewrite the latter equation as

$$\nabla^2_\perp \tilde{E} + 2\mathrm{i}\frac{\omega}{c}\frac{\partial \tilde{E}}{\partial z} + \frac{\omega^2}{c^2}\chi(\boldsymbol{r}_\perp)\tilde{E} = 0 \ ,$$

where $\chi = (n')^2 - 1$. We seek the solution of this equation in the form of (4.2). The eigenfunction of the waveguide, $\Phi(\boldsymbol{r}_\perp)$, is the solution of the equation

$$\nabla^2_\perp \Phi(\boldsymbol{r}_\perp) + 2\mathrm{i}\frac{\omega}{c}\Lambda\Phi(\boldsymbol{r}_\perp) + \frac{\omega^2}{c^2}\chi(\boldsymbol{r}_\perp)\Phi(\boldsymbol{r}_\perp) = 0 \ . \tag{4.3}$$

Now let us consider the amplification process in the undulator. The electric field of the wave in the helical undulator is subjected to the integro-differential equation

$$\nabla^2_\perp \tilde{E} + 2\mathrm{i}\frac{\omega}{c}\frac{\partial \tilde{E}}{\partial z} = \mathrm{const.} j_0(\boldsymbol{r}_\perp) \int_0^z \mathrm{d}z' \tilde{E}(z', \boldsymbol{r}_\perp) G(z - z') \ ,$$

where $j_0(\boldsymbol{r}_\perp)$ is the beam current density. Substituting (4.2) into the latter equation, in the limit of

$$|\exp(\Lambda z)| \gg 1 \ ,$$

we obtain the following equation:

$$\nabla^2_\perp \Phi(\boldsymbol{r}_\perp) + 2\mathrm{i}\frac{\omega}{c}\Lambda\Phi(\boldsymbol{r}_\perp) + \frac{\omega^2}{c^2}\chi_{\mathrm{eff}}(\boldsymbol{r}_\perp, \Lambda)\Phi(\boldsymbol{r}_\perp) = 0 \ . \tag{4.4}$$

At first glance, this equation is similar to (4.3). Nevertheless, there is a principal difference. In the case of an optical waveguide, the function $\chi(\boldsymbol{r}_\perp)$ (see (4.3)) takes the same values for all modes, while $\chi_{\mathrm{eff}}(\boldsymbol{r}_\perp, \Lambda)$ appearing in (4.4) depends on the eigenvalue Λ. As a result, the eigenfunctions of the beam radiation modes are not orthogonal.

The chapter is organized as follows. We begin with a general derivation of the self-consistent equations describing the linear mode of FEL amplifier operation. They are obtained by means of the self-consistent solution of the Vlasov equation and Maxwell's equations written down in the paraxial approximation. The self-consistent equations are obtained in two equivalent forms: for the field amplitude and for the first harmonic of the beam current density. In principle, these self-consistent equations can be solved numerically to obtain specific results at specific parameters of the FEL amplifier. Nevertheless, even though they give us the possibility of obtaining specific

158   4. Diffraction Effects in the FEL Amplifier

results, numerical methods do not provide deep insight into FEL physics. In this chapter we concentrate mainly on those physical situations which can be investigated analytically.

We start the analytical investigation with the simplest model of the FEL amplifier with a sheet electron beam. This model is important from the methodological point of view, since it allows one to investigate the physics of the FEL amplifier with simple mathematics. In particular, expressions for the eigenvalue equations and the beam radiation modes in the case of a stepped profile of the electron beam contain elementary functions. Despite its simplicity, such a model allows one to study the effect of optical guiding and other important physical effects influencing the operation of the FEL amplifier. Then we complicate the considerations with the analysis of electron beams with a gradient profile of the current density. Analytical study is performed using the model of the electron beam with the Epstein profile of the beam current density. It is interesting to note that even in this complicated case the eigenvalue equation and the expressions for the beam radiation modes are written in terms of elementary functions. The next profile allowing analytical solution is a parabolic profile. Finally, we analyze the FEL amplifier with an arbitrary gradient profile. Such an analysis is performed with the multilayer approximation method. This semi-analytical method allows one to perform a complete description of the FEL amplifier with an arbitrary gradient profile of the electron beam. Analytical solutions obtained for the Epstein profile are used to check the accuracy of the multilayer approximation method.

The second part of the chapter is devoted to the description of the FEL amplifier with an axisymmetric electron beam. Taking into account the greater practical importance of this model, we perform a more detailed analysis. As in the case of a sheet electron beam, we begin our study with a stepped profile of the electron beam. The eigenvalue problem and the initial-value problem are solved analytically. The optimal conditions of the input radiation focusing on the electron beam are found for a Gaussian laser beam.

Then the consideration is complicated with the case of an FEL amplifier with an arbitrary gradient profile of the electron beam. The case of the parabolic profile is studied in detail, allowing analytical solution of the eigenvalue problem. The general case of an arbitrary gradient profile is studied by means of the multilayer approximation method. The accuracy of the multilayer method is controlled by means of analytical results for the parabolic profile. The important practical case of a Gaussian beam profile is studied numerically. We finish the study of the linear mode of FEL amplifier operation by presenting an algorithm for the numerical solution of the initial-value problem.

We complete the study of the FEL amplifier with the analysis of the nonlinear mode of operation. The macroparticle method, developed in Chap. 2, is extended to the three-dimensional case. Further extension of the similarity techniques is discussed and the reduced self-consistent system of the FEL amplifier equations is formulated, taking into account the space charge fields, energy spread, and diffraction effects. Then, using the results of numerical calculations presented in reduced form, we analyze various features of the FEL amplifier in the nonlinear mode.

4.1 Self-Consistent Equations

Let us consider a relativistic electron beam moving along the z axis in the field of a helical undulator. We neglect the transverse variation of the undulator field and assume the electrons move along constrained helical trajectories in parallel with the z axis. The electron rotation angle is considered to be small and the longitudinal electron velocity v_z is close to the velocity of light c ($v_z \simeq c$). The electric field of the amplified wave may be represented in the complex form of (4.1). We describe the electron motion using energy-phase variables $\mathcal{E}$ and $\psi = k_\mathrm{w} z + \omega(z/c - t)$. In Chap. 2 we started the study with the derivation of the Hamiltonian and then used it for writing down the canonical equations of motion. Here we begin the study by writing down the equations of motion. The relevant value of ψ is that at the location of the particle. Hence, the total derivative of the phase ψ is given by

$$\frac{d\psi}{dz} = \frac{\partial \psi}{\partial z} + \frac{\partial \psi}{\partial t}\frac{dt}{dz} \ .$$

The explicit form for this derivative is

$$\frac{d\psi}{dz} = k_\mathrm{w} + \frac{\omega}{c} - \omega \frac{dt}{dz} = k_\mathrm{w} + \frac{\omega}{c} - \frac{\omega}{v_z(\mathcal{E})} \ .$$

When writing down this equation, we assumed that the radiation field does not disturb the transverse constrained motion of the electron in the undulator. If the energy $\mathcal{E}$ does not differ significantly from the nominal value $\mathcal{E}_0$, we can write the following expression for the total derivative:

$$\frac{d\psi}{dz} = k_\mathrm{w} + \frac{\omega}{c} - \frac{\omega}{v_z(\mathcal{E}_0)} + \frac{\omega}{v_z^2(\mathcal{E}_0)}\frac{dv_z}{d\mathcal{E}}(\mathcal{E} - \mathcal{E}_0) \ . \tag{4.5}$$

The rate of electron energy change is given by the expression:

$$\frac{d\mathcal{E}}{dt} = -e\boldsymbol{v}\cdot\boldsymbol{E} \ ,$$

so, we can write the total derivative $d\mathcal{E}/dz$ in the following form:

$$\frac{d\mathcal{E}}{dz} = -e\frac{\boldsymbol{v}\cdot\boldsymbol{E}}{v_z} = -eE_z - e\theta_\mathrm{s}\tilde{E}\mathrm{e}^{i\psi} + \mathrm{C.C.} \ , \tag{4.6}$$

where $E_z(\psi, z, \boldsymbol{r}_\perp)$ is the longitudinal space charge field and $\theta_\mathrm{s} = eH_\mathrm{w}/(\mathcal{E}_0 k_\mathrm{w})$ is the rotation angle of the electron with nominal energy.

Analyzing (4.5) and (4.6), we can conclude that these equations are Hamilton's canonical equations corresponding to the following Hamiltonian:

4. Diffraction Effects in the FEL Amplifier

$$H(P, \psi, z) = CP + \frac{\omega}{2c\gamma_z^2 \mathcal{E}_0} P^2 - \left(U e^{i\psi} + U^* e^{-i\psi}\right) + \int d\psi e E_z , \qquad (4.7)$$

where $P = \mathcal{E} - \mathcal{E}_0$, $C = k_w - \omega/(2c\gamma_z^2)$ is the detuning of the electron with the nominal energy $\mathcal{E}_0$, $U = -e\theta_s \tilde{E}(z, \boldsymbol{r}_\perp)/(2i)$ is the complex amplitude of the effective potential of the particle interaction with the electromagnetic wave, and

$$\gamma_z^{-2} = \gamma^{-2} + \theta_s^2 , \qquad \gamma = \mathcal{E}_0/(m_e c^2) .$$

The effective Hamiltonian (4.7) seems to be similar to that obtained in Chap. 2 (see (2.6)), but we should note some diffferences. In the present consideration we have neglected the influence of the radiation field on the transverse constrained motion of the electron in the undulator. This is equivalent to neglecting the term $(Ue^{i\psi} + \text{C.C.})P/\mathcal{E}_0$ in (2.6) which is valid for the low-efficiency approximation. In this chapter we do not go beyond the margins of this approximation. Another important difference is that now the radiation field and the space charge field depend on the transverse coordinate $\boldsymbol{r}_\perp$. So, in the present consideration these fields are three-dimensional. At the same time, the dynamical equations (4.5) and (4.6) corresponding to the Hamiltonian (4.7) are one-dimensional. This means that from the point of view of dynamics the transverse coordinate $\boldsymbol{r}_\perp$ is a parameter, but not a dynamical variable.

We describe the evolution of the electron beam distribution function by the Vlasov equation

$$\frac{\partial f}{\partial z} + \frac{\partial H}{\partial P} \frac{\partial f}{\partial \psi} - \frac{\partial H}{\partial \psi} \frac{\partial f}{\partial P} = 0 , \qquad (4.8)$$

where $f(\psi, P, z, \boldsymbol{r}_\perp)$ now depends on the parameter $\boldsymbol{r}_\perp$ which is averaged over the undulator period transverse coordinate. We have already mentioned above that the transverse coordinated is not treated as a dynamical variable in this model.

For the electron beam with small density perturbation we seek the solutions for f and E_z in the form:

$$f = f_0 + \tilde{f}_1 e^{i\psi} + \tilde{f}_1^* e^{-i\psi} , \qquad E_z = \tilde{E}_z e^{i\psi} + \tilde{E}_z^* e^{-i\psi} ,$$

and obtain the following linear equation for the amplitude $\tilde{f}_1$:

$$\frac{\partial \tilde{f}_1}{\partial z} + i\left[C + \frac{\omega P}{c\gamma_z^2 \mathcal{E}_0}\right] \tilde{f}_1 + \left(iU - e\tilde{E}_z\right) \frac{\partial f_0}{\partial P} = 0 . \qquad (4.9)$$

In what follows we use the following assumptions:

- complex amplitudes of the electric field and of the first harmonic of the beam current density are slowly varying functions;
- the transverse size of the electron beam is large, i.e. $r_b^2 \gg \gamma_z^2 c^2/\omega^2$;
- the electron beam at the undulator entrance is modulated neither in velocity nor density, i.e.

$$\tilde{f}_1|_{z=0} = 0, \qquad f_0 = n_0(\boldsymbol{r}_\perp)F(P) \,, \tag{4.10}$$

where the initial distribution function in energy, $F(P)$, is normalized to unity. In the ultrarelativistic approximation $v_z \simeq c$, so the beam current density and the distribution function are related as:

$$j_z = -j_0(\boldsymbol{r}_\perp) + \tilde{j}_1 e^{i\psi} + \text{C.C.} \,, \qquad \tilde{j}_1 \simeq -ec \int \tilde{f}_1 \mathrm{d}P \,,$$

where $-j_0(\boldsymbol{r}_\perp) \simeq -ecn_0(\boldsymbol{r}_\perp)$ is the longitudinal component of the beam current density at the undulator entrance at $z = 0$.

In the framework of the accepted limitation on the transverse size of the electron beam, $r_\mathrm{b}^2 \gg c^2\gamma_z^2/\omega^2$, the longitudinal space charge field is defined by the local modulation of the beam density (see Appendix A.2):

$$\tilde{E}_z = -\mathrm{i}\frac{4\pi}{\omega}\tilde{j}_1(z,\boldsymbol{r}_\perp) \,. \tag{4.11}$$

The complex amplitude $\tilde{j}_1$ can be obtained by integration of the kinetic equation (4.9). Using the initial conditions (4.10) and expression (4.11), we find

$$\tilde{j}_1(z,\boldsymbol{r}_\perp) = \mathrm{i}j_0(\boldsymbol{r}_\perp)\int_0^z \mathrm{d}z' \left[U(z',\boldsymbol{r}_\perp) + 4\pi e\tilde{j}_1(z',\boldsymbol{r}_\perp)/\omega\right]$$

$$\times \int_{-\infty}^{\infty} \mathrm{d}P\frac{\mathrm{d}F(P)}{\mathrm{d}P} \exp\left[\mathrm{i}\left(C + \omega P/(c\gamma_z^2\mathcal{E}_0)\right)(z'-z)\right] \,. \tag{4.12}$$

Here it is important to note the following. When the condition $r_\mathrm{b}^2 \gg \gamma_z^2 c^2/\omega^2$ is satisfied, the radius of the electron rotation in the undulator, $r_\mathrm{w} = \theta_\mathrm{s}/k_\mathrm{w}$, is always much less than the transverse size of the electron beam. Indeed, taking into account that $\theta_\mathrm{s} = K/\gamma$, we can write

$$\frac{r_\mathrm{b}^2}{r_\mathrm{w}^2} \simeq \frac{1+K^2}{K^2}\left(\frac{r_\mathrm{b}^2\omega^2}{c^2\gamma_z^2}\right) \gg 1 \,.$$

Now we should consider the electrodynamic problem. Using Maxwell's equations, we can write the equation for the electric field:

$$\boldsymbol{\nabla} \times (\boldsymbol{\nabla} \times \boldsymbol{E}) = -\frac{1}{c^2}\frac{\partial^2 \boldsymbol{E}}{\partial t^2} - \frac{4\pi}{c^2}\frac{\partial \boldsymbol{j}}{\partial t} \,.$$

Then we should take into account that

$$\boldsymbol{\nabla} \times (\boldsymbol{\nabla} \times \boldsymbol{E}) = \boldsymbol{\nabla}(\boldsymbol{\nabla} \cdot \boldsymbol{E}) - \boldsymbol{\nabla}^2 \boldsymbol{E} \,.$$

The value of $\boldsymbol{\nabla} \cdot \boldsymbol{E}$ can be found from the Poisson equation. Finally, we come to the inhomogeneous wave equation for $\boldsymbol{E}$:

$$\boldsymbol{\nabla}^2 \boldsymbol{E} - \frac{1}{c^2}\frac{\partial^2 \boldsymbol{E}}{\partial t^2} = 4\pi\boldsymbol{\nabla}\rho_\mathrm{e} + \frac{4\pi}{c^2}\frac{\partial \boldsymbol{j}}{\partial t} \,. \tag{4.13}$$

This equation allows one to calculate the electric field $\boldsymbol{E}(\boldsymbol{r},t)$ for given charge and current sources, $\rho_\mathrm{e}(\boldsymbol{r},t)$ and $\boldsymbol{j}(\boldsymbol{r},t)$. Since in the paraxial approximation the radiation field has only transverse components, we are interested in the transverse component of (4.13). Let us show that the second term (or, the current term) in the right-hand side of (4.13) provides the main contribution to the value of the radiation field. The contribution of the charge term $\nabla_\perp \rho_e$ in (4.13) is negligibly small when

$$|\nabla_\perp \rho_e| z^{(c)} \left[\omega c^{-2} |\bm{j}_\perp| z^{(j)} \right]^{-1} \ll 1 , \tag{4.14}$$

where $z^{(c)}$ and $z^{(j)}$ are typical formation lengths of the field produced by the charge and the current sources, respectively. The current term is a resonant one, while the charge term is not. So, the typical formation length of the radiation produced by the current term, $z^{(j)}$, is of about the gain length, $l_g = (\mathrm{Re}\, \Lambda)^{-1}$ (see (4.2)):

$$z^{(j)} \simeq l_g = (\mathrm{Re}\, \Lambda)^{-1} .$$

The typical formation length of the radiation produced by the charge sources, $z^{(c)}$, is of the order of

$$z^{(c)} \simeq \lambda/(1 - v_z/c) \simeq k_w^{-1} .$$

Taking into account that

$$|\nabla_\perp \rho_e| \simeq \rho_e/r_b , \quad \omega |\bm{j}_\perp|/c^2 \simeq \theta_s \omega \rho_e/c ,$$

we find that condition (4.14) is equivalent to

$$(k_w l_g)^{-1} (r_b \omega \theta_s/c)^{-1} \ll 1 . \tag{4.15}$$

It is relevant to remember that the self-consistent equations have been derived using the following assumptions:

$$\gamma_z^2 \gg 1 , \quad k_w l_g \gg 1 , \quad \omega^2 r_b^2/(c^2 \gamma_z^2) \gg 1 . \tag{4.16}$$

It can be shown that when these conditions are fulfilled, condition (4.15) is fulfilled, too. Thus, the charge term can be neglected in (4.13). Let us perform simple estimates for the important practical case of the value of the undulator parameter $K \gtrsim 1$. Taking into account that $\gamma_z \simeq \theta_s^{-1}$ in this case, we find that (4.15) can be written as

$$\left(\frac{1}{k_w l_g} \right) \left(\frac{\gamma_z c}{r_b \omega} \right) \ll 1 . \tag{4.17}$$

Taking into account the limitations (4.16), we find that each of the two terms in the left-hand side of (4.17) is small. This means that we can neglect the charge term in (4.13). Similar estimates can be made for the case of small values of the undulator parameters, $K \to 0$. Here the problem is that the gain length, l_g, tends to infinity, and we come to the limit $0 \times \infty$ in relation (4.15). To resolve this indeterminacy, we need more specific knowledge about the dependence of the gain length on the problem parameters. This question can be answered after reading this chapter, since the diffraction effects are essential in the limit of $K \to 0$. The reader can find that condition (4.15) is always valid under the limitations (4.16).

Thus, we consider the following wave equation:

$$c^2 \nabla^2 \boldsymbol{E}_\perp - \partial^2 \boldsymbol{E}_\perp / \partial t^2 = 4\pi \partial \boldsymbol{j}_\perp / \partial t , \qquad (4.18)$$

where the perturbed transverse current density is given by

$$\boldsymbol{j}_\perp = \theta_{\rm s} \left[\boldsymbol{e}_x \cos(k_{\rm w} z) - \boldsymbol{e}_y \sin(k_{\rm w} z) \right] \left(\tilde{j}_1 e^{i\psi} + {\rm C.C.} \right) . \qquad (4.19)$$

The electric field, $\boldsymbol{E}_\perp$, is also presented in the complex form (4.1). To simplify (4.18), we assume that the complex amplitude $\tilde{j}_1$ is a slowly varying function on the scale of the undulator period. Then, for $k_{\rm w} z \gg 1$ we can neglect rapidly oscillating terms in the right-hand side, and omit the second derivative of $\tilde{E}$ with respect to z in the left-hand side of (4.18). Finally, we obtain the equation for the slowly varying amplitudes $\tilde{j}_1$ and $\tilde{E}$:

$$c^2 \left[\nabla^2_\perp + 2i(\omega/c) \partial/\partial z \right] \tilde{E} = -4\pi i \theta_{\rm s} \omega \tilde{j}_1 , \qquad (4.20)$$

where $\nabla^2_\perp$ is the Laplace operator in transverse coordinates. When the right-hand side of (4.20) is equal to zero, it transforms to the well-known paraxial wave equation in optics.

So, we have obtained the system of self-consistent field equations (4.12) and (4.20). This system can be solved by two methods. First, we can substitute (4.12) into the right-hand side of (4.20). Then, expressing the function $\tilde{j}_1(z', \boldsymbol{r}_\perp)$ in terms of (4.20), we obtain an integro-differential equation for the field amplitude $\tilde{E}$:

$$\nabla^2_\perp \tilde{E} + 2i \frac{\omega}{c} \frac{\partial \tilde{E}}{\partial z} = i j_0(\boldsymbol{r}_\perp) \int_0^z dz' \left\{ \frac{2\pi e}{c^2} \theta_{\rm s}^2 \omega \tilde{E}(z', \boldsymbol{r}_\perp) \right. $$

$$\left. + \frac{4\pi e}{\omega} \left[\nabla^2_\perp \tilde{E} + 2i \frac{\omega}{c} \frac{\partial \tilde{E}}{\partial z'} \right] \right\}$$

$$\times \int_{-\infty}^\infty dP \frac{dF}{dP} \exp \left[i \left(C + \frac{\omega}{\gamma_z^2 \mathcal{E}_0 c} P \right) (z' - z) \right] . \qquad (4.21)$$

Another method consists in deriving the integral equation for the first harmonic $\tilde{j}_1(z, \boldsymbol{r}_\perp)$. In this case we solve (4.20) with respect to $\tilde{E}(z', \boldsymbol{r}_\perp)$, then substitute this solution into (4.12), and obtain an integral equation for the first harmonic of the beam current density $\tilde{j}_1(z, \boldsymbol{r}_\perp)$.

Let us represent the radiation field as a sum of external and radiated waves: $\tilde{E} = \tilde{E}_{\rm ext} + \tilde{E}_{\rm i}$. Solving (4.20), we find the field of the radiated wave $\tilde{E}_{\rm i}$:

$$\tilde{E}_{\rm i}(z, \boldsymbol{r}_\perp) = \frac{i\theta_{\rm s} \omega}{c^2} \int_0^z \frac{dz'}{z - z'} \int d\boldsymbol{r}'_\perp \tilde{j}_1(z', \boldsymbol{r}'_\perp) \exp \left[\frac{i\omega |\boldsymbol{r}_\perp - \boldsymbol{r}'_\perp|^2}{2c(z - z')} \right] . \qquad (4.22)$$

Substituting (4.22) into (4.12), we obtain an integral equation for the first harmonic of the beam current density $\tilde{j}_1$:

$$\tilde{j}_1(z, \boldsymbol{r}_\perp) = \mathrm{i} j_0(\boldsymbol{r}_\perp) \int_0^z \mathrm{d}z' \left\{ -\frac{e\theta_s}{2\mathrm{i}} \tilde{E}_\mathrm{ext}(z', \boldsymbol{r}_\perp) + \frac{4\pi e}{\omega} \tilde{j}_1(z', \boldsymbol{r}_\perp) \right.$$

$$-\frac{e\theta_s^2 \omega}{2c^2} \int_0^{z'} \frac{\mathrm{d}z''}{z' - z''} \int \mathrm{d}\boldsymbol{r}'_\perp \tilde{j}_1(z'', \boldsymbol{r}'_\perp) \exp\left[\frac{\mathrm{i}\omega|\boldsymbol{r}_\perp - \boldsymbol{r}'_\perp|^2}{2c(z' - z'')}\right] \right\}$$

$$\times \int_{-\infty}^\infty \mathrm{d}P \frac{\mathrm{d}F}{\mathrm{d}P} \exp\left[\mathrm{i}\left(C + \frac{\omega}{\gamma_z^2 \mathcal{E}_0 c} P\right)(z' - z)\right]. \tag{4.23}$$

When the energy spread is negligibly small ($F(P) \to \delta(P)$, where $\delta(P)$ is the delta function), this equation is reduced to:

$$\frac{\mathrm{d}^2 \tilde{j}_1}{\mathrm{d}z^2} + 2\mathrm{i}C \frac{\mathrm{d}\tilde{j}_1}{\mathrm{d}z} + \left[\frac{4\pi e}{c\gamma_z^2 \mathcal{E}_0} j_0(\boldsymbol{r}_\perp) - C^2\right] \tilde{j}_1$$

$$= \frac{\omega}{c\gamma_z^2 \mathcal{E}_0} j_0(\boldsymbol{r}_\perp) \left\{ \frac{e\theta_s}{2\mathrm{i}} \tilde{E}_\mathrm{ext}(z, \boldsymbol{r}_\perp) \right.$$

$$\left. + \frac{e\theta_s^2 \omega}{2c^2} \int_0^z \frac{\mathrm{d}z'}{z - z'} \int \mathrm{d}\boldsymbol{r}'_\perp \tilde{j}_1(z', \boldsymbol{r}'_\perp) \exp\left[\frac{\mathrm{i}\omega|\boldsymbol{r}_\perp - \boldsymbol{r}'_\perp|^2}{2c(z - z')}\right] \right\}. \tag{4.24}$$

So, in the linear approximation we can write down the self-consistent equation either for the field amplitude of the amplified wave (4.21), or for the modulation amplitude of the beam current density (4.23). We will illustrate below that both ways lead to the same results. We should stress that the equation for the wave field is preferable for an analytical solution, since the mathematical techniques are always connected with more conventional differential equations. The situation with computer simulations is proved to be reversed, and the method using the equation for the modulation amplitude of the beam current density is more convenient.

4.2 Power Balance

Let us show that the output radiation power of the FEL amplifier is equal to the power losses of the electron beam. In the paraxial approximation the diffraction angles are small, the vectors of the electric and magnetic field are equal in absolute value and are perpendicular to each other.

Thus, the expression for the radiation power, W, can be written in the form:

$$W = \frac{c}{4\pi} \int \langle |\boldsymbol{E}_\perp|^2 \rangle \mathrm{d}\boldsymbol{r}_\perp ,$$

4.2 Power Balance

where $\langle\ldots\rangle$ denotes averaging in time. The total electromagnetic field in the undulator is the sum of the external and radiated waves, so the total radiation power consists of three summands: $W = W_1 + W_2 + W_3$. The summand W_1 refers to the radiated wave:

$$W_1 = \frac{c}{4\pi} \int \langle|\boldsymbol{E}_\text{i}|^2\rangle \mathrm{d}\boldsymbol{r}_\perp . \quad (4.25\text{a})$$

The summand W_2 refers to the external wave and is equal to the power of the external wave:

$$W_2 = \frac{c}{4\pi} \int \langle|\boldsymbol{E}_\text{ext}|^2\rangle \mathrm{d}\boldsymbol{r}_\perp = W_\text{ext} . \quad (4.25\text{b})$$

The interference summand W_3 is equal to:

$$W_3 = \frac{c}{2\pi} \int \langle \boldsymbol{E}_\text{i} \cdot \boldsymbol{E}_\text{ext} \rangle \mathrm{d}\boldsymbol{r}_\perp. \quad (4.25\text{c})$$

Let us consider the summand W_1. Since the radiation field in the near zone has the form (4.22), then W_1 is given by:

$$W_1 = \frac{c}{4\pi} \int \langle|\boldsymbol{E}_\text{i}|^2\rangle \mathrm{d}\boldsymbol{r}_\perp$$

$$= \frac{\omega^2 \theta_\text{s}^2}{4\pi c^3} \int \mathrm{d}\boldsymbol{r}_\perp \left\{ \int_0^z \frac{\mathrm{d}z'}{z-z'} \int \mathrm{d}\boldsymbol{r}'_\perp \tilde{j}_1(z',\boldsymbol{r}'_\perp) \exp\left[\frac{i\omega|\boldsymbol{r}_\perp - \boldsymbol{r}'_\perp|^2}{2c(z-z')}\right] \right\}$$

$$\times \left\{ \int_0^z \frac{\mathrm{d}z''}{z-z''} \int \mathrm{d}\boldsymbol{r}''_\perp \tilde{j}_1^*(z'',\boldsymbol{r}''_\perp) \exp\left[-\frac{i\omega|\boldsymbol{r}_\perp - \boldsymbol{r}''_\perp|^2}{2c(z-z'')}\right] \right\} . \quad (4.26)$$

The products of integrals over z' and z'' can be represented as

$$\int_0^z \Phi(z')\mathrm{d}z' \int_0^z \Phi^*(z'')\mathrm{d}z'' = \int_0^z \Phi(z')\mathrm{d}z' \int_0^{z'} \Phi^*(z'')\mathrm{d}z'' + \text{C.C.} \quad (4.27)$$

The integral over the transverse coordinate $\boldsymbol{r}_\perp$ is equal to:

$$\int \mathrm{d}\boldsymbol{r}_\perp \exp\left\{\frac{i\omega|\boldsymbol{r}_\perp - \boldsymbol{r}'_\perp|^2}{2c(z-z')} - \frac{i\omega|\boldsymbol{r}_\perp - \boldsymbol{r}''_\perp|^2}{2c(z-z'')}\right\}$$

$$= \int_{-\infty}^{\infty} \mathrm{d}x \int_{-\infty}^{\infty} \mathrm{d}y \exp\left\{\frac{i\omega}{2c}\frac{(x-x')^2 + (y-y')^2 - (x-x'')^2 - (y-y'')^2}{z-z''}\right\}$$

$$= \frac{2\pi i c}{\omega} \frac{(z-z')(z-z'')}{z'-z''} \exp\left\{\frac{-i\omega|\boldsymbol{r}'_\perp - \boldsymbol{r}''_\perp|^2}{2c(z'-z'')}\right\} . \quad (4.28)$$

As a result, expression (4.26) can be written in the form:

4. Diffraction Effects in the FEL Amplifier

$$W_1 = \frac{i\omega\theta_s^2}{2c^2} \int_0^z dz' \int_0^{z'} \frac{dz''}{z'-z''} \int d\bm{r}'_\perp \int d\bm{r}''_\perp \tilde{j}_1^*(z'',\bm{r}''_\perp)\tilde{j}_1(z',\bm{r}'_\perp)$$

$$\times \exp\left\{\frac{-i\omega|\bm{r}'_\perp - \bm{r}''_\perp|^2}{2c(z'-z'')}\right\} + \text{C.C.} \qquad (4.29)$$

The procedure for the calculation of the interference summand, W_3, is more complicated. First, we write the two-dimensional Fourier transform of the field of the external wave at $z = 0$:

$$[(E_x)_{\text{ext}} + i(E_y)_{\text{ext}}]_{z=0} = \tilde{E}_{\text{ext}}(0,\bm{r}_\perp)e^{-i\omega t}$$

$$= e^{-i\omega t} \int d\bm{k}_\perp A(\bm{k}_\perp) \exp(i\bm{k}_\perp \cdot \bm{r}_\perp) . \qquad (4.30)$$

Each Fourier component is a plane wave,

$$\tilde{E}_{\text{ext}}(z,\bm{r}_\perp)e^{i\omega(z/c-t)} = e^{-i\omega t} \int d\bm{k}_\perp A(\bm{k}_\perp) \exp(i\bm{k}_\perp \cdot \bm{r}_\perp + ik_z z) ,$$

where

$$k_z = \frac{1}{c}\sqrt{\omega^2 - k_\perp^2 c^2} \simeq \frac{\omega}{c} - \frac{ck_\perp^2}{2\omega} .$$

Therefore, the complex amplitude $\tilde{E}_{\text{ext}}(z,\bm{r}_\perp)$ of the external field is given by

$$\tilde{E}_{\text{ext}}(z,\bm{r}_\perp) = \int d\bm{k}_\perp A(\bm{k}_\perp) \exp\left\{i\bm{k}_\perp \cdot \bm{r}_\perp - \frac{ik_\perp^2 c}{2\omega}z\right\} . \qquad (4.31)$$

Substituting this expression for $\tilde{E}_{\text{ext}}(z,\bm{r}_\perp)$ into (4.25c), we obtain:

$$W_3 = \frac{c}{4\pi} \int (\tilde{E}_{\text{ext}}\tilde{E}_i^* + \tilde{E}_{\text{ext}}^*\tilde{E}_i)d\bm{r}_\perp$$

$$= -i\frac{\theta_s \omega}{4\pi c} \int d\bm{r}_\perp \int d\bm{k}_\perp A(\bm{k}_\perp) \exp\left\{i\bm{k}_\perp \cdot \bm{r}_\perp - \frac{ik_\perp^2 c}{2\omega}z\right\}$$

$$\times \int_0^z \frac{dz'}{z-z'} \int d\bm{r}'_\perp \tilde{j}_1^*(z',\bm{r}'_\perp) \exp\left\{-\frac{i\omega|\bm{r}_\perp - \bm{r}'_\perp|^2}{2c(z-z')}\right\}$$

$$+ \text{C.C.} \qquad (4.32)$$

The integral over the transverse coordinates $\bm{r}_\perp$ can be calculated as:

$$\int d\bm{r}_\perp \exp\left\{i\bm{k}_\perp \cdot \bm{r}_\perp - \frac{i\omega|\bm{r}_\perp - \bm{r}'_\perp|^2}{2c(z-z')}\right\}$$

$$= \int_{-\infty}^{\infty} d\zeta \exp\left\{ik_x\zeta - \frac{i\omega(\zeta-x')^2}{2c(z-z')}\right\} \int_{-\infty}^{\infty} d\xi \exp\left\{ik_y\xi - \frac{i\omega(\xi-y')^2}{2c(z-z')}\right\}$$

$$= -\frac{2\pi ic}{\omega}(z-z') \exp\left\{i\bm{k}_\perp \cdot \bm{r}'_\perp + \frac{ik_\perp^2 c}{2\omega}(z-z')\right\} . \qquad (4.33)$$

As a result, we obtain:

$$W_3 = -\frac{1}{2}\theta_s \int_0^z dz' \int dr'_\perp \tilde{j}_1^*(z', r'_\perp) \int dk_\perp A(k_\perp)$$

$$\times \exp\left\{ik_\perp \cdot r'_\perp - \frac{ik_\perp^2 c}{2\omega}z'\right\} + \text{C.C.}$$

$$= -\frac{\theta_s}{2} \int_0^z dz' \int dr'_\perp \tilde{j}_1^*(z', r'_\perp) \tilde{E}_{\text{ext}}(z', r'_\perp) + \text{C.C.} \qquad (4.34)$$

The expressions for W_1 and W_3 can be written in terms of the effective potentials of the radiated field, U_i, and of the external field, U_{ext}:

$$U_i = -\frac{e\theta_s \tilde{E}_i(z, r_\perp)}{2i}, \qquad U_{\text{ext}} = -\frac{e\theta_s \tilde{E}_{\text{ext}}(z, r_\perp)}{2i}.$$

Using (4.22), (4.29) and (4.34) we obtain the expression for the sum $W_1 + W_3$:

$$W_1 + W_3 = \frac{1}{e}\int_0^z dz' \left\{ i\int dr'_\perp [U_i(z', r'_\perp) \right.$$

$$\left. + U_{\text{ext}}(z', r'_\perp)] \tilde{j}_1^*(z', r'_\perp) + \text{C.C.} \right\}. \qquad (4.35)$$

The radiated power, $W_1 + W_3$, must be equal to the difference of the electron beam power at the exit and the entrance of the undulator. The rate of energy change of a single electron is given by

$$\frac{d\mathcal{E}}{dz} = -\frac{\partial H}{\partial \psi} = i(U_i + U_{\text{ext}} + 4\pi e\tilde{j}_1/\omega)e^{i\psi} + \text{C.C.}$$

To obtain the mean power loss by the electron beam, we should multiply this value by the particle flux density

$$-j_z(z, \psi, r_\perp)/e ,$$

perform an averaging over the phase ψ, and integrate over the beam cross-section and the undulator length. Finally, we obtain the following result:

$$\Delta W_e = -\frac{1}{2\pi e}\int_0^{2\pi} d\psi \int dz' dr'_\perp \int_0^z j_z(\psi, z', r'_\perp)\frac{d\mathcal{E}}{dz'}$$

$$= -\frac{i}{2\pi e}\int_0^z dz' \int_0^{2\pi} d\psi \int dr'_\perp (U_i + U_{\text{ext}} + 4\pi e\tilde{j}_1/\omega)j_z e^{i\psi} + \text{C.C.}$$

$$= -\frac{i}{e}\int_0^z dz' \int dr'_\perp (U_i + U_{\text{ext}})\tilde{j}_1^* + \text{C.C.} \qquad (4.36)$$

Comparing this expression with (4.35), we see that the radiation power and the change in the electron beam power have equal absolute values and are opposite in sign, i.e. $\Delta W_e + W_1 + W_3 = 0$. So, power balance takes place.

After reading this section the reader can formulate the reasonable question: why is it necessary to go through all this derivation, since there exists the well-known Poynting theorem stating that the energy conservation law always holds in the system of particles and fields. Here we direct the attention of the reader to the fact that in the FEL theory we use the paraxial approximation for the wave equation and for the Poynting vector. The study, presented in this section, shows that power balance holds strictly mathematically. From a practical point of view, this fact is important, for instance, for testing the accuracy of numerical simulation codes.

4.3 Linear Theory of the FEL Amplifier with a Sheet Electron Beam

In this section we study a model of the FEL amplifier with a sheet electron beam. The approximations of the linear theory are similar to those described in the previous section. Taking into account the planar geometry of the electron beam, we assume the undulator to be planar. The axis of the undulator is directed along the z axis of the Cartesian coordinate system (x, y, z). The electron beam has infinitely large width along the y axis. The whole system is symmetrical with respect to the plane $x = 0$. The electron beam moves along the z axis in the field of the planar undulator:

$$\boldsymbol{H}_\ell(z) = \boldsymbol{e}_x H_\ell \cos(k_w z) ,$$

where $\boldsymbol{e}_x$ is the unit vector directed along the x axis. We assume that electrons move along identical sinusoidal trajectories parallel to the z axis. The oscillation amplitude of the transverse velocity of the electrons is considered to be small and the longitudinal velocity of the electrons v_z is close to the velocity of light. Only the linearly polarized electromagnetic wave can be amplified in this FEL amplifier. The vector of the electric field of the amplified wave can be written in the form:

$$\boldsymbol{E}_\perp = \boldsymbol{e}_y \tilde{E}_y(z, x) \exp\left[\mathrm{i}\omega(z/c - t)\right] + \text{C.C.}$$

For the electron beam with a small density perturbation, the distribution function can be written in the form:

$$f(\mathcal{E}, \psi, z, x) = f_0 + \tilde{f}_1 \exp(\mathrm{i}\psi) + \tilde{f}_1^* \exp(-\mathrm{i}\psi) .$$

The evolution of the perturbation of the distribution function, $\tilde{f}_1$, is governed by (2.66). We consider the case of an unmodulated electron beam at the undulator entrance,

$$\tilde{f}_1|_{z=0} = 0, \quad f_0 = n_0(x)F(P) .$$

The complex amplitude of the first harmonic of the longitudinal beam current density, $\tilde{j}_1$, and the perturbation of the distribution function, $\tilde{f}_1$, are connected by the relation

$$\tilde{j}_1 \simeq -ec \int \tilde{f}_1 \mathrm{d}P \ .$$

It is convenient to write complex amplitude $\tilde{j}_1$ in the form:

$$\tilde{j}_1 = \tilde{j}_\mathrm{a}(z,\ x) \exp\left[-\mathrm{i}Q \sin\left(2k_\mathrm{w} z\right)\right] \ ,$$

where

$$Q = \theta_\ell^2 \omega/(8ck_\mathrm{w}) \ .$$

Expanding the function

$$\exp\left[\mathrm{i}Q \sin\left(2k_\mathrm{w} z\right)\right]$$

in a Fourier series and neglecting rapidly oscillating terms, we obtain the equation for $\tilde{j}_\mathrm{a}(z,\ x)$ (see (2.72)):

$$\tilde{j}_\mathrm{a}(z,\ x) = \mathrm{i}j_0(x) \int_0^z \mathrm{d}z' \left\{ -\frac{A_{\mathrm{JJ}} e \theta_\ell}{2} \tilde{E}_y(z',\ x) + \frac{4\pi e}{\omega} \tilde{j}_\mathrm{a}(z',\ x) \right\}$$

$$\times \int \mathrm{d}P F'(P) \exp\left\{ \mathrm{i}\left[C + \frac{\omega P}{c\gamma_\ell^2 \mathcal{E}_0}\right](z'-z) \right\} \ , \qquad (4.37\mathrm{a})$$

where

$$A_{\mathrm{JJ}} = [J_0(Q) - J_1(Q)] \ , \quad F'(P) = \mathrm{d}F(P)/\mathrm{d}P \ , \quad C = k_\mathrm{w} - \omega/(2c\gamma_\ell^2) \ ,$$

and $-j_0(x) \simeq -ecn_0(x)$ is the longitudinal component of the beam current density at the undulator entrance. When the energy spread is negligibly small, this equation is reduced to

$$\frac{\mathrm{d}^2 \tilde{j}_\mathrm{a}}{\mathrm{d}z^2} + 2\mathrm{i}C \frac{\mathrm{d}\tilde{j}_\mathrm{a}}{\mathrm{d}z} + \left[\frac{4\pi e}{c\gamma_\ell^2 \mathcal{E}_0} j_0(x) - C^2\right] \tilde{j}_\mathrm{a} = \frac{e\omega \theta_\ell A_{\mathrm{JJ}}}{2c\gamma_\ell^2 \mathcal{E}_0} j_0(x) \tilde{E}_y \ . \qquad (4.37\mathrm{b})$$

One more relation connecting the complex amplitudes $\tilde{j}_\mathrm{a}$ and $\tilde{E}_y$ follows from Maxwell's equations. The electromagnetic field of the amplified wave is subjected to the wave equation (4.18). Substituting the perturbation of the transverse component of the beam current density,

$$\mathbf{j}_\perp = -\mathbf{e}_y \theta_\ell \sin\left(k_\mathrm{w} z\right) \left[\tilde{j}_1 \exp(\mathrm{i}\psi) + \mathrm{C.C.}\right] \ ,$$

into (4.18) and neglecting rapidly oscillating terms, we obtain the following equation for the complex amplitudes $\tilde{E}_y$ and $\tilde{j}_\mathrm{a}$:

$$\left[\frac{\partial^2}{\partial x^2} + 2\mathrm{i}\frac{\omega}{c}\frac{\partial}{\partial z}\right] \tilde{E}_y = -2\pi \theta_\ell c^{-2} \omega A_{\mathrm{JJ}} \tilde{j}_\mathrm{a} \ . \qquad (4.38\mathrm{a})$$

Here we have taken into account that in this two-dimensional case $\partial \tilde{E}_y/\partial y = 0$. In what follows it is convenient to write down the solution of this equation in the form:

$$\tilde{E}_y = \frac{\theta_\ell \omega A_{\rm JJ}}{2c^2} \int_{-\infty}^{\infty} dx' \int_{-\infty}^{\infty} dy' \int_0^z dz' \frac{\tilde{j}_a(z', x')}{z - z'}$$

$$\times \exp\left\{\frac{i\omega |\mathbf{r}_\perp - \mathbf{r}'_\perp|^2}{2c(z-z')}\right\}. \qquad (4.38b)$$

The self-consistent equations (4.37) and (4.38) describe the amplification of the radiation field in the FEL amplifier with a sheet electron beam and a planar undulator.

4.3.1 Eigenvalue Problem for a Stepped Profile

Let us consider the electron beam with a stepped profile of the beam current density. The electron beam is symmetrical with respect to the plane $x = 0$. We use the self-consistent equations (4.37a) and (4.38a) to obtain a unique integro-differential equation for the radiation field amplitude. It is convenient to write this equation in the following dimensionless form:

Region 1, ($|\hat{x}| < 1$):

$$\left(\frac{\partial^2}{\partial \hat{x}^2} + 2\mathrm{i}B\frac{\partial}{\partial \hat{z}}\right)\tilde{E}_y(\hat{z}, \hat{x})$$

$$= \mathrm{i}\int_0^{\hat{z}} d\hat{z}' \left[2 + \hat{A}_{\rm p}^2\left(\frac{\partial^2}{\partial \hat{x}^2} + 2\mathrm{i}B\frac{\partial}{\partial \hat{z}'}\right)\right]\tilde{E}_y(\hat{z}', \hat{x})$$

$$\times \int_{-\infty}^{\infty} d\hat{P}\hat{F}'(\hat{P}) \exp\left[\mathrm{i}(\hat{P} + \hat{C})(\hat{z}' - \hat{z})\right], \qquad (4.39a)$$

Region 2, ($1 < |\hat{x}|$)

$$\left(\frac{\partial^2}{\partial \hat{x}^2} + 2\mathrm{i}B\frac{\partial}{\partial \hat{z}}\right)\tilde{E}_y(\hat{z}, \hat{x}) = 0. \qquad (4.39b)$$

Here the following notation is introduced: $2d$ is the width of the electron beam, $\hat{x} = x/d$, $\hat{z} = \Gamma z$, $\hat{A}_{\rm p}^2 = A_{\rm p}^2/\Gamma^2 = 8c^2\left(\omega\theta_\ell A_{\rm JJ}d\right)^{-2}$ is the space charge parameter, $\hat{C} = C/\Gamma$ is the detuning parameter, $B = \Gamma\omega d^2/c$ is the diffraction parameter,

$$\Gamma = \left[\pi A_{\rm JJ}^2 \omega^2 \theta_\ell^2 \bar{j}_0 d \left(4c^2\gamma_\ell^2\gamma I_{\rm A}\right)^{-1}\right]^{1/2}$$

is the gain parameter, $\bar{j}_0 = 2dj_0$, j_0 is the beam current density, $I_{\rm A} = m_e c^3/e \simeq 17$ kA is Alfven current and

$$\Lambda_{\rm p} = \left[2\pi\bar{j}_0\left(\gamma_\ell^2\gamma I_{\rm A}d\right)^{-1}\right]^{1/2}$$

is the wavenumber of the longitudinal plasma oscillations. The distribution function for the normalized energy deviation, $\hat{F}(\hat{P})$, is normalized to unity:

$$\int \hat{F}(\hat{P})\mathrm{d}\hat{P} = 1 ,$$

where

$$\hat{P} = (\mathcal{E} - \mathcal{E}_0)/(\rho\mathcal{E}_0)$$

is the reduced energy deviation. The efficiency parameter ρ is given by

$$\rho = \gamma_\ell^2 c\Gamma/\omega .$$

For a Gaussian energy spread in the electron beam given by (2.20a), the explicit form of the function $\hat{F}(\hat{P})$ is

$$\hat{F}(\hat{P}) = \left(2\pi\hat{\Lambda}_\mathrm{T}^2\right)^{-1/2} \exp\left(-\frac{\hat{P}^2}{2\hat{\Lambda}_\mathrm{T}^2}\right) .$$

The energy spread parameter, $\hat{\Lambda}_\mathrm{T}^2$, and the energy spread in the electron beam, $\langle(\Delta\mathcal{E})^2\rangle$, are connected by the relation

$$\hat{\Lambda}_\mathrm{T}^2 = \langle(\Delta\mathcal{E})^2\rangle/\left(\rho^2\mathcal{E}_0^2\right) .$$

The reader can show that we use notation similar to that introduced in Chap. 2. Despite the normalizing factors and the reduced parameters are expressed in terms of physical parameters in a different way with respect to Chap. 2, their physical interpretation is similar. Also, an additional parameter of the problem, the diffraction parameter B, appears in the theory, since we take into account diffraction effects. The physical significance of this parameter will become clear after the solution of the eigenvalue equation.

In the high-gain linear regime we seek a solution for the field amplitude, $\tilde{E}_y(\hat{z}, \hat{x})$, in the form:

$$\tilde{E}_y = \Phi(\hat{x})\exp(\hat{\Lambda}\hat{z}), \qquad \mathrm{Re}\,\hat{\Lambda} > 0 .$$

According to (4.39), the function $\Phi(\hat{x})$ is subjected to the following equations:

Region 1, ($|\hat{x}| < 1$)

$$\frac{\mathrm{d}^2\Phi(\hat{x})}{\mathrm{d}\hat{x}^2} + \left[\frac{-2\mathrm{i}\hat{D}}{1 - \mathrm{i}\hat{\Lambda}_\mathrm{p}^2\hat{D}} + 2\mathrm{i}B\hat{\Lambda}\right]\Phi(\hat{x}) = 0 , \qquad (4.40\mathrm{a})$$

Region 2, ($1 < |\hat{x}|$)

$$\frac{\mathrm{d}^2\Phi(\hat{x})}{\mathrm{d}\hat{x}^2} + (2\mathrm{i}B\hat{\Lambda})\Phi(\hat{x}) = 0 . \qquad (4.40\mathrm{b})$$

The function $\hat{D}$ is given by the expression:

$$\int_{-\infty}^{\infty} \mathrm{d}\hat{P}\,\frac{\hat{F}'(\hat{P})}{\hat{\Lambda} + \mathrm{i}\hat{C} + \mathrm{i}\hat{P}} .$$

In particular, for a Gaussian energy spread we have

$$\hat{D} = i \int_0^\infty \xi \exp\left[-\frac{\hat{\Lambda}_T^2 \xi^2}{2} - (\hat{\Lambda} + i\hat{C})\xi\right] d\xi .$$

When the energy spread is negligibly small, i.e. $\hat{\Lambda}_T^2 \to 0$, we have $\hat{D} = i(\hat{\Lambda} + i\hat{C})^{-2}$.

The function $\Phi(\hat{x})$ must meet the requirement of quadratical integrability. Taking into account that the boundary conditions

$$\Phi(x) \to 0 \quad \text{for} \quad x \to \pm\infty ,$$

and equations (4.40) are invariant as $x \to -x$, we conclude that the solutions for $\Phi(x)$ must be either even or odd. This statement can be proved in the following way. First, we show that there is no degeneracy of the radiation modes. Let $\Phi_1(\hat{x})$ and $\Phi_2(\hat{x})$ be different eigenfunctions corresponding to the same eigenvalue. Using (4.40a) and (4.40b), we write

$$\Phi_2 \frac{d^2 \Phi_1}{d\hat{x}^2} - \Phi_1 \frac{d^2 \Phi_2}{d\hat{x}^2} = 0 .$$

It follows from the latter equation that

$$\Phi_2 \frac{d\Phi_1}{d\hat{x}} - \Phi_1 \frac{d\Phi_2}{d\hat{x}} = \text{const.}$$

Since

$$\lim_{|\hat{x}| \to \infty} \Phi_{1,2}(\hat{x}) = 0$$

as $|\hat{x}| \to \infty$, then

$$\Phi_2 \frac{d\Phi_1}{d\hat{x}} - \Phi_1 \frac{d\Phi_2}{d\hat{x}} = 0 ,$$

and we can write

$$\Phi_1(\hat{x}) = \text{const.} \times \Phi_2(\hat{x}) .$$

Each function can be represented as a superposition of two functions, an even function

$$\Phi^e(\hat{x}) = \Phi^e(-\hat{x}) ,$$

and an odd function,

$$\Phi^o(\hat{x}) = -\Phi^o(-\hat{x}) .$$

Thus, we can write

$$\Phi(\hat{x}) = A\Phi^e(\hat{x}) + B\Phi^o(\hat{x}) ,$$
$$\Phi(-\hat{x}) = A\Phi^e(\hat{x}) - B\Phi^o(\hat{x}) .$$

If $\Phi(\hat{x})$ is an eigenfunction, then $\Phi(-\hat{x})$ is also an eigenfunction. Since there is no degeneration, we have

$\Phi(-\hat{x}) = \text{const.} \times \Phi(\hat{x})$,

$A\Phi^e(\hat{x}) - B\Phi^o(\hat{x}) = \text{const.} \times [A\Phi^e(\hat{x}) + B\Phi^o(\hat{x})]$.

Since the functions $\Phi^e(\hat{x})$ and $\Phi^o(\hat{x})$ are linearly independent, the latter expression is valid only when (const. = 1, $B = 0$) or (const. = -1, $A = 0$).

It follows from (4.40a) that inside the electron beam the even solution for $\Phi(\hat{x})$ is

$$\Phi(\hat{x}) = C_1 \cos(\nu\hat{x}), \quad \text{for} \quad |\hat{x}| < 1.$$

According to (4.40b), we have the following solution outside the electron beam:

$$\Phi(\hat{x}) = C_2 \exp(-q|\hat{x}|), \quad \text{for} \quad |\hat{x}| > 1,$$

where C_1 and C_2 are constants, and

$$\nu^2 = \frac{-2\mathrm{i}\hat{D}}{1 - \mathrm{i}\hat{\Lambda}_\mathrm{p}^2 \hat{D}} + 2\mathrm{i}B\hat{\Lambda}, \qquad q^2 = -2\mathrm{i}B\hat{\Lambda}.$$

To be specific, we assume that $\operatorname{Re} q > 0$. The solutions inside and outside the electron beam must be sewn together at the beam boundary. So, the next problem is to derive the boundary conditions. This can be done in the following way. Equations (4.40a) and (4.40b) can be written in the form

$$\frac{\mathrm{d}^2\Phi(\hat{x})}{\mathrm{d}\hat{x}^2} + g(\hat{x})\Phi(\hat{x}) = 0,$$

where the function $g(\hat{x})$ has a break at the beam boundary. Formally, the stepped profile of the electron beam assumes that the beam current density also has a break at the beam boundary. In an actual physical situation there is no break of the current density, but it drops very rapidly to zero. Thus, we can consider $j_0(\hat{x})$ and $g(\hat{x})$ as continuous functions of $\hat{x}$. Integrating the latter equation in the vicinity of $\hat{x} = 1$, we get:

$$\int_{1-\Delta}^{1+\Delta} \Phi''(\hat{x})\mathrm{d}\hat{x} + \int_{1-\Delta}^{1+\Delta} g(\hat{x})\Phi(\hat{x})\mathrm{d}\hat{x} = 0.$$

It follows from this expression that

$$\Phi'(\hat{x})|_{\hat{x}=1-\Delta}^{\hat{x}=1+\Delta} = -\int_{1-\Delta}^{1+\Delta} g(\hat{x})\Phi(\hat{x})\mathrm{d}\hat{x}.$$

The limiting transition $\Delta \to 0$ gives us the following boundary condition for the derivative of the eigenfunction:

$$\Phi'(\hat{x})|_{\hat{x}=1-0}^{\hat{x}=1+0} = 0.$$

Using the same considerations, we obtain the boundary condition for the eigenfunction:

$$\Phi(\hat{x})|_{\hat{x}=1-0}^{\hat{x}=1+0} = 0.$$

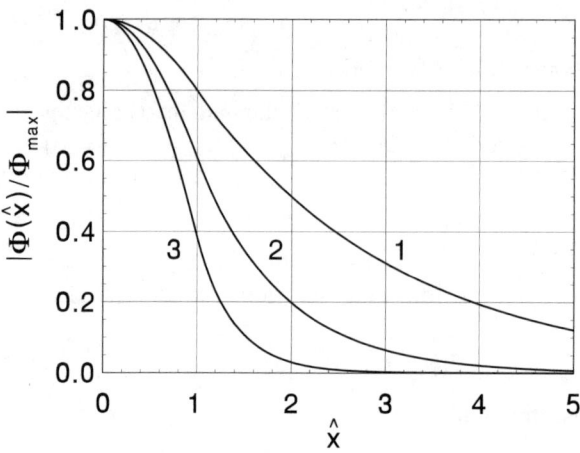

Fig. 4.1. Sheet electron beam with stepped profile: transverse distribution of the field amplitude for the first even mode. Here $\hat{C} = 0$, $\hat{\Lambda}_p^2 \to 0$, and $\hat{\Lambda}_T^2 = 0$. Curve 1: $B = 0.1$. Curve 2: $B = 1$. Curve (3): $B = 10$

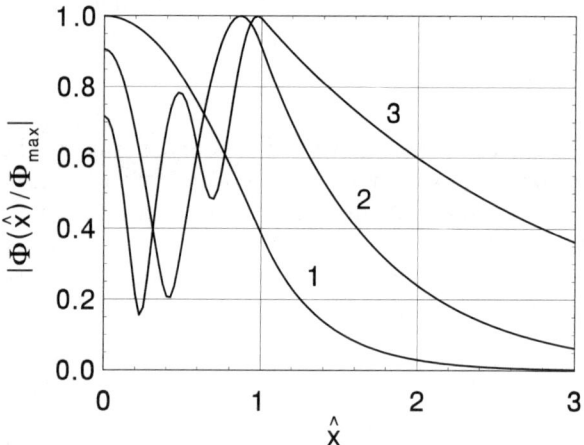

Fig. 4.2. Sheet electron beam with stepped profile: transverse distribution of the field amplitude for the even modes. Here $B = 10$, $\hat{\Lambda}_p^2 \to 0$, and $\hat{\Lambda}_T^2 = 0$. Curve 1: the first even mode ($\hat{C} = 0$). Curve 2: the second even mode ($\hat{C} = 0.4$). Curve 3: the third even mode ($\hat{C} = 1.5$)

Using the continuity conditions of $\Phi(\hat{x})$ and $\Phi'(\hat{x})$, we obtain the following eigenvalue equation for the even beam radiation modes:

$$\nu \tan(\nu) = q \ . \tag{4.41}$$

An explicit expression for the even eigenfunction has the form:

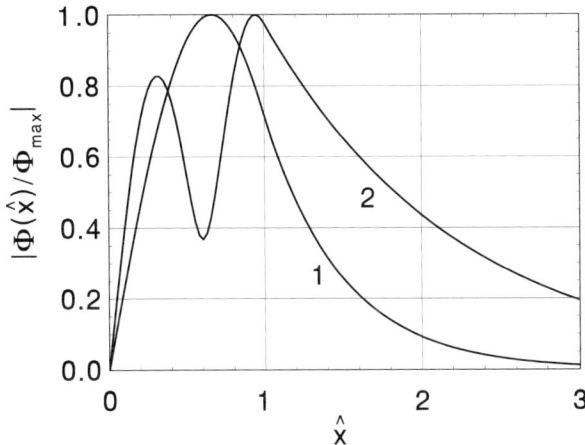

Fig. 4.3. Sheet electron beam with stepped profile: transverse distribution of the field amplitude for the odd modes. Here $B = 10$, $\hat{\Lambda}_{\rm p}^2 \to 0$, and $\hat{\Lambda}_{\rm T}^2 = 0$. Curve 1: the first odd mode ($\hat{C} = 0.15$). Curve 2: the second odd mode ($\hat{C} = 0.8$)

$$\Phi(\hat{x}) = \begin{cases} \cos(\nu \hat{x}), & \text{for } |\hat{x}| < 1 \\ \cos(\nu) \exp(q - q|\hat{x}|), & \text{for } |\hat{x}| > 1. \end{cases} \quad (4.42)$$

The eigenvalue equation and the expression for the odd beam radiation modes are as follows:

$$\nu \cot(\nu) = -q, \quad (4.43)$$

$$\Phi(\hat{x}) = \begin{cases} \sin(\nu \hat{x}), & \text{for } |\hat{x}| < 1 \\ {\rm sgn}(\hat{x}) \sin(\nu) \exp(q - q|\hat{x}|), & \text{for } |\hat{x}| > 1, \end{cases} \quad (4.44)$$

where

$${\rm sgn}(\hat{x}) = 1 \quad \text{for } \hat{x} > 0,$$
$${\rm sgn}(\hat{x}) = -1 \quad \text{for } \hat{x} < 0.$$

So, we have solved the eigenvalue problem starting from the self-consistent equation for the radiation field. The same result can be obtained from the solution of the integral equation for the first harmonic of the beam current density. Let us illustrate the application of the latter method for the case of a negligibly small energy spread, $\hat{\Lambda}_{\rm T}^2 \to 0$. In the high-gain limit we seek the solution for $\tilde{j}_{\rm a}(\hat{z}, \hat{x})$ in the form

$$\tilde{j}_{\rm a} = a(\hat{x}) \exp(\hat{\Lambda}\hat{z}).$$

Substituting (4.38b) into (4.37b), we obtain the equation for the eigenfunction of the beam modulation, $a(\hat{x})$:

$$\left[(\hat{A} + i\hat{C}^2) + \hat{A}_p^2\right] a(\hat{x})$$

$$= \frac{1}{2\pi} \int_{-1}^{1} d\hat{x}' \int_{-\infty}^{\infty} d\hat{y}' \int_{0}^{\infty} d\zeta \frac{a(\hat{x}')}{\zeta}$$

$$\times \exp\left\{-\hat{A}\zeta + \frac{iB\left[(\hat{x} - \hat{x}')^2 + (\hat{y} - \hat{y}')^2\right]}{2\zeta}\right\}, \qquad (4.45)$$

where $\hat{y} = y/d$ and $\hat{A} = A/\Gamma$. Here the normalization procedure is identical to that used in the beginning of this section. The integral over the longitudinal coordinate in the latter equation is calculated analytically:

$$\int_{0}^{\infty} d\zeta \frac{\exp(-\zeta - \beta/\zeta)}{\zeta} = 2K_0(2\sqrt{\beta}),$$

which leads to the following result:

$$\left[(\hat{A} + i\hat{C}^2) + \hat{A}_p^2\right] a(\hat{x})$$

$$= \frac{1}{\pi} \int_{-1}^{1} d\hat{x}' a(\hat{x}') \int_{-\infty}^{\infty} d\hat{y}' K_0\left(q\sqrt{(\hat{x} - \hat{x}')^2 + (\hat{y} - \hat{y}')^2}\right), \qquad (4.46)$$

where $q = \sqrt{-2iB\hat{A}}$ and $\operatorname{Re} q > 0$. The integral over $\hat{y}'$ is calculated as follows

$$\int_{-\infty}^{\infty} d\hat{y}' K_0\left(q\sqrt{(\hat{x} - \hat{x}')^2 + (\hat{y} - \hat{y}')^2}\right) = \frac{\pi}{q} \exp(-q|\hat{x} - \hat{x}'|).$$

Finally, we obtain a linear homogeneous Fredholm equation of the second kind:

$$\left[(\hat{A} + i\hat{C}^2) + \hat{A}_p^2\right] a(\hat{x}) = \frac{1}{q} \int_{-1}^{1} d\hat{x}' a(\hat{x}') \exp(-q|\hat{x} - \hat{x}'|). \qquad (4.47)$$

It follows from (4.47) that the function $a(\hat{x})$ is subjected to the following differential equation:

$$\frac{d^2 a(\hat{x})}{d\hat{x}^2} + \left[\frac{2}{(\hat{A} + i\hat{C})^2 + \hat{A}_p^2} - q^2\right] a(\hat{x}) = 0.$$

Solutions of this equation are separated into even and odd functions. The even solution for the beam modulation is $a(\hat{x}) = \text{const.} \times \cos(\nu \hat{x})$, where the parameter ν is given by

$$\nu^2 = \left[\frac{2}{(\hat{A} + i\hat{C})^2 + \hat{A}_p^2} - q^2\right].$$

Substituting this solution for $a(\hat{x})$ into the integral equation (4.47), we have

$$\frac{\cos(\nu\hat{x})}{\nu^2+q^2} = \frac{1}{2q}\int_{-1}^{\hat{x}} d\hat{x}'\cos(\nu\hat{x}')\exp[-q(\hat{x}-\hat{x}')]$$

$$+\frac{1}{2q}\int_{\hat{x}}^{1} d\hat{x}'\cos(\nu\hat{x}')\exp[q(\hat{x}-\hat{x}')] \ . \tag{4.48}$$

The integrals appearing in (4.48) are calculated in the following way:

$$\int d\hat{x}\cos(\nu\hat{x})\exp(q\hat{x}) = \frac{\exp(q\hat{x})}{\nu^2+q^2}[q\cos(\nu\hat{x})+\nu\sin(\nu\hat{x})] \ .$$

Finally, we obtain the eigenvalue equation for the even radiation modes,

$$q = \nu\tan(\nu) \ ,$$

which is identical to (4.41) obtained from the solution of the self-consistent equation for the field amplitude.

In a similar way, $a(\hat{x}) = \text{const.} \times \sin(\nu\hat{x})$ for the odd solution. Substituting $a(\hat{x})$ into (4.47), we have

$$\frac{\sin(\nu\hat{x})}{\nu^2+q^2} = \frac{1}{2q}\int_{-1}^{\hat{x}} d\hat{x}'\sin(\nu\hat{x}')\exp[-q(\hat{x}-\hat{x}')]$$

$$+\frac{1}{2q}\int_{\hat{x}}^{1} d\hat{x}'\sin(\nu\hat{x}')\exp[q(\hat{x}-\hat{x}')] \ . \tag{4.49}$$

These integrals are calculated analytically:

$$\int d\hat{x}\sin(\nu\hat{x})\exp(q\hat{x}) = \frac{\exp(q\hat{x})}{\nu^2+q^2}[q\sin(\nu\hat{x})-\nu\cos(\nu\hat{x})] \ ,$$

which leads to

$$\exp(-q)\left[\sin(\nu)+\frac{\nu}{q}\cos(\nu)\right]\sinh(q\hat{x}) = 0.$$

The latter equation can be rewritten in the form

$$q = -\nu\cot(\nu) \ ,$$

which is identical to (4.43) for the odd beam radiation modes obtained from the differential equation for the radiation field.

It should be noted that we did not use the continuity conditions for the eigenfunction $\Phi(\hat{x})$ and its derivative at the beam boundary when deriving the eigenvalue equation by means of the integral equation method. Since the eigenfunction is proportional to the radiation field amplitude, these boundary conditions are automatically satisfied as follows from (4.38b).

4.3.2 Analysis of the Beam Radiation Modes

Transverse profiles of the radiation field for the first even mode are presented in Fig. 4.1. It is seen that the field is concentrated mainly inside the electron beam at large values of the diffraction parameter B. When the value of the diffraction parameter is decreased, the field begins to expand outside the electron beam. Figure 4.2 illustrates the field distribution of the first, second, and third even radiation modes. The field distribution of the first and second odd modes is shown in Fig. 4.3. Since the eigenfunctions are given by complex functions of the transverse coordinate, the wavefront of each eigenfunction is not a plane. It is interesting to trace the variation of the field phase across the electron beam. Figures. 4.4 and 4.5 show the distribution of the amplitude and of the phase of the radiation beam for two even modes.

The solution of the eigenvalue problem allows one to find the transverse distribution of the radiation field at the exit of the undulator, at $z = l_w$. From a practical point of view it is necessary to know the field distribution in the space after the undulator, at $z > l_w$. When the radiation field leaves the undulator, it is subjected to the parabolic equation

$$\left[\frac{\partial^2}{\partial x^2} + 2\mathrm{i}\frac{\omega}{c}\frac{\partial}{\partial z}\right]\tilde{E}_y(z,x) = 0 \ .$$

It follows from the latter equation that the field amplitude in the space after the undulator and the field amplitude at the undulator exit are connected by

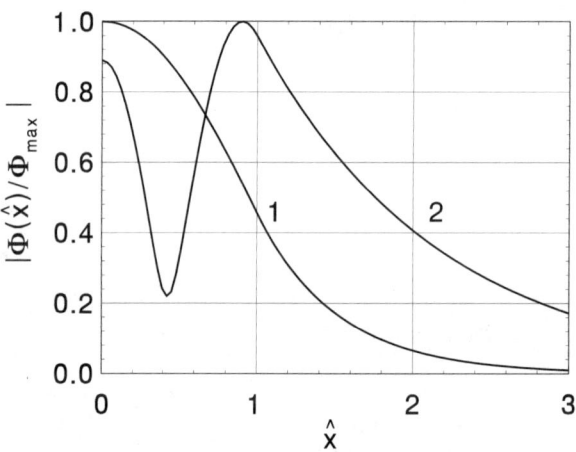

Fig. 4.4. Sheet electron beam with stepped profile: transverse distribution of the field amplitude for the even modes. Here $B = 5$, $\hat{\Lambda}_\mathrm{p}^2 \to 0$, and $\hat{\Lambda}_\mathrm{T}^2 = 0$. Curve 1: the first even mode ($\hat{C} = 0.05$). Curve 2: the second even mode ($\hat{C} = 0.6$)

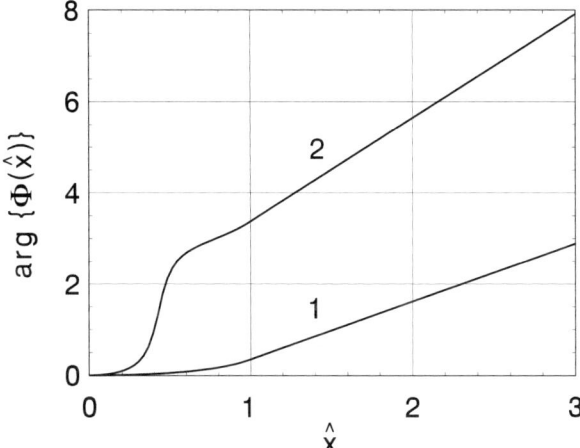

Fig. 4.5. The stepped beam profile. The field phase distribution of the first two even modes versus the reduced coordinate $\hat{x}$. Here $B = 5$, $\hat{\Lambda}_{\mathrm{p}}^2 \to 0$, and $\hat{\Lambda}_{\mathrm{T}}^2 = 0$. Curve 1: the first even mode ($\hat{C} = 0.05$). Curve 2: the second even mode ($\hat{C} = 0.6$)

$$\tilde{E}_y(z,\, x) = \left[\frac{\mathrm{i}\omega}{2\pi c\, (z - l_{\mathrm{w}})}\right]^{1/2} \int_{-\infty}^{\infty} \mathrm{d}x'\, \tilde{E}_y(l_{\mathrm{w}},\, x')$$

$$\times \exp\left[\frac{\mathrm{i}\omega\, (x - x')^2}{2c\, (z - l_{\mathrm{w}})}\right] . \qquad (4.50)$$

The subject of particular interest is the angular distribution of the radiation intensity. The radiation field at the undulator exit may be presented as a superposition of plane waves, all with the same wavenumber $k = \omega/c$. The value of k_x/k gives the sine of the angle between the z axis and the direction of propagation of the plane wave. In the paraxial approximation $k_x/k = \sin\theta \simeq \theta$. The angular distribution of the radiation intensity, $I(\theta)$, can be expressed as follows:

$$\frac{I(\theta)}{I(0)} = \left|\frac{\Xi(\theta)}{\Xi(0)}\right|^2 ,$$

where $\Xi(\theta)$ is the spatial Fourier transform of the complex amplitude of the radiation field, $\Phi(x')$, at the exit of the undulator,

$$\Xi(\theta) = \int_{-\infty}^{\infty} \Phi(x') \exp(-\mathrm{i}\omega x'\theta/c)\mathrm{d}x' . \qquad (4.51)$$

Let us consider the FEL amplifier operating at an even beam radiation mode. Substituting (4.42) into (4.51), we obtain:

4. Diffraction Effects in the FEL Amplifier

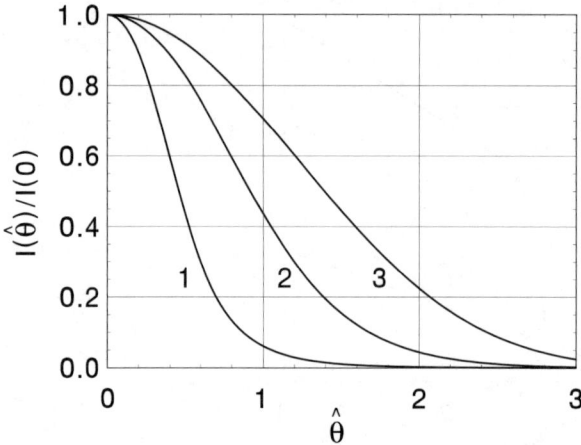

Fig. 4.6. Sheet electron beam with stepped profile. Angular distribution of the radiation intensity for the first even mode. Here $\hat{C} = 0$, $\hat{\Lambda}_p^2 \to 0$, and $\hat{\Lambda}_T^2 = 0$. Curve 1: $B = 0.1$. Curve 2: $B = 1$. Curve 3: $B = 10$

$$\Xi(\hat{\theta}) = \left(\hat{\theta}^2 - \nu^2\right)^{-1}\left[\hat{\theta}\sin(\hat{\theta})\cos(\nu) - \nu\sin(\nu)\cos(\hat{\theta})\right]$$
$$+ \left(q^2 + \hat{\theta}^2\right)^{-1}\cos(\nu)\left[q\cos(\hat{\theta}) - \hat{\theta}\sin(\hat{\theta})\right],$$

where $\hat{\theta} = \theta\omega d/c$. Using (4.41), we find the expression for the angular distribution of the radiation intensity,

$$\frac{I(\theta)}{I(0)} = \left|\frac{\Xi(\hat{\theta})}{\Xi(0)}\right|^2 = \left|\frac{\cos(\hat{\theta}) - (\hat{\theta}/q)\sin(\hat{\theta})}{\left(1 + \hat{\theta}^2/q^2\right)\left(1 - \hat{\theta}^2/\nu^2\right)}\right|^2. \tag{4.52}$$

At large values of the diffraction parameter B, the region of applicability for the far radiation zone is given by the relation $(z - l_w) \gg \omega d^2/c$. When the value of the diffraction parameter is about equal to or less than unity, the latter condition transforms to $(z - l_w) \gg |\Lambda|^{-1} \simeq \Gamma^{-1}$. In Fig. 4.6 we present the angular distributions of the radiation intensity for the FEL amplifier operating in the first even mode. It is seen that the output radiation is concentrated in a small cone. In the limit of large diffraction parameter B, the FWHM of the distribution of the radiation intensity over the normalized angle does not depend on the value of B and is equal to $\Delta\theta \simeq 3.2c/(\omega d)$. On the other hand, for small values of the diffraction parameter B, the width of the angular distribution is significantly less than $c/(\omega d)$. The reason for this is that the effective size of the radiation source is significantly larger than the transverse size of the electron beam (see Fig. 4.1).

It follows from (4.41) that the reduced field growth rate, $\operatorname{Re}\hat{\Lambda}$, is a universal function of four dimensionless parameters:

$$\operatorname{Re}\hat{\Lambda} = F(\hat{C}, \hat{\Lambda}_p^2, \hat{\Lambda}_T^2, B).$$

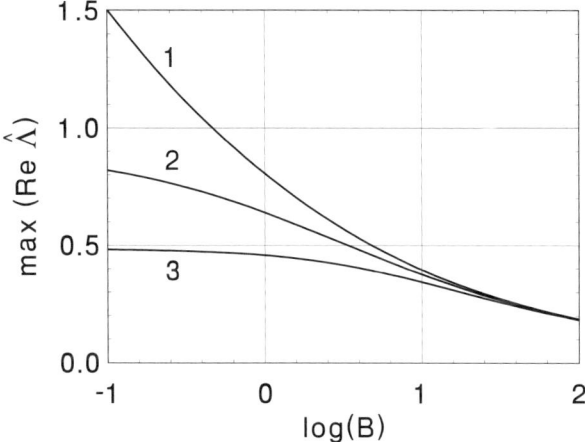

Fig. 4.7. The stepped beam profile. The dependence of the maximal reduced growth rate $\max(\operatorname{Re}\hat{\Lambda})$ on the diffraction parameter B. Here $\hat{\Lambda}_{\mathrm{p}}^2 \to 0$, $\hat{\Lambda}_{\mathrm{T}}^2 = 0$. Curve 1: the first even mode. Curve 2: the first odd mode. Curve 3: the second even mode

Let us consider the case of negligibly small energy spread and space charge effects, $\hat{\Lambda}_{\mathrm{p}}^2 \to 0$ and $\hat{\Lambda}_{\mathrm{T}}^2 \to 0$. In this case for each mode and for each value of the diffraction parameter B there is an optimal detuning parameter at which the growth rate achieves its maximum. The dependence of the maximal field growth rate on the value of B for different modes is presented in Fig. 4.7. The dependence of the reduced field growth rate on the detuning parameter is presented in Fig. 4.8.

Let us study the asymptotic behavior of the field growth rate for the even beam radiation modes at large values of the diffraction parameter B. It follows from (4.41) that $|q| \to \infty$ and $\nu \to \pi k/2$ as $B \to \infty$ ($k = 1, 3, \ldots$). Thus, we obtain

$$\hat{\Lambda}(\hat{\Lambda} + i\hat{C})^2 = \frac{i}{B} . \tag{4.53}$$

Redetermination of the gain parameter as

$$\Gamma B^{-1/3} \to \Gamma = \left[\pi A_{\mathrm{JJ}}^2 \omega \theta_\ell^2 j_0 \left(2c^2 \gamma_\ell^2 \gamma I_{\mathrm{A}}\right)^{-1}\right]^{1/3} ,$$

leads to the cubic eigenvalue equation of the one-dimensional model of the FEL amplifier (here $j_0 = \bar{j}_0/(2d)$). A similar asymptote exists for the odd beam radiation modes. Indeed, we obtain from (4.43) that $|q| \to \infty$ and $\nu \to \pi k$ as $B \to \infty$ ($k = 1, 2, \ldots$), and we obtain the cubic eigenvalue equation (4.53).

Analysis of the curves in Fig. 4.8 shows that the maxima of the field growth rate of different modes are achieved at different positive values of the detuning parameter. This is due to the fact that the phase velocity of

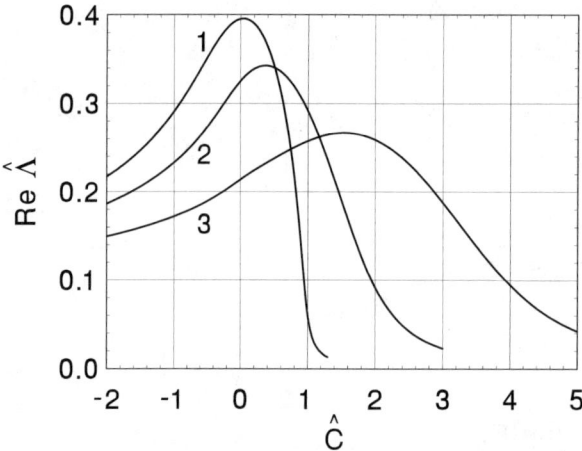

Fig. 4.8. Sheet electron beam with stepped profile. The reduced field growth rate Re $\hat{\Lambda}$ versus the detuning parameter $\hat{C}$. Here $B = 10$, $\hat{\Lambda}_{\mathrm{p}}^2 \to 0$, $\hat{\Lambda}_{\mathrm{T}}^2 = 0$. Curve 1: the first even mode. Curve 2: the second even mode. Curve 3: the third even mode

the radiation field of the mode is larger than the velocity of light. Indeed, the radiation field of the mode can be represented as a superposition of plane waves propagating at different angles. At large values of the diffraction parameter, $B \gg 1$, the typical value of the transverse wave vector of the first even mode is

$$k_x \simeq \pi/(2d) \ .$$

Remembering that

$$k_x^2 + k_z^2 = \omega^2/c^2 \ ,$$

we estimate the value of the longitudinal wavenumber,

$$k_z \simeq \frac{\omega}{c} - \frac{ck_x^2}{2\omega} \simeq \frac{\omega}{c} - \frac{c\pi^2}{8\omega d^2} \ .$$

Therefore, the phase velocity is

$$v_{\mathrm{ph}} \simeq c + \frac{c^3 \pi^2}{8\omega^2 d^2} \ .$$

When the phase velocity differs from the velocity of light, the resonance condition (1.8) can be rewritten as

$$k_{\mathrm{w}} + \omega\left(\frac{1}{v_{\mathrm{ph}}} - \frac{1}{v_z}\right) = 0 \ . \tag{4.54}$$

Using the definition of the detuning parameter

$$\hat{C} = \frac{k_{\mathrm{w}}}{\Gamma} + \frac{\omega}{\Gamma}\left(\frac{1}{v_z} - \frac{1}{c}\right) \ ,$$

we obtain that the resonance condition (4.54) for the first even mode occurs at $\hat{C}_m \simeq \pi^2/(8B)$. This estimation provides good accuracy for the value of the diffraction parameter $B \gtrsim 10$.

In discussing the phase velocity of the beam radiation mode, we note that at each longitudinal z coordinate, the radiation field can be represented as a superposition of plane waves. The reader may wonder why the spatial Fourier transform of the radiation field does not depend on the z coordinate. The answer lies on the effect of the self-reproduction of the radiation field. Analysis of the field distributions at different longitudinal coordinates shows that they are identical with an accuracy of some complex factor which depends only on the longitudinal coordinate along the undulator. The modulus of this factor is the amplification of the amplitude, and its phase gives the common phase shift of the radiation field. This is a very interesting physical effect. Indeed, the radiation, propagating in free space, does not reproduce its configuration and expands due to diffraction effects. Diffraction of the radiation occurs in the FEL amplifier, too, but the electron beam also radiates the wave. The sum of these fields always gives a self-reproducing field configuration when the FEL amplifier operates in the high-gain linear regime.

Now let us study the asymptote of a thin electron beam. In this case $B \to 0$, $|q| \to 0$ and for the first even mode we have asymptotically:

$$q = \nu \tan(\nu) \simeq \nu^2 \simeq 2\left(\hat{A} + i\hat{C}\right)^{-2}.$$

The eigenvalue equation (4.41) takes the form:

$$\hat{A}\left(\hat{A} + i\hat{C}\right)^4 = \frac{2i}{B}. \tag{4.55}$$

Here one should remember that only one root of (4.55) has physical significance and satisfies the condition

$$\operatorname{Re} q = \operatorname{Re}(\hat{A} + i\hat{C})^{-2} > 0.$$

The accuracy of the asymptotes of the wide and the thin electron beam, (4.53) and (4.55), is illustrated with the plots presented in Figs. 4.9 and 4.10.

In the limit of a thin electron beam, $B \to 0$, the maximal growth rate tends to a constant value for the first odd mode:

$$|q| \to 0, \quad \nu \to \pi/2, \quad \max(\operatorname{Re} \hat{A}) \to 2\sqrt{2}/\pi.$$

The accuracy of this approximation is about 2% at $B = 0.01$ when tuning the FEL amplifier to the maximum of the field growth rate.

Now we go over to the analysis of the space charge effects. For simplicity we assume the energy spread to be negligibly small, $\hat{A}_T^2 \to 0$. In the same way as was done above, we find that in the limit of a wide electron beam, $B \to \infty$, and as $\hat{A}_T^2 \to 0$, the eigenvalue equation (4.41) reduces to the eigenvalue equation of the one-dimensional model:

$$\hat{A}\left[\left(\hat{A} + i\hat{C}\right)^2 + \hat{A}_p^2\right] = \frac{i}{B}. \tag{4.56}$$

184   4. Diffraction Effects in the FEL Amplifier

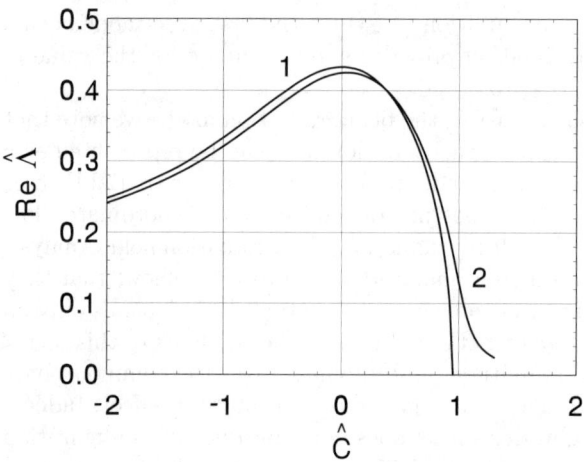

Fig. 4.9. Sheet electron beam with stepped profile. The reduced field growth rate $\operatorname{Re}\hat{\Lambda}$ versus the detuning parameter $\hat{C}$. Here $B = 8$, $\hat{\Lambda}_p^2 \to 0$, and $\hat{\Lambda}_T^2 = 0$. Curve 1 is the solution of the asymptotic equation (4.53). Curve 2 is the solution of (4.41) for the first even mode

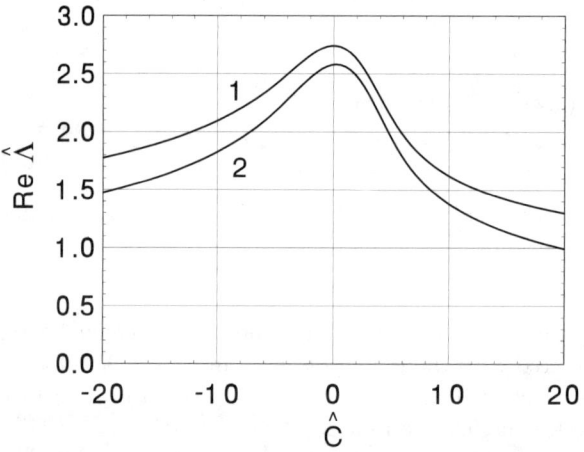

Fig. 4.10. Sheet electron beam with stepped profile. The reduced field growth rate $\operatorname{Re}\hat{\Lambda}$ versus the detuning parameter $\hat{C}$. Here $B = 0.01$, $\hat{\Lambda}_p^2 \to 0$, $\hat{\Lambda}_T^2 = 0$. Curve 1 is the solution of the asymptotic equation (4.55). Curve 2 is the solution of (4.41)

Figure 4.11 illustrates the accuracy of approximation (4.56).

In the limit of a thin electron beam, $B \to 0$, and as $\hat{\Lambda}_T^2 \to 0$, the eigenvalue equation (4.41) for the first even radiation mode takes the form:

$$\hat{\Lambda}\left[\left(\hat{\Lambda}+\mathrm{i}\hat{C}\right)^2 + \hat{\Lambda}_p^2\right]^2 = \frac{2\mathrm{i}}{B} \quad \text{for} \quad \operatorname{Re}\left[(\hat{\Lambda}+\mathrm{i}\hat{C})^2 + \hat{\Lambda}_p^2\right]^{-1} > 0 \,. \quad (4.57)$$

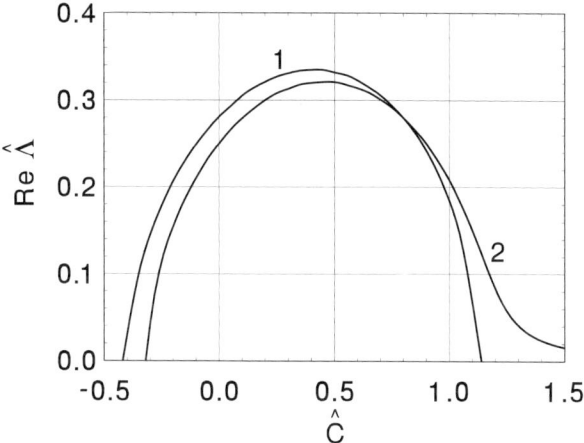

Fig. 4.11. The stepped beam profile. The reduced growth rate $\operatorname{Re}\hat{\Lambda}$ versus the reduced detuning $\hat{C}$. Here $B = 8$, $\hat{\Lambda}_P^2 = 0.25$, and $\hat{\Lambda}_T^2 = 0$. Curve 1 is the solution of the asymptotic equation (4.56). Curve 2 is the solution of (4.41) for the first even mode

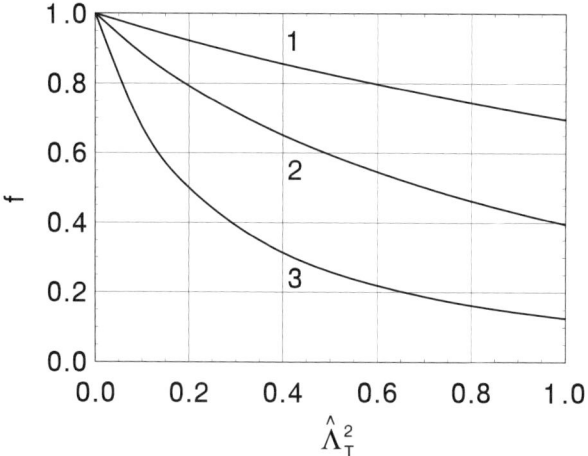

Fig. 4.12. Sheet electron beam with stepped profile. The dependence of the function f for the first even mode on the energy spread parameter $\hat{\Lambda}_T^2$. Curve 1: $B = 0.1$. Curve 2: $B = 1$. Curve 3: $B = 10$

Now let us study the influence of the energy spread on FEL amplifier operation. To clarify the analysis, we consider the case of a negligibly small space charge field, $\hat{\Lambda}_P^2 \to 0$. The maximal growth rate is a function of two parameters, the energy spread parameter and the diffraction parameter, and can be written as

$$\max(\operatorname{Re}\hat{\Lambda}) = \max(\operatorname{Re}\hat{\Lambda})|_{\hat{\Lambda}_T^2 \to 0} \times f(\hat{\Lambda}_T^2, B) \ .$$

186   4. Diffraction Effects in the FEL Amplifier

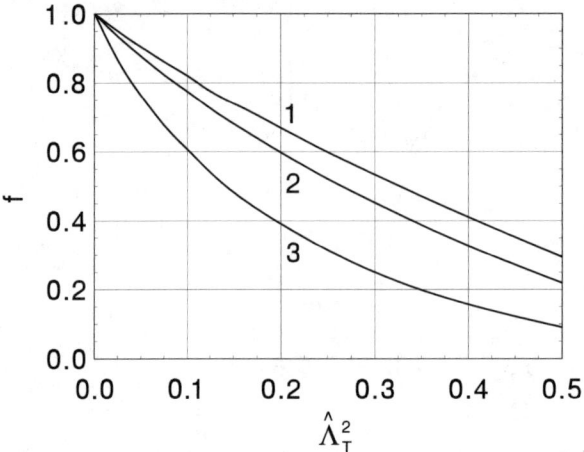

Fig. 4.13. Sheet electron beam with stepped profile. The dependence of the function f for the first odd mode on the energy spread parameter $\hat{\Lambda}_T^2$. Curve 1: $B = 0.1$. Curve 2: $B = 1$. Curve 3: $B = 10$

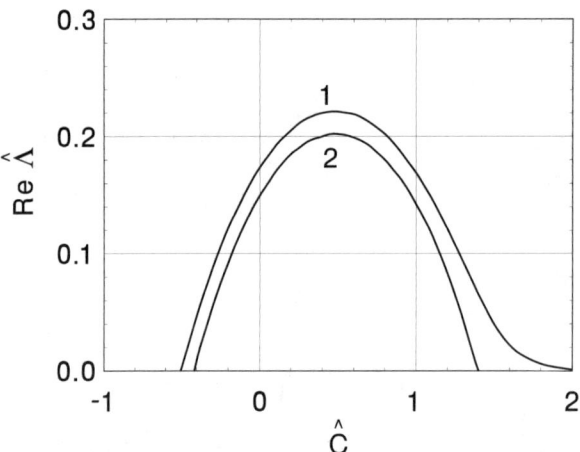

Fig. 4.14. Sheet electron beam with stepped profile. The reduced field growth rate $\mathrm{Re}\,\hat{\Lambda}$ versus the detuning parameter $\hat{C}$. Here $B = 8$, $\hat{\Lambda}_p^2 \to 0$, and $\hat{\Lambda}_T^2 = 0.25$. Curve 1 is the solution of the asymptotic equation (4.58). Curve 2 is the solution of (4.41) for the first even mode

The plots of the function f for the first even and the first odd modes are presented in Figs. 4.12 and 4.13, respectively. It is seen that the energy spread has a greater effect on the odd mode. More detailed study shows that the higher-order modes are suppressed even more strongly with increase of the energy spread. This means that the energy spread serves as a mode selector.

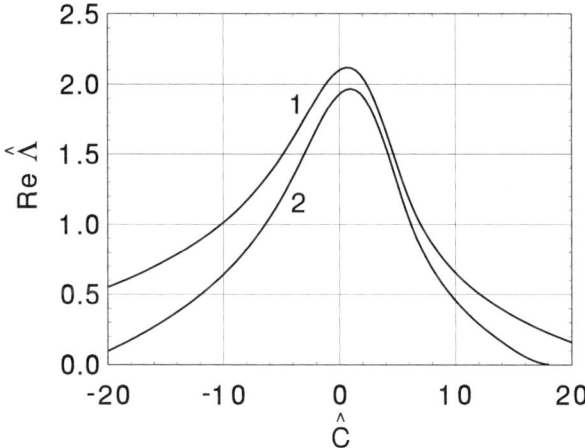

Fig. 4.15. Sheet electron beam with stepped profile. The reduced field growth rate $\operatorname{Re}\hat{\Lambda}$ versus the detuning parameter $\hat{C}$. Here $B = 0.01$, $\hat{\Lambda}_p^2 \to 0$, $\hat{\Lambda}_T^2 = 0.5$. Curve 1 is the solution of the asymptotic equation (4.59). Curve 2 is the solution of (4.41)

Let us study the asymptote of the first even mode for a wide electron beam, $B \to \infty$. It follows from (4.41) that $q \to \infty$ and $\nu \to \pi/2$, which leads to the eigenvalue equation of the one-dimensional approximation:

$$\int_0^\infty \xi \exp\left[-\frac{\hat{\Lambda}_T^2 \xi^2}{2} - \left(\hat{\Lambda} + i\hat{C}\right)\xi\right] d\xi = -iB\hat{\Lambda}. \quad (4.58)$$

In the limit of a thin electron beam, as $B \to 0$, we have for the first even mode (as $\hat{\Lambda}_p^2 \to 0$):

$$\frac{2i}{B}\left\{\int_0^\infty \xi \exp\left[-\frac{\hat{\Lambda}_T^2 \xi^2}{2} - \left(\hat{\Lambda} + i\hat{C}\right)\xi\right] d\xi\right\}^2 = \hat{\Lambda}. \quad (4.59)$$

The roots of the latter equation must satisfy the condition

$$\operatorname{Re}\left\{\int_0^\infty \xi \exp\left[-\frac{\hat{\Lambda}_T^2 \xi^2}{2} - \left(\hat{\Lambda} + i\hat{C}\right)\xi\right] d\xi\right\} > 0.$$

Figures 4.14 and 4.15 illustrate the accuracy of approximations (4.58) and (4.59).

4.3.3 Initial-Value Problem for a Stepped Profile

In the previous section we presented the solution of the eigenvalue problem. The next problem consists in describing the evolution of the radiation field

4. Diffraction Effects in the FEL Amplifier

in the undulator for given initial conditions at the undulator entrance. This problem is usually referred to as an initial-value problem.

In this section we present the solution of the initial-value problem for an FEL amplifier with a sheet electron beam with stepped profile of the current density. We consider the specific case of initial conditions when an external electromagnetic wave of amplitude $\tilde{E}_{\text{ext}}(x)$ and unmodulated electron beam are fed to the undulator entrance. Under the accepted initial conditions the field amplitude $\tilde{E}_y(z,x)$ is subject to (4.39). Equation (4.39) is an integro-differential equation with kernel depending on the difference of two variables, and it can be solved by the Laplace transformation technique. We multiply (4.39) by a factor of $\exp(-p\hat{z})$ (with $\operatorname{Re} p > 0$) and integrate over $\hat{z}$ in the limits from 0 to ∞. The Laplace transforms of (4.39a) and (4.39b) are as follows:

Region 1, ($|\hat{x}| < 1$)

$$\left[\frac{\partial^2}{\partial \hat{x}^2} + \bar{\nu}^2\right]\bar{E} = \hat{f}(\hat{x}) , \qquad (4.60\text{a})$$

Region 2, ($1 < |\hat{x}|$)

$$\left[\frac{\partial^2}{\partial \hat{x}^2} - \bar{q}^2\right]\bar{E} = \hat{f}(\hat{x}) , \qquad (4.60\text{b})$$

where notation

$$\bar{E}(p,\hat{x}) = \int_0^\infty \exp(-p\hat{z})\tilde{E}_y(\hat{z},\hat{x})\mathrm{d}\hat{z} ,$$

$$\bar{q}^2 = -2\mathrm{i}Bp , \qquad \bar{\nu}^2 = \frac{-2\mathrm{i}\hat{D}}{1 - \mathrm{i}\hat{A}_p^2 \hat{D}} - \bar{q}^2 ,$$

$$\hat{D} = \int_{-\infty}^{\infty} \mathrm{d}\hat{P}\, \frac{\hat{F}'(\hat{P})}{p + \mathrm{i}\left(\hat{P} + \hat{C}\right)} , \qquad \hat{f}(\hat{x}) = 2\mathrm{i}B\tilde{E}_{\text{ext}}(\hat{x}) .$$

is introduced.

We use the Green's function method to solve the nonhomogeneous differential equations (4.60). For the region inside the electron beam we seek the solution of (4.60a) for the Green's function $\hat{G}(\hat{x},\hat{x}')$ which satisfies the homogeneous equation

$$\frac{\mathrm{d}^2\hat{G}}{\mathrm{d}\hat{x}^2} + \bar{\nu}^2\hat{G} = 0 \qquad (4.61)$$

for all points except $\hat{x} = \hat{x}'$. The following conditions must be fulfilled at the latter point:

$$\hat{G}|_{\hat{x}=\hat{x}'+0} - \hat{G}|_{\hat{x}=\hat{x}'-0} = 0 , \qquad \frac{\mathrm{d}\hat{G}}{\mathrm{d}\hat{x}}\bigg|_{\hat{x}=\hat{x}'+0} - \frac{\mathrm{d}\hat{G}}{\mathrm{d}\hat{x}}\bigg|_{\hat{x}=\hat{x}'-0} = 1 . \qquad (4.62)$$

We limit our study to the case of an even function $\hat{f}(\hat{x})$, so we consider only positive values of $\hat{x}$ and $\hat{x}'$. In this case the Green's function must satisfy the following condition:

$$\frac{d\hat{G}}{d\hat{x}}\bigg|_{\hat{x}=0} = 0 . \tag{4.63}$$

Let $\chi(\hat{x})$ and $\psi(\hat{x})$ be two linearly independent solutions of (4.61) satisfying the normalization condition

$$\psi \frac{d\chi}{d\hat{x}} - \chi \frac{d\psi}{d\hat{x}} = 1 . \tag{4.64}$$

Then the Green's function

$$\hat{G}(\hat{x}, \hat{x}') = \begin{cases} \psi(\hat{x})\chi(\hat{x}'), & \hat{x} < \hat{x}' \\ \psi(\hat{x}')\chi(\hat{x}), & \hat{x} > \hat{x}' \end{cases} \tag{4.65}$$

is the solution of (4.61) satisfying conditions (4.62). If

$$\frac{d\psi}{d\hat{x}}\bigg|_{\hat{x}=0} = 0 , \tag{4.66}$$

then condition (4.63) is fulfilled. The solutions of the homogeneous equation (4.61) satisfying conditions (4.64) and (4.66) are as follows:

$$\psi(\hat{x}) = \frac{1}{\sqrt{\bar{\nu}}} \cos(\bar{\nu}\hat{x}), \qquad \chi(\hat{x}) = \frac{1}{\sqrt{\bar{\nu}}} \sin(\bar{\nu}\hat{x}).$$

We obtain from (4.65) that

$$\hat{G}(\hat{x}, \hat{x}') = \begin{cases} (\bar{\nu})^{-1} \cos(\bar{\nu}\hat{x}) \sin(\bar{\nu}\hat{x}'), & \hat{x} < \hat{x}' \\ (\bar{\nu})^{-1} \cos(\bar{\nu}\hat{x}') \sin(\bar{\nu}\hat{x}), & \hat{x} > \hat{x}' . \end{cases} \tag{4.67}$$

Thus, we can write the general form for the even solution of (4.60a) inside the beam:

$$\bar{E} = C_1 \cos(\bar{\nu}\hat{x}) + \frac{1}{\bar{\nu}} \sin(\bar{\nu}\hat{x}) \int_0^{\hat{x}} \cos(\bar{\nu}\zeta) \hat{f}(\zeta) d\zeta$$

$$+ \frac{1}{\bar{\nu}} \cos(\bar{\nu}\hat{x}) \int_{\hat{x}}^1 \sin(\bar{\nu}\zeta) \hat{f}(\zeta) d\zeta . \tag{4.68}$$

To find the solution outside the electron beam, we must find the Green's function satisfying the equation

$$\frac{d^2 \hat{G}}{d\hat{x}^2} - \bar{q}^2 \hat{G} = 0 . \tag{4.69}$$

To provide the quadratic integrability of the solution, the Green's function must satisfy the boundary condition

$$\hat{G}(\hat{x}, \hat{x}') \to 0 \quad \text{as} \quad \hat{x} \to \infty . \tag{4.70}$$

The solutions of the homogeneous equation satisfying the normalization condition (4.64) are as follows:

$$\psi(\hat{x}) = \frac{1}{\mathrm{i}\sqrt{2\bar{q}}} \exp(\bar{q}\hat{x}), \qquad \chi(\hat{x}) = \frac{1}{\mathrm{i}\sqrt{2\bar{q}}} \exp(-\bar{q}\hat{x}),$$

where $\mathrm{Re}\,\bar{q} > 0$. Using (4.65), we find that the Green's function satisfying boundary condition (4.70) is given by the expression:

$$\hat{G}(\hat{x}, \hat{x}') = \begin{cases} -(2\bar{q})^{-1} \exp\left[\bar{q}(\hat{x} - \hat{x}')\right], & \hat{x} < \hat{x}' \\ -(2\bar{q})^{-1} \exp\left[-\bar{q}(\hat{x} - \hat{x}')\right], & \hat{x} > \hat{x}' . \end{cases} \tag{4.71}$$

Finally, the general solution of (4.60b) outside the electron beam satisfying the condition $\bar{E} \to 0$ as $\hat{x} \to \infty$ is as follows

$$\bar{E} = C_2 \exp(-\bar{q}\hat{x}) - \frac{1}{2\bar{q}} \int_{\hat{x}}^{\infty} \exp\left[\bar{q}(\hat{x} - \zeta)\right] \hat{f}(\zeta) \mathrm{d}\zeta$$

$$- \frac{1}{2\bar{q}} \int_{1}^{\hat{x}} \exp\left[-\bar{q}(\hat{x} - \zeta)\right] \hat{f}(\zeta) \mathrm{d}\zeta . \tag{4.72}$$

The coefficients C_1 and C_2 are defined from the continuity condition for $\bar{E}$ and its derivative at the beam boundary:

$$\bar{E}|_{\hat{x}=1+0} = \bar{E}|_{\hat{x}=1-0}, \qquad \left.\frac{\mathrm{d}\bar{E}}{\mathrm{d}\hat{x}}\right|_{\hat{x}=1+0} = \left.\frac{\mathrm{d}\bar{E}}{\mathrm{d}\hat{x}}\right|_{\hat{x}=1-0} . \tag{4.73}$$

Using (4.68), we find from (4.73) that

$$C_1 \cos(\bar{\nu}) + \frac{1}{\bar{\nu}} \sin(\bar{\nu}) \int_0^1 \cos(\bar{\nu}\zeta) \hat{f}(\zeta) \mathrm{d}\zeta$$

$$= C_2 \exp(-\bar{q}) - \frac{1}{2\bar{q}} \int_1^{\infty} \exp\left[\bar{q}(1-\zeta)\right] \hat{f}(\zeta) \mathrm{d}\zeta ,$$

$$-C_1 \bar{\nu} \sin(\bar{\nu}) + \cos(\bar{\nu}) \int_0^1 \cos(\bar{\nu}\zeta) \hat{f}(\zeta) \mathrm{d}\zeta$$

$$= -C_2 \bar{q} \exp(-\bar{q}) - \frac{1}{2} \int_1^{\infty} \exp\left[\bar{q}(1-\zeta)\right] \hat{f}(\zeta) \mathrm{d}\zeta . \tag{4.74}$$

Solutions of (4.74) for C_1 and C_2 are as follows:

$$C_1 = \frac{1}{\cos(\bar{\nu})[\bar{\nu}\tan(\bar{\nu}) - \bar{q}]} \left[\int_1^\infty \exp[\bar{q}(1-\zeta)]\hat{f}(\zeta)d\zeta \right.$$

$$\left. + \left[\cos(\bar{\nu}) + \frac{\bar{q}}{\bar{\nu}}\sin(\bar{\nu})\right] \int_0^1 \cos(\bar{\nu}\zeta)\hat{f}(\zeta)d\zeta \right],$$

$$C_2 = \frac{\exp(\bar{q})}{\bar{\nu}\tan(\bar{\nu}) - \bar{q}} \left[\frac{1}{\cos(\bar{\nu})} \int_0^1 \cos(\bar{\nu}\zeta)\hat{f}(\zeta)d\zeta \right.$$

$$\left. + \frac{1}{2\bar{q}}[\bar{q} + \bar{\nu}\tan(\bar{\nu})] \int_1^\infty \exp[\bar{q}(1-\zeta)]\hat{f}(\zeta)d\zeta \right]. \tag{4.75}$$

Substituting these expressions in (4.68) and (4.72) we obtain $\bar{E}(p,\hat{x})$. The value of the field amplitude $\tilde{E}(\hat{z},\hat{x})$ is obtained by means of the inverse Laplace transform of $\bar{E}(p,\hat{x})$:

$$\tilde{E}(\hat{z},\hat{x}) = \frac{1}{2\pi\mathrm{i}} \int_{\gamma'-\mathrm{i}\infty}^{\gamma'+\mathrm{i}\infty} d\lambda \bar{E}(\lambda,\hat{x})\exp(\lambda\hat{z}) . \tag{4.76}$$

The integration path on the complex plane λ is parallel to the imaginary axis. The constant γ' is real and positive with a value larger than all the real parts of the singularities of the integrand. According to (4.60), (4.72) and (4.75), the origin of coordinates is the branch point for the integrand in (4.76). For the function $\bar{E}(\lambda,\hat{x})$ to be single-valued in the variable λ, it is necessary to make a cross-cut of the complex plane along the negative part of the imaginary axis. Let us consider a closed path composed of the line $(\gamma' - \mathrm{i}\infty, \gamma' + \mathrm{i}\infty)$, the semicircle of infinite radius on the left half-plane, the bipartite cross-cut along the negative part of the imaginary axis, and the circle of infinitely small radius with center at the origin of coordinates. If $\bar{E}(\lambda,\hat{x})$ satisfies Jordan's lemma, then it follows from Cauchy's residue theorem that the integral along $(\gamma' - \mathrm{i}\infty, \gamma' + \mathrm{i}\infty)$ can be written as a sum of the residues located inside the integration contour, and of integrals along the cross-cuts (it follows from (4.68), (4.72), and (4.75) that the integral along the circle of infinitely small radius with center at the origin of coordinates is equal to zero).

Let us consider the specific case of a high gain. If the undulator is sufficiently long, the contributions of the residues proportional to $\exp(\lambda_j \hat{z})$ with $\mathrm{Re}(\lambda_j) > 0$ are larger than all the other terms, and the contributions of the latter can be neglected. In this case the even solution for the field amplitude $\tilde{E}(\hat{z},\hat{x})$ takes the form:

$$\tilde{E}(\hat{z},\hat{x}) = \begin{cases} \sum_j u_j [\cos(\nu_j)]^{-1} \cos(\nu_j \hat{x})\exp(\lambda_j \hat{z}), & |\hat{x}| < 1 \\ \sum_j u_j \exp(q_j)\exp(-q_j|\hat{x}|)\exp(\lambda_j \hat{z}), & |\hat{x}| > 1 , \end{cases} \tag{4.77}$$

where λ_j is the jth root of the eigenvalue equation ($\mathrm{Re}\,\lambda_j > 0$):

$$\bar{\nu}(\lambda_j)\tan\bar{\nu}(\lambda_j) - \bar{q}(\lambda_j) = 0 \ . \tag{4.78}$$

It follows from (4.77), (4.78), (4.41), and (4.42) that in the high gain limit the field of the amplified wave can be presented as a set of beam radiation modes with amplitude coefficients

$$u_j = \left[\frac{\mathrm{d}}{\mathrm{d}\lambda}\left(\bar{\nu}\tan(\bar{\nu}) - \bar{q}\right)\right]_{\lambda=\lambda_j}^{-1} \left\{\int_0^1 \frac{\cos(\nu_j\zeta)}{\cos(\nu_j)}\hat{f}(\zeta)\mathrm{d}\zeta \right.$$
$$\left. + \int_1^\infty \exp(q_j - q_j\zeta)\hat{f}(\zeta)\mathrm{d}\zeta\right\} \ . \tag{4.79}$$

The parameters ν_j and q_j are given by

$$\nu_j^2 = \bar{\nu}^2(\lambda_j) = -2\mathrm{i}\hat{D}(\lambda_j)\left[1 - \mathrm{i}\hat{\Lambda}_\mathrm{p}^2\hat{D}(\lambda_j)\right]^{-1} - q_j^2 \ ,$$
$$q_j^2 = \bar{q}^2(\lambda_j) = -2\mathrm{i}B\lambda_j \ ,$$
$$\hat{D}(\lambda_j) = \int_{-\infty}^\infty \mathrm{d}\hat{P}\,\frac{\hat{F}'(\hat{P})}{\lambda_j + \mathrm{i}(\hat{P} + \hat{C})} \ .$$

For a Gaussian energy spread in the electron beam, the function $\hat{D}(\lambda_j)$ is given by

$$\hat{D}(\lambda_j) = \mathrm{i}\int_0^\infty \xi\exp\left[-\frac{\hat{\Lambda}_\mathrm{T}^2\xi^2}{2} - (\lambda_j + \mathrm{i}\hat{C})\xi\right]\mathrm{d}\xi \ .$$

It should be noted that the conditions of Jordan's lemma are not fulfilled for the case of a Gaussian energy spread in the electron beam. So, it is impossible to calculate the integral (4.76) by closing the integration loop in the left half-plane. Nevertheless, it can be shown that all the asymptotic high-gain formulae (4.77)-(4.79) are also valid for the case of a Gaussian energy spread.

Let us study the asymptotic behavior of the solution of the initial-value problem obtained at large values of the diffraction parameter B. We have shown above that in the limit of a wide electron beam the eigenvalue equation is reduced to the eigenvalue equation of the one-dimensional model. We can also expect that the asymptotic solution of the initial-value problem, (4.77) and (4.78), should reduce to the corresponding one-dimensional asymptote. In the limit of $B \to \infty$ it follows from (4.78) that $q_j \to \infty$ and $\nu_j \to (2j-1)\pi/2$. Thus, we have at $B \to \infty$, $\hat{\Lambda}_\mathrm{p}^2 \to 0$, $\hat{\Lambda}_\mathrm{T}^2 = 0$, and $\hat{C} = 0$:

$$\nu_j^2 = 2\lambda_j^{-2} + 2\mathrm{i}B\lambda_j \simeq \pi^2(2j-1)^2/4 \ . \tag{4.80}$$

Taking into account that $q_j \propto B^{1/3}$ and $\cos(\nu_j) \propto B^{-1/3}$ as $B \to \infty$, we obtain:

$$\left[\frac{\mathrm{d}}{\mathrm{d}\lambda}\left(\bar{\nu}\tan(\bar{\nu})-\bar{q}\right)\right]_{\lambda=\lambda_j} \simeq \left[\mathrm{i}B - 2\lambda_j^{-3}\right]\left(\cos^2(\nu_j)\right)^{-1} .$$

It follows from (4.80) that

$$\lambda_j^3 \simeq \frac{\mathrm{i}}{B}, \qquad \left[\frac{\mathrm{d}}{\mathrm{d}\lambda}\left(\bar{\nu}\tan(\bar{\nu})-\bar{q}\right)\right]_{\lambda=\lambda_j} \simeq \frac{3\mathrm{i}B}{\cos^2(\nu_j)} .$$

Let the input signal be a plane wave $\tilde{E}_{\text{ext}} = E_{\text{ext}} = \text{const.}$ at the undulator entrance. It follows from the latter expression that

$$\begin{aligned}
u_j &= \left[\frac{\mathrm{d}}{\mathrm{d}\lambda}\left(\bar{\nu}\tan(\bar{\nu})-\bar{q}\right)\right]_{\lambda=\lambda_j}^{-1} \int_0^1 \frac{\cos(\nu_j\zeta)}{\cos(\nu_j)} \hat{f}(\zeta)\mathrm{d}\zeta \\
&\simeq \frac{2E_{\text{ext}}}{3\nu_j}\sin(\nu_j)\cos(\nu_j) \\
&\simeq \frac{4E_{\text{ext}}}{3\pi(2j-1)}(-1)^{j+1}\cos(\nu_j) .
\end{aligned} \qquad (4.81)$$

For $B \gg 1$ the asymptotic expression for the field amplitude takes the form:

$$\begin{aligned}
\tilde{E} &= \sum_j u_j \frac{\cos(\nu_j\hat{x})}{\cos(\nu_j)}\exp(\lambda_j\hat{z}) \\
&\simeq \frac{4}{3\pi}E_{\text{ext}}\exp\left[(\sqrt{3}+\mathrm{i})\frac{\Gamma z}{2\sqrt[3]{B}}\right] \\
&\quad \times \sum_j \frac{(-1)^{j+1}}{(2j-1)}\cos\left[\frac{(2j-1)\pi}{2}\hat{x}\right] .
\end{aligned} \qquad (4.82)$$

The sum in (4.82) is equal to $\pi/4$, which leads to the asymptotic formula of the one-dimensional model:

$$\tilde{E} = \frac{1}{3}E_{\text{ext}}\exp\left[(\sqrt{3}+\mathrm{i})\frac{\Gamma z}{2\sqrt[3]{B}}\right] .$$

An important characteristic of the FEL amplifier is the power gain G. In the paraxial approximation the power gain is given by the expression

$$G(\hat{z}) = \int_{-\infty}^{\infty} |\tilde{E}(\hat{z},\hat{x})|^2 \mathrm{d}\hat{x} \left[\int_{-\infty}^{\infty} |\tilde{E}_{\text{ext}}(\hat{x})|^2 \mathrm{d}\hat{x}\right]^{-1} .$$

In the region of the diffraction parameter $B \lesssim 10$ the field growth rate of the first even mode visibly exceeds the growth rates of the higher modes (see Fig. 4.8). If the undulator is long enough, the contribution of the first even mode in (4.77) significantly exceeds the contributions of all the higher modes. For this high-gain limit we can write the following asymptotic expression for the power gain:

$$G = 4B^2 \exp\left[2\operatorname{Re}\lambda_1 \hat{z}\right] \left| \left\{ \frac{\mathrm{d}}{\mathrm{d}\lambda} \left(\bar{\nu}\tan(\bar{\nu}) - \bar{q}\right) \right\}_{\lambda=\lambda_1} \right|^{-2}$$

$$\times \left\{ \int_0^\infty \left|\tilde{E}_{\text{ext}}\right|^2 \mathrm{d}\zeta \right\}^{-1} \left| \int_0^1 \frac{\cos(\nu_1\zeta)}{\cos(\nu_1)} \tilde{E}_{\text{ext}}(\zeta)\mathrm{d}\zeta \right.$$

$$\left. + \int_1^\infty \exp(q_1 - q_1\zeta)\tilde{E}_{\text{ext}}(\zeta)\mathrm{d}\zeta \right|^2$$

$$\times \left\{ \int_0^1 \left|\frac{\cos(\nu_1\zeta)}{\cos(\nu_1)}\right|^2 \mathrm{d}\zeta + \int_1^\infty |\exp(q_1 - q_1\zeta)|^2 \mathrm{d}\zeta \right\}. \tag{4.83}$$

This asymptotic expression provides an accuracy of about a few per cent for the power gain $G \gtrsim 20$ dB and the value of the diffraction parameter $B \lesssim 10$.

4.3.4 Epstein Profile

In the previous section we presented the linear theory of the FEL amplifier with a stepped profile of the electron beam. Such a simplified model provides the possibility of completing the analytical description of the linear mode of FEL amplifier operation (the eigenvalue problem and initial-value problem). In reality the electron beam has a gradient distribution of the beam current density. In the following sections we extend the study of the eigenvalue problem for this general case. We solve the eigenvalue problem in two steps. First, we study electron beams with gradient profiles allowing analytical solutions. These profiles are the Epstein and the parabolic profiles. Then we

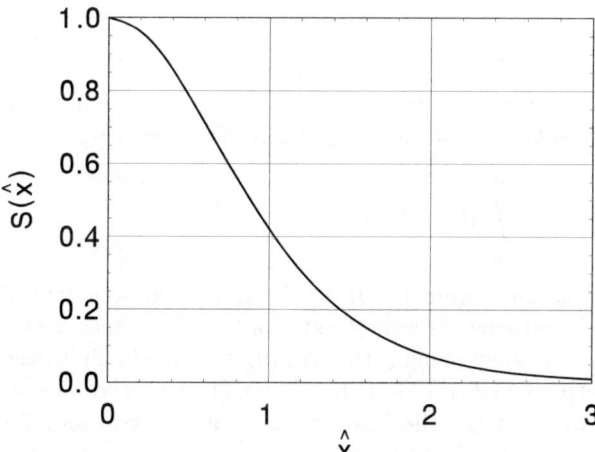

Fig. 4.16. Function of the Epstein profile, $S(\hat{x}) = \cosh^{-2}(\hat{x})$

introduce a semi-analytical multilayer approximation method allowing one to describe any arbitrary gradient profile. Analytical solutions serve as primary standards for testing the approximate methods.

The general form of the transverse distribution of the beam current density of the sheet electron beam is

$$j_0(x) = \bar{j}_0 S(ax) \left(\int_{-\infty}^{\infty} S(ax) \mathrm{d}x \right)^{-1}, \qquad (4.84)$$

where $S(ax)$ is the a function of the profile, a is a normalizing factor, and $\bar{j}_0$ is the linear current density. We study only profiles symmetric with respect to the plane $x = 0$ of the Cartesian coordinate system (x, y, z). To be specific, we normalize the function $S(ax)$ in such a way that $S(0) = 1$.

Using (4.37a) and (4.38a), we can write the following dimensionless self-consistent equation for the field amplitude:

$$\left(\frac{\partial^2}{\partial \hat{x}^2} + 2\mathrm{i}B \frac{\partial}{\partial \hat{z}} \right) \tilde{E}_y(\hat{z}, \hat{x})$$

$$= \mathrm{i} S(\hat{x}) \int_0^{\hat{z}} \mathrm{d}\hat{z}' \left[2 + \hat{\Lambda}_\mathrm{p}^2 \left(\frac{\partial^2}{\partial \hat{x}^2} + 2\mathrm{i}B \frac{\partial}{\partial \hat{z}'} \right) \right] \tilde{E}_y(\hat{z}', \hat{x})$$

$$\times \int_{-\infty}^{\infty} \hat{F}'(\hat{P}) \exp\left[\mathrm{i}(\hat{P} + \hat{C})(\hat{z}' - \hat{z}) \right] \mathrm{d}\hat{P} . \qquad (4.85)$$

Here the following notation is introduced:

$$\hat{x} = ax, \quad \hat{z} = \Gamma z, \quad \hat{C} = C/\Gamma, \quad B = \Gamma \omega/(ca^2),$$

$$\hat{\Lambda}_\mathrm{p}^2 = 8c^2 a^2 \left(A_{\mathrm{JJ}}^2 \theta_\ell^2 \omega^2 \right)^{-1},$$

$$\Gamma = \left[\pi A_{\mathrm{JJ}}^2 \theta_\ell^2 \omega^2 \bar{j}_0 \left(4c^2 a \gamma_\ell^2 \gamma I_\mathrm{A} \int_0^\infty S(\zeta) \mathrm{d}\zeta \right)^{-1} \right]^{1/2}.$$

For a Gaussian energy spread in the electron beam, the distribution function $\hat{F}(\hat{P})$ has the following form:

$$\hat{F}(\hat{P}) = \left(2\pi \hat{\Lambda}_\mathrm{T}^2 \right)^{-1/2} \exp\left(-\frac{\hat{P}^2}{2\hat{\Lambda}_\mathrm{T}^2} \right).$$

where the energy spread parameter $\hat{\Lambda}_\mathrm{T}^2$ is given by

$$\hat{\Lambda}_\mathrm{T}^2 = \langle (\Delta \mathcal{E})^2 \rangle / (\rho^2 \mathcal{E}_0^2), \qquad \rho = c\gamma_\ell^2 \Gamma/\omega.$$

In the linear high-gain limit we seek the solution for the field amplitude $\tilde{E}_y(\hat{z}, \hat{x})$ in the form:

196   4. Diffraction Effects in the FEL Amplifier

$$\tilde{E}_y = \Phi(\hat{x})\exp(\hat{\Lambda}\hat{z}), \qquad \mathrm{Re}\,\hat{\Lambda} > 0 .$$

According to (4.85), the function $\Phi(\hat{x})$ is subject to the equation

$$\frac{\mathrm{d}^2\Phi(\hat{x})}{\mathrm{d}\hat{x}^2} + \left[\frac{-2\mathrm{i}S(\hat{x})\hat{D}}{1 - \mathrm{i}\hat{\Lambda}_\mathrm{p}^2 S(\hat{x})\hat{D}} + 2\mathrm{i}B\hat{\Lambda}\right]\Phi(\hat{x}) = 0 , \qquad (4.86)$$

where the function $\hat{D}$ is given by

$$\hat{D} = \int_{-\infty}^{\infty} \mathrm{d}\hat{P}\,\frac{\hat{F}'(\hat{P})}{\left(\hat{\Lambda} + \mathrm{i}\hat{C}\right) + \mathrm{i}\hat{P}} .$$

For the Gaussian energy spread we have:

$$\hat{D} = \mathrm{i}\int_0^\infty \xi\exp\left[-\frac{\hat{\Lambda}_\mathrm{T}^2\xi^2}{2} - (\hat{\Lambda} + \mathrm{i}\hat{C})\xi\right]\mathrm{d}\xi .$$

One should remember also that the function $\Phi(\hat{x})$ must satisfy the requirement of quadratic integrability.

Let us consider the specific case of the Epstein profile allowing analytical solution of (4.86) in the limit of a negligibly small space charge influence, $\hat{\Lambda}_\mathrm{p}^2 \to 0$. The beam current density in the case of the Epstein profile is given by

$$j_0(x) = \bar{j}_0 a\left[2\cosh^2(ax)\right]^{-1} .$$

The profile function is given by

$$S(\hat{x}) = \cosh^{-2}(\hat{x}) ,$$

where $\hat{x} = ax$ (see Fig. 4.16). The diffraction parameter, B, is given by

$$B = \Gamma\frac{\omega}{ca^2} = \left[\pi A_{\mathrm{JJ}}^2 \theta_\ell^2 \omega^2 \bar{j}_0 \left(4c^2\gamma_\ell^2\gamma I_A a\right)^{-1}\right]^{1/2}\frac{\omega}{ca^2} .$$

To transform (4.86) to standard form, we introduce the following notation:

$$\chi^2 = -2\mathrm{i}B\hat{\Lambda}, \qquad \epsilon(\epsilon - 1) = -2\mathrm{i}\hat{D} .$$

To be specific, we assume that $\mathrm{Re}\,\chi > 0$ and $\mathrm{Re}\,\epsilon > 0$. As a result, (4.86) takes the form:

$$\frac{\mathrm{d}^2\Phi(\hat{x})}{\mathrm{d}\hat{x}^2} + \left[\frac{\epsilon(\epsilon - 1)}{\cosh^2(\hat{x})} - \chi^2\right]\Phi(\hat{x}) = 0 . \qquad (4.87)$$

Formally this equation is similar to that used in the theory of optical waveguides for finding the eigenmodes of the waveguide with the Epstein profile of the refractive index. Nevertheless, there is a principal difference. In the case of an optical waveguide, only the parameter χ depends on the eigenvalue Λ, while in the case of the FEL amplifier the parameter ϵ is also function of the eigenvalue. The consequence is that the eigenmodes of the dielectric waveguide are orthogonal, while the beam radiation modes of the FEL amplifier are not orthogonal.

Equation (4.87) can be transformed to a hypergeometric equation. The standard form of the hypergeometric equation for the function $u(\zeta)$ is:

$$\frac{d^2u}{d\zeta^2} - \frac{(a+b+1)\zeta - c}{\zeta(1-\zeta)} \frac{du}{d\zeta} - \frac{ab}{\zeta(1-\zeta)} u = 0 ,$$

where a, b, and c are constants. Using the substitution

$$u = v(\hat{x})\Phi(\hat{x}) , \quad \zeta = Y(\hat{x}) ,$$

where $Y(\hat{x})$ is an arbitrary function and

$$v(\hat{x}) = \zeta^{-c/2} (1-\zeta)^{(c-a-b-1)/2} \left(\frac{d\zeta}{d\hat{x}}\right)^{1/2} ,$$

we reduce the hypergeometric equation to the form:

$$\frac{d^2\Phi(\hat{x})}{d\hat{x}^2} + g(\hat{x})\Phi(\hat{x}) = 0 .$$

The function $g(\hat{x})$ is given by the expression:

$$g(\hat{x}) = \frac{1}{2}\frac{d^2}{d\hat{x}^2}\left[\ln\left(\frac{dY}{d\hat{x}}\right)\right] - \frac{1}{4}\left\{\frac{d}{d\hat{x}}\left[\ln\left(\frac{dY}{d\hat{x}}\right)\right]\right\}^2$$
$$- \left[\frac{d}{d\hat{x}}[\ln(Y)]\right]^2 \left[D_1 + D_2\frac{Y}{1-Y} + D_3\frac{Y}{(1-Y)^2}\right] , \quad (4.88)$$

where

$$4D_1 = c(c-2) , \quad 4D_2 = 1 - (a-b)^2 + c(c-2) ,$$
$$4D_3 = (a+b-c)^2 - 1 .$$

Let the function $Y(\hat{x})$ be

$$Y(\hat{x}) = -\exp(m\hat{x}) .$$

Then the function $g(\hat{x})$ takes the form:

$$g(\hat{x}) = -m^2\left[D_1 + \frac{1}{4}\right] + m^2 D_2 \frac{\exp(m\hat{x})}{1+\exp(m\hat{x})}$$
$$+ m^2 D_3 \frac{\exp(m\hat{x})}{[1+\exp(m\hat{x})]^2} . \quad (4.89)$$

Choosing constants D_1, D_2, D_3, and m to be equal to

$$D_1 = \chi^2/4 - 1/4 , \quad D_2 = 0 , \quad D_3 = \epsilon(\epsilon-1) , \quad m = -2 ,$$

we obtain the equation

$$\frac{d^2\Phi(\hat{x})}{d\hat{x}^2} + g(\hat{x})\Phi(\hat{x}) = 0 ,$$

where

$$g(\hat{x}) = \frac{\epsilon(\epsilon-1)}{\cosh^2(\hat{x})} - \chi^2 .$$

The parameters χ and ϵ can be expressed in terms of the parameters of the hypergeometric equation a, b, and c:

$$2a = 1 + \sqrt{1+4D_1} + \sqrt{1+4D_3} - \sqrt{1+4(D_1-D_2)} = 2\epsilon \,,$$

$$2b = 1 + \sqrt{1+4D_1} + \sqrt{1+4D_3} + \sqrt{1+4(D_1-D_2)} = 2(\epsilon+\chi) \,,$$

$$c = 1 + \sqrt{1+4D_1} = 1 + \chi \,.$$

The general solution of the hypergeometric equation in the region $|\zeta| < 1$ is given by the expression:

$$u = C_1 \,_2F_1(a,\,b,\,c,\,\zeta) + C_2 \zeta^{1-c} \,_2F_1(a+1-c,\,b+1-c,\,2-c,\,\zeta) \,,$$

where $_2F_1(a,\,b,\,c,\,\zeta)$ is the hypergeometric function. It can be written as a series in the region $|\zeta| < 1$:

$$_2F_1(a,\,b,\,c,\,\zeta) = \frac{\Gamma(c)}{\Gamma(a)\Gamma(b)} \sum_{n=0}^{\infty} \frac{\Gamma(a+n)\Gamma(b+n)}{\Gamma(c+n)n!} \zeta^n \,.$$

Remembering that $\Phi(\hat{x}) = u/v$ and expressing the coefficients of the hypergeometric equations via the coefficients χ and ϵ, we find that the general solution of (4.87) for $0 < \hat{x} < \infty$ is given by

$$\begin{aligned}\Phi(\hat{x}) = &\, C_1 \exp(-\chi\hat{x}) \,_2F_1\left(\epsilon, \epsilon+\chi, 1+\chi, -\exp(-2\hat{x})\right) \\ &\times [1+\exp(-2\hat{x})]^{\epsilon} \\ &+ C_2 \exp(\chi\hat{x}) \,_2F_1\left(\epsilon-\chi, \epsilon, 1-\chi, -\exp(-2\hat{x})\right) \\ &\times [1+\exp(-2\hat{x})]^{\epsilon} \,.\end{aligned} \qquad (4.90)$$

As $\hat{x} \to \infty$, the function $\Phi(\hat{x})$ tends asymptotically to

$$\Phi(\hat{x}) \simeq C_1 \exp(-\chi\hat{x}) + C_2 \exp(\chi\hat{x}) \,.$$

Taking into account the requirement of quadratic integrability of $\Phi(\hat{x})$ and remembering that $\mathrm{Re}\,\chi > 0$, we must set the coefficient C_2 equal to zero. Using the representation of the spherical functions in terms of the hypergeometric functions,

$$P_\beta^\alpha(\xi) = \frac{1}{\Gamma(1-\alpha)} \left(\frac{\xi+1}{\xi-1}\right)^{\alpha/2} \,_2F_1\left(-\beta,\,\beta+1,\,1-\alpha,\,(1-\xi)/2\right) \,,$$

and the relation between the hypergeometric functions,

$$_2F_1(a,\,b,\,c,\,\zeta) = (1-\zeta)^{-a} \,_2F_1\left(a,\,c-b,\,c,\,\zeta/(\zeta-1)\right) \,,$$

we obtain

$$\Phi(\hat{x}) = C_1 \Gamma(1+\chi) P_{\epsilon-1}^{-\chi}(\tanh(\hat{x})) \,, \qquad \text{for} \quad 0 < \hat{x} < \infty \,. \qquad (4.91)$$

The differential equation (4.87) and the boundary conditions are invariant with respect to inversion $\hat{x} \to -\hat{x}$, so the solutions of (4.91) are separated into even and odd functions. The derivative of the even eigenfunction $\Phi(\hat{x})$

must be equal to zero at the center of the beam at $\hat{x} = 0$. Using the derivative of the spherical function at the zero value of the argument,

$$\frac{\sqrt{\pi}}{2^{\alpha+1}} \left[\frac{dP_\alpha^\beta(\xi)}{d\xi}\right]_{\xi=0} = \sin\left[\frac{\pi(\alpha+\beta)}{2}\right] \Gamma\left(1 + \frac{\alpha+\beta}{2}\right)$$
$$\times \left[\Gamma\left(\frac{\beta-\alpha+1}{2}\right)\right]^{-1}, \qquad (4.92)$$

and writing the sine as a product of gamma functions,

$$\frac{\pi}{\Gamma(\alpha)\Gamma(1-\alpha)} = \sin(\pi\alpha),$$

we obtain the eigenvalue equation for the even modes:

$$\Gamma(1+\chi)\left[\Gamma\left(\frac{\epsilon+\chi}{2}\right)\Gamma\left(\frac{1+\chi-\epsilon}{2}\right)\right]^{-1} = 0. \qquad (4.93)$$

Remembering that $\operatorname{Re}\chi > 0$ and $\operatorname{Re}\epsilon > 0$, we find that the arguments of the gamma functions,

$$\Gamma(1+\chi), \quad \Gamma\left(\frac{\chi+\epsilon}{2}\right)$$

are located in the right half-plane, and the gamma functions never take values equal to zero or infinity. As a result, (4.93) is simplified to

$$\Gamma\left(\frac{1+\chi-\epsilon}{2}\right) = \infty.$$

The poles of the gamma functions are negative integer numbers and zero, so the eigenvalue equation for the even modes takes the form:

$$\chi = \epsilon + 1 - 2n, \quad n = 1, 2, \ldots \qquad (4.94)$$

The odd eigenfunction must take a zero value in the center of the beam, $\Phi(0) = 0$. Using the expression for the spherical function of zero argument,

$$P_\beta^\alpha(0) = \sqrt{\pi}2^\alpha \left[\Gamma\left(\frac{\beta-\alpha}{2}+1\right)\Gamma\left(\frac{1-\alpha-\beta}{2}\right)\right]^{-1},$$

we obtain the eigenvalue equation for the odd beam radiation modes:

$$\Gamma(1+\chi)\left[\Gamma\left(\frac{1+\chi+\epsilon}{2}\right)\Gamma\left(1+\frac{\chi-\epsilon}{2}\right)\right]^{-1} = 0. \qquad (4.95)$$

Since $\operatorname{Re}\chi > 0$ and $\operatorname{Re}\epsilon > 0$ in the case under study, so (4.95) is transformed to

$$\Gamma\left(1 + \frac{\chi-\epsilon}{2}\right) = \infty.$$

Finally, the eigenvalue equation for the odd modes can be written in the form:

$$\chi = \epsilon - 2n. \qquad (4.96)$$

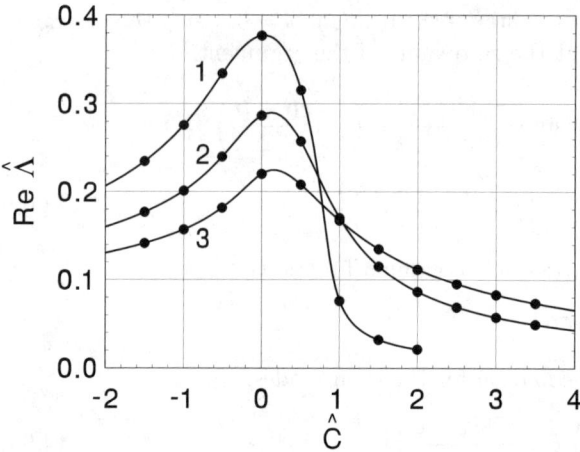

Fig. 4.17. Epstein beam profile. The reduced field growth rate Re $\hat{\Lambda}$ versus $\hat{C}$. Here $B = 10$, $\hat{\Lambda}_\mathrm{p}^2 \to 0$, $\hat{\Lambda}_\mathrm{T}^2 = 0$. Curve 1: the first even mode. Curve 2: the second even mode. Curve 3: the third even mode. The curves show analytical results and the circles are calculated with the multilayer approximation method

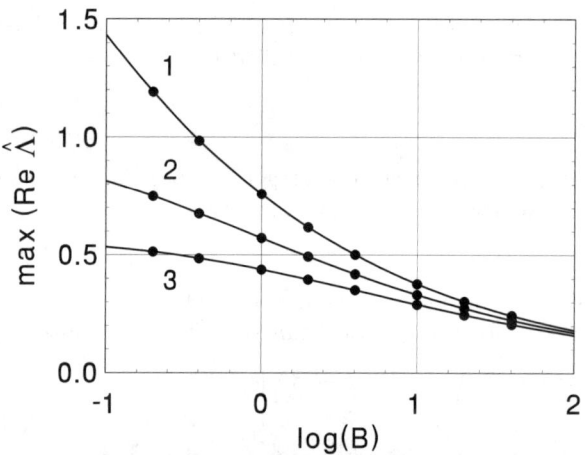

Fig. 4.18. Epstein beam profile. The dependence of the maximal reduced growth rate on the diffraction parameter. Here $\hat{\Lambda}_\mathrm{p}^2 \to 0$, $\hat{\Lambda}_\mathrm{T}^2 = 0$. Curve 1: the first even mode. Curve 2: the first odd mode. Curve 3: the second even mode. The curves show analytical results and the circles are calculated with the multilayer approximation method

Using (4.91) and (4.94), and omitting inessential factors, we can write the expressions for the even radiation modes:

$$\Phi_n^\mathrm{e}(\hat{x}) = \mathrm{P}_{2n-2+\chi}^{-\chi}\left(\tanh(\hat{x})\right) \ .$$

The spherical function $P_{2n-2+\chi}^{-\chi}(\xi)$ can be expressed via elementary functions. For the first even mode we have the following explicit expression:

$$\Phi_1^e(\hat{x}) = P_\chi^{-\chi}(\tanh(\hat{x})) = \text{const.} \times \left[\tanh^2(\hat{x}) - 1\right]^{\chi/2} = \frac{\text{const.}}{[\cosh(\hat{x})]^\chi} . \quad (4.97)$$

The higher-order eigenfunctions can be calculated using the relations:

$$\xi P_\alpha^{-\alpha}(\xi) = P_{\alpha+1}^{-\alpha}(\xi) ,$$

$$\xi(2\beta + 1)P_\beta^\alpha(\xi) = (\beta - \alpha + 1)P_{\beta+1}^\alpha(\xi) + (\beta + \alpha)P_{\beta-1}^\alpha(\xi) .$$

For instance, an explicit expression for the second even mode is

$$\Phi_2^e(\hat{x}) = \left[(2\chi + 3)\tanh^2(\hat{x}) - 1\right][\cosh(\hat{x})]^{-\chi} .$$

Expressions for the odd radiation modes follow from (4.91) and (4.96):

$$\Phi_n^o(\hat{x}) = P_{2n-1+\chi}^{-\chi}(\tanh(\hat{x})) .$$

Explicit expressions for the first and second odd eigenfunctions are as follows:

$$\Phi_1^o(\hat{x}) = \frac{\tanh(\hat{x})}{[\cosh(\hat{x})]^\chi} ,$$

$$\Phi_2^o(\hat{x}) = \frac{\tanh(\hat{x})}{[\cosh(\hat{x})]^\chi} \left[(4\chi^2 + 16\chi + 15)\tanh^2(\hat{x}) - 6\chi + 9\right] .$$

The eigenvalue equations can be solved numerically. Figure 4.17 presents plots of the field growth rate, Re $\hat{\Lambda}$, for the first, second, and third even radiation modes. For each mode and for each value of the diffraction parameter B there is an optimal detuning parameter at which the growth rate of the mode achieves its maximum. In Fig. 4.18 we present the dependencies of the maximal growth rate for different modes on the value of the diffraction parameter, B.

The transverse distribution of the field for the first even mode at different values of the diffraction parameter B is presented in Fig. 4.19. Figure 4.20 illustrates the transverse distribution of the field for the first three even modes at a fixed value of the diffraction parameter $B = 10$.

The next problem of interest is the angular distribution of the radiation intensity. The radiation field at the undulator exit may be written as a superposition of plane waves. The spatial Fourier transform of the complex amplitude of the radiation field, $\Phi(\hat{x})$, at the exit of the undulator, is given by

$$\Xi(\hat{\theta}) = \int_{-\infty}^{\infty} \Phi(\zeta) \exp(-i\hat{\theta}\zeta) d\zeta , \quad (4.98)$$

where $\hat{\theta} = \omega\theta/(ca)$ and $\theta \simeq k_x/k$ is the angle between the wave vector $\mathbf{k}$ and the z axis.

Let us calculate $\Xi(\hat{\theta})$ for the first even radiation mode. Substituting the expression for $\Phi_1^e(\hat{x})$ from (4.97) into (4.98), we obtain:

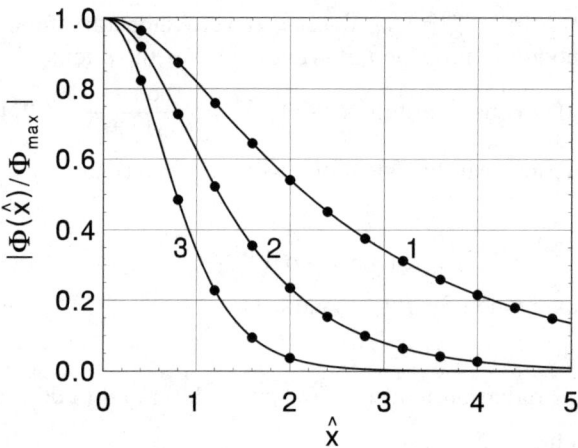

Fig. 4.19. Epstein beam profile. The field distribution of the first even mode versus the reduced coordinate $\hat{x}$. Here $\hat{C} = 0$, $\hat{\Lambda}_{\rm p}^2 \to 0$, $\hat{\Lambda}_{\rm T}^2 = 0$. Curve (1): $B = 0.1$, curve (2): $B = 1$ and curve (3): $B = 10$. The curves show analytical results and the circles are calculated with the multilayer approximation method

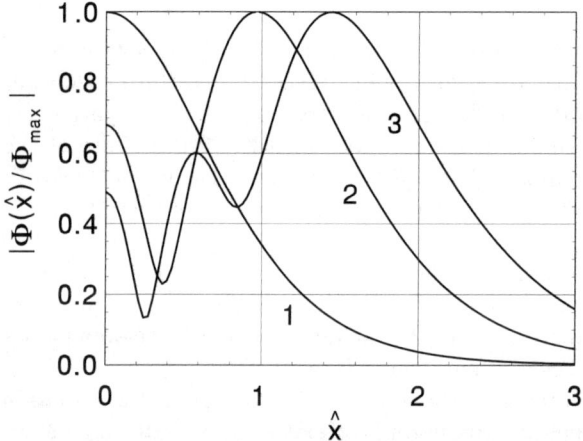

Fig. 4.20. Epstein beam profile. The field distribution of the first three even modes versus the reduced coordinate $\hat{x}$. Here $B = 10$, $\hat{\Lambda}_{\rm p}^2 \to 0$, $\hat{\Lambda}_{\rm T}^2 = 0$ and $\hat{C} = 0$. Curve (1): the first even mode, curve (2): the second even mode, curve (3): the third even mode

$$\Xi(\hat{\theta}) = \int_{-\infty}^{\infty} \cos(\hat{\theta}\zeta) \left[\cosh(\zeta)\right]^{-\chi} d\zeta \ .$$

Taking into account that

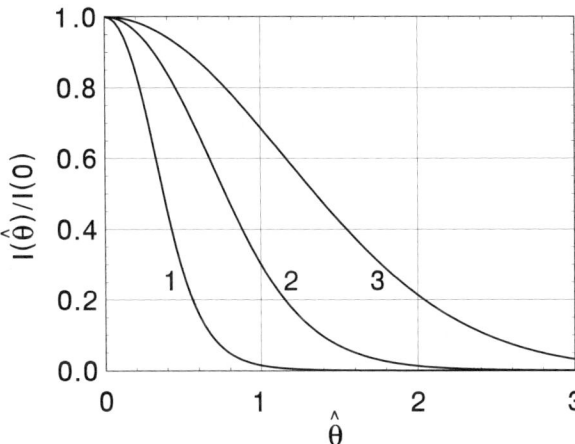

Fig. 4.21. Epstein beam profile. Angular distribution of the radiation intensity for the first even mode. Here $\hat{C} = 0$, $\hat{\Lambda}_p^2 \to 0$, and $\hat{\Lambda}_T^2 = 0$. Curve 1: $B = 0.1$. Curve 2: $B = 1$. Curve 3: $B = 10$

$$\int_0^\infty d\zeta \, \frac{\cos(\alpha\zeta)}{[\cosh(\beta\zeta)]^\nu} = \frac{2^{\nu-2}}{\beta \Gamma(\nu)} \Gamma\left(\frac{\nu}{2} + \frac{i\alpha}{2\beta}\right) \Gamma\left(\frac{\nu}{2} - \frac{i\alpha}{2\beta}\right),$$

for $\operatorname{Re}\nu$, $\operatorname{Re}\beta$, and $\operatorname{Re}\alpha > 0$, and writing the modulus of the gamma function as an infinite product,

$$|\Gamma(x + iy)| = \Gamma(x) \prod_{n=0}^{\infty} \left[1 + \frac{y^2}{(x+n)^2}\right]^{-1/2},$$

we obtain the expression for the angle distribution of the radiation intensity:

$$\frac{I(\hat{\theta})}{I(0)} = \left|\frac{\Xi(\hat{\theta})}{\Xi(0)}\right|^2 = \prod_{n=0}^{\infty} \left[(\chi_1 + 2n)^2 + \chi_2^2\right]^2 \left[(\chi_1 + 2n)^2 + (\chi_2 + \hat{\theta})^2\right]^{-1}$$
$$\times \left[(\chi_1 + 2n)^2 + (\chi_2 - \hat{\theta})^2\right]^{-1}, \qquad (4.99)$$

where $\chi_1 = \operatorname{Re}\chi$ and $\chi_2 = \operatorname{Im}\chi$. Figure 4.21 presents the angular distribution of the radiation intensity in the far zone calculated with (4.99).

Let us study the asymptotic behavior of the field growth rate in the limits of a wide and thin electron beam. To simplify the consideration, we assume the energy spread to be negligibly small, $\hat{\Lambda}_T^2 \to 0$. The parameter $|\chi| \to \infty$ in the limit of a wide electron beam, $B \to \infty$, and we obtain the following asymptote for the even and odd modes:

$$\chi^2 = -2iB\hat{\Lambda} \simeq \epsilon(\epsilon - 1) = 2\left(\hat{\Lambda} + i\hat{C}\right)^{-2}. \qquad (4.100)$$

Redetermination of the gain parameter as $\Gamma B^{-1/3} \to \Gamma$ leads to the cubic eigenvalue equation of the one-dimensional approximation:

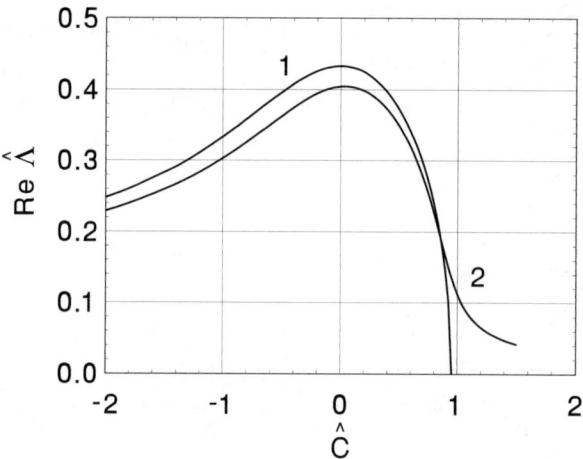

Fig. 4.22. Epstein beam profile. The reduced field growth rate $\operatorname{Re}\hat{\Lambda}$ versus the detuning parameter $\hat{C}$. Here $B = 8$, $\hat{\Lambda}_p^2 \to 0$, and $\hat{\Lambda}_T^2 = 0$. Curve 1 is the solution of the asymptotic equation (4.100). Curve 2 is the solution of (4.94) for the first even mode

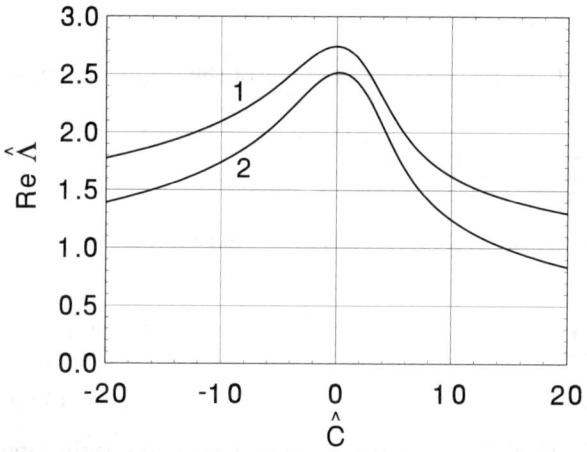

Fig. 4.23. Epstein beam profile. The reduced field growth rate $\operatorname{Re}\hat{\Lambda}$ versus the detuning parameter $\hat{C}$. Here $B = 0.01$, $\hat{\Lambda}_p^2 \to 0$, $\hat{\Lambda}_T^2 = 0$. Curve 1 is the solution of the asymptotic equation (4.101). Curve 2 is the solution of (4.94)

$$\hat{\Lambda}\left(\hat{\Lambda} + i\hat{C}\right)^2 = i \,.$$

The next asymptote is that of a thin electron beam, $B \to 0$. In this case $|\chi| \to 0$, and for the first even mode we can use the following approximation:

$$\chi = \epsilon - 1 = \left[\frac{1}{4} + \frac{2}{\left(\hat{\Lambda} + i\hat{C}\right)^2}\right]^{1/2} - \frac{1}{2} \simeq \frac{2}{\left(\hat{\Lambda} + i\hat{C}\right)^2} \,.$$

As a result, the eigenvalue equation takes the form:

$$\left(\hat{\Lambda} + i\hat{C}\right)^4 \hat{\Lambda} = \frac{2i}{B}, \qquad \mathrm{Re}\left(\hat{\Lambda} + i\hat{C}\right)^{-2} > 0 \,. \tag{4.101}$$

Figures 4.22 and 4.23 illustrate the accuracy of approximations (4.100) and (4.101).

In the limit of a thin electron beam, as $B \to 0$, for the first odd mode we obtain:

$$\chi = \epsilon - 2 = \left[\frac{1}{4} + \frac{2}{\left(\hat{\Lambda} + i\hat{C}\right)^2}\right]^{1/2} - \frac{3}{2} \to 0 \,,$$

and the maximal field growth rate tends asymptotically to $\max(\mathrm{Re}\,\hat{\Lambda}) \to 1$.

4.3.5 Parabolic Profile

The bounded parabolic profile also allows one to obtain analytical solutions of (4.86) in the limit of a negligibly small space charge parameter, $\hat{\Lambda}_p^2 \to 0$. The current density of the electron beam with bounded parabolic profile is given by

$$\bar{j}_0(x) = \frac{\bar{j}_0}{2d} \frac{(1 - x^2/d_1^2)}{[1 - d^2/(3d_1^2)]} \qquad \text{for } |x| < d \,,$$

$$\bar{j}_0(x) = 0 \qquad \text{for } |x| > d \,,$$

where $d < d_1$. The profile function has the form

$$S(\hat{x}) = 1 - k_1^2 \hat{x}^2 \,,$$

where $\hat{x} = x/d$ and $k_1 = d/d_1$ (see Fig. 4.24). The diffraction parameter, B, is given by

$$B = \Gamma \frac{\omega d^2}{c} = \left[\frac{\pi A_{\mathrm{JJ}}^2 \omega^2 \theta_\ell^2 \bar{j}_0 d}{4c^2 I_A \gamma_\ell^2 \gamma (1 - k_1^2/3)}\right]^{1/2} \frac{\omega d^2}{c} \,.$$

To reduce (4.86) to standard form, we introduce the following notation:

$$\nu^2 = -2i\hat{D} - q^2, \qquad q^2 = -2iB\hat{\Lambda}, \qquad \delta^2 = -2i\hat{D}k_1^2 \,.$$

To be specific, we assume that $\mathrm{Re}\,\delta > 0$ and $\mathrm{Re}\,q > 0$. As a result, we obtain the following equations from (4.86):

$$\frac{\mathrm{d}^2 \Phi(\hat{x})}{\mathrm{d}\hat{x}^2} + \left[\nu^2 - \delta^2 \hat{x}^2\right] \Phi(\hat{x}) = 0, \qquad \text{for } |\hat{x}| < 1 \,, \tag{4.102a}$$

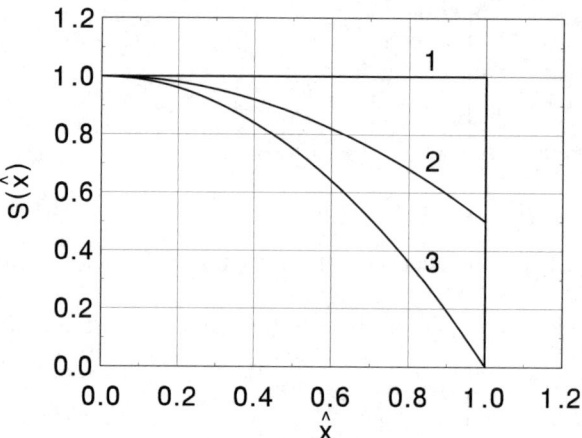

Fig. 4.24. Parabolic beam profile function $S(\hat{x}) = 1 - k_1^2 \hat{x}^2$. Curve 1: $k_1^2 = 0$. Curve 2: $k_1^2 = 0.5$. Curve (3): $k_1^2 = 1$

$$\frac{d^2 \Phi(\hat{x})}{d\hat{x}^2} - q^2 \Phi(\hat{x}) = 0, \qquad \text{for} \quad |\hat{x}| > 1. \tag{4.102b}$$

Introducing the variable $\zeta = \delta \hat{x}^2$, we transform (4.102a) into

$$\zeta \frac{d^2 \Phi(\zeta)}{d\zeta^2} + \frac{1}{2} \frac{d\Phi(\zeta)}{d\zeta} + \frac{1}{4} \left[\frac{\nu^2}{\delta} - \zeta \right] \Phi(\zeta) = 0. \tag{4.103}$$

We seek the solution for $\Phi(\zeta)$ in the form

$$\Phi(\zeta) = \exp(-\zeta/2) u(\zeta).$$

Substituting the latter expression into (4.103), we obtain the Kummer equation for $u(\zeta)$:

$$\zeta \frac{d^2 u}{d\zeta^2} + \left(\frac{1}{2} - \zeta \right) \frac{du}{d\zeta} - \epsilon u = 0, \tag{4.104}$$

where $\epsilon = 1/4 - \nu^2/(4\delta)$. The solutions of (4.104) are separated into even and odd functions. The even solution is

$$u(\zeta) = {}_1F_1(\epsilon, 1/2, \zeta),$$

where ${}_1F_1(\epsilon, \chi, \zeta)$ is the confluent hypergeometric function given by the series

$${}_1F_1(\epsilon, \chi, \zeta) = \frac{\Gamma(\chi)}{\Gamma(\epsilon)} \sum_{n=0}^{\infty} \frac{\Gamma(\epsilon + n)}{\Gamma(\chi + n)} \frac{\zeta^n}{n!}.$$

The odd solution of (4.104) is

$$\zeta^{1/2} {}_1F_1(\epsilon + 1/2, 3/2, \zeta).$$

The even eigenfunction can be written in the form ($\text{Re}\,\delta > 0$ and $\text{Re}\,q > 0$):

$$\Phi(x) = \begin{cases} C_1 \exp(-\delta \hat{x}^2/2) \, _1F_1(\epsilon, \, 1/2, \, \delta \hat{x}^2), & \text{for } |\hat{x}| < 1 \\ C_2 \exp(-q|\hat{x}|), & \text{for } |\hat{x}| > 1. \end{cases} \quad (4.105)$$

The continuity requirement for $\Phi(\hat{x})$ and $\mathrm{d}\Phi(\hat{x})/\mathrm{d}\hat{x}$ at the beam boundary, at $\hat{x} = 1$, leads to the eigenvalue equation for the even modes:

$$(q - \delta) \, _1F_1(\epsilon, \, 1/2, \, \delta) + 4\epsilon\delta \, _1F_1(\epsilon + 1, \, 3/2, \, \delta) = 0. \quad (4.106)$$

When deriving this equation we used the following relation for the derivative of the confluent hypergeometric function:

$$\frac{\mathrm{d}}{\mathrm{d}\zeta} \, _1F_1(\epsilon, \, \chi, \, \zeta) = \frac{\epsilon}{\chi} \, _1F_1(\epsilon + 1, \, \chi + 1, \, \zeta),$$

The explicit expression for the even eigenfunction is

$$\Phi(x) = \begin{cases} \exp(-\delta \hat{x}^2/2) \, _1F_1(\epsilon, \, 1/2, \, \delta \hat{x}^2), & \text{for } |\hat{x}| < 1 \\ \exp(-\delta/2) \, _1F_1(\epsilon, \, 1/2, \, \delta) \exp(q - q|\hat{x}|), & \text{for } |\hat{x}| > 1. \end{cases} \quad (4.107)$$

In a similar way we obtain the eigenvalue equation for the odd radiation modes:

$$\frac{3(\delta - 1 - q)}{(4\epsilon + 2)\delta} \, _1F_1(\epsilon + 1/2, \, 3/2, \, \delta) = \, _1F_1(\epsilon + 3/2, \, 5/2, \, \delta), \quad (4.108)$$

and the explicit expression for the odd eigenfunction:

Region 1, ($|\hat{x}| < 1$):

$$\Phi(\hat{x}) = \exp(-\delta \hat{x}^2/2) \hat{x} \delta^{1/2} \, _1F_1(\epsilon + 1/2, \, 3/2, \, \delta \hat{x}^2), \quad (4.109a)$$

Region 2, ($1 < |\hat{x}|$)

$$\Phi(\hat{x}) = \mathrm{sgn}(\hat{x}) \exp(-\delta/2) \delta^{1/2}$$
$$\times \, _1F_1(\epsilon + 1/2, \, 3/2, \, \delta) \exp(q - q|\hat{x}|). \quad (4.109b)$$

In Fig. 4.25 we present the dependencies of the field growth rate on the detuning parameter calculated with (4.106). The transverse distribution of the radiation field (4.107) is presented in Fig. 4.26.

Let us show that the solutions obtained provide correct asymptotics for the stepped profile of the electron beam. The bounded parabolic profile tends asymptotically to a stepped profile as $k_1 \to 0$ ($d_1 \to \infty$ in this case). The parameter δ tends to zero as $k_1 \to 0$, thus, taking into account the limits

$$\lim_{\delta \to 0} \, _1F_1\left(-\frac{\nu^2}{4\delta}, \, \beta, \, \delta\right) = \Gamma(\beta) \left(\frac{\nu}{2}\right)^{1-\beta} J_{\beta-1}(\nu),$$

we obtain

$$\lim_{\delta \to 0} \, _1F_1(\epsilon, \, 1/2, \, \delta) = \cos(\nu),$$

$$\lim_{\delta \to 0} \, _1F_1(\epsilon + 1, \, 3/2, \, \delta) = \frac{1}{\nu} \sin(\nu).$$

208   4. Diffraction Effects in the FEL Amplifier

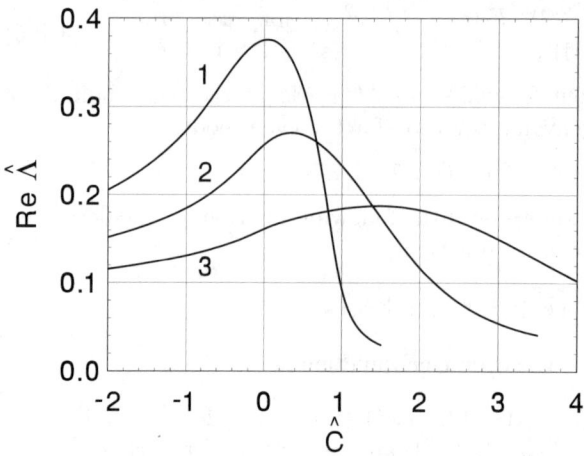

Fig. 4.25. The parabolic beam profile. The reduced field growth rate $\operatorname{Re}\hat{\Lambda}$ versus the detuning parameter $\hat{C}$. Here $k_1 = 1$, $B = 10$, $\hat{\Lambda}_p^2 \to 0$, $\hat{\Lambda}_T^2 = 0$. Curve 1: the first even mode. Curve 2: the second even mode. Curve 3: the third even mode

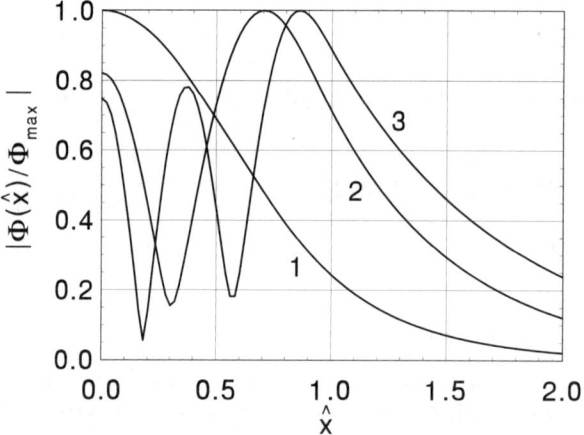

Fig. 4.26. The parabolic beam profile. The field distribution of the first three even modes versus the reduced coordinate $\hat{x}$. Here $k_1 = 1$, $B = 10$, $\hat{\Lambda}_p^2 \to 0$, $\hat{\Lambda}_T^2 = 0$, and $\hat{C} = 0$. Curve 1: the first even mode. Curve 2: the second even mode. Curve 3: the third even mode

Substituting these relations into (4.106), we obtain the eigenvalue equation of the FEL amplifier with a stepped profile for the electron beam (4.41):

$$\nu \tan(\nu) = q \ .$$

Now let us study the asymptotics of the solutions obtained in the limit of a wide and a thin electron beam. In the limit of a wide electron beam $d \to \infty$ at $k_1 = $ const, and the diffraction parameter tends to infinity, $B \to \infty$.

One can obtain that $|q| \to \infty$ and $|\delta| \to \infty$ in this case, and we can use the asymptotic expansion of the hypergeometric functions entering (4.106). The asymptotic expansion of the confluent hypergeometric function at large values of the argument is

$${}_1F_1(\epsilon, \chi, \zeta) \to \frac{\Gamma(\chi)}{\Gamma(\epsilon)} \zeta^{\epsilon-\chi} \exp(\zeta) \quad \text{as} \quad \text{Re}\,\zeta \to \infty .$$

Taking into account (4.107), we find that the eigenfunction grows exponentially in this case. This divergence does not take place only when

$$\epsilon \to -j, \quad j = 0, \ 1, \ 2, \ \ldots$$

which leads to

$$\nu^2 = \delta^2/k_1^2 - q^2 \to (1+4j)\delta \quad \text{as} \quad \delta \to \infty .$$

As a result, the eigenvalue equation (4.106) is simplified to

$$\int_0^\infty \zeta \exp\left[-\frac{\hat{\Lambda}_T^2 \zeta^2}{2} - (\hat{\Lambda} + i\hat{C})\zeta\right] d\zeta = -iB\hat{\Lambda} . \tag{4.110}$$

Rescaling the gain parameter as $\Gamma B^{-1/3} \to \Gamma$, in the limit of a negligibly small energy spread we obtain a cubic equation of the one-dimensional model.

Now let us turn to the asymptote of a thin electron beam, $B \to 0$. One can find that $|q|, |\delta| \to 0$ for the first even mode, and the value of $|q|$ decreases faster than $|\delta|$, i.e. $\nu^2 \simeq \delta^2/k_1^2$ and $\nu^2/\delta \to 0$. So, we can expand the confluent hypergeometric function at small values of the argument:

$${}_1F_1(\epsilon, \chi, \zeta) \simeq 1 + \frac{\epsilon}{\chi}\zeta .$$

As a result, (4.106) is reduced to

$$q = \left(\frac{1}{k_1^2} - \frac{1}{3}\right)\delta^2, \quad \text{Re}\,q > 0 . \tag{4.111}$$

In the limit of a negligibly small energy spread, we obtain from the latter equation:

$$\hat{\Lambda}(\hat{\Lambda} + i\hat{C})^4 = \frac{2i}{B}\left(1 - \frac{k_1^2}{3}\right)^2, \quad \text{Re}\left(\hat{\Lambda} + i\hat{C}\right)^{-2} > 0 .$$

4.3.6 Arbitrary Gradient Profile

It has been shown in the previous sections that analytical techniques allow one to study the eigenvalue problem for a limited number of electron beam profiles (the stepped, the Epstein, and the parabolic profile). To investigate the general case of an arbitrary gradient profile, approximate methods should be used. In this section we describe a multilayer approximation method to solve the eigenvalue problem. The idea of this method is similar to that used in the theory of optical waveguides. The electron beam is divided into a number of layers. The beam current density is assumed to be constant within each layer. The self-consistent field equations are solved within each layer to find the eigenfunctions. The requirements for the eigenfunction and its derivative to be continuous at the boundaries leads to a system of linear equations. Solution of this system allows one to find the eigenvalues and eigenfunctions. Our experience has shown that application of the multilayer method provides good results as for beams with a sharp edge, as well for beams with a smooth distribution of the beam current density (for instance, for a Gaussian profile). To obtain the required accuracy in the latter case, the actual beam is represented as a beam with a sharp boundary. It is always possible to choose the parameters of the divisions (the coordinate of the beam boundary and the number of layers) to provide the required accuracy of the calculations. It is important that multilayer method can be tested with the rigorous solutions obtained above for the Epstein and the parabolic profile.

Let $\hat{x}_b$ be the coordinate of the beam boundary. We divide the interval $(0, \hat{x}_b)$ into K layers with size $(\hat{x}_{j-1}, \hat{x}_j)$. The coordinate of the boundary between the jth and $(j+1)$th layers is

$$\hat{x}_j = j\hat{x}_b/K; \quad j = 1, 2, \ldots, K .$$

The beam current density is assumed to be constant within each layer. We let this value be equal to the beam current density in the middle of the division and denote it as $S_{j-\frac{1}{2}} = S(\hat{x}_{j-\frac{1}{2}})$, where $\hat{x}_{j-\frac{1}{2}} = \hat{x}_b(j-1/2)/K$. The solution of (4.86) in each layer has the form:

$$\Phi_j(\hat{x}) = A_j \cos(\nu_j \hat{x}) + D_j \sin(\nu_j \hat{x}) ,$$

where $(j-1)\hat{x}_b/K < \hat{x} < j\hat{x}_b/K$, and A_j and D_j are constants. The values of ν_j are given by

$$\nu_j^2 = -2\mathrm{i}\hat{D}S_{j-\frac{1}{2}}\left(1 - \mathrm{i}\hat{\Lambda}_p^2 \hat{D}S_{j-\frac{1}{2}}\right)^{-1} - q^2 ,$$
$$q^2 = -2\mathrm{i}B\hat{\Lambda}, \quad \mathrm{Re}\, q > 0 .$$

The solutions of (4.86) are divided into even and odd functions for a symmetric beam profile. The derivative of the eigenfunction of the even solution must be equal to zero in the center of the beam, $\mathrm{d}\Phi(\hat{x})/\mathrm{d}\hat{x} = 0$ at $\hat{x} = 0$. This requirement defines the value of the coefficient, $D_1 = 0$. For the odd solution we have $\Phi(0) = 0$, therefore $A_1 = 0$ in this case. All the other coefficients can be found using the continuity conditions of the eigenfunction and its derivative, $\Phi(\hat{x})$ and $\mathrm{d}\Phi(\hat{x})/\mathrm{d}\hat{x}$, at the boundaries between the layers and the beam boundary. The continuity condition at the boundary between the jth and $(j+1)$th layer is

$$A_j c_j + D_j s_j = A_{j+1} c_{j+1} + D_{j+1} s_{j+1} ,$$
$$\nu_j[-A_j s_j + D_j c_j] = \nu_{j+1}[-A_{j+1} s_{j+1} + D_{j+1} c_{j+1}] ,$$

where

$$c_j = \cos(\nu_j \hat{x}_j) , \quad s_j = \sin(\nu_j \hat{x}_j) ,$$

$$c_{j+1} = \cos(\nu_{j+1}\hat{x}_j), \quad s_{j+1} = \sin(\nu_{j+1}\hat{x}_j).$$

We can write the equations for the coefficients A_{j+1} and D_{j+1} in the matrix form:

$$\begin{pmatrix} A_{j+1} \\ D_{j+1} \end{pmatrix} = T_j \begin{pmatrix} A_j \\ D_j \end{pmatrix}, \tag{4.112}$$

with the elements of the matrix T_j given by

$$(T_j)_{11} = c_j c_{j+1} + \frac{\nu_j}{\nu_{j+1}} s_j s_{j+1},$$

$$(T_j)_{12} = s_j c_{j+1} - \frac{\nu_j}{\nu_{j+1}} c_j s_{j+1},$$

$$(T_j)_{21} = c_j s_{j+1} - \frac{\nu_j}{\nu_{j+1}} s_j c_{j+1},$$

$$(T_j)_{22} = s_j s_{j+1} + \frac{\nu_j}{\nu_{j+1}} c_j c_{j+1}. \tag{4.113}$$

According to (4.86), the solution for the eigenfunction outside the electron beam, at $\hat{x} > \hat{x}_b$, is

$$\Phi(\hat{x}) = F_1 \exp(-q\hat{x}) + F_2 \exp(q\hat{x}).$$

The boundary condition $\Phi(\hat{x}) \to 0$ at $\hat{x} \to \infty$ defines the coefficient $F_2 = 0$. The continuity conditions for the eigenfunction and its derivative at the beam boundary, at $\hat{x} = \hat{x}_b$, give the relations:

$$A_K c_K + D_K s_K = F_1 \exp(-q\hat{x}_b),$$

$$-\nu_K A_K s_K + \nu_K D_K c_K = -q F_1 \exp(-q\hat{x}_b),$$

where

$$c_K = \cos(\nu_K \hat{x}_b), \quad s_K = \sin(\nu_K \hat{x}_b).$$

These relations can be rewritten in the matrix form:

$$T_K \begin{pmatrix} A_K \\ D_K \end{pmatrix} = \exp(q\hat{x}_b) \begin{pmatrix} c_K & s_K \\ \nu_K s_K/q & -\nu_K c_K/q \end{pmatrix} \times \begin{pmatrix} A_K \\ D_K \end{pmatrix}$$

$$= F_1 \begin{pmatrix} 1 \\ 1 \end{pmatrix}. \tag{4.114}$$

Using (4.112), we can express the coefficient F_1 in terms of A_1 and D_1, which gives us the following matrix equation:

$$T_K \times T_{K-1} \times \ldots \times T_1 \begin{pmatrix} A_1 \\ D_1 \end{pmatrix} = T \begin{pmatrix} A_1 \\ D_1 \end{pmatrix} = F_1 \begin{pmatrix} 1 \\ 1 \end{pmatrix}, \tag{4.115}$$

where the matrix T is a function of the unknown quantity $\hat{\Lambda}$. It should be noted that (4.115) is valid for the even modes, as well as for the odd modes. As we mentioned above, the coefficient D_1 must be equal to zero for the even mode. A_1 may be chosen arbitrarily, so without loss of generality, we set

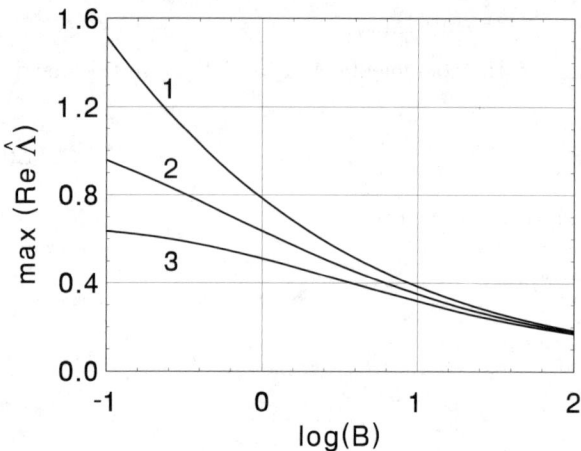

Fig. 4.27. Gaussian beam profile. The maximal reduced growth rate $\max(\operatorname{Re}\hat{\Lambda})$ versus the diffraction parameter B. Here $\hat{\Lambda}_p^2 \to 0$, $\hat{\Lambda}_T^2 = 0$. Curve 1: the first even mode. Curve 2: the first odd mode. Curve 3: the second even mode

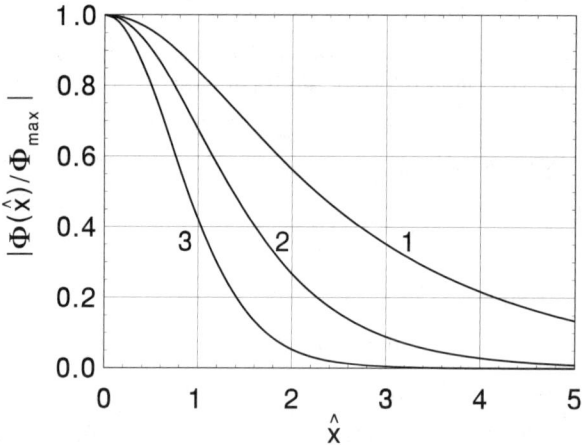

Fig. 4.28. Gaussian beam profile. The field distribution of the first even mode versus the reduced coordinate $\hat{x}$. Here $\hat{C} = 0$, $\hat{\Lambda}_p^2 \to 0$, $\hat{\Lambda}_T^2 = 0$. Curve 1: $B = 0.1$. Curve 2: $B = 1$. Curve 3: $B = 10$

$A_1 = 1$. Then, excluding F_1 from (4.115), it is easy to obtain the eigenvalue equation for the even modes:

$$(T)_{11} = (T)_{21} \,. \tag{4.116}$$

A similar procedure can be applied for the odd modes. Now $A_1 = 0$ and $D_1 = 1$ in this case, and the eigenvalue equation takes the form:

$$(T)_{12} = (T)_{22} \,. \tag{4.117}$$

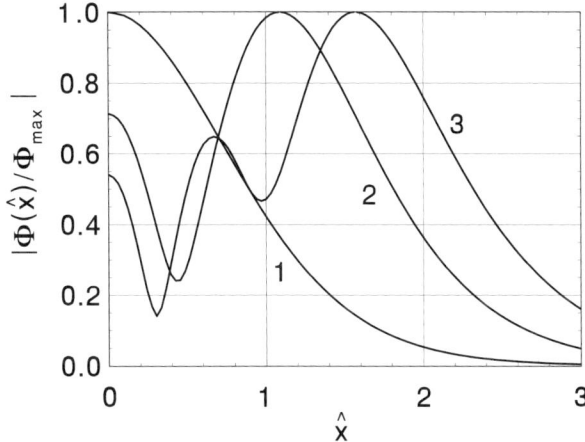

Fig. 4.29. Gaussian beam profile. The field distribution of the first three even modes versus the reduced coordinate $\hat{x}$. Here $B = 10$, $\hat{\Lambda}_p^2 \to 0$, $\hat{\Lambda}_T^2 = 0$, and $\hat{C} = 0$. Curve 1: the first even mode. Curve 2: the second even mode. Curve 3: the third even mode

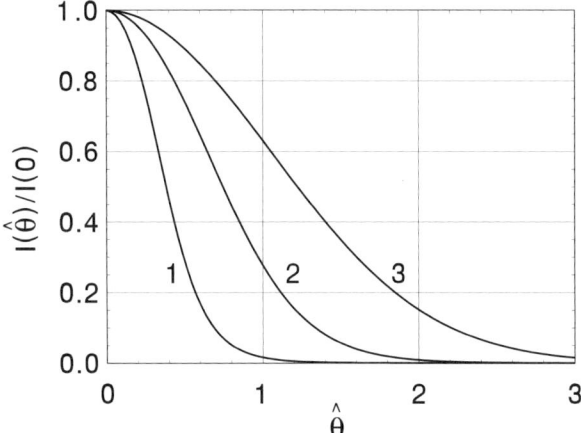

Fig. 4.30. Gaussian beam profile. Angular distribution of the radiation intensity for the first even mode. Here $\hat{C} = 0$, $\hat{\Lambda}_p^2 \to 0$, and $\hat{\Lambda}_T^2 = 0$. Curve 1: $B = 0.1$. Curve 2: $B = 1$. Curve 3: $B = 10$

The beam radiation eigenmode is calculated using (4.112) and (4.114).

Let us illustrate the multilayer method by applying it to a Gaussian profile of the beam current density:

$$j_0(x) = \frac{\bar{j}_0}{\sqrt{2\pi}\sigma} \exp\left(-\frac{x^2}{2\sigma^2}\right) .$$

The transverse coordinate is normalized to σ^{-1}, i.e. $\hat{x} = x/\sigma$. The corresponding profile function is

$$S(\hat{x}) = \exp(-\hat{x}^2/2) \, .$$

Dimensionless parameters are defined as follows:

$$B = \Gamma \frac{\omega \sigma^2}{c} = \left(\frac{\sqrt{\pi} A_{JJ}^2 \omega^2 \theta_\ell^2 \bar{j}_0 \sigma}{\sqrt{8} c^2 I_A \gamma_\ell^2 \gamma} \right)^{1/2} \frac{\omega \sigma^2}{c} \, ,$$

$$\hat{\Lambda}_p^2 = \frac{8c^2}{\sigma^2 \omega^2 \theta_\ell^2 A_{JJ}^2} \, .$$

The reduced observation angle in the far zone is given by $\hat{\theta} = \theta \omega \sigma / c$.

The results of the calculations are presented in Figs. 4.27-4.30. The number of layers is equal to $K = 40$. Testing of the multilayer method was performed with the analytical solutions for the Epstein profile (see Figs. 4.17-4.19). When the number of layers is $K = 40$ the difference between the numerical and analytical results is less than 0.1%.

4.4 Linear Theory of the FEL Amplifier with an Axisymmetric Electron Beam

In this section we present the linear theory of the FEL amplifier with an axisymmetric electron beam. We start with the simplest model of a stepped profile. This allows one to perform a complete analytical study of the eigenvalue problem. Also, the initial-value problem can be solved analytically in the high-gain limit. Then we complicate the considerations with the case of a parabolic profile which allows one to obtain an analytical solution of the eigenvalue problem. The eigenvalue problem for an arbitrary gradient profile is solved by means of the multilayer approximation method. Finally, we present an algorithm for the numerical solution of the initial-value problem.

4.4.1 Eigenvalue Problem for a Stepped Profile

Let us consider a homogeneous axisymmetric electron beam with radius r_0 moving in the magnetic field of the helical undulator. Using cylindrical coordinates (r, φ, z), in the high-gain limit we seek the solution of (4.21) in the form:

$$\tilde{E}(z, r, \varphi) = \Phi_n(r) e^{\Lambda z} \begin{pmatrix} \sin(n\varphi) \\ \cos(n\varphi) \end{pmatrix} , \qquad (4.118)$$

where n is an integer, $n \geq 0$. Substituting (4.118) into (4.21) we get the Bessel equations:

$$\hat{r}^2 \frac{d^2 \Phi_n}{d\hat{r}^2} + \hat{r} \frac{d\Phi_n}{d\hat{r}} + \left(\mu^2 \hat{r}^2 - n^2 \right) \Phi_n = 0 \qquad \text{for } \hat{r} < 1 \, , \qquad (4.119a)$$

$$\hat{r}^2 \frac{d^2 \Phi_n}{d\hat{r}^2} + \hat{r} \frac{d\Phi_n}{d\hat{r}} - \left(g^2 \hat{r}^2 + n^2\right) \Phi_n = 0, \qquad \text{for } \hat{r} > 1 , \tag{4.119b}$$

where $\hat{r} = r/r_0$, $\hat{C} = C/\Gamma$ is the detuning parameter, $\hat{\Lambda} = \Lambda/\Gamma$ is the reduced eigenvalue, $g^2 = -2\mathrm{i}B\hat{\Lambda}$,

$$\mu^2 = \frac{-2\mathrm{i}\hat{D}}{1 - \mathrm{i}\hat{\Lambda}_\mathrm{p}^2 \hat{D}} - g^2 , \qquad \hat{D} = \int_{-\infty}^{\infty} d\hat{P} \, \frac{d\hat{F}(\hat{P})/d\hat{P}}{\hat{\Lambda} + \mathrm{i}\hat{C} + \mathrm{i}\hat{P}} .$$

The reduced energy distribution function, $\hat{F}(\hat{P})$, is normalized to unity, $\int \hat{F}(\hat{P}) d\hat{P} = 1$, where $\hat{P} = (\mathcal{E} - \mathcal{E}_0)/(\rho \mathcal{E}_0)$ is the reduced energy deviation. The gain parameter, Γ, the diffraction parameter, B, the space charge parameter, $\hat{\Lambda}_\mathrm{p}^2$, and the efficiency parameter, ρ, are defined by the formulae:

$$\Gamma = \left[\frac{I_0 \omega^2 \theta_\mathrm{s}^2}{I_\mathrm{A} c^2 \gamma_z^2 \gamma}\right]^{1/2} , \qquad B = \Gamma r_0^2 \omega / c ,$$

$$\hat{\Lambda}_\mathrm{p}^2 = 4c^2/(\omega^2 r_0^2 \theta_\mathrm{s}^2) , \qquad \rho = c\gamma_z^2 \Gamma/\omega ,$$

where $I_0 = \pi r_0^2 j_0$ is the beam current and $I_\mathrm{A} = m_\mathrm{e} c^3/e \simeq 17$ kA is the Alfven current. When the energy spread in the electron beam is Gaussian, the function $\hat{D}$ is given by

$$\hat{D} = \mathrm{i} \int_0^\infty \xi \exp\left[-\hat{\Lambda}_\mathrm{T}^2 \xi^2/2 - (\hat{\Lambda} + \mathrm{i}\hat{C})\xi\right] d\xi ,$$

where $\hat{\Lambda}_\mathrm{T}^2 = \langle (\Delta \mathcal{E})^2 \rangle / (\rho^2 \mathcal{E}_0^2)$ is the energy spread parameter. When the energy spread is negligibly small, $\hat{\Lambda}_\mathrm{T}^2 \to 0$, we have $\hat{D} = \mathrm{i}(\hat{\Lambda} + \mathrm{i}\hat{C})^{-2}$.

To avoid the singularity at $\hat{r} = 0$, the solution for $\Phi_n(\hat{r})$ inside the beam should be chosen to be of the form:

$$\Phi_n(\hat{r}) = C_1 J_n(\mu \hat{r}) \qquad \text{for } \hat{r} < 1 ,$$

where J_n is the Bessel function of the first kind of order n. As the field must vanish as $r \to \infty$, we should choose the solution for $\Phi_n(\hat{r})$ outside the beam in the form (to be specific, we assume here that $\operatorname{Re} g > 0$):

$$\Phi_n(\hat{r}) = C_2 K_n(g\hat{r}) \qquad \text{for } \hat{r} > 1 ,$$

where K_n is the modified Bessel function. The continuity conditions of Φ_n and $d\Phi_n/d\hat{r}$ at the beam boundary give us a system of two linear equations:

$$\begin{aligned} C_1 J_n(\mu) &= C_2 K_n(g) , \\ \mu C_1 J_n'(\mu) &= g C_2 K_n'(g) . \end{aligned} \tag{4.120}$$

From the compatibility condition of this system we obtain the eigenvalue equation for the FEL amplifier with a homogeneous axisymmetric electron beam:

$$\mu J_{n+1}(\mu) K_n(g) = g J_n(\mu) K_{n+1}(g) . \tag{4.121}$$

When deriving this equation we used the following relations between the Bessel functions:

$$K'_n(\zeta) = nK_n(\zeta)/\zeta - K_{n+1}(\zeta) , \qquad J'_n(\zeta) = nJ_n(\zeta)/\zeta - J_{n+1}(\zeta) .$$

The field mode eigenfunction (i.e. the transverse field distribution inside the undulator) is given by the expressions:

$$\Phi_n(\hat{r}) = \begin{cases} J_n(\mu\hat{r}) & \text{for } \hat{r} < 1 \\ J_n(\mu)K_n(g\hat{r})/K_n(g) & \text{for } \hat{r} > 1 . \end{cases} \qquad (4.122)$$

It follows from (4.118) and (4.121) that all the eigenvalues are doubly degenerate because two different eigenfunctions (4.118) correspond to each value of the azimuthal index n.

The angular distribution of the radiation intensity is one of the important characteristics of the FEL amplifier. In the axisymmetric case the spatial Fourier transform of the radiation field is given by

$$\Xi(\hat{\theta}) = \int_0^\infty \Phi_0(\hat{r}) J_0(\hat{\theta}\hat{r}) \hat{r} d\hat{r} , \qquad (4.123)$$

where $\hat{\theta} = \theta r_0 \omega/c$, θ is the observation angle, $\Phi_0(\hat{r})$ is the complex amplitude of the axisymmetric radiation mode at the amplifier exit and J_0 is the Bessel function of the first kind. Using (4.122), we can write the following expression for the function $\Xi(\theta)$:

$$\Xi(\hat{\theta}) = \int_0^1 \zeta J_0(\hat{\theta}\zeta) J_0(\mu\zeta) d\zeta + \int_1^\infty d\zeta \zeta J_0(\mu) J_0(\hat{\theta}\zeta) K_0(g\zeta)/K_0(g)$$

$$= \frac{1}{\hat{\theta}^2 - \mu^2} \left[\hat{\theta} J_1(\hat{\theta}) J_0(\mu) - \mu J_1(\mu) J_0(\hat{\theta}) \right]$$

$$- \frac{1}{g^2 + \hat{\theta}^2} \left[\hat{\theta} J_1(\hat{\theta}) J_0(\mu) - g J_0(\hat{\theta}) J_0(\mu) K_1(g)/K_0(g) \right] . \qquad (4.124)$$

Taking into account (4.121), we get the following expression for the angular distribution of the radiation intensity:

$$\frac{I(\hat{\theta})}{I(0)} = \left| \frac{\Xi(\hat{\theta})}{\Xi(0)} \right|^2 = \left| \frac{J_0(\hat{\theta}) - \hat{\theta} J_1(\hat{\theta}) J_0(\mu)/(\mu J_1(\mu))}{(1 + \hat{\theta}^2/g^2)(1 - \hat{\theta}^2/\mu^2)} \right|^2 . \qquad (4.125)$$

At large values of the diffraction parameter B the Fraunhofer diffraction approximation may be used when $cR_i/(r_0^2\omega) \gg 1$, where R_i is the distance between the observation point and the amplifier exit. When $B \lesssim 1$ the above condition changes to $|\Lambda| R_i \simeq \Gamma R_i \gg 1$.

4.4.2 Analysis of the Solutions

The field distribution of the fundamental symmetric TEM$_{00}$ mode is presented in Fig. 4.31. At large values of the diffraction parameter B the radiation field is mainly concentrated inside and in the vicinity of the electron beam. When the parameter B decreases, the radiation field expands out of the electron beam. Figures 4.32 and 4.33 illustrate the field distribution of the higher radiation modes TEM_{01}, TEM_{02}, TEM_{10}, and TEM_{11}. It should be noticed that the phase front of the beam radiation mode is not a plane one. Figures 4.34 and 4.35 show transverse distributions of the amplitude and of the phase of the TEM_{00} and TEM_{01} modes.

Figure 4.36 presents the angular distribution of the radiation intensity for the fundamental TEM_{00} mode. One can see that the radiation power is mainly concentrated in the small angle near the z axis. At large values of the diffraction parameter B the FWHM of the distribution is approximately equal to $\Delta\theta \simeq 3.6c/(r_0\omega)$. At small values of the parameter B the width of the distribution is much less than $c/(r_0\omega)$. This is due to the fact that in this case the transverse dimension of the beam radiation mode at the amplifier exit significantly exceeds the size of the electron beam (see Fig. 4.31).

According to (4.121), the growth rate is a function of four reduced parameters:

$$\mathrm{Re}\,\hat{\Lambda} = F(\hat{C}, \hat{\Lambda}_{\mathrm{p}}^2, \hat{\Lambda}_{\mathrm{T}}^2, B) \ .$$

To be specific, we assume the energy spread in the electron beam to be Gaussian. We start with the case of a negligibly small space charge field and energy spread, $\hat{\Lambda}_{\mathrm{p}}^2 \to 0$ and $\hat{\Lambda}_{\mathrm{T}}^2 \to 0$. In this region of the parameters the growth rate is a function of only two parameters, the detuning and the diffraction parameter. At some fixed value of the diffraction parameter there is always detuning when the growth rate reaches the maximal value (see

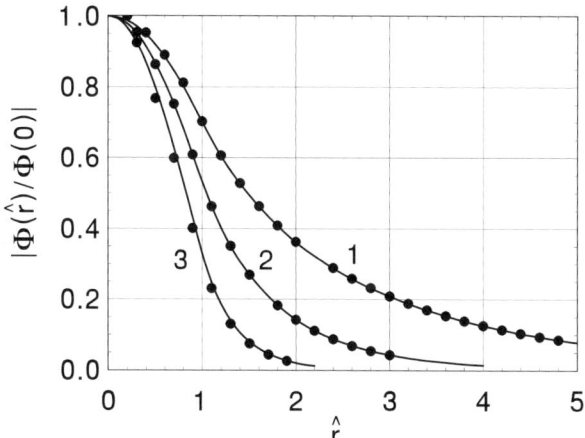

Fig. 4.31. Field distribution of the TEM_{00} mode versus the reduced radius $\hat{r}$. Here $\hat{C} = 0$, $\hat{\Lambda}_{\mathrm{p}}^2 \to 0$, $\hat{\Lambda}_{\mathrm{T}}^2 = 0$. Curve 1: $B = 0.1$. Curve 2: $B = 1$. Curve (3): $B = 10$. The curves show analytical results (TEM_{00} mode) and the circles are calculated with nonlinear simulation code in the linear stage

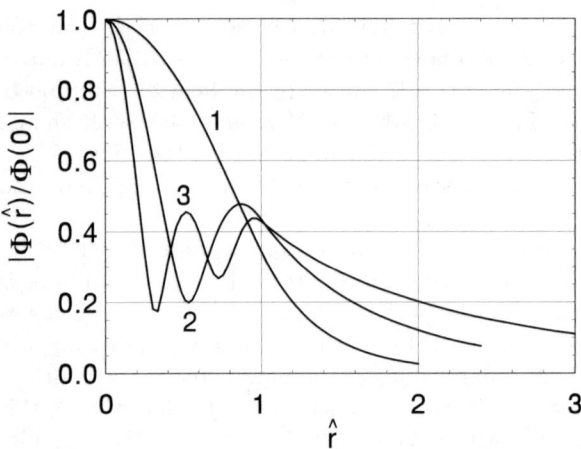

Fig. 4.32. The field distribution of TEM_{00}, TEM_{01} and TEM_{02} modes versus the reduced radius $\hat{r}$. Here $B = 10$, $\hat{\Lambda}_p^2 \to 0$, and $\hat{\Lambda}_T^2 = 0$. Curve 1: TEM_{00} mode ($\hat{C} = 0.1$). Curve 2: TEM_{01} mode ($\hat{C} = 0.7$). Curve 3: TEM_{02} mode ($\hat{C} = 2.0$)

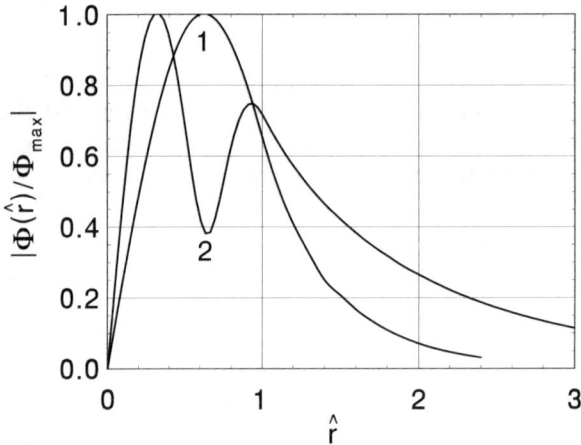

Fig. 4.33. The field distribution of TEM_{10} and TEM_{11} modes versus the reduced radius $\hat{r}$. Here $B = 10$, $\hat{\Lambda}_p^2 \to 0$, $\hat{\Lambda}_T^2 = 0$. Curve 1: TEM_{10} mode ($\hat{C} = 0.2$). Curve 2: TEM_{11} mode ($\hat{C} = 1.1$)

Fig 4.37). It is interesting to trace the dependence of the maximal growth rate as a function of the diffraction parameter. The corresponding dependencies for TEM_{00}, TEM_{01}, and TEM_{02} modes are presented in Fig. 4.38. It is seen from this plot that the fundamental TEM_{00} mode has an advantage over the higher modes. Also, one can see that the asymptotical behavior of the maximal growth rate is quite different for $B \ll 1$ and $B \gg 1$.

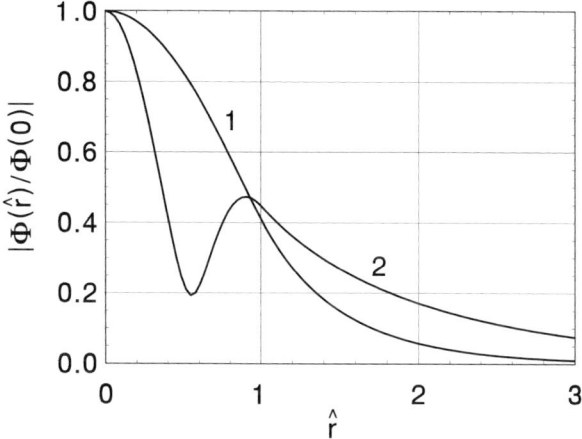

Fig. 4.34. The field distribution of TEM$_{00}$ and TEM$_{01}$ modes versus the reduced coordinate $\hat{r}$. Here $B = 5$, $\hat{\Lambda}_{\rm p}^2 \to 0$, and $\hat{\Lambda}_{\rm T}^2 = 0$. Curve 1: TEM$_{00}$ mode ($\hat{C} = 0.2$). Curve 2: TEM$_{01}$ mode ($\hat{C} = 0.9$)

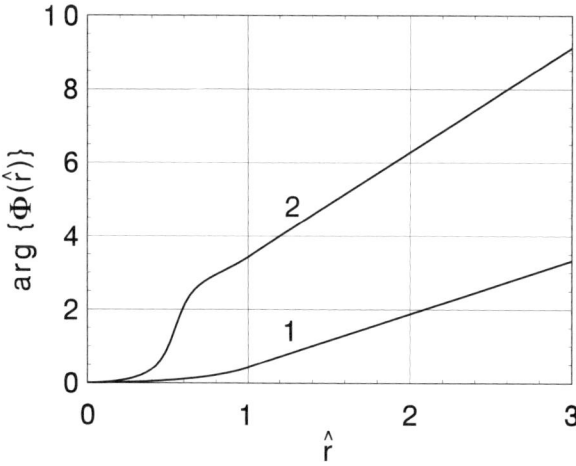

Fig. 4.35. The field phase distribution of TEM$_{00}$ and TEM$_{01}$ modes versus the reduced coordinate $\hat{r}$. Here $B = 5$, $\hat{\Lambda}_{\rm p}^2 \to 0$, and $\hat{\Lambda}_{\rm T}^2 = 0$. Curve 1: TEM$_{00}$ mode ($\hat{C} = 0.2$). Curve 2: TEM$_{01}$ mode ($\hat{C} = 0.9$)

Let us study the asymptote for large values of the diffraction parameter, $B \gg 1$. In this case we let $|g| \gg 1$ and $K_n(g) \simeq K_{n+1}(g)$. Hence, we get from (4.121) that $J_n(\mu) \simeq 0$ which is possible only when $\mu \simeq \nu_{ni}$, where ν_{ni} is the ith root of the Bessel function of order n. As a result, we get asymptotically as $B \to \infty$, $\hat{\Lambda}_{\rm p}^2 \to 0$ and $\hat{\Lambda}_{\rm T}^2 \to 0$:

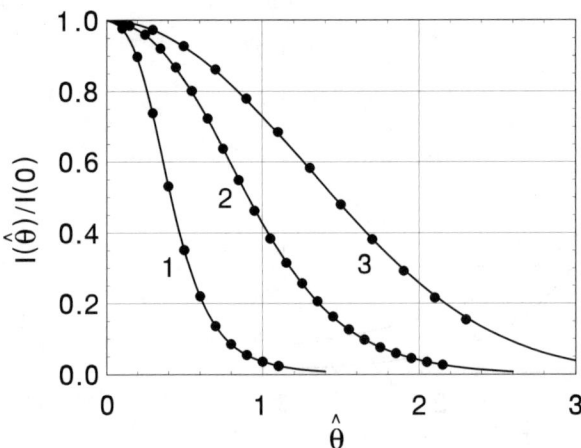

Fig. 4.36. Angular distribution of the radiation intensity for the TEM$_{00}$ mode. Here $\hat{C} = 0$, $\hat{\Lambda}_{\rm p}^2 \to 0$, and $\hat{\Lambda}_{\rm T}^2 = 0$. Curve 1: $B = 0.1$. Curve 2: $B = 1$. Curve 3: $B = 10$. The curves show analytical results (TEM$_{00}$ mode) and the circles are calculated with nonlinear simulation code at the linear stage

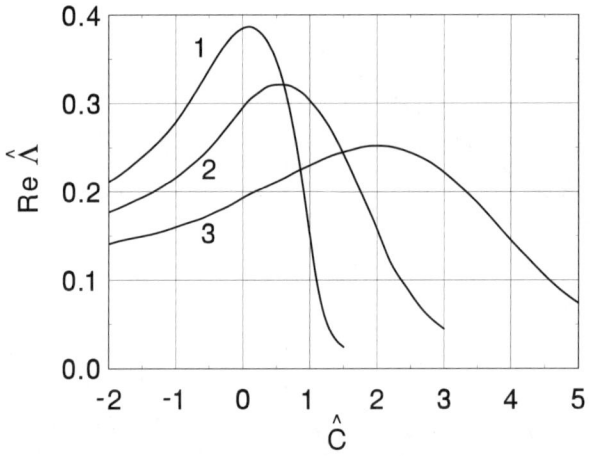

Fig. 4.37. The reduced growth rate $\mathrm{Re}\,\hat{\Lambda}$ versus the reduced detuning $\hat{C}$. Here $B = 10$, $\hat{\Lambda}_{\rm p}^2 \to 0$, $\hat{\Lambda}_{\rm T}^2 = 0$. Curve 1: TEM$_{00}$ mode. Curve 2: TEM$_{01}$ mode. Curve 3: TEM$_{02}$ mode

$$2/(\hat{\Lambda} + i\hat{C})^2 \simeq g^2 = -2iB\hat{\Lambda}\ . \tag{4.126}$$

Redetermination of the gain parameter as

$$\Gamma B^{-1/3} \to \Gamma = \left[\frac{I_0\omega\theta_{\rm s}^2}{I_{\rm A}c\gamma\gamma_z^2 r_0^2}\right]^{1/3}, \tag{4.127}$$

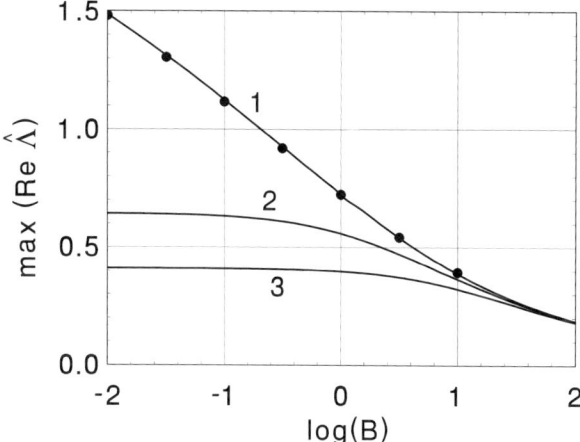

Fig. 4.38. The dependence of the maximal reduced growth rate on the diffraction parameter. Here $\hat{\Lambda}_p^2 \to 0$, $\hat{\Lambda}_T^2 = 0$. Curve 1: TEM$_{00}$ mode. Curve 2: TEM$_{10}$ mode. Curve 3: TEM$_{01}$ mode. The curves show analytical results and the circles are calculated with nonlinear simulation code in the linear stage

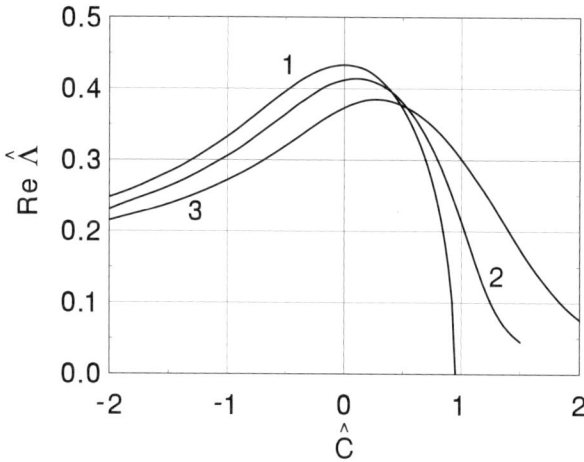

Fig. 4.39. The reduced growth rate Re $\hat{\Lambda}$ versus the reduced detuning $\hat{C}$. Here $B = 8$, $\hat{\Lambda}_p^2 \to 0$ and $\hat{\Lambda}_T^2 = 0$. Curve 1 is the solution of the asymptotic equation (2.36). Curve 2 – (TEM$_{00}$ mode) and curve 3 – (TEM$_{10}$ mode) are the solutions of (4.121)

corresponds to the definition of the gain parameter in the one-dimensional approximation. Then we see that (4.126) becomes identical to the eigenvalue equation (2.36) of the one-dimensional theory. The accuracy of this approximation at the value of the diffraction parameter $B = 8$ is illustrated in Fig. 4.39.

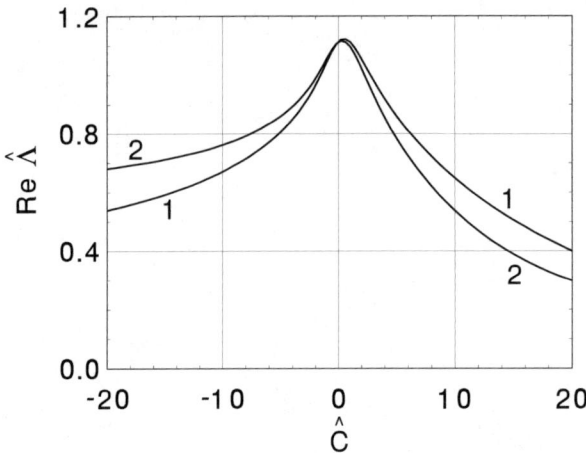

Fig. 4.40. The reduced growth rate $\mathrm{Re}\,\hat{\Lambda}$ versus the reduced detuning $\hat{C}$. Here $B = 0.1$, $\hat{\Lambda}_{\mathrm{p}}^2 \to 0$, $\hat{\Lambda}_{\mathrm{T}}^2 = 0$. Curve 1 is the solution of (4.121) and curve 2 is the solution of the asymptotic equation (4.128)

The asymptote of a thin beam corresponds to small values of the diffraction parameter, $B \to 0$. The argument of the modified Bessel function tends to null in this case, and we can use the expansion for small arguments. For the fundamental TEM$_{00}$ mode we have:

$$K_0(g) \simeq -\ln(g/2) - \gamma_{\mathrm{E}}\,, \qquad K_1(g) \simeq 1/g\,,$$

where $\gamma_{\mathrm{E}} \simeq 0.577$ is Euler's constant. Substituting this approximation into (4.121), we get:

$$\mu J_1(\mu)/J_0(\mu) \simeq -\left[\ln(g/2) + \gamma_{\mathrm{E}}\right]^{-1}\,.$$

Assuming the value of $\ln(1/B)$ to be large, we find with double logarithmic accuracy the eigenvalue equation for the fundamental TEM$_{00}$ mode of the thin beam:

$$2(\hat{\Lambda} + \mathrm{i}\hat{C})^2 = -\ln(-\mathrm{i}B\hat{\Lambda}) + (\ln 2 - 2\gamma_{\mathrm{E}} + 1/2)\,. \tag{4.128}$$

The sum of the last three terms in the right-hand side of (4.128) is equal to 0.03 and may be neglected. The accuracy of this asymptotic is illustrated in Fig. 4.40.

Let us now study the influence of the space charge field. In the limit of $\hat{\Lambda}_{\mathrm{T}}^2 \to 0$ and at fixed parameters $\hat{\Lambda}_{\mathrm{p}}^2$ and B, there is always a value of the detuning parameter $\hat{C}_{\mathrm{m}}$ at which the growth rate achieves its maximal value. This maximal growth rate is a function of two parameters, the space charge parameter $\hat{\Lambda}_{\mathrm{p}}^2$ and the diffraction parameter B, and may be represented in the form:

$$\max(\mathrm{Re}\,\hat{\Lambda}) = \max(\mathrm{Re}\,\hat{\Lambda})|_{\hat{\Lambda}_{\mathrm{p}}^2 \to 0} \times f_1(\hat{\Lambda}_{\mathrm{p}}^2, B)\,.$$

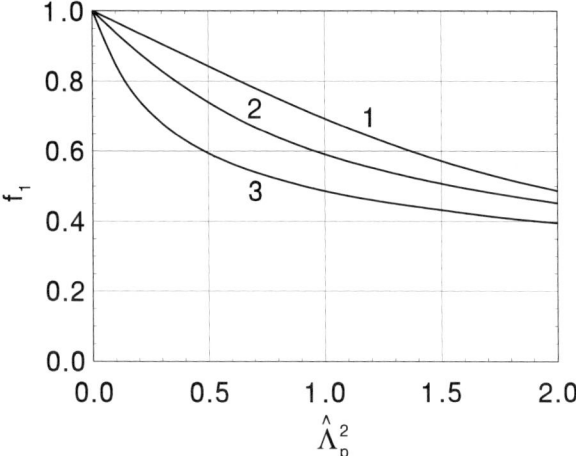

Fig. 4.41. The dependence of the function f_1 for the TEM$_{00}$ mode on the space charge parameter $\hat{\Lambda}_p^2$. Curve 1: $B = 0.1$. Curve 2: $B = 1$. Curve 3: $B = 10$

The plots of the function f_1 for the fundamental TEM$_{00}$ mode are presented in Fig. 4.41.

In the limit of $B \to \infty$ and $\hat{\Lambda}_T^2 \to 0$, (4.121) transforms to the eigenvalue equation of the one-dimensional approximation:

$$\left[(\hat{\Lambda} + i\hat{C})^2 + \hat{\Lambda}_p^2\right] \hat{\Lambda} = i/B \ .$$

Taking into account the redetermination procedure (4.127) for the gain parameter, we find that this equation is identical to the corresponding eigenvalue equation (2.32) of the one-dimensional model.

In the limit of $\hat{\Lambda}_T^2 \to 0$ and $B \to 0$, the eigenvalue equation (4.121) for the fundamental TEM$_{00}$ mode takes the form:

$$2\left[(\hat{\Lambda} + i\hat{C})^2 + \hat{\Lambda}_p^2\right] = -\ln(-iB\hat{\Lambda}) \ .$$

Let us now study the influence of the energy spread on FEL amplifier operation. In the limit of $\hat{\Lambda}_p^2 \to 0$ and at fixed parameters $\hat{\Lambda}_T^2$ and B, there is always a value of the detuning parameter $\hat{C}_m$ when the growth rate achieves its maximum. This maximal growth rate is a function of only two parameters, the energy spread parameter $\hat{\Lambda}_T^2$ and the diffraction parameter B:

$$\max(\mathrm{Re}\,\hat{\Lambda}) = \max(\mathrm{Re}\,\hat{\Lambda})|_{\hat{\Lambda}_T^2 \to 0} \times f_2(\hat{\Lambda}_T^2, B) \ .$$

The plots of the function f_2 for the first two axisymmetric modes are presented in Figs. 4.42 and 4.43. One can find from these plots that the energy spread acts as a strong radiation mode selector. Even at relatively small values of the energy spread parameter, $\hat{\Lambda}_T^2 \simeq 0.1$, the maximal growth rate of the TEM$_{01}$ mode is decreased drastically with respect to the fundamental TEM$_{00}$ mode.

224   4. Diffraction Effects in the FEL Amplifier

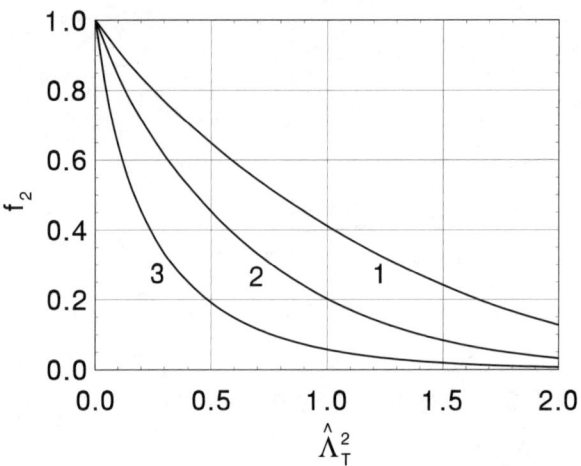

Fig. 4.42. The dependence of the function f_2 for the TEM$_{00}$ mode on the energy spread parameter $\hat{\Lambda}_{\mathrm{T}}^2$. Curve 1: $B = 0.1$. Curve 2: $B = 1$. Curve 3: $B = 10$

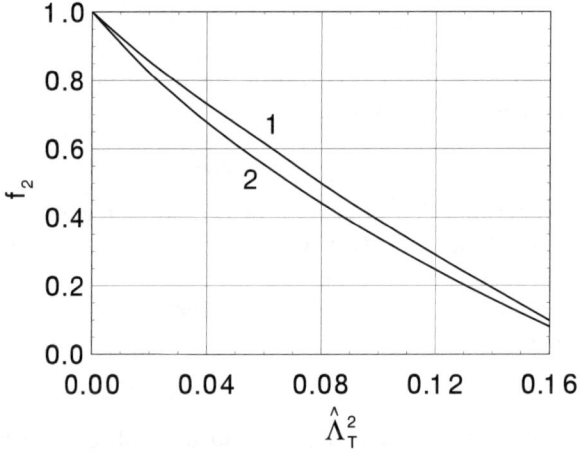

Fig. 4.43. The dependence of the function f_2 for the TEM$_{01}$ mode on the energy spread parameter $\hat{\Lambda}_{\mathrm{T}}^2$. Curve 1: $B = 0.1$. Curve 2: $B = 10$

In the limit of $B \to \infty$ and $\hat{\Lambda}_{\mathrm{p}}^2 \to 0$, the eigenvalue equation (4.121) transforms to:

$$\int_0^\infty \xi \exp\left\{-\hat{\Lambda}_{\mathrm{T}}^2 \xi^2/2 - (\hat{\Lambda} + \mathrm{i}\hat{C})\xi\right\} \mathrm{d}\xi = -\mathrm{i}B\hat{\Lambda} \ .$$

After the redetermination procedure (4.127) of the gain parameter, this equation is reduced to the eigenvalue equation (2.50) of the one-dimensional model.

As $B \to 0$ and $\hat{\Lambda}_p^2 \to 0$, the eigenvalue equation (4.121) for the TEM$_{00}$ mode transforms to:

$$\int_0^\infty \xi \exp\left\{-\hat{\Lambda}_T^2 \xi^2/2 - (\hat{\Lambda} + i\hat{C})\xi\right\} d\xi = -2/\ln(-iB\hat{\Lambda}).$$

4.4.3 Initial-Value Problem for a Stepped Profile

To find the evolution of the electric field of the amplified wave $\tilde{E}(z, r, \varphi)$ one should solve the self-consistent field equations under the given conditions at the undulator entrance. In this chapter we consider the specific, but important practical case of the following initial conditions:

- the electron beam is modulated neither in velocity nor density at the undulator entrance;
- the electric field amplitude $\tilde{E}$ takes the value $\tilde{E}_{\text{ext}}(r, \varphi)$ at the undulator entrance.

In this case the evolution of the complex amplitude $\tilde{E}(z, r, \varphi)$ of the amplified wave is defined by (4.21). We consider the case of an axisymmetric electron beam with stepped profile of the current density. Using the notation introduced in Sect. 4.4.1 we write (4.21) in the following reduced form:

Region 1, ($\hat{r} < 1$):

$$\left[\frac{\partial^2}{\partial \hat{r}^2} + \frac{1}{\hat{r}}\frac{\partial}{\partial \hat{r}} + \frac{1}{\hat{r}^2}\frac{\partial^2}{\partial \varphi^2} + 2iB\frac{\partial}{\partial \hat{z}}\right]\tilde{E}(\hat{z}, \hat{r}, \varphi)$$

$$= i\int_0^{\hat{z}} d\hat{z}' \left[2\tilde{E}(\hat{z}', \hat{r}, \varphi)\right.$$

$$+ \hat{\Lambda}_p^2 \left[\frac{\partial^2}{\partial \hat{r}^2} + \frac{1}{\hat{r}}\frac{\partial}{\partial \hat{r}} + \frac{1}{\hat{r}^2}\frac{\partial^2}{\partial \varphi^2} + 2iB\frac{\partial}{\partial \hat{z}'}\right]\tilde{E}(\hat{z}', \hat{r}, \varphi)\right]$$

$$\times \int_{-\infty}^\infty d\hat{P}(d\hat{F}(\hat{P})/d\hat{P})\exp\left[i(\hat{P} + \hat{C})(\hat{z}' - \hat{z})\right], \quad (4.129a)$$

Region 2, ($1 < \hat{r}$):

$$\left[\frac{\partial^2}{\partial \hat{r}^2} + \frac{1}{\hat{r}}\frac{\partial}{\partial \hat{r}} + \frac{1}{\hat{r}^2}\frac{\partial^2}{\partial \varphi^2} + 2iB\frac{\partial}{\partial \hat{z}}\right]\tilde{E}(\hat{z}, \hat{r}, \varphi) = 0. \quad (4.129b)$$

226   4. Diffraction Effects in the FEL Amplifier

We represent $\tilde{E}$ as a Fourier series in the angle φ:[1]

$$\tilde{E}(\hat{z}, \hat{r}, \varphi) = \sum_{n=-\infty}^{n=+\infty} \tilde{E}^{(n)}(\hat{z}, \hat{r}) e^{-in\varphi} .$$

The Laplace transforms of the Fourier coefficients $\tilde{E}^{(n)}$,

$$\bar{E}^{(n)}(p, \hat{r}) = \int_0^\infty e^{-p\hat{z}} \tilde{E}^{(n)}(\hat{z}, \hat{r}) \mathrm{d}\hat{z} ,$$

satisfy the following equations:

$$\left[\frac{\mathrm{d}^2}{\mathrm{d}\hat{r}^2} + \frac{1}{\hat{r}} \frac{\mathrm{d}}{\mathrm{d}\hat{r}} - \frac{n^2}{\hat{r}^2} + \bar{\mu}^2 \right] \bar{E}^{(n)}(p, \hat{r}) = \hat{f}^{(n)}(\hat{r}) \quad \text{for} \quad \hat{r} < 1 , \quad (4.130\mathrm{a})$$

$$\left[\frac{\mathrm{d}^2}{\mathrm{d}\hat{r}^2} + \frac{1}{\hat{r}} \frac{\mathrm{d}}{\mathrm{d}\hat{r}} - \frac{n^2}{\hat{r}^2} - \bar{g}^2 \right] \bar{E}^{(n)}(p, \hat{r}) = \hat{f}^{(n)}(\hat{r}) \quad \text{for} \quad \hat{r} > 1 , \quad (4.130\mathrm{b})$$

where:

$$\bar{\mu}^2 = -2\mathrm{i}\hat{D}\left[1 - \mathrm{i}\hat{A}_\mathrm{p}^2 \hat{D}\right]^{-1} - \bar{g}^2 , \qquad \bar{g}^2 = -2\mathrm{i}Bp ,$$

$$\hat{D} = \int_{-\infty}^\infty \mathrm{d}\hat{P} \frac{\mathrm{d}\hat{F}(\hat{P})/\mathrm{d}\hat{P}}{p + \mathrm{i}(\hat{P} + \hat{C})} , \qquad \hat{f}^{(n)}(\hat{r}) = 2\mathrm{i}B\tilde{E}_\mathrm{ext}^{(n)}(\hat{r}) .$$

So, using the Fourier expansion and the Laplace transform allows us to go over from the partial integro-differential equations (4.129) to the ordinary differential equations (4.130). This leads to a significant simplification of the problem. To find $\bar{E}^{(n)}$, one must solve (4.130) with the following boundary and continuity conditions:

$$\bar{E}^{(n)}(p, \hat{r}) \to 0 \quad \text{for} \quad \hat{r} \to \infty ,$$

$$\bar{E}^{(n)}|_{\hat{r}=1+0} = \bar{E}^{(n)}|_{\hat{r}=1-0} , \qquad \left.\frac{\mathrm{d}\bar{E}^{(n)}}{\mathrm{d}\hat{r}}\right|_{\hat{r}=1+0} = \left.\frac{\mathrm{d}\bar{E}^{(n)}}{\mathrm{d}\hat{r}}\right|_{\hat{r}=1-0} .$$

We use the Green's function method to solve the inhomogeneous equations (4.130). First, we consider the region inside the beam. We seek the Green's function $\hat{G}(\hat{r}, \hat{r}')$ satisfying the homogeneous equation,

$$\hat{r} \frac{\mathrm{d}}{\mathrm{d}\hat{r}} \left[\hat{r} \frac{\mathrm{d}\hat{G}}{\mathrm{d}\hat{r}} \right] + (\bar{\mu}^2 \hat{r}^2 - n^2) \hat{G} = 0 , \qquad (4.131)$$

for all $\hat{r}$ except $\hat{r} = \hat{r}'$. The following conditions must hold at the latter point:

---

[1] This is an expansion in rotating field configurations. The field components of a mode with index n have $|n|$ variations over the azimuthal angle ϕ. The field vectors rotate in time around the z axis at a fixed longitudinal coordinate. The direction of rotation coincides with the electron rotation direction in the undulator at positive values of n, and vice versa.

$$\hat{G}|_{\hat{r}=\hat{r}'+0} - \hat{G}|_{\hat{r}=\hat{r}'-0} = 0 ,$$

$$\left.\frac{\mathrm{d}\hat{G}}{\mathrm{d}\hat{r}}\right|_{\hat{r}=\hat{r}'+0} - \left.\frac{\mathrm{d}\hat{G}}{\mathrm{d}\hat{r}}\right|_{\hat{r}=\hat{r}'-0} = \frac{1}{\hat{r}} . \tag{4.132}$$

In addition, the Green's function must be finite at $\hat{r} = 0$. If two linearly independent solutions of (4.130a), $\chi(\hat{r})$ and $\psi(\hat{r})$, are normalized as

$$\psi\frac{\mathrm{d}\chi}{\mathrm{d}\hat{r}} - \chi\frac{\mathrm{d}\psi}{\mathrm{d}\hat{r}} = \frac{1}{\hat{r}} , \tag{4.133}$$

then the Green's function is

$$\hat{G}(\hat{r},\hat{r}') = \begin{cases} \psi(\hat{r})\chi(\hat{r}') & \text{for } \hat{r} < \hat{r}' , \\ \psi(\hat{r}')\chi(\hat{r}) & \text{for } \hat{r} > \hat{r}' . \end{cases}$$

This satisfies (4.131) and conditions (4.132). Moreover, if the function $\psi(\hat{r})$ is finite at $\hat{r} = 0$, then the requirement for the Green's function to be finite is fulfilled. The following solutions of the homogeneous equations satisfy all these conditions:

$$\psi = (\pi/2)^{1/2} J_n(\bar{\mu}\hat{r}) , \qquad \chi = (\pi/2)^{1/2} N_n(\bar{\mu}\hat{r}) ,$$

where N_n is the Bessel function of the second kind of order n. Thus, we obtain

$$\hat{G} = \begin{cases} (\pi/2) J_n(\bar{\mu}\hat{r}) N_n(\bar{\mu}\hat{r}') & \text{for } \hat{r} < \hat{r}' , \\ (\pi/2) J_n(\bar{\mu}\hat{r}') N_n(\bar{\mu}\hat{r}) & \text{for } \hat{r} > \hat{r}' . \end{cases}$$

Finally, we find the general form of the solution of the inhomogeneous equation (4.130a):

$$\bar{E}^{(n)} = C_1 J_n(\bar{\mu}\hat{r}) + \frac{\pi}{2} N_n(\bar{\mu}\hat{r}) \int_0^{\hat{r}} \mathrm{d}\zeta\, J_n(\bar{\mu}\zeta)\zeta \hat{f}^{(n)}(\zeta)$$

$$+ \frac{\pi}{2} J_n(\bar{\mu}\hat{r}) \int_{\hat{r}}^{1} \mathrm{d}\zeta\, N_n(\bar{\mu}\zeta)\zeta \hat{f}^{(n)}(\zeta) . \tag{4.134}$$

To find the solution in the region outside the beam, at $\hat{r} > 1$, we also seek the Green's function $\hat{G}(\hat{r},\hat{r}')$ of the homogeneous equation corresponding to the inhomogeneous equation (4.130b). To provide boundedness of the solution as $\hat{r} \to \infty$, the following condition must be fulfilled:

$$\hat{G} \to 0 \qquad \text{for } \hat{r} \to \infty . \tag{4.135}$$

We choose the following solutions of the homogeneous equation satisfying the normalization conditions (4.133):

$$\psi = \mathrm{i} I_n(\bar{g}\hat{r}) , \qquad \chi = \mathrm{i} K_n(\bar{g}\hat{r}) ,$$

and obtain the following Green's function:

228   4. Diffraction Effects in the FEL Amplifier

$$\hat{G} = \begin{cases} -I_n(\bar{g}\hat{r})K_n(\bar{g}\hat{r}') & \text{for } \hat{r} < \hat{r}', \\ -I_n(\bar{g}\hat{r}')K_n(\bar{g}\hat{r}) & \text{for } \hat{r} > \hat{r}'. \end{cases} \quad (4.136)$$

Since $K_n(\bar{g}\hat{r})$ tends to zero as $\hat{r} \to \infty$, the boundary condition (4.135) for the Green's function is fulfilled. Thus, the general solution of the inhomogeneous equation (4.130b) has the form:

$$\bar{E}^{(n)} = C_2 K_n(\bar{g}\hat{r}) - K_n(\bar{g}\hat{r}) \int_1^{\hat{r}} d\zeta I_n(\bar{g}\zeta) \zeta \hat{f}^{(n)}(\zeta)$$

$$- I_n(\bar{g}\hat{r}) \int_{\hat{r}}^{\infty} d\zeta K_n(\bar{g}\zeta) \zeta \hat{f}^{(n)}(\zeta). \quad (4.137)$$

The continuity conditions of $\bar{E}^{(n)}(p,\hat{r})$ and $d\bar{E}^{(n)}(p,\hat{r})/d\hat{r}$ at the beam boundary $\hat{r} = 1$ give us the following equations:

$$C_1 J_n(\bar{\mu}) + \frac{\pi}{2} N_n(\bar{\mu}) \int_0^1 d\zeta J_n(\bar{\mu}\zeta) \zeta \hat{f}^{(n)}(\zeta)$$

$$= C_2 K_n(\bar{g}) - I_n(\bar{g}) \int_1^{\infty} d\zeta K_n(\bar{g}\zeta) \zeta \hat{f}^{(n)}(\zeta),$$

$$C_1 \bar{\mu} J_{n+1}(\bar{\mu}) + \frac{\pi}{2} N_{n+1}(\bar{\mu}) \int_0^1 d\zeta J_n(\bar{\mu}\zeta) \zeta \hat{f}^{(n)}(\zeta)$$

$$= C_2 \bar{g} K_{n+1}(\bar{g}) + \bar{g} I_{n+1}(\bar{g}) \int_1^{\infty} d\zeta K_n(\bar{g}\zeta) \zeta \hat{f}^{(n)}(\zeta). \quad (4.138)$$

Solving this system of equations and using the relations

$$N_{n+1}(\xi) J_n(\xi) - N_n(\xi) J_{n+1}(\xi) = -\frac{2}{\pi \xi},$$

$$I_{n+1}(\xi) K_n(\xi) + I_n(\xi) K_{n+1}(\xi) = \frac{1}{\xi},$$

we obtain

$$C_1 = \frac{1}{\bar{\mu} J_{n+1}(\bar{\mu}) K_n(\bar{g}) - \bar{g} K_{n+1}(\bar{g}) J_n(\bar{\mu})} \int_1^{\infty} d\zeta K_n(\bar{g}\zeta) \zeta \hat{f}^{(n)}(\zeta)$$

$$+ \frac{\pi}{2} \frac{\bar{g} K_{n+1}(\bar{g}) N_n(\bar{\mu}) - \bar{\mu} K_n(\bar{g}) N_{n+1}(\bar{\mu})}{\bar{\mu} J_{n+1}(\bar{\mu}) K_n(\bar{g}) - \bar{g} K_{n+1}(\bar{g}) J_n(\bar{\mu})} \int_0^1 d\zeta J_n(\bar{\mu}\zeta) \zeta \hat{f}^{(n)}(\zeta),$$

$$C_2 = \frac{1}{\bar{\mu} J_{n+1}(\bar{\mu}) K_n(\bar{g}) - \bar{g} K_{n+1}(\bar{g}) J_n(\bar{\mu})} \int_0^1 d\zeta\, J_n(\bar{\mu}\zeta) \zeta \hat{f}^{(n)}(\zeta)$$

$$+ \frac{\bar{g} J_n(\bar{\mu}) I_{n+1}(\bar{g}) + \bar{\mu} J_{n+1}(\bar{\mu}) I_n(\bar{g})}{\bar{\mu} J_{n+1}(\bar{\mu}) K_n(\bar{g}) - \bar{g} K_{n+1}(\bar{g}) J_n(\bar{\mu})} \int_1^\infty d\zeta\, K_n(\bar{g}\zeta) \zeta \hat{f}^{(n)}(\zeta) \;. \quad (4.139)$$

Substituting C_1 and C_2 into (4.134) and (4.137) we get the solution for $\bar{E}^{(n)}$. To find $\tilde{E}^{(n)}(\hat{z},\hat{r})$, we use the inverse Laplace transformation:

$$\tilde{E}^{(n)}(\hat{z},\hat{r}) = \frac{1}{2\pi\mathrm{i}} \int_{\gamma'-\mathrm{i}\infty}^{\gamma'+\mathrm{i}\infty} d\lambda\, \bar{E}^{(n)}(\lambda,\hat{r}) \mathrm{e}^{\lambda z} \;, \quad (4.140)$$

where the integration path in the complex plane λ is parallel to the imaginary axis. The real constant γ' is larger than the real parts of all the singularities of $\bar{E}^{(n)}(\lambda,\hat{r})$. We shall consider only the high-gain limit. In this case the exponentially growing solutions are given by the residues of the integrand in (4.140) lying in the right-hand half of the complex plane λ. Using (4.134), (4.137), (4.139) and the relations

$$J_n(\xi) = \mathrm{i}^n I_n(-\mathrm{i}\xi) \;,$$

$$K_n(\xi) = \frac{\pi}{2}\mathrm{i}^n \left[J_n(\mathrm{i}\xi) + \mathrm{i} N_n(\mathrm{i}\xi)\right] \;,$$

we may write:

$$\tilde{E}^{(n)} = \sum_j u_j J_n(\mu_j \hat{r}) \exp(\lambda_j \hat{z}) \qquad \text{for} \quad \hat{r} < 1 \;, \quad (4.141\mathrm{a})$$

$$\tilde{E}^{(n)} = \sum_j u_j \frac{J_n(\mu_j)}{K_n(g_j)} K_n(g_j \hat{r}) \exp(\lambda_j \hat{z}) \qquad \text{for} \quad \hat{r} > 1 \;. \quad (4.141\mathrm{b})$$

Here λ_j is the jth root of the equation ($\mathrm{Re}\,\lambda_j > 0$):

$$\bar{\mu}(\lambda_j) J_{n+1}(\bar{\mu}(\lambda_j)) K_n(\bar{g}(\lambda_j)) - \bar{g}(\lambda_j) K_{n+1}(\bar{g}(\lambda_j)) J_n(\bar{\mu}(\lambda_j)) = 0 \;,$$

and

$$u_j = \frac{K_n(g_j)[J_n(\mu_j)]^{-1} \int_0^1 d\zeta\, J_n(\mu_j \zeta) \zeta \hat{f}^{(n)}(\zeta) + \int_1^\infty d\zeta\, K_n(g_j \zeta) \zeta \hat{f}^{(n)}(\zeta)}{d/d\lambda \left[\bar{\mu} J_{n+1}(\bar{\mu}) K_n(\bar{g}) - \bar{g} K_{n+1}(\bar{g}) J_n(\bar{\mu})\right]|_{\lambda=\lambda_j}} \;,$$

$$\mu_j^2 = \frac{-2\mathrm{i}\hat{D}_j}{1 - \mathrm{i}\hat{\Lambda}_\mathrm{p}^2 \hat{D}_j} - g_j^2 \;, \quad g_j^2 = -2\mathrm{i}B\lambda_j \;, \quad \hat{D}_j = \int_0^\infty d\hat{P}\, \frac{d\hat{F}(\hat{P})/d\hat{P}}{\lambda_j + \mathrm{i}(\hat{P} + \hat{C})} \;.$$

Each term on the right-hand sides of (4.141) corresponds to an individual radiation mode and is characterized by a unique amplitude factor, growth rate, and eigenfunction.

4. Diffraction Effects in the FEL Amplifier

In the paraxial approximation, the power gain coefficient G of the radiation field with azimuthal index n is given by the expression:

$$G = \int_0^\infty r|\tilde{E}^{(n)}(z,r)|^2 dr \left[\int_0^\infty r|\tilde{E}^{(n)}_{\text{ext}}(r)|^2 dr\right]^{-1}.$$

When the diffraction parameter B is not very large, $B \lesssim 10$, one can find that the growth rate of the TEM_{00} mode is visibly larger than the growth rates of higher modes (see Figs. 4.38 and 4.37). When the undulator is sufficiently long, the contribution of the TEM_{00} mode in (4.141) significantly exceeds the contributions of all other modes. In this case we may use the single-mode approximation and write:

$$G = 4B^2 \exp[2\,\text{Re}\,\lambda_1 \hat{z}] \left|\frac{d}{d\lambda}[\bar{\mu}J_1(\bar{\mu})K_0(\bar{g}) - \bar{g}K_1(\bar{g})J_0(\bar{\mu})]\right|_{\lambda=\lambda_1}^{-2}$$

$$\times \left\{\int_0^\infty \hat{r}|\tilde{E}_{\text{ext}}(\hat{r})|^2 d\hat{r}\right\}^{-1}$$

$$\times \left\{\int_0^1 \hat{r}|J_0(\mu_1\hat{r})|^2 d\hat{r} + \int_1^\infty \hat{r}\left|\frac{J_0(\mu_1)K_0(g_1\hat{r})}{K_0(g_1)}\right|^2 d\hat{r}\right\}$$

$$\times \left|\frac{K_0(g_1)}{J_0(\mu_1)}\int_0^1 \hat{r}J_0(\mu_1\hat{r})\tilde{E}_{\text{ext}}(\hat{r})d\hat{r} + \int_1^\infty \hat{r}K_0(g_1\hat{r})\tilde{E}_{\text{ext}}(\hat{r})d\hat{r}\right|^2 , \quad (4.142)$$

where λ_1 is the reduced eigenvalue of the TEM_{00} mode.

Optimal Focusing of the Master Oscillator Radiation. Let us study the problem of optimal focusing of the external radiation on the electron beam at the undulator entrance. We consider the case when the circularly polarized radiation of the master oscillator has the form of a Gaussian laser beam:

$$E_x + iE_y = \tilde{E}_{\text{ext}}(z,r)\exp[i\omega(z/c - t)]$$
$$= \frac{-iE_g w^2(\omega/c)e^{-i\omega t}}{2(z-z_0) - iw^2\omega/c}$$
$$\times \exp\left\{i\frac{\omega}{c}(z-z_0) + \frac{2i(\omega/c)(z-z_0)r^2 - (rw\omega/c)^2}{4(z-z_0)^2 + (w^2\omega/c)^2}\right\}. \quad (4.143)$$

Here z_0 and w are the position of the focus and the waist size in the focus of the Gaussian laser beam, respectively. At $z = z_0$ the Gaussian laser beam has a plane phase front and Gaussian distribution of the amplitude:

$$\tilde{E}_{\text{ext}}(z_0, r)\exp[i\omega(z_0/c - t)] = E_g e^{-i\omega t}\exp(-r^2/w^2). \quad (4.144)$$

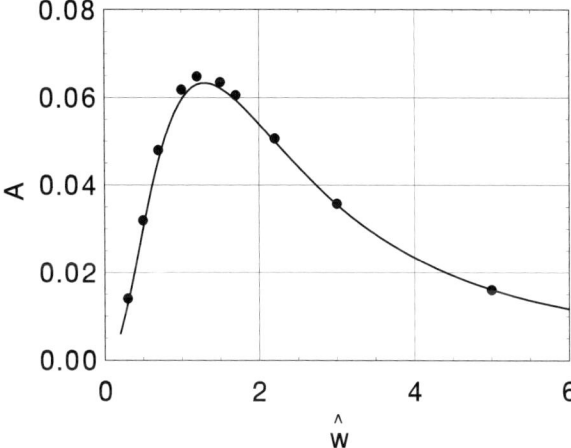

Fig. 4.44. Dependence of the input coupling factor A on the reduced Gaussian laser beam waist $\hat{w}$. Here $B = 1$, $\hat{C} = 0$, $\hat{z}_0 = 0$, $\hat{\Lambda}_p^2 \to 0$, $\hat{\Lambda}_T^2 = 0$. The curve is calculated with (4.121), (4.142), and (4.143) and the circles are the numerical solution of the initial problem

One can easily see that the complex amplitude $\tilde{E}_{\text{ext}}(z,r)$ appearing in (4.143) is an exact solution of the paraxial wave equation

$$[\partial^2/\partial r^2 + r^{-1}\partial/\partial r + 2\mathrm{i}c^{-1}\omega\partial/\partial z]\, \tilde{E}_{\text{ext}}(z,r) = 0\ .$$

When the undulator is sufficiently long, the output power grows exponentially with undulator length, so the power gain, $G = W_{\text{out}}/W_{\text{ext}}$, can be written as

$$G = A \exp\left[2\,\mathrm{Re}\,\lambda_1 \hat{z}\right]\ .$$

In the linear regime the power gain does not depend on the input power W_{ext}, so the input coupling factor A is a function of six reduced parameters: $B, \hat{C}$, $\hat{\Lambda}_p^2$, $\hat{\Lambda}_T^2$, $\hat{w} = w/r_0$, and $\hat{z}_0 = \Gamma z_0$.

Let us consider the operation of the FEL amplifier at exact FEL resonance, ($\hat{C} = 0$). To simplify the considerations, we assume the space charge field and the energy spread to be negligibly small, $\hat{\Lambda}_p^2 \to 0$ and $\hat{\Lambda}_T^2 \to 0$. In this case for each value of the diffraction parameter B there are always optimal values of the Gaussian laser beam parameters, $\hat{w}$ and $\hat{z}_0$, when the input coupling factor A achieves its maximum. To simplify the optimization problem we will not perform the variation of $\hat{z}_0$ and set it equal to zero. We will show below that such a choice of $\hat{z}_0$ is close to the optimum. Figure 4.44 presents the dependence of the input coupling factor A on the reduced laser beam waist $\hat{w}$ at a fixed value of the parameter B. The maximal value of the factor A and the laser beam waist $\hat{w}$ corresponding to this maximum are universal functions only of the diffraction parameter B. The plots of these functions are presented in Figs. 4.45 and 4.46. Figure 4.47 presents the de-

232   4. Diffraction Effects in the FEL Amplifier

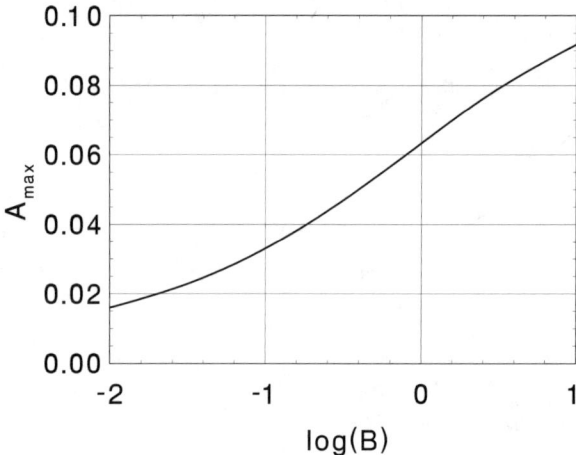

Fig. 4.45. Maximal input coupling factor A versus the diffraction parameter B. Here $\hat{C} = 0$, $\hat{z}_0 = 0$, $\hat{A}_p^2 \to 0$, $\hat{A}_T^2 = 0$

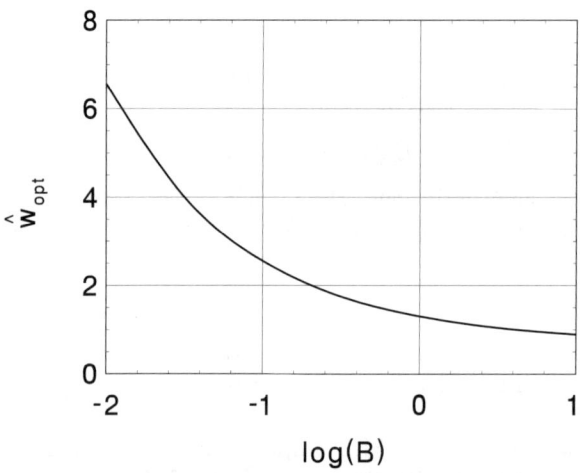

Fig. 4.46. Optimum of the reduced Gaussian laser beam waist $\hat{w}$ versus the diffraction parameter B. Here $\hat{C} = 0$, $\hat{z}_0 = 0$, $\hat{A}_p^2 \to 0$, $\hat{A}_T^2 = 0$

pendence of the input coupling factor A on the reduced laser beam focus coordinate, $\hat{z}_0$. It is clearly seen that the value of A at $\hat{z}_0 = 0$ does not differ significantly from its maximal value. The plots in Figs. 4.44-4.47 allow one to maximize the output power of the FEL amplifier operating in the linear regime at a fixed power of the master laser.

An important characteristic of the FEL amplifier is the amplification bandwidth. Figure 4.48 presents a specific example of the dependence of the

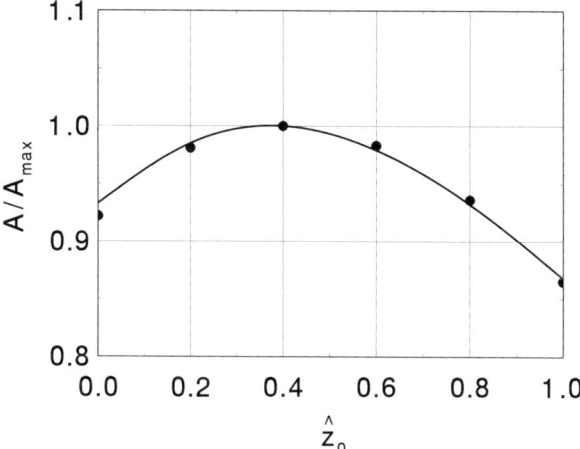

Fig. 4.47. The dependence of the input coupling factor A on the position of the Gaussian laser beam focus. Here $B = 1$, $\hat{w} = 1.3$, $\hat{C} = 0$, $\hat{\Lambda}_p^2 \to 0$, $\hat{\Lambda}_T^2 = 0$. The curve is calculated with (4.121), (4.142), and (4.143) and the circles are the numerical solution of the initial problem

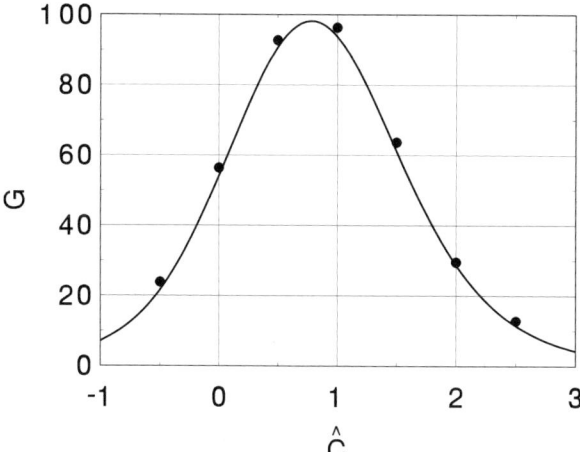

Fig. 4.48. The dependence of the power gain G on the reduced detuning $\hat{C}$. Here $B = 1$, $\hat{w} = 1.3$, $\hat{z} = 4.7$, $\hat{z}_0 = 0$, $\hat{\Lambda}_p^2 \to 0$, $\hat{\Lambda}_T^2 = 0$. The curve is calculated with (4.121), (4.142), and (4.143) and the circles are the numerical solution of the initial problem

power gain on the detuning parameter. The plots in Fig. 4.49 can be used to find the FWHM bandwidth for an amplifier with different values of the power gain G and diffraction parameter B.

234   4. Diffraction Effects in the FEL Amplifier

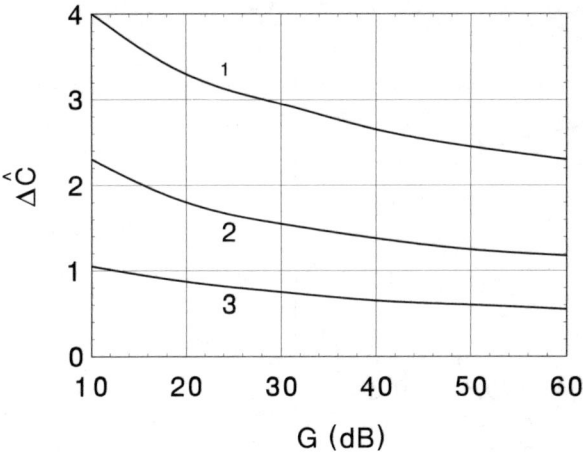

Fig. 4.49. The amplifier reduced bandwidth $\Delta \hat{C}$ versus the power gain G. Here $\hat{z}_0 = 0$, $\hat{\Lambda}_p^2 \to 0$, $\hat{\Lambda}_T^2 = 0$. The reduced laser beam waist is optimized at $\hat{C} = 0$. Curve 1: $B = 0.1$ and $\hat{w} = 2.55$. Curve 2: $B = 1$ and $\hat{w} = 1.3$. Curve 3: $B = 10$ and $\hat{w} = 0.89$.

4.4.4 Parabolic Profile

In the following sections we study the FEL amplifier with a gradient profile of the beam current density. In the axisymmetric case the transverse distribution of the beam current density is given by

$$j_0(r) = I_0 S(r/r_0) \left[2\pi \int_0^\infty r S(r/r_0) \mathrm{d}r \right]^{-1} , \qquad (4.145)$$

where r_0 is the beam profile parameter (typical transverse size of the beam) and I_0 is the beam current. To be specific, we set $S(0) = 1$. Using (4.21) and (4.145), we obtain the dimensionless equations for the field amplitude $\tilde{E}(z, r, \varphi)$:

$$\left[\frac{\partial^2}{\partial \hat{r}^2} + \frac{1}{\hat{r}} \frac{\partial}{\partial \hat{r}} + \frac{1}{\hat{r}^2} \frac{\partial^2}{\partial \varphi^2} + 2\mathrm{i}B \frac{\partial}{\partial \hat{z}} \right] \tilde{E}(\hat{z}, \hat{r}, \varphi)$$

$$= \mathrm{i}S(\hat{r}) \int_0^{\hat{z}} \mathrm{d}\hat{z}' \left\{ 2 + \hat{\Lambda}_p^2 \left[\frac{\partial^2}{\partial \hat{r}^2} + \frac{1}{\hat{r}} \frac{\partial}{\partial \hat{r}} + \frac{1}{\hat{r}^2} \frac{\partial^2}{\partial \varphi^2} + 2\mathrm{i}B \frac{\partial}{\partial \hat{z}'} \right] \right\} \tilde{E}(\hat{z}', \hat{r}, \varphi)$$

$$\times \int_{-\infty}^\infty \mathrm{d}\hat{P} \mathrm{d}\hat{F}(\hat{P})/\mathrm{d}\hat{P} \exp\left[\mathrm{i}(\hat{P} + \hat{C})(\hat{z}' - \hat{z}) \right] , \qquad (4.146)$$

where

$$\hat{z} = \Gamma z, \quad \hat{r} = r/r_0, \quad \hat{C} = C/\Gamma, \quad B = r_0^2 \Gamma \omega / c ,$$

$$\hat{\Lambda}_{\rm p}^2 = 4c^2(\theta_{\rm s} r_0 \omega)^{-2}, \quad \Gamma = \left[I_0 \omega^2 \theta_{\rm s}^2 \left(2I_A c^2 \gamma_z^2 \gamma \int_0^\infty \zeta S(\zeta) {\rm d}\zeta\right)^{-1}\right]^{1/2}.$$

In the high-gain limit we seek the solution for the field amplitude in the form:

$$\tilde{E}(\hat{z},\ \hat{r},\ \varphi) = \Phi_n(\hat{r}) \binom{\sin(n\varphi)}{\cos(n\varphi)} \exp(\hat{\Lambda}\hat{z}),$$

where n is an integer, $n \geq 0$. According to (4.146), the function $\Phi_n(\hat{r})$ satisfies the equations:

$$\left[\frac{{\rm d}^2}{{\rm d}\hat{r}^2} + \frac{1}{\hat{r}} \frac{\rm d}{{\rm d}\hat{r}} - \frac{2{\rm i}\hat{D}S(\hat{r})}{1 - {\rm i}\hat{\Lambda}_{\rm p}^2 \hat{D}S(\hat{r})} + 2{\rm i}B\hat{\Lambda} - \frac{n^2}{\hat{r}^2}\right]\Phi_n(\hat{r}) = 0, \quad (4.147)$$

where

$$\hat{D} = {\rm i} \int_0^\infty \exp\left[-\frac{\hat{\Lambda}_{\rm T}^2 \xi^2}{2} - \left(\hat{\Lambda} + {\rm i}\hat{C}\right)\xi\right] \xi {\rm d}\xi.$$

The energy spread parameter is defined as $\hat{\Lambda}_{\rm T}^2 = \langle(\Delta\mathcal{E})^2\rangle/(\rho^2 \mathcal{E}_0^2)$ where $\rho = c\gamma_z^2 \Gamma/\omega$.

The bounded parabolic profile is the only gradient profile allowing an analytical solution of (4.147) in the limit of $\hat{\Lambda}_{\rm p}^2 \to 0$. The bounded parabolic profile of the beam current density is given by the expression:

$$j_0(r) = \frac{I_0(1 - r^2/r_1^2)}{\pi r_0^2 \left[1 - r_0^2/(2r_1^2)\right]} \quad \text{for} \quad r < r_0,$$

$$j_0(r) = 0 \quad \text{for} \quad r > r_0. \quad (4.148)$$

where $r_0 < r_1$. The corresponding profile function is

$$S(\hat{r}) = 1 - k_1^2 \hat{r}^2,$$

where $\hat{r} = r/r_0$ and $k_1 = r_0/r_1$. Using (4.147), we obtain the following equation for $\Phi_n(\hat{r})$ inside the electron beam:

$$\left[\frac{{\rm d}^2}{{\rm d}\hat{r}^2} + \frac{1}{\hat{r}}\frac{\rm d}{{\rm d}\hat{r}} - 2{\rm i}\hat{D}(1 - k_1^2\hat{r}^2) + 2{\rm i}B\hat{\Lambda} - \frac{n^2}{\hat{r}^2}\right]\Phi_n(\hat{r}) = 0,$$

where $B = \Gamma r_0^2 \omega/c$ and

$$\Gamma = \left[I_0\omega^2\theta_{\rm s}^2 \left(c^2 I_A \gamma_z^2 \gamma(1 - k_1^2/2)\right)^{-1}\right]^{1/2}.$$

Using notation

$$\mu^2 = -2{\rm i}\hat{D} + 2{\rm i}B\hat{\Lambda}, \quad \delta^2 = -2{\rm i}\hat{D}k_1^2,$$

we rewrite the latter equation in the standard form:

$$\frac{{\rm d}^2\Phi_n}{{\rm d}\hat{r}^2} + \frac{1}{\hat{r}}\frac{{\rm d}\Phi_n}{{\rm d}\hat{r}} + \left[\mu^2 - \delta^2\hat{r}^2 - \frac{n^2}{\hat{r}^2}\right]\Phi_n = 0.$$

Introducing the new variable $\zeta = \delta \hat{r}^2$, we obtain the equation:

$$\zeta \frac{d^2 \Phi_n}{d\zeta^2} + \frac{d\Phi_n}{d\zeta} + \frac{1}{4}\left[\frac{\mu^2}{\delta} - \zeta - \frac{n^2}{\zeta}\right] \Phi_n = 0 .$$

Introducing the notation

$$\Phi_n = \zeta^{n/2} \exp(-\zeta/2) u_n(\zeta) ,$$

we obtain the Kummer equation for the function $u_n(\zeta)$:

$$\zeta u_n'' + (\chi - \zeta) u_n' - \epsilon u_n = 0 , \qquad (4.149)$$

where

$$\chi = n + 1, \qquad \epsilon = (n+1)/2 - \mu^2/(4\delta) .$$

The nondivergent solution of (4.149) has the form:

$$\Phi_n(\hat{r}) = C_1 \hat{r}^n \exp(-\delta \hat{r}^2 / 2) \,_1F_1(\epsilon, \, n+1, \, \delta \hat{r}^2) \qquad \text{for} \quad \hat{r} < 1 .$$

The solution of (4.147) for $\Phi_n(\hat{r})$ outside the electron beam has the form:

$$\Phi_n(\hat{r}) = C_2 K_n(g\hat{r}), \qquad \text{for} \quad \hat{r} > 1 ,$$

where $g^2 = -2\mathrm{i}B\hat{\Lambda}$, $\operatorname{Re} g > 0$. The continuity condition for the eigenfunction and its derivative, $\Phi_n(\hat{r})$ and $\mathrm{d}\Phi_n(\hat{r})/\mathrm{d}\hat{r}$, at the beam boundary leads to the eigenvalue equation:

$$\delta K_n(g) \left[2\epsilon(n+1)^{-1} \,_1F_1(\epsilon+1, \, n+2, \, \delta) - \,_1F_1(\epsilon, \, n+1, \, \delta)\right]$$
$$+ g K_{n+1}(g) \,_1F_1(\epsilon, \, n+1, \, \delta) = 0 . \qquad (4.150)$$

An explicit expression for the eigenfunction has the form:

$$\Phi_n(\hat{r}) = \hat{r}^n \exp(-\delta \hat{r}^2/2) \,_1F_1(\epsilon, \, n+1, \, \delta \hat{r}^2) , \qquad \text{for} \quad \hat{r} < 1 .$$

$$\Phi_n(\hat{r}) = \exp(-\delta/2) \,_1F_1(\epsilon, \, n+1, \, \delta) \frac{K_n(g\hat{r})}{K_n(g)} , \qquad \text{for} \quad \hat{r} > 1. \qquad (4.151)$$

The results of calculations of the eigenvalues and the eigenfunctions are presented in Figs. 4.50 and 4.51.

Let us find the asymptote of (4.150) and (4.151) in the limit of a stepped profile, $k_1 \to 0$. When $k_1 \to 0$, the parameter δ tends to zero. Using the relation

$$\lim_{\alpha \to \infty} \,_1F_1\left(\alpha, \, \beta, \, -\frac{\nu^2}{4\alpha}\right) = \Gamma(\beta) \left(\frac{\nu}{2}\right)^{1-\beta} J_{\beta-1}(\nu) ,$$

we obtain:

$$\lim_{\delta \to 0} \,_1F_1(\epsilon, \, n+1, \, \delta) = 2^n \mu^{-n} n! J_n(\mu) ,$$

$$\lim_{\delta \to 0} \,_1F_1(\epsilon+1, \, n+2, \, \delta) = 2^{n+1} \mu^{-n-1} (n+1)! J_{n+1}(\mu) .$$

As a result, (4.150) is reduced to the eigenvalue equation for a stepped profile of the electron beam:

Linear Theory of the FEL Amplifier with an Axisymmetric Electron Beam 237

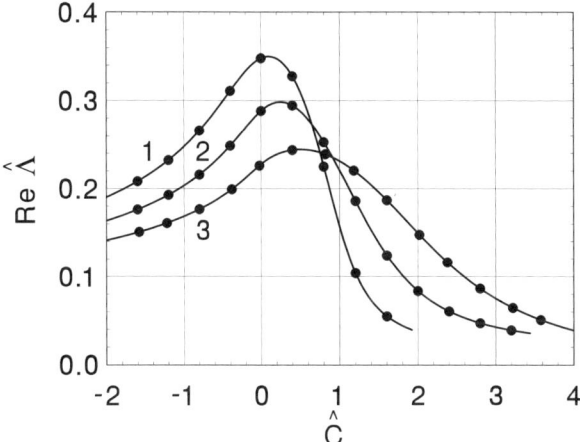

Fig. 4.50. The parabolic beam profile. The reduced growth rate Re $\hat{\Lambda}$ versus $\hat{C}$. Here $k_1 = 1$, $B = 10$, $\hat{\Lambda}_p^2 \to 0$, $\hat{\Lambda}_T^2 = 0$. Curve 1: TEM$_{00}$ mode. Curve 2: TEM$_{01}$ mode. Curve 3: TEM$_{02}$ mode. The curves show analytical results and the circles are calculated with the multilayer approximation method

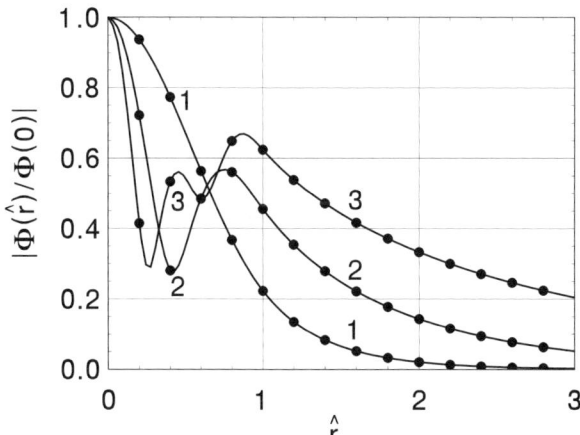

Fig. 4.51. The parabolic beam profile. The field distribution versus the reduced radius $\hat{r}$. Here $k_1 = 1$, $B = 10$, $\hat{\Lambda}_p^2 \to 0$, $\hat{\Lambda}_T^2 = 0$, and $\hat{C} = 0$. Curve 1: TEM$_{00}$. Curve 2: TEM$_{01}$ mode. Curve 3: TEM$_{02}$ mode. The curves show analytical results and the circles are calculated with the multilayer approximation method

$$gK_{n+1}(g)J_n(\mu) - \mu J_{n+1}(\mu)K_n(g) = 0 \ .$$

Now let us consider the asymptote of (4.150) in the limit of a wide electron beam, as $B \to \infty$. The parameter $|\delta| \to \infty$ in this case, and the eigenfunction becomes divergent for all values of ϵ except of $\epsilon \simeq -j$, where $j = 0, 1, 2, \ldots$. So, we have asymptotically:

$$\mu^2 = \delta^2/k_1^2 - g^2 \simeq 2(n+1+2j)\delta \quad \text{as} \quad |\delta| \to \infty,$$

and (4.150) takes the simple form, $\hat{D} = B\hat{\Lambda}$.

In the limit of a thin electron beam, as $B \to 0$, for the fundamental TEM$_{00}$ mode we find that the argument of $K_0(g)$ and $K_1(g)$ tends to zero and we can use the expansion:

$$K_0(g) \simeq -\ln(g/2) - \gamma_E, \quad K_1(g) \simeq g^{-1}.$$

Substituting this expansion into (4.150), we obtain:

$$\delta\left[1 - 2\epsilon\frac{{}_1F_1(\epsilon+1,\,2,\,\delta)}{{}_1F_1(\epsilon,\,1,\,\delta)}\right] = \frac{1}{[\ln(g/2)+\gamma_E]}.$$

Assuming the right-hand part of this equation is small, and expanding the hypergeometric functions in a series, we obtain

$$\left[\frac{\mu^2}{2}\left(1-\frac{k_1^2}{2}\right) + \frac{\mu^4}{16}\left(1-\frac{2k_1^2}{3}+\frac{k_1^4}{8}\right)\right]\left[\ln\left(\frac{g}{2}\right)+\gamma_E\right] = -1.$$

The latter equation, written down with double logarithmic accuracy, has the form:

$$\frac{\mu^2}{2}\left(1-\frac{k_1^2}{2}\right)\left[\ln\left(\frac{g}{2}\right)+0.577-\frac{1-2k_1^2/3+k_1^4/8}{4-4k_1^2+k_1^4}\right] = -1.$$

4.4.5 Arbitrary Gradient Profile

The eigenvalue problem for the case of an arbitrary gradient axisymmetric profile is solved by means of the multilayer approximation method. Let r_0 be the beam profile parameter and r_b be the radius of the beam boundary. The transverse coordinate is normalized as $\hat{r} = r/r_0$, so the normalized radius of the beam boundary is $\hat{r}_b = r_b/r_0$. We divide the region $0 < \hat{r} < \hat{r}_b$ into K layers. The beam current density is assumed to be constant within each layer. According to (4.147), the solution for the eigenfunction within each layer is

$$\Phi_n^{(j)} = A_j J_n(\mu_j \hat{r}) + D_j N_n(\mu_j \hat{r}),$$

where $(j-1)/K < \hat{r} < j\hat{r}_b/K$, A_j and D_j are constants, and

$$\mu_j^2 = -\frac{2\mathrm{i}\hat{D}S_{j-1/2}}{1-\mathrm{i}\hat{\Lambda}_p^2\hat{D}S_{j-1/2}} - g^2, \quad g^2 = -2\mathrm{i}B\hat{\Lambda},$$

where $S_{j-1/2} = S(\hat{r}_{j-1/2})$ and $\hat{r}_{j-1/2} = \hat{r}_b(j-1/2)/K$. To avoid a singularity of the eigenfunction at $\hat{r} = 0$, we should let $D_1 = 0$. All the other coefficients are obtained from the continuity conditions for the eigenfunction and its derivative at the boundaries between the layers. These equations can be written in the matrix form:

$$\begin{pmatrix}A_{j+1}\\D_{j+1}\end{pmatrix} = T_j \begin{pmatrix}A_j\\D_j\end{pmatrix}, \quad j=1,\,2,\,\ldots,\,K-1, \tag{4.152}$$

where the coefficients T_j are given by ($\hat{r}_j = \hat{r}_b j/K$):

$$(T_j)_{11} = (\pi/2)\hat{r}_j \left[\mu_j J_{n+1}(\mu_j \hat{r}_j) N_n(\mu_{j+1}\hat{r}_j) \right.$$
$$\left. - \mu_{j+1} J_n(\mu_j \hat{r}_j) N_{n+1}(\mu_{j+1}\hat{r}_j) \right] ,$$
$$(T_j)_{12} = (\pi/2)\hat{r}_j \left[\mu_j N_{n+1}(\mu_j \hat{r}_j) N_n(\mu_{j+1}\hat{r}_j) \right.$$
$$\left. - \mu_{j+1} N_n(\mu_j \hat{r}_j) N_{n+1}(\mu_{j+1}\hat{r}_j) \right] ,$$
$$(T_j)_{21} = -(\pi/2)\hat{r}_j \left[\mu_j J_{n+1}(\mu_j \hat{r}_j) J_n(\mu_{j+1}\hat{r}_j) \right.$$
$$\left. - \mu_{j+1} J_n(\mu_j \hat{r}_j) J_{n+1}(\mu_{j+1}\hat{r}_j) \right] ,$$
$$(T_j)_{22} = -(\pi/2)\hat{r}_j \left[\mu_j N_{n+1}(\mu_j \hat{r}_j) J_n(\mu_{j+1}\hat{r}_j) \right.$$
$$\left. - \mu_{j+1} N_n(\mu_j \hat{r}_j) J_{n+1}(\mu_{j+1}\hat{r}_j) \right] . \tag{4.153}$$

According to (4.147), the solution for the eigenfunction outside the beam, $\hat{r} > \hat{r}_{\mathrm{b}}$, satisfying the condition of quadratic integrability is

$$\Phi_n(\hat{r}) = F_1 K_n(g\hat{r}) , \quad \operatorname{Re} g > 0 .$$

At the beam boundary, at $\hat{r} = \hat{r}_{\mathrm{b}}$, the continuity condition gives the following relations:

$$A_K J_n(\mu_K \hat{r}_{\mathrm{b}}) + D_K N_n(\mu_K \hat{r}_{\mathrm{b}}) = F_1 K_n(g\hat{r}_{\mathrm{b}}) ,$$
$$\mu_K A_K J_{n+1}(\mu_K \hat{r}_{\mathrm{b}}) + \mu_K D_K N_{n+1}(\mu_K \hat{r}_{\mathrm{b}}) = g F_1 K_{n+1}(g\hat{r}_{\mathrm{b}}) ,$$

which can also be written in the matrix form:

$$T_K \begin{pmatrix} A_K \\ D_K \end{pmatrix} = F_1 \begin{pmatrix} 1 \\ 1 \end{pmatrix} . \tag{4.154}$$

The coefficient F_1 can be expressed in terms of the coefficient A_1 by multiple use of (4.152). The coefficient A_1 may be chosen arbitrarily, so without loss of generality we let $A_1 = 1$. Then we can write the following matrix equation:

$$T_K \times T_{K-1} \times \ldots \times T_1 \begin{pmatrix} 1 \\ 0 \end{pmatrix} = T \begin{pmatrix} 1 \\ 0 \end{pmatrix} = F_1 \begin{pmatrix} 1 \\ 1 \end{pmatrix} , \tag{4.155}$$

where the matrix T depends on the unknown quantity $\hat{\Lambda}$. Another unknown quantity in (4.155) is the coefficient F_1, which can be easily excluded. Thus, we obtain the eigenvalue equation

$$(T)_{11} = (T)_{21} , \tag{4.156}$$

which allows one to find the eigenvalue $\hat{\Lambda}$. The eigenfunction is calculated using (4.152) and (4.154).

The accuracy of the multilayer approximation method can be tested with analytical results for the bounded parabolic profile. It is seen from Figs. 4.50 and 4.51 that there is good agreement between the numerical and analytical results. The difference is less than 0.1% when the number of layers is $K = 40$.

Now let us consider a beam with Gaussian distribution of the current density:

$$j_0(r) = \frac{I_0}{2\pi\sigma^2} \exp\left(-\frac{r^2}{2\sigma^2} \right) .$$

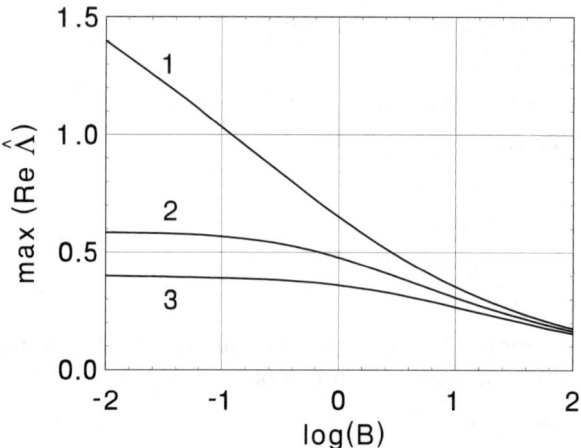

Fig. 4.52. Gaussian beam profile. The dependence of the maximal reduced growth rate max(Re $\hat{\Lambda}$) on the diffraction parameter B. Here $\hat{\Lambda}_p^2 \to 0$, $\hat{\Lambda}_T^2 = 0$. Curve 1: TEM$_{00}$ mode. Curve 2: TEM$_{10}$ mode. Curve 3: TEM$_{01}$ mode

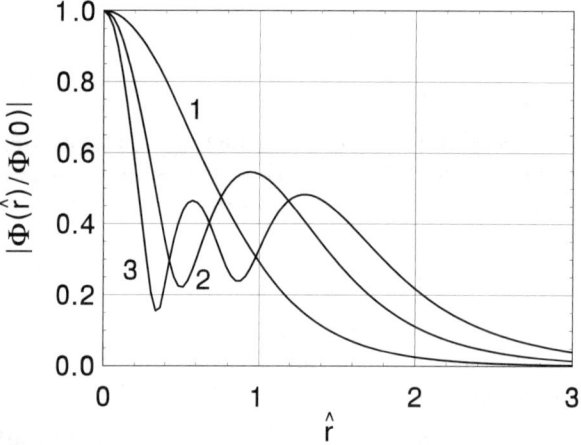

Fig. 4.53. Gaussian beam profile. The field distribution of TEM$_{00}$, TEM$_{01}$, and TEM$_{02}$ modes versus the reduced radius $\hat{r}$. Here $B = 10$, $\hat{\Lambda}_p^2 \to 0$, $\hat{\Lambda}_T^2 = 0$, and $\hat{C} = 0$. Curve 1: TEM$_{00}$ mode. Curve 2: TEM$_{01}$ mode. Curve 3: TEM$_{02}$ mode

It is reasonable to choose the rms radius, $\sqrt{2}\sigma$, as the profile parameter r_0. The profile function is

$$S(\hat{r}) = \exp\left(-\hat{r}^2\right),$$

where $\hat{r} = r/(\sqrt{2}\sigma)$. The normalizing factors and reduced parameters in the case of the Gaussian profile are as follows

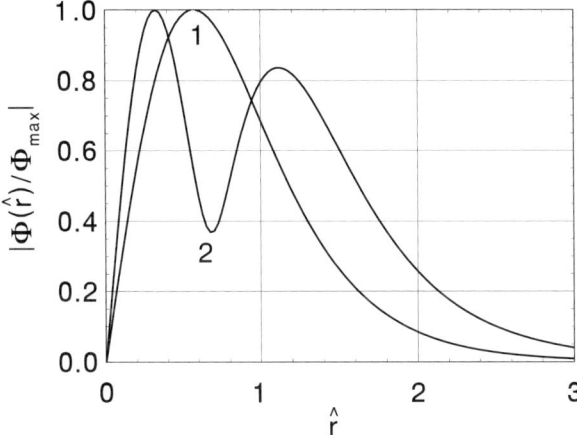

Fig. 4.54. Gaussian beam profile. The field distribution of TEM_{10} and TEM_{11} modes versus the reduced radius $\hat{r}$. Here $B = 10$, $\hat{\Lambda}_\text{p}^2 \to 0$, $\hat{\Lambda}_\text{T}^2 = 0$, $\hat{C} = 0.2$. Curve 1: TEM_{10} mode. Curve 2: TEM_{11} mode

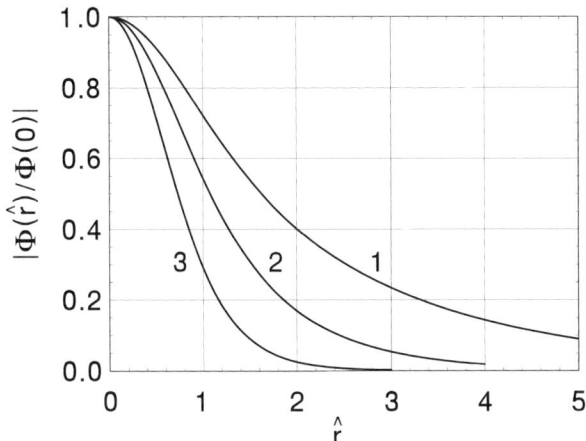

Fig. 4.55. Gaussian beam profile. The field distribution of the TEM_{00} mode versus the reduced radius $\hat{r}$. Here $\hat{C} = 0$, $\hat{\Lambda}_\text{p}^2 \to 0$, $\hat{\Lambda}_\text{T}^2 = 0$. Curve 1: $B = 0.1$. Curve 2: $B = 1$. Curve 3: $B = 10$

$$\Gamma = \left[I_0\omega^2\theta_\text{s}^2\left(I_\text{A}c^2\gamma_z^2\gamma\right)^{-1}\right]^{1/2}, \qquad B = 2\sigma^2\Gamma\omega/c,$$

$$\hat{\Lambda}_\text{p}^2 = 2c^2(\theta_\text{s}\sigma\omega)^{-2}, \qquad \hat{\Lambda}_\text{T}^2 = \langle(\Delta\mathcal{E})^2\rangle/(\rho^2\mathcal{E}^2), \qquad \rho = c\gamma_z^2\Gamma/\omega.$$

Let us present some results of calculations with the multilayer approximation method for the Gaussian current profile. We start with the case of a negligibly small space charge field and energy spread, $\hat{\Lambda}_\text{p}^2 \to 0$ and $\hat{\Lambda}_\text{T}^2 \to 0$. In this region of the parameters the growth rate is a function of only two

242   4. Diffraction Effects in the FEL Amplifier

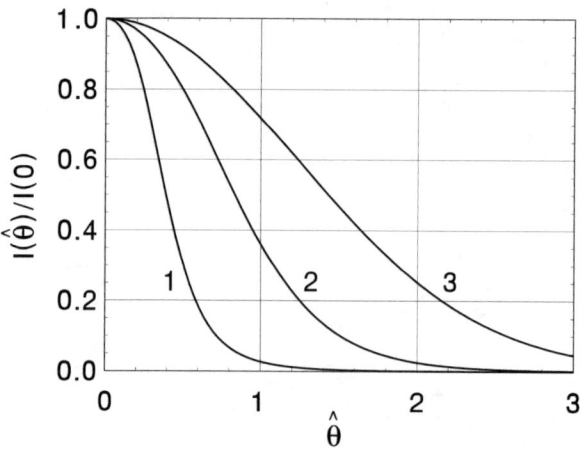

Fig. 4.56. Gaussian beam profile. Angular distribution of the radiation intensity for the TEM$_{00}$ mode. Here $\hat{C} = 0$, $\hat{A}_p^2 \to 0$, and $\hat{A}_T^2 = 0$. Curve 1: $B = 0.1$. Curve 2: $B = 1$. Curve 3: $B = 10$

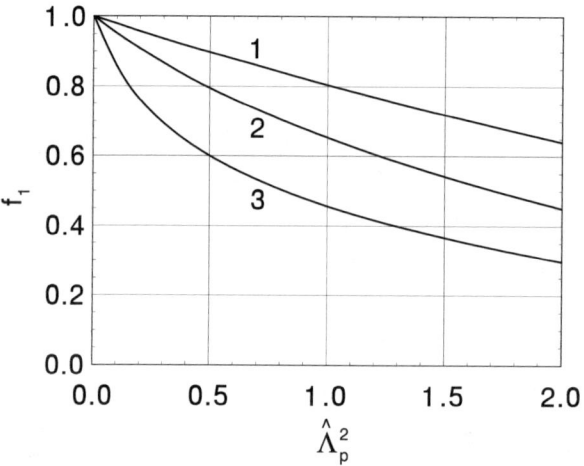

Fig. 4.57. Gaussian beam profile. The dependence of the function f_1 for the TEM$_{00}$ mode on the space charge parameter $\hat{A}_p^2$. Curve 1: $B = 0.1$. Curve 2: $B = 1$. Curve 3: $B = 10$

parameters, the detuning and the diffraction parameter. At some fixed value of the diffraction parameter there is always detuning when the growth rate reaches its maximal value. The maximal growth rate as a function of the diffraction parameter is plotted in Fig. 4.52.

The field distributions of different radiation modes are presented in Figs. 4.53-4.55. The angular distribution of the radiation intensity for the

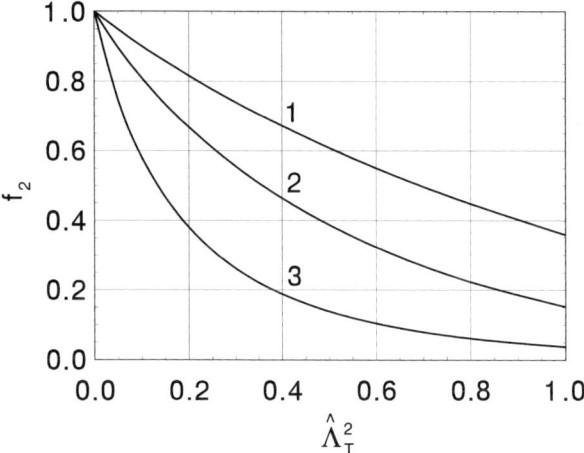

Fig. 4.58. Gaussian beam profile. The dependence of the function f_2 for the TEM$_{00}$ mode on the energy spread parameter $\hat{\Lambda}_T^2$. Curve 1: $B = 0.1$. Curve 2: $B = 1$. Curve 3: $B = 10$

fundamental TEM$_{00}$ mode is shown in Fig. 4.56. The normalized angle, $\hat{\theta}$, and the observation angle, θ, are connected by the relation $\hat{\theta} = \sqrt{2}\sigma\theta\omega/c$.

When the space charge influences the FEL process, the maximal growth rate is a function of two parameters, the space charge parameter $\hat{\Lambda}_p^2$ and the diffraction parameter B, and may be presented in the form:

$$\max(\operatorname{Re}\hat{\Lambda}) = \max(\operatorname{Re}\hat{\Lambda})|_{\hat{\Lambda}_p^2 \to 0} \times f_1(\hat{\Lambda}_p^2, B) \ .$$

The plots of the function f_1 for the fundamental TEM$_{00}$ mode are presented in Fig. 4.57.

Now let us consider the influence of the energy spread on FEL operation assuming that $\hat{\Lambda}_p^2 \to 0$. The maximal growth rate is a function of two parameters, the energy spread parameter and the diffraction parameter, and can be written as

$$\max(\operatorname{Re}\hat{\Lambda}) = \max(\operatorname{Re}\hat{\Lambda})|_{\hat{\Lambda}_T^2 \to 0} \times f_2(\hat{\Lambda}_T^2, B) \ .$$

The function f_2 for the fundamental mode is plotted in Fig. 4.58.

4.4.6 Numerical Solution of Initial-Value Problem

We finish the study of the linear mode of FEL amplifier operation with presenting an algorithm for numerical solution of the initial-value problem. The self-consistent field method in the linear approximation enables one to get from the kinetic equation and Maxwell's equations a unique equation either for the field amplitude of the amplified wave, or for the modulation amplitude of the beam current density. Both ways lead to the same results. For

244   4. Diffraction Effects in the FEL Amplifier

analytical calculations it is more convenient to use the equation for the wave field (4.21). The situation with computer simulations is proved to be reversed, and the method using the equation for the modulation amplitude of the beam current density (4.24) is more convenient.

In this section we present an algorithm for the numerical solution of the initial-value problem using (4.24). To simplify the considerations, we write all the formulae for the case of a stepped profile of the electron beam and axially symmetric radiation modes. The field of the master oscillator has the form of a Gaussian laser beam (4.143) with amplitude

$$E_g = \left[8W_{\text{ext}}/(w^2 c)\right]^{1/2} ,$$

where W_{ext} is the total power of the master oscillator. In the case of a negligibly small energy spread, using the standard normalization procedure, we rewrite (4.24) in the form:

$$\frac{d^2 \hat{a}_1}{d\hat{z}^2} + 2i\hat{C}\frac{d\hat{a}_1}{d\hat{z}} + \left[\hat{\Lambda}_p^2 - \hat{C}^2\right]\hat{a}_1$$

$$= \int_0^{\hat{z}} \frac{d\hat{z}'}{\hat{z} - \hat{z}'} \int_0^1 d\hat{r}' \hat{r}' \hat{a}_1(\hat{z}', \hat{r}') J_0\left[\frac{B\hat{r}\hat{r}'}{\hat{z} - \hat{z}'}\right]$$

$$\times \exp\left\{\frac{iB(\hat{r}^2 + \hat{r}'^2)}{2(\hat{z} - \hat{z}')}\right\} - \hat{U}_{\text{ext}}(\hat{z}, \hat{r}) , \qquad (4.157)$$

where

$$\hat{a}_1(z, r) = \tilde{j}_1(z, r)\pi r_0^2 / I_0 , \qquad \hat{W}_{\text{ext}} = W_{\text{ext}}/W_0 ,$$

$$\hat{E}_g = \left[8\hat{W}_{\text{ext}}/(B\hat{w}^2)\right]^{1/2} , \qquad W_0 = I_0 \mathcal{E}_0 \rho / e ,$$

$$\hat{U}_{\text{ext}}(\hat{z}, \hat{r}) = \frac{B\hat{w}^2 \hat{E}_g}{2\left[2(\hat{z} - \hat{z}_0) - iB w^2\right]} \exp\left[\frac{2iB(\hat{z} - \hat{z}_0)\hat{r}^2 - (B\hat{w}\hat{r})^2}{4(\hat{z} - \hat{z}_0)^2 + (B\hat{w}^2)^2}\right] .(4.158)$$

The power gain G is calculated as

$$G = 1 + \left\{\frac{2i}{\hat{W}_{\text{ext}}} \int_0^{\hat{z}} d\hat{z}' \int_0^{\hat{z}'} d\hat{z}'' \int_0^1 d\hat{r}' \int_0^1 d\hat{r}'' \hat{a}_1(\hat{z}', \hat{r}') \frac{\hat{r}'\hat{r}''}{\hat{z}' - \hat{z}''}\right.$$

$$\left. \times \hat{a}_1^*(\hat{z}'', \hat{r}'') J_0\left[\frac{B\hat{r}'\hat{r}''}{\hat{z}' - \hat{z}''}\right] \exp\left\{-\frac{iB(\hat{r}'^2 + \hat{r}''^2)}{2(\hat{z}' - \hat{z}'')}\right\} + \text{C.C.}\right\}$$

$$- \left\{\frac{2i}{\hat{W}_{\text{ext}}} \int_0^{\hat{z}} d\hat{z}' \int_0^1 d\hat{r}' \hat{U}_{\text{ext}}^*(\hat{z}', \hat{r}') \hat{a}_1(\hat{z}', \hat{r}') \hat{r}' + \text{C.C.}\right\} . \qquad (4.159)$$

Figures 4.44, 4.47, and 4.48 present the results of the numerical solution of (4.157). Comparison with analytical results shows that in the high-gain limit there is good agreement between the numerical and analytical results.

4.5 Nonlinear Mode of Operation

Saturation effects have been studied thoroughly in Chap. 2 in the framework of the one-dimensional model. The saturation mechanism remains mainly the same in the three-dimensional case. To find the FEL characteristics at saturation, it is necessary to solve the equations of the nonlinear theory of the FEL amplifier. Analytical methods provide limited tools for the description of saturation effects, and numerical simulation codes are widely used for the calculation of FEL amplifiers.

In this book we decided to limit our considerations mainly to including the fundamental effects: the amplification mechanism itself, space charge effects, and the diffraction effects. An important nonfundamental effect of the energy spread in the electron is included in the considerations, too. The approach presented here to numerical simulations satisfies the requirements for reliability and clear physical interpretation of the results obtained. First, the model approximations allow one to check the linear stage of amplification with the rigorous solutions of the linear theory. Second, when writing down the final equations we use similarity techniques. This enables one not only to reduce the number of problem parameters but also to go over to variables possessing a clear physical interpretation. Each physical factor influencing FEL operation (diffraction, space charge, energy spread, etc.) is matched by its own reduced parameter. For the effect under study this reduced parameter is a measure of the corresponding physical effect. When some effect becomes less important for FEL amplifier operation, it falls out of the number of problem parameters. The results of numerical simulations, rescaled with application of similarity techniques, possess a high degree of generality and provide a deep insight into FEL physics.

In this model we use a three-dimensional treatment of the radiation field, while the motion of particles is considered to be one-dimensional. This model proved to be very fruitful, since it allows one to take into account such effects as the radiation diffraction, space charge fields, and energy spread of the electrons in the beam. To simplify the considerations, we consider the specific case of the initial conditions when the radiation from the master oscillator (the Gaussian laser beam) and an unmodulated electron beam are fed to the undulator entrance. The derived numerical algorithm enables one to calculate the frequency, amplitude, and current characteristics of the FEL amplifier, and the field distributions in the near and far zone. The code allows one to calculate all these characteristics for constant undulator parameters as well as for tapering ones.

4.5.1 Nonlinear Simulation Algorithm

In this section we briefly describe nonlinear simulation code for the FEL amplifier with an axisymmetric electron beam. To simplify the considerations, we present all the formulae for the stepped profile of the beam current density and for an axisymmetric radiation field. The reader can see that the algorithm is transparent and can be easily extended to an arbitrary transverse profile of the electron beam and a three-dimensional representation of the radiation field. The electron beam moves along the axis of the helical undulator. The electron motion is described in the energy-phase variables with the phase

$$\psi = \int k_\mathrm{w}(z)\mathrm{d}z - \omega(t - z/c)$$

as canonical coordinate and energy $\mathcal{E}$ as canonical momentum. When the energy $\mathcal{E}$ of a particle does not differ significantly from the nominal value $\mathcal{E}_0$, we can use the Hamiltonian (4.7). The only difference is that now the detuning can be a function of z if there is undulator tapering.

Then we perform the usual normalization procedure. The diffraction parameter B, the gain parameter Γ, space charge parameter $\hat{\Lambda}_\mathrm{p}^2$, and energy spread parameter $\hat{\Lambda}_\mathrm{T}^2$ are defined the same way as in Sect. 4.4.1. The equations of motion corresponding to the Hamiltonian (4.7) can be written in the following reduced form:

$$\frac{\mathrm{d}\hat{P}}{\mathrm{d}\hat{z}} = -2\,\mathrm{Im}\left[\mathrm{e}^{\mathrm{i}\psi}\hat{U} + \hat{\Lambda}_\mathrm{p}^2\hat{U}_\mathrm{c}\right],$$

$$\frac{\mathrm{d}\psi}{\mathrm{d}\hat{z}} = \hat{C} + \hat{P}. \qquad (4.160)$$

Here $\hat{z} = \Gamma z$, $\hat{C} = C/\Gamma$, $\hat{P} = P/(\rho\mathcal{E}_0)$, $P = \mathcal{E} - \mathcal{E}_0$, and $\rho = c\gamma_z^2\Gamma/\omega$ is the efficiency parameter.

The expression for the reduced complex amplitude of the effective potential of the axisymmetric radiation field $\hat{U}$ is given by

$$\hat{U} = \hat{U}_\mathrm{ext} - \int_0^{\hat{z}}\frac{\mathrm{d}\hat{z}'}{\hat{z}-\hat{z}'}\int_0^1 \mathrm{d}\hat{r}'\hat{r}'\hat{a}_1(\hat{z}',\hat{r}')$$

$$\times \exp\left[\frac{\mathrm{i}B(\hat{r}^2+\hat{r}'^2)}{2(\hat{z}-\hat{z}')}\right]J_0\left[\frac{B\hat{r}\hat{r}'}{\hat{z}-\hat{z}'}\right], \qquad (4.161)$$

where $\hat{U}_\mathrm{ext}$ is given by (4.158).

The complex amplitude $\hat{a}_1 = |\hat{a}_1|\exp(\mathrm{i}\psi_1)$ entering (4.161) is calculated with the local macroparticle ensemble:

$$|\hat{a}_1| = \frac{1}{N}\left[\left[\sum_{k=1}^N \cos(\psi_{(k)})\right]^2 + \left[\sum_{k=1}^N \sin(\psi_{(k)})\right]^2\right]^{1/2},$$

$$\psi_1 = -\mathrm{arctg}\left[\frac{\sum_{k=1}^{N}\sin(\psi_{(k)})}{\sum_{k=1}^{N}\cos(\psi_{(k)})}\right] . \quad (4.162)$$

Under the limitation $r_0^2 \gg \gamma_z^2 c^2/\omega^2$, the expression for the reduced effective potential of the space charge fields $\hat{U}_c$ takes the following simple form:

$$\hat{U}_c = \sum_{n=1}^{\infty} e^{in\psi} \frac{\hat{a}_n(\hat{r},\hat{z})}{n} . \quad (4.163)$$

The complex amplitudes $\hat{a}_n = |\hat{a}_n|\exp(i\psi_n)$ are calculated as follows:

$$|\hat{a}_n| = \frac{1}{N}\left[\left[\sum_{k=1}^{N}\cos(n\psi_{(k)})\right]^2 + \left[\sum_{k=1}^{N}\sin(n\psi_{(k)})\right]^2\right]^{1/2} ,$$

$$\psi_n = -\mathrm{arctg}\left[\frac{\sum_{k=1}^{N}\sin(n\psi_{(k)})}{\sum_{k=1}^{N}\cos(n\psi_{(k)})}\right] . \quad (4.164)$$

It has been shown in Sect. 2.2.5 that the Fourier series (4.163) of the space charge harmonics can be reduced to the sum of the trigonometric series, and the latter results in a simple algebraic function.

The power gain coefficient G is calculated using (4.159). We should mention that a simpler way to calculate the gain can be used which is based on the energy conservation law. The increase of the electromagnetic wave power due to radiation by the electron beam may be expressed as

$$\Delta W = -I_0 \langle P \rangle / e ,$$

where $\langle P \rangle$ is the average change of the energy of the particles in the beam. Performing the normalization, we get

$$G = 1 + \Delta W / W_\mathrm{ext} = 1 - \langle \hat{P} \rangle / \hat{W}_\mathrm{ext} ,$$

where $\hat{W}_\mathrm{ext} = W_\mathrm{ext}/W_0$ and $W_0 = \rho I_0 \mathcal{E}_0/e$. The efficiency η is defined as

$$\eta = -\langle P \rangle / \mathcal{E}_0 .$$

Normalizing the latter expression, we obtain

$$\hat{\eta} = \eta/\rho = -\langle \hat{P} \rangle .$$

The field distribution in the near zone (i.e. inside the undulator) is given by (4.161). At a large distance of z from the undulator exit, the radiation has the form of a spherical wave with amplitude $\Xi(\theta)$ depending on the observation angle θ (we assume here the Fraunhofer diffraction approximation):

$$\Xi(\hat{\theta}) = \int_0^{\hat{l}_w} d\hat{z} \int_0^1 d\hat{r}\,\hat{r}\,\hat{a}_1(\hat{z},\hat{r})\exp\left(i\hat{\theta}^2\hat{z}/B\right)J_0(\hat{\theta}\hat{r})$$

$$-\frac{\hat{E}_g \hat{w}^2 B}{4}\exp\left[-\frac{i\hat{\theta}^2 \hat{z}_0}{2B} - \frac{\hat{w}^2\hat{\theta}^2}{4}\right] , \quad (4.165)$$

248   4. Diffraction Effects in the FEL Amplifier

where $\hat{\theta}$, $\hat{E}_\mathrm{g}$, $\hat{w}$, and $\hat{z}_0$ were defined in the previous section.

The simulation is performed with the macroparticle method. The macroparticle ensemble is prepared as follows: the electron beam is divided into M layers over the radius and in each layer we distribute uniformly N macroparticles over the phase ψ from 0 to 2π. The initial energy spread is simulated with the additional distribution of the particles according to the Gaussian law (see Chap. 2 for more details):

$$\hat{F}(\hat{P}) = \frac{1}{\sqrt{2\pi\hat{\Lambda}_\mathrm{T}^2}} \exp\left[-\frac{\hat{P}^2}{2\hat{\Lambda}_\mathrm{T}^2}\right].$$

As a result, we get a system of $2 \times N \times M$ ordinary differential equations (4.160) which is integrated with the Runge Kutta technique. It should be noted that standard numerical quadratures are not effective for the calculation of the integral over $\hat{z}'$ in (4.161), since the integrand has a singularity at $\hat{z}' \to \hat{z}$. To calculate this integral, we have developed a special algorithm. The integration interval $(0, \hat{z})$ is divided into a number of subintervals. The Bessel function $J_0(\zeta)$, where $\zeta = B\hat{r}\hat{r}'/(\hat{z}-\hat{z}')$, is approximated with polynomials at small values of ζ, and at large values of ζ we use the asymptotic expansion. As a result the calculation of the integral (4.161) over $\hat{z}'$ is reduced to the sum of special functions: Fresnel integrals, integral sine and cosine. The integral over the transverse coordinate is calculated with standard quadrature formulae.

It should be noted that the algorithm provides high speed and high accuracy of calculations. For instance, a version of this code, adapted for a modern personal computer, allows one to perform a simulation run of an actual physical device within a few seconds. This feature of the code is mainly connected with fast and precise calculations of the radiation fields. The high accuracy of the calculations is provided by using a rigorous solution of the electrodynamic problem. Thus, we avoid the problems which occur in the codes based on direct numerical solution of the wave equation. The high speed of calculations is provided by using an effective procedure for numerical calculations of the integrals.

4.5.2 Some Results of Numerical Simulations

First, we present some test results for the code in the linear stage of operation. The characteristics of the FEL amplifier, calculated in the high-gain linear regime, must correspond to those calculated analytically in Sect. 4.4. Figure 4.38 presents the calculations of the field growth rate. One can see that even when the number of radial mesh divisions is $N = 5$ the difference between the simulation and analytical results is less than 1% in a wide range of the diffraction parameter B. Figures 4.31 and 4.36 present comparative results of the field distributions. Figure 4.59 shows the power gain in the initial stage of amplification. Thorough testing of the simulation code has shown that the code is stable and provides the required accuracy of calculations for the correct choice of simulation parameters (number of radial mesh divisions, number of macroparticles, integration step, etc.).

In Chap. 2 we presented a comprehensive analysis of the nonlinear effects in the framework of the one-dimensional approximation. Some of the features show similar behavior when we go over to the three-dimensional case. This refers to the dependence of the saturation efficiency on the detuning, the space charge, and the energy spread. On the other hand, diffraction effects play a significant role in the nonlinear regime and we meet novel effects with respect to the one-dimensional model.

We start with the case of an untapered undulator. In this case the maximal output radiation power is achieved at the saturation point when most of the electrons fall into the accelerating phase of the effective potential. When the external input signal power is small, $W_{\text{ext}} \ll W_0$, the FEL amplifier output characteristics at the saturation depend on neither the input signal power nor the interaction length and are functions of four reduced parameters: $\hat{C}$, B, $\hat{\Lambda}_{\text{p}}^2$, and $\hat{\Lambda}_{\text{T}}^2$. Let us now illustrate the characteristic features of the FEL amplifier operating at saturation.

Figure 4.60 illustrates the simulation results of the reduced FEL amplifier efficiency $\hat{\eta}$ versus the undulator length at the value $B = 1$ of the diffraction parameter. It is clearly seen from this plot that the growth of the output power ceases at the saturation point when most of the electrons fall into the accelerating phase of the effective potential.

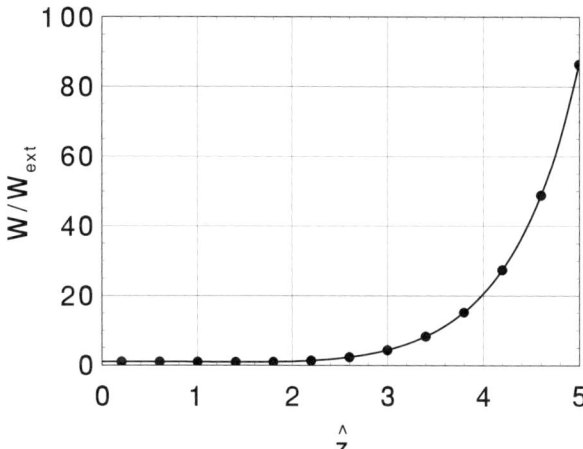

Fig. 4.59. The power gain in the linear stage of amplification. The curve is calculated with the linear simulation code (see (4.157)) and the circles are calculated with the nonlinear simulation code. Here $B = 1$, $\hat{C} = 0$, $\hat{\Lambda}_{\text{p}}^2 \to 0$, $\hat{\Lambda}_{\text{T}}^2 = 0$ and $\hat{w} = 1.2$

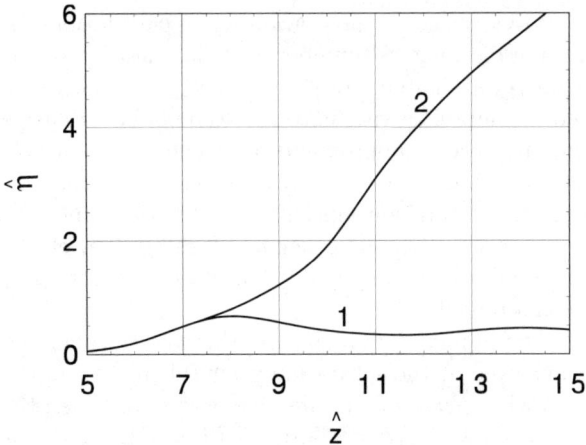

Fig. 4.60. The reduced efficiency $\hat{\eta}$ versus the interaction length. Here $B = 1$, $\hat{C} = 0$, $\hat{\Lambda}_p^2 \to 0$, $\hat{\Lambda}_T^2 = 0$, $\hat{w} = 1.2$, and $\hat{W}_{\text{ext}} = 10^{-3}$. Curve 1: without tapering. Curve 2: tapering according to (4.166)

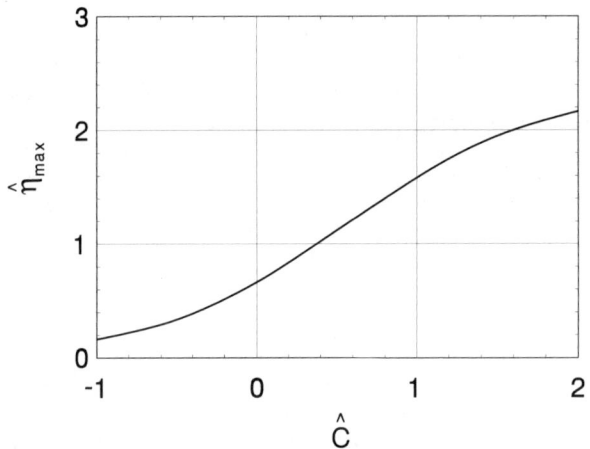

Fig. 4.61. The reduced efficiency $\hat{\eta}$ at saturation versus the detuning parameter $\hat{C}$. Here $B = 1$, $\hat{\Lambda}_p^2 \to 0$, and $\hat{\Lambda}_T^2 = 0$

Figure 4.61 presents the dependence of the maximal reduced efficiency $\hat{\eta}$ on the detuning parameter $\hat{C}$. One can see from this plot that the amplifier efficiency is an increasing function of the detuning parameter $\hat{C}$. This is explained by the fact that when the detuning parameter is increased, the electrons interact with the wave for a longer distance (one should remember that this takes place only when the detuning is inside the amplification bandwidth).

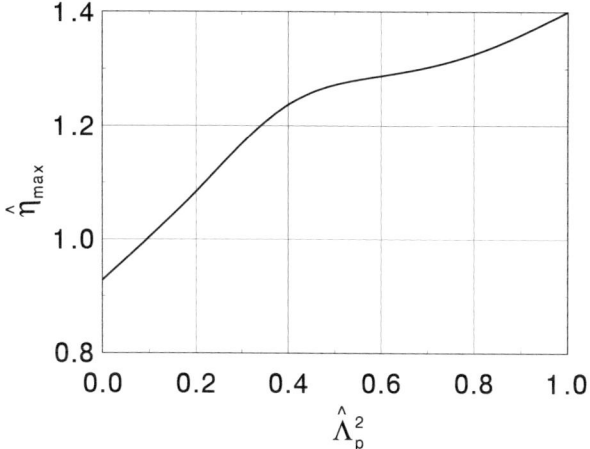

Fig. 4.62. The reduced efficiency $\hat{\eta}$ at saturation versus the space charge parameter $\hat{\Lambda}_p^2$. Here $B = 1$ and $\hat{\Lambda}_T^2 = 0$. (The detuning parameter corresponds to the maximal growth rate in the linear high-gain limit.)

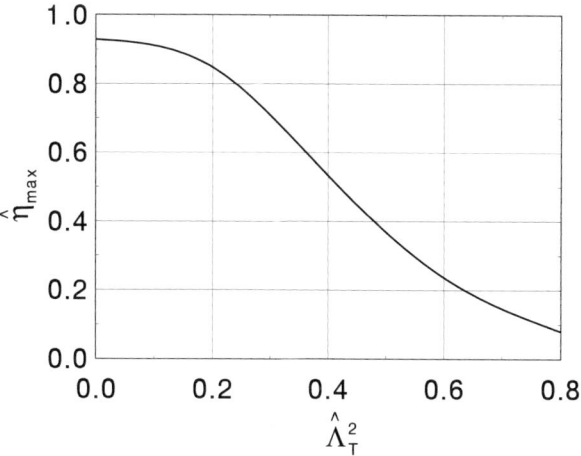

Fig. 4.63. The reduced efficiency $\hat{\eta}$ at saturation versus the energy spread parameter $\hat{\Lambda}_T^2$. Here $B = 1$ and $\hat{\Lambda}_p^2 \to 0$. (The detuning parameter corresponds to the maximum field growth rate in the linear high-gain limit.)

Figure 4.62 presents the dependence of the maximal reduced efficiency on the space charge parameter $\hat{\Lambda}_p^2$. It is clearly seen that the efficiency of the FEL amplifier is an increasing function of the space charge parameter. This is a consequence of the fact that the space charge fields prevent beam overmodulation near the saturation point and the interaction of the modulated electron beam with the wave is prolonged.

252   4. Diffraction Effects in the FEL Amplifier

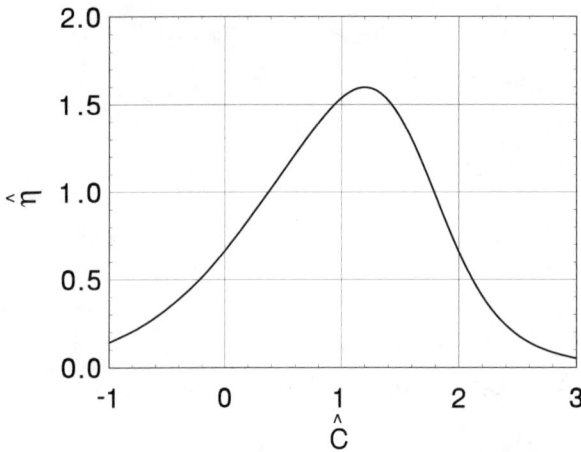

Fig. 4.64. The reduced efficiency $\hat{\eta}$ versus the detuning parameter. At $\hat{C} = 0$ the FEL amplifier operates at saturation with the power gain equal to 30 dB. Here $B = 1$, $\hat{\Lambda}_p^2 \to 0$, $\hat{\Lambda}_T^2 = 0$

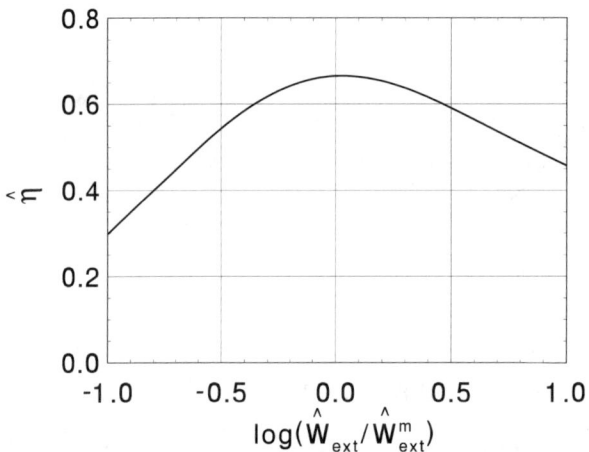

Fig. 4.65. The reduced efficiency $\hat{\eta}$ versus the deviation of the input radiation power $\hat{W}_{ext}$ from the nominal value $\hat{W}_{ext}^m$. The FEL amplifier operates at saturation. Here $B = 1$, $\hat{C} = 0$, $\hat{\Lambda}_p^2 \to 0$, $\hat{\Lambda}_T^2 = 0$, $\hat{w} = 1.2$, and $\hat{W}_{ext}^m = 10^{-3}$

Figure 4.63 shows the dependence of the maximal reduced efficiency on the energy spread parameter. One can see that the amplifier efficiency is decreased drastically with energy spread. Thus, we find that there are several general features of the operation of the FEL amplifier operating in the saturation regime which are the same as in the one-dimensional model: namely, the saturation efficiency can be controlled by an appropriate adjustment of

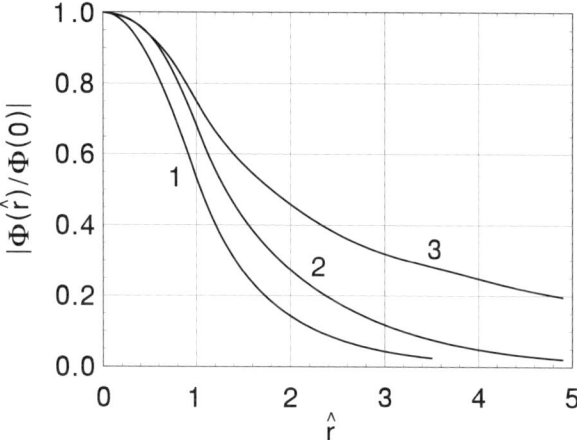

Fig. 4.66. The field distribution in the undulator versus radius. Curve 1: at the linear stage (analytical results). Curve 2: at saturation. Curve 3: at $\hat{z} = 15$ with tapering according to (4.166). Here $B = 1$, $\hat{C} = 0$, $\hat{\Lambda}_\mathrm{p}^2 \to 0$, $\hat{\Lambda}_\mathrm{T}^2 = 0$, $\hat{w} = 1.2$, and $\hat{W}_\mathrm{ext} = 10^{-3}$

the detuning. Also, the influence of space charge effects on FEL amplifier operation has the same physical origin. As for the energy spread in the electron beam, there is the same influence on the saturation efficiency – it drops drastically with an increase of the energy spread. It is important to note that an additional consequence of the energy spread in the linear regime consists of a strong suppression of the higher radiation modes.

In Figs. 4.64 and 4.65 one can see the basic output characteristics of the FEL amplifier at saturation: the reduced resonance characteristic (the dependence of the efficiency on the detuning parameter) and the amplitude characteristic (the dependence of the efficiency versus the deviation of the input power from the nominal value). Using the reduced resonance characteristic one can find the reduced amplification bandwidth $\Delta\hat{C}$. It is connected with the physical parameters by the simple relations: $\Delta\omega/\omega = 2\rho\Delta\hat{C}$, $\Delta\mathcal{E}/\mathcal{E} = \rho\Delta\hat{C}$, and $\Delta H_\mathrm{w}/H_\mathrm{w} = \rho(1+K^2)\Delta\hat{C}/K^2$.

Since we consider the three-dimensional case, we can calculate distributions of the radiation field in the far and the near diffraction zones. The corresponding plots are presented in Figs. 4.66 and 4.67. Figure 4.66 shows the transverse distribution of the radiation field when the FEL amplifier operates at saturation. The field distribution of the fundamental symmetric TEM$_{00}$ mode (linear stage) is presented in this figure, too. One can see that the field distribution at saturation is wider than that at the linear stage and the radiation field expands out of the electron beam.

Figure 4.67 presents the angular distribution of the radiation power for the FEL amplifier operating at saturation and for the fundamental TEM$_{00}$ mode. It is clearly seen from these plots that the width of the field distribu-

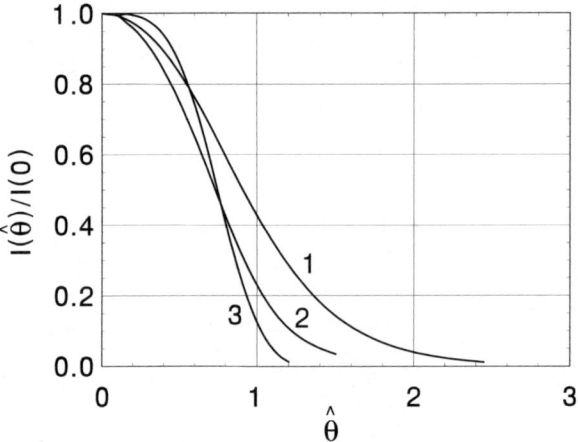

Fig. 4.67. Angular distribution of the radiation intensity. Curve 1: at the linear stage (analytical results). Curve 2: at saturation. Curve 3: at $\hat{z} = 15$ with tapering according to (4.166). Here $B = 1$, $\hat{C} = 0$, $\hat{\Lambda}_{\mathrm{P}}^2 \to 0$, $\hat{\Lambda}_{\mathrm{T}}^2 = 0$, $\hat{w} = 1.2$, and $\hat{W}_{\mathrm{ext}} = 10^{-3}$

tion at saturation is less than at the linear stage. This is a consequence of the fact that the effective size of the radiation spot at the amplifier exit is larger at saturation with respect to the linear stage (see Fig. 4.66). Also, the width of the angular distribution at saturation is a monotonic function of the diffraction parameter. The width of the distribution is always increased with an increase of the diffraction parameter.

All the numerical simulations, presented above, have illustrated common features of the FEL amplifier for fixed value of the diffraction parameter, $B = 1$. It would be interesting to trace with Fig. 4.68 the dependence of the reduced efficiency on the value of the diffraction parameter B. From a practical point of view, this plot covers all the region of interest for the range of the diffraction parameter.

Now we should like to draw the attention of the reader to novel nonlinear effects which do not occur in the framework of the one-dimensional model (see Chap. 2). The analysis of the linear mode of FEL amplifier operation shows that at the value $B \simeq 10$ of the diffraction parameter the difference between the TEM_{00} mode growth rate and one-dimensional model growth rate does not exceed a few per cent (see Fig. 4.39). It is natural to suppose that similar behavior may occur at the nonlinear stage, too. We remember that in the framework of the one-dimensional model, the value of the reduced efficiency at saturation is $\hat{\eta}_{\mathrm{1D}} \simeq 1.37$. This value, recalculated in terms of three-dimensional parameters, gives $\hat{\eta}_{\mathrm{3D}} \simeq 1.37/B^{1/3}$. Let us study the plot in Fig. 4.68. It is natural that the one-dimensional model gives a significant overestimation of the efficiency at the value $B \lesssim 1$ of the diffraction param-

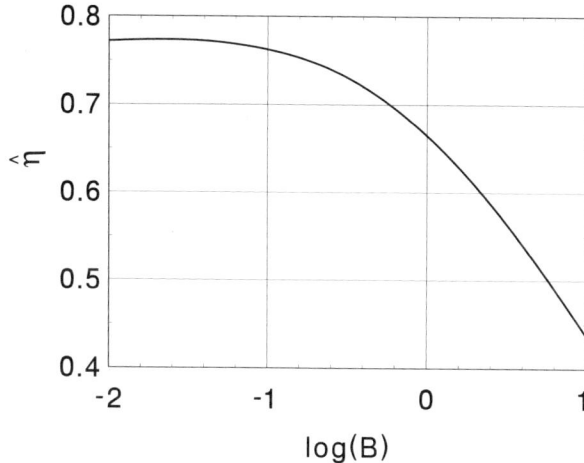

Fig. 4.68. The reduced efficiency $\hat{\eta}$ at saturation versus the diffraction parameter B. Here $\hat{C} = 0$, $\hat{\Lambda}_p^2 \to 0$, and $\hat{\Lambda}_T^2 = 0$

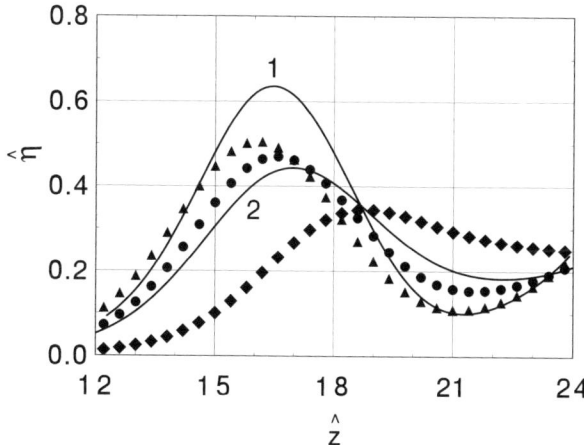

Fig. 4.69. The reduced efficiency $\hat{\eta}$ versus $\hat{z}$. The power gain at saturation is $G = 40$ dB. Curve 1: one-dimensional simulations. Two-dimensional simulation at the value of diffraction parameter $B = 10$. Curve 2: the total efficiency; ($\triangle$): the efficiency at $\hat{r} = 0$; (o): the efficiency at $\hat{r} = 0.5$; and ($\lozenge$): the efficiency at $\hat{r} = 1$. Here $\hat{C} = 0$, $\hat{\Lambda}_p^2 \to 0$, and $\hat{\Lambda}_T^2 = 0$

eter. The reason for this is that at small values of the diffraction parameter the size of the radiation mode is wider than the size of the electron beam even in the linear regime.

On the other hand, it seems to be surprising that there is a significant difference in the efficiency calculations with these two models at large values of the diffraction parameter. Even at the value $B = 10$ of the diffraction parameter the difference is still of about 30%. To explain this phenomenon, we should analyze the distribution of the electron energy losses over the radial coordinate. One can see from Fig. 4.69 that the electron energy losses are smaller for particles located closer to the beam boundary. This is a consequence of the nonuniform distribution of the radiation field over the beam cross-section. Such a nonuniformity has its origin in diffraction effects. Though the value of the TEM$_{00}$ mode growth rate is close to the one-dimensional asymptote at $B \simeq 10$, the growth rates of the higher modes: TEM$_{01}$, TEM$_{02}$, etc., are visibly less. When the FEL power gain is large, only the TEM$_{00}$ radiation mode survives. One can see from Fig. 4.31 that the nonuniformity of the TEM$_{00}$ mode increases with the value of the diffraction parameter. In conclusion, we should emphasize that the results of the one-dimensional nonlinear theory should be used carefully because even at large values of the diffraction parameter diffraction effects could play a significant role. This happens when the gain is large enough to provide mode selection.

Another important subject for discussion is undulator tapering, allowing one to increase the FEL amplifier efficiency. A similar problem has been studied in Chap. 2 in the framework of the one-dimensional model. It has been shown that optimal undulator tapering occurs when the variation of the undulator parameters is a quadratic law. The situation changes drastically if we take into account diffraction effects. In particular, the quadratic law of tapering no longer works in the three-dimensional case, and a linear law should be used.

Let us begin with a qualitative analysis of the FEL amplifier with a tapered undulator. The bunched electron beam in the FEL amplifier can be considered as a sequence of periodically spaced oscillators. The radiation of these oscillators always interferes coherently at zero angle with respect to the undulator axis. In the limit of small size of the electron beam this interference will be constructive within an angle of about

$$\Delta\theta_c \simeq \sqrt{c/(z_{\text{tap}}\omega)},$$

where z_{tap} is the length of the tapered section. In the limit of large size of the electron beam, the angle of coherence is about

$$\Delta\theta_c \simeq c/(r_b\omega).$$

The boundary between these two asymptotes is about

$$r_{\text{dif}} \simeq \sqrt{cz_{\text{tap}}/\omega}.$$

We can estimate the power radiated by the bunched electron beam in the tapered section of length z_{tap}:

$$W \simeq c|\tilde{E}|^2 S_{\text{rad}},$$

where the square of the optical beam is

$$S_{\text{rad}} \simeq r_{\text{dif}}^2 \qquad \text{for} \qquad r_b^2 \ll r_{\text{dif}}^2,$$

4.5 Nonlinear Mode of Operation

$$S_{\text{rad}} \simeq r_b^2 \qquad \text{for} \qquad r_{\text{dif}}^2 \ll r_b^2 \ .$$

Assuming that a significant fraction of the particles is trapped in the regime of coherent deceleration, we can estimate the power loss by the electron beam in the tapered section as:

$$W \simeq I_0 |\tilde{E}| \theta_s z_{\text{tap}} \ .$$

In the latter expression we used the following estimation for the radiation field:

$$\int_0^{z_{\text{tap}}} \tilde{E}(z) \mathrm{d}z \simeq \tilde{E}(z_{\text{tap}}) z_{\text{tap}} \ .$$

Thus, we have the following power balance:

1. Thin electron beam ($r_b^2 \ll c z_{\text{tap}}/\omega$):

   $$S_{\text{rad}} \simeq c z_{\text{tap}}/\omega \ ,$$
   $$W \simeq I_0 |\tilde{E}| \theta_s z_{\text{tap}} \simeq |\tilde{E}|^2 c^2 z_{\text{tap}}/\omega \ \to$$
   $$W \simeq I_0^2 \theta_s^2 z_{\text{tap}} \omega / c^2 \ , \qquad |\tilde{E}| \simeq I_0 \theta_s \omega / c^2 \ .$$

2. Wide electron beam ($r_b^2 \gg c z_{\text{tap}}/\omega$):

   $$S_{\text{rad}} \simeq r_b^2 \ ,$$
   $$W \simeq I_0 |\tilde{E}| \theta_s z_{\text{tap}} \simeq |\tilde{E}|^2 c r_b^2 \ \to$$
   $$W \simeq I_0^2 \theta_s^2 z_{\text{tap}}^2 / (c r_b^2) \ , \qquad |\tilde{E}| \simeq I_0 \theta_s z_{\text{tap}} / (c r_b^2) \ .$$

Our estimates show that in the case of a thin electron beam, the radiation field, acting on the electrons, is almost constant along the undulator axis. The radiation power grows linearly with the length of the tapered section, z_{tap}. Thus, we can conclude that the regime of coherent deceleration of the particles should take place only for a linear law of undulator tapering, i.e. the detuning $\hat{C}(\hat{z})$ should change linearly with the z coordinate.

Let us consider the case of a large value of the diffraction parameter,

$$B = \Gamma \omega r_b^2 / c \gg 1 \ .$$

At the beginning of the tapering section, when $\Gamma z_{\text{tap}} \lesssim B$, we deal with the case of a wide electron beam and most of the radiation overlaps with the electron beam. When the length of the tapered section increases, the radiation expands out of the electron beam. When $\Gamma z_{\text{tap}} \gg B$ we always fall in the region when diffraction effects are important, i.e. the electron beam becomes thin with respect to the radiation beam. Thus, we come to the conclusion that the asymptotically stable regime of coherent deceleration should occur only for a linear law of undulator tapering.

The results of numerical simulations confirm these simple physical considerations. Let us present a specific numerical example for the value of the diffraction parameter $B = 1$. In Figs. 4.60, 4.66, and 4.67 we present the simulation results of the FEL amplifier with tapered undulator. The tapering has been performed at a fixed value of undulator parameter K according to the law

$$\hat{C}(\hat{z}) = \begin{cases} 0 & \text{for} \quad \hat{z} < 7, \\ \hat{z} - 7 & \text{for} \quad \hat{z} > 7. \end{cases} \tag{4.166}$$

The parameters of the tapering have been optimized in order to achieve maximal efficiency at the exit of the tapered section. The radiation power in the tapered section grows linearly with undulator length (see Fig. 4.60). The radiation spot size at the amplifier exit is larger than that at saturation (see Fig. 4.66). As a result, the width of the angle distribution becomes narrower (see Fig. 4.67).

4.5.3 Planar Undulator

All the results of the linear and nonlinear theory of the FEL amplifier with helical undulator and circularly polarized radiation can also be used for the case of a planar undulator and linearly polarized radiation for the following redetermination of the parameters (for comparison, see Sect. 4.4.4):

$$\Gamma = \left[A_{\mathrm{JJ}}^2 I_0 \omega^2 \theta_\ell^2 \left(4 I_A c^2 \gamma_\ell^2 \gamma \int_0^\infty S(\zeta)\zeta \mathrm{d}\zeta \right)^{-1} \right]^{1/2},$$

$$B = r_0^2 \Gamma \omega / c,$$

$$\hat{C} = \left[k_\mathrm{w} - \omega/(2c\gamma_\ell^2) \right]/\Gamma,$$

$$\hat{\Lambda}_\mathrm{p}^2 = 8c^2(\omega^2 r_0^2 \theta_\ell^2 A_{\mathrm{JJ}}^2)^{-1},$$

$$\hat{\Lambda}_\mathrm{T}^2 = \langle(\Delta\mathcal{E})^2\rangle/(\rho^2 \mathcal{E}_0^2),$$

$$\rho = \gamma_\ell^2 \Gamma c/\omega,$$

$$W_0 = I_0 \mathcal{E}_0 \rho/e,$$

$$\theta_\ell = e H_\ell/(\mathcal{E}_0 k_\mathrm{w}),$$

$$\gamma_\ell^{-2} = \gamma^{-2} + \theta_\ell^2/2.$$

The factor A_{JJ} is given by the formula

$$A_{\mathrm{JJ}} = [J_0(Q) - J_1(Q)],$$

where $Q = \theta_\ell^2 \omega/(8ck_\mathrm{w})$, and J_0 and J_1 are Bessel functions.

4.6 Concluding Remarks

The model of the FEL amplifier considered in this chapter is based on a full three-dimensional description of the electromagnetic field, but the electron motion is considered to be one-dimensional. It is assumed that the electrons move along constrained helical (or sinusoidal) trajectories in parallel with the undulator axis. In any real magnetic system of an FEL amplifier the electrons also execute free oscillations in the transverse direction, the so-called betatron oscillations. The typical period of such oscillations is much larger than the period of constrained motion and is defined by focusing properties of the undulator field (and/or of the external focusing magnetic field). The properties of the focusing field in the x and y planes are described by the beta functions $\beta_{x,y}$. Let us consider the simplest case of azimuthally symmetric focusing without axial variation of the beta function, $\beta_x = \beta_y = \beta = \text{const}$. For instance, an intrinsic feature of the helical undulator is that it provides natural uniform focusing in both transverse planes with the beta function equal to $\beta = \sqrt{2}/(k_\mathrm{w}\theta_\mathrm{s})$.

The rms transverse emittance of an axisymmetric electron beam, matched with the uniform focusing system, is given by the expression:

$$\epsilon = (\langle x^2\rangle\langle (x')^2\rangle)^{1/2} = (\langle y^2\rangle\langle (y')^2\rangle)^{1/2} \ .$$

Here $\langle \ldots \rangle$ means averaging over the electrons in the beam. The prime denotes the derivative with respect to z. The following simple relations hold:

$$\langle x^2\rangle = \epsilon\beta \ , \qquad \langle (x')^2\rangle = \frac{\epsilon}{\beta} \ . \tag{4.167}$$

We note two effects which can influence the operation of the FEL amplifier. The first one is connected with the motion of the particle across the beam. As a rule, the radiation field in the FEL amplifier has a complicated transverse profile, so the particle interacts with different fields when crossing the beam. This can lead, in principle, to a decrease of the field growth rate, especially for nonsymmetric modes. This effect is negligible when the beta function is much larger than the gain length. Indeed, the typical longitudinal scale for the amplification process is the gain length. When the beta function is larger than the gain length, the position of the particle in the beam changes only slightly. The larger the beta function, the better the model approximation of constrained motion along the z axis. The criterion for neglecting this effect may be approximately written as $\beta\Gamma \gg 1$, where Γ is the gain parameter. Another situation when we can neglect this effect is the case of a thin electron beam. The transverse variation of the radiation field across the electron beam is small in this case, so all the particles interact with the same field.

The second effect of betatron oscillations, which can influence the operation of the FEL amplifier, has its origin in an additional longitudinal velocity spread. Particles with equal energies, but with different betatron angles, have different longitudinal velocities. In other words, on top of the longitudinal

4. Diffraction Effects in the FEL Amplifier

velocity spread due to the energy spread, there is an additional source of velocity spread. To estimate the power of this effect, we should calculate the dispersion of the longitudinal velocities due to both effects. Let us write the expression for the longitudinal velocity $\beta_z = v_z/c$ of the particle averaged over period of focusing system:[2]

$$\beta_z \simeq 1 - \frac{1}{\gamma_z^2} - \frac{\beta_\perp^2}{2} \, ,$$

where $\gamma_z^2 = (1 + K^2)/\gamma^2$, $\gamma = \mathcal{E}/(m_e c^2)$, K is the undulator parameter, $\beta_\perp^2 = (x')^2 + (y')^2$, $x' = \sqrt{A_x/\beta_x}$, $y' = \sqrt{A_y/\beta_y}$, and A_x and A_y are Courant-Snyder invariants. A particle with zero amplitude of the betatron oscillations and with nominal energy $\gamma_0 = \mathcal{E}_0/(m_e c^2)$ has the following nominal longitudinal velocity:

$$\beta_{z0} \simeq 1 - \frac{1}{2\gamma_{z0}^2} \, .$$

The deviation of the longitudinal velocity from the nominal value is

$$\Delta\beta_z = \frac{1}{\gamma_{z0}^2} \frac{\Delta\gamma}{\gamma_0} - \frac{\beta_\perp^2}{2} \, .$$

We consider the case of a symmetric energy distribution with respect to γ_0. Assuming $\Delta\gamma$, x', and y' be uncorrelated, we obtain in the axisymmetric case:

$$\langle \beta_z^2 \rangle - \langle \beta_z \rangle^2 = \langle (\Delta\beta_z)^2 \rangle - \langle \Delta\beta_z \rangle^2$$
$$= \frac{1}{\gamma_{z0}^4} \frac{\langle (\Delta\gamma)^2 \rangle}{\gamma_0^2} + \frac{1}{2} \left[\langle (x')^4 \rangle - \langle (x')^2 \rangle^2 \right] \, . \quad (4.168)$$

Let us consider the Gaussian energy distribution (2.20a) and the Gaussian distribution in transverse phase space. The normalized distribution for the betatron angles is

$$f(x') = \frac{1}{\sqrt{2\pi \langle (x')^2 \rangle}} \exp\left[-\frac{(x')^2}{2\langle (x')^2 \rangle} \right] \, .$$

Then we obtain that

$$\langle (x')^4 \rangle = 3 \langle (x')^2 \rangle^2 \, ,$$

and, using (4.167), rewrite (4.168) in the form:

$$\langle (\Delta\beta_z)^2 \rangle - \langle \Delta\beta_z \rangle^2 = \frac{1}{\gamma_z^4} \frac{\langle (\Delta\mathcal{E})^2 \rangle}{\mathcal{E}_0^2} + \frac{\epsilon^2}{\beta^2} \, . \quad (4.169)$$

Now we can write the expression for the effective energy spread which provides the same dispersion of the longitudinal velocities, as betatron motion:

---

[2] Here we consider the case when the period of focusing system is much shorter than the betatron wavelength. Within the framework of this smooth approximation the longitudinal velocity of the particle, averaged over period of the focusing system, does not depend on the phase of betatron oscillation, and is a function of only Courant-Snyder invariant.

$$\frac{\langle(\Delta\mathcal{E})^2\rangle_{\text{eff}}}{\mathcal{E}_0^2} = \frac{\gamma_z^4 \epsilon^2}{\beta^2} \ . \tag{4.170}$$

This effective energy spread can be used for the calculation of the effective energy spread parameter:

$$\left(\hat{\Lambda}_{\text{T}}^2\right)_{\text{eff}} = \frac{\langle(\Delta\mathcal{E})^2\rangle_{\text{eff}}}{\rho^2 \mathcal{E}_0^2} \ .$$

One can find that the influence of betatron oscillations on the longitudinal velocity spread can be neglected when $\left(\hat{\Lambda}_{\text{T}}^2\right)_{\text{eff}} \ll 1$. When the following conditions are fulfilled,

$$\beta\Gamma \gg 1 \ , \qquad \frac{\gamma_z^4 \epsilon^2}{\beta^2} \ll \rho^2 \ ,$$

the model of constrained motion, described in this chapter, always provides a correct description of the FEL amplifier. If only the first condition is violated, the model provides correct results for the case of a thin electron beam, $B \ll 1$. In all other cases a more general model should be used, carefully incorporating the betatron motion into the FEL theory and the simulation codes. Relevant references can be found in the suggested bibliography.

4.7 Suggested Bibliography

Here we present selected references to the original papers devoted to the study of diffraction effects in the theory of the FEL amplifier. The first consideration of diffraction effects in the FEL amplifier was done in [4.1]. The asymptotic solution for a thin electron beam (small values of the diffraction parameter) was obtained. The rigorous solution of the eigenvalue problem for an axisymmetric electron beam with a stepped profile was obtained in [4.2]. Later the initial-value problem was solved in [4.3] in the framework of the same model of the electron beam. A more general study of the eigenvalue and initial-value problems was reported in [4.4–4.9]. Different complications have been taken into account, such as gradient profile of the electron beam, space charge and energy spread effects. The influence of betatron oscillations on the operation of the FEL amplifier has been investigated in [4.10, 4.11]. Obtaining a rigorous solution is difficult in this case, so the approximate expressions for the field growth rate were derived. Finally, we mention [4.12–4.17] describing numerical algorithms for the simulation of an FEL amplifier.

5. Waveguide FELs

In the previous chapter we presented the analysis of the FEL amplifier using the approximation of an open electron beam. Such an approximation describes well FEL amplifiers operating in the visible down to X-ray wavelength ranges. To describe FEL amplifiers operating in the millimeter or far infrared wavelength ranges, one should take into account the influence of the waveguide walls on the amplification process. Also, a more general consideration allows us to find the region of applicability for the model of an open electron beam.

The present study assumes the waveguide to be overmoded. Such an assumption does not reduce significantly the practical applicability of the results obtained. Indeed, the FEL amplifier has an advantage over conventional vacuum tube devices only when the undulator period λ_w is much larger than the radiation wavelength λ. The FEL resonance condition is

$$\frac{\lambda_w}{v_z} = \frac{\lambda}{v_{ph} - v_z},$$

where v_z is the longitudinal velocity of the electrons and v_{ph} is the phase velocity of the electromagnetic wave:

$$v_{ph} = c\left(1 - c^2 k_\perp^2/\omega^2\right)^{-1/2}.$$

Here $k_\perp$ is the transverse wavenumber of the wave. It is obvious that the parameter $c^2 k_\perp^2/\omega^2$ should be much less than unity when $\lambda_w \gg \lambda$, i.e. the waveguide should be overmoded.

The linear mode of operation of the FEL amplifier with an overmoded waveguide can be described by two different methods. The first one is based on the paraxial wave equation and approximate boundary conditions on the waveguide walls. Using this method, we derive self-consistent field equations which can be solved analytically in a number of cases. Another approach is based on the method of Green's function. To some extent, this approach is more consistent. Indeed, we start with the rigorous solutions for the eigenfunctions of a passive waveguide. Using these eigenfunctions, we find the Green's function. Finally, the Green's function is simplified using the paraxial approximation. Application of the Green's function method allows one to describe the linear mode of FEL amplifier operation by means of integro-differential equations for the first harmonic of the beam current density. The obtained results are fully compatible with those obtained from the paraxial wave equation and approximate boundary conditions.

Thus, we find that the radiation field in an overmoded waveguide is described adequately using the paraxial approximation. The reader can show that the same approximation was used in the previous chapter for the analysis of diffraction effects in the FEL amplifier. So, the theory of the waveguide FEL, presented here, is a natural extension of the theory of the open electron beam. The main ideas and methods used in the previous chapter can be simply extended to the case of the waveguide FEL.

When presenting the linear theory, we use the power of analytical techniques. In particular, this refers to the FEL amplifier with an axisymmetric electron beam, helical undulator, and circular waveguide. Rigorous analytical solutions of the eigenvalue problem can be obtained for the case of a stepped and parabolic profile of the beam current density. Analysis of the rigorous solutions provides the possibility of studying different physical effects in the clearest form. For instance, we find that the azimuthal symmetry of the electron beam current density modulation does not lead to azimuthal symmetry of the radiation fields, and vise versa. Physically this means that azimuthal symmetry is violated by the helicity of the undulator. Indeed, the electron beam moving in the helical magnetic field may be considered as a gyrotropic active medium. In the case of the open beam this results in the circular polarization of the radiation field only. The presence of the waveguide walls significantly complicates the situation, because the boundary conditions on the walls are different for the radial and azimuthal components of the electric field. As a result, double degeneration of azimuthal modes is eliminated and this leads to the difference mentioned above between the azimuthal dependencies of the electron beam density modulation and radiation field.

Analytical techniques provide the possibility of describing the operation of the FEL amplifier only in a limited number of cases. More specific information can be obtained by means of numerical simulation codes. In this chapter we describe an approach for developing the numerical simulation codes for the FEL amplifier with a waveguide. The radiation fields are calculated using the Green's function method. The analytical results are used as primary standards for testing the numerical simulation codes.

It is worth mentioning that application of similarity techniques helps significantly in a clear physical interpretation of the numerical results. In this chapter we continue discussion of similarity techniques in the FEL theory. Despite the fact that the initial equations of the waveguide FEL amplifier are much more complicated than in the case of an open electron beam, we find in the end that all these complications result in the appearance of only one additional dimensionless parameter, namely the waveguide diffraction parameter.

The structure for presenting the material is as follows. We start with the derivation of the self-consistent equations describing the linear mode of operation of the FEL amplifier with waveguide. In the following sections we present a detailed investigation of the FEL amplifier with an axisymmetric electron beam, helical undulator, and circular waveguide. We find analytical solutions of the eigenvalue problem for the case of a stepped and parabolic beam profile. The case of an arbitrary gradient profile of the electron beam is studied by means of the multilayer approximation method. The initial-value problem is also solved by two methods: analytically (for the stepped profile) and numerically (for an arbitrary gradient profile). Finally, we investigate the nonlinear mode of operation of the FEL amplifier with circular waveguide and helical undulator.

At the end of the chapter we deal with extensions to the theory of the waveguide FEL. First, we present an approach for developing the numerical simulation codes for the FEL amplifier with rectangular waveguide and helical undulator. Second, the generalization of the theory for the case of finite conductivity of the walls is presented. It is based on the application of the Green's function method and Leontovich's boundary conditions.

5.1 Self-Consistent Equations

Let us consider a helical undulator. Electrons in the helical undulator move along the constrained helical trajectory parallel to the z axis. The electron rotation angle $\theta_s = K/\gamma$ is assumed to be small and the longitudinal electron velocity v_z is close to the velocity of light, $v_z \simeq c$. A waveguide is placed inside the undulator. The electromagnetic wave propagates in the waveguide in the same direction as the electron beam. The transverse electric field of the wave can be written in the form:

$$\boldsymbol{E}_\perp = \left[\boldsymbol{e}_x \tilde{E}_x(z,\,\boldsymbol{r}_\perp) + \boldsymbol{e}_y \tilde{E}_y(z,\,\boldsymbol{r}_\perp) \right] \exp\left[\mathrm{i}\omega(z/c - t) \right] + \mathrm{C.C.} \;, \qquad (5.1)$$

where $\boldsymbol{e}_x$ and $\boldsymbol{e}_y$ are unit vectors of the Cartesian coordinate system and ω is the frequency of the amplified wave.

For an electron beam with a small density perturbation, the distribution function can be written in the form
$$f(z, \psi, \mathcal{E}, \boldsymbol{r}_\perp) = f_0(\mathcal{E}, \boldsymbol{r}_\perp) + \tilde{f}_1(z, \mathcal{E}, \boldsymbol{r}_\perp) \exp(\mathrm{i}\psi) + \mathrm{C.C.}$$
The evolution of the perturbation to the distribution function, $\tilde{f}_1$, is described by the Vlasov equation (4.9). In the case of a helical undulator, the amplitude of the effective potential, U, appearing in (4.9) is connected with the complex amplitude of the electromagnetic wave, $\tilde{E}_{x,y}$, by

$$U = -\frac{e\theta_s}{2\mathrm{i}} \left[\tilde{E}_x + \mathrm{i}\tilde{E}_y \right] \;. \qquad (5.2)$$

In the following we assume also that the transverse size of the electron beam is rather large, $r_b^2/\gamma_z^2 \gg c^2/\omega^2$. In particular, it follows from this assumption that we can neglect the reduction of the plasma wavenumber due to the presence of the waveguide walls.

5. Waveguide FELs

We consider the initial conditions when an external electromagnetic wave and unmodulated electron beam are fed to the undulator entrance:

$$\tilde{f}_1|_{z=0} = 0 , \quad f_0 = n_0(\boldsymbol{r}_\perp)F(P) ,$$

where $P = \mathcal{E} - \mathcal{E}_0$, $\int dP F(P) = 1$. Integration of the Vlasov equation gives us the relation between the longitudinal component of the beam current density, j_z, and the amplitude of the wave:

$$j_z = -j_0(\boldsymbol{r}_\perp) + \tilde{j}_1 \exp(i\psi) + \text{C.C.} ,$$

$$\tilde{j}_1(z, \boldsymbol{r}_\perp) \simeq -ec \int dP \tilde{f}_1$$

$$= -ij_0(\boldsymbol{r}_\perp) \int_0^z dz' \left\{ \frac{e\theta_s}{2i} \left[\tilde{E}_x(z', \boldsymbol{r}_\perp) \right.\right.$$

$$\left.\left. + i\tilde{E}_y(z', \boldsymbol{r}_\perp) \right] - \frac{4\pi e}{\omega} \tilde{j}_1(z', \boldsymbol{r}_\perp) \right\}$$

$$\times \int_{-\infty}^{\infty} dP F'(P) \exp\left\{ i \left[\frac{\omega P}{c\gamma_z^2 \mathcal{E}_0} + C \right](z' - z) \right\} , \quad (5.3)$$

where $-j_0(\boldsymbol{r}_\perp) \simeq -ecn_0(\boldsymbol{r}_\perp)$ is the longitudinal component of the beam current density at the undulator entrance.

It follows from the wave equation that

$$e^{i\omega(z/c-t)} \left[\nabla_\perp^2 \begin{pmatrix} \tilde{E}_x \\ \tilde{E}_y \end{pmatrix} + \frac{\partial^2}{\partial z^2} \begin{pmatrix} \tilde{E}_x \\ \tilde{E}_y \end{pmatrix} + 2i \frac{\omega}{c} \frac{\partial}{\partial z} \begin{pmatrix} \tilde{E}_x \\ \tilde{E}_y \end{pmatrix} \right] + \text{C.C.}$$

$$= \frac{4\pi}{c^2} \frac{\partial}{\partial t} \begin{pmatrix} j_x \\ j_y \end{pmatrix}$$

$$= -4\pi \frac{\omega}{c^2} \theta_s \begin{pmatrix} \cos(k_w z) \\ -\sin(k_w z) \end{pmatrix} (i\tilde{j}_1 \exp(i\psi) + \text{C.C.}) . \quad (5.4)$$

In the framework of the accepted limitations, (5.4) may be reduced to an equation for the slowly varying amplitudes:

$$\nabla_\perp^2 \begin{pmatrix} \tilde{E}_x \\ \tilde{E}_y \end{pmatrix} + \frac{2i\omega}{c} \frac{\partial}{\partial z} \begin{pmatrix} \tilde{E}_x \\ \tilde{E}_y \end{pmatrix} = -\begin{pmatrix} i \\ 1 \end{pmatrix} 2\pi \frac{\omega}{c^2} \theta_s \tilde{j}_1(z, \boldsymbol{r}_\perp) . \quad (5.5)$$

5.1.1 Integro-Differential Equation for the Field

The self-consistent equations (5.3) and (5.5) can be solved in two ways. The first way is to substitute (5.3) into the right-hand side of (5.5), which leads to a system of two coupled integro-differential equations for $\tilde{E}_x$ and $\tilde{E}_y$:

5.1 Self-Consistent Equations

$$\nabla_\perp^2 \begin{pmatrix} \tilde{E}_x \\ \tilde{E}_y \end{pmatrix} + \frac{2i\omega}{c} \frac{\partial}{\partial z} \begin{pmatrix} \tilde{E}_x \\ \tilde{E}_y \end{pmatrix}$$

$$= j_0(\mathbf{r}_\perp) \int_0^z dz' \left\{ \begin{pmatrix} i \\ 1 \end{pmatrix} \pi \frac{\omega \theta_s^2 e}{c^2} \left(\tilde{E}_x + i\tilde{E}_y \right) \right.$$

$$\left. + i \frac{4\pi e}{\omega} \left[\nabla_\perp^2 \begin{pmatrix} \tilde{E}_x \\ \tilde{E}_y \end{pmatrix} + \frac{2i\omega}{c} \frac{\partial}{\partial z'} \begin{pmatrix} \tilde{E}_x \\ \tilde{E}_y \end{pmatrix} \right] \right\}$$

$$\times \int_{-\infty}^{\infty} dP F'(P) \exp\left\{ i \left[\frac{\omega P}{c \gamma_z^2 \mathcal{E}_0} + C \right] (z' - z) \right\} . \qquad (5.6)$$

These equations can be rewritten as

$$\left(\nabla_\perp^2 + \frac{2i\omega}{c} \frac{\partial}{\partial z} \right) \left(\tilde{E}_x + i\tilde{E}_y \right)$$

$$= ij_0(\mathbf{r}_\perp) \int_0^z dz' \left\{ 2\pi \frac{\omega e}{c^2} \theta_s^2 \left(\tilde{E}_x + i\tilde{E}_y \right) \right.$$

$$\left. + i \frac{4\pi e}{\omega} \left(\nabla_\perp^2 + \frac{2i\omega}{c} \frac{\partial}{\partial z'} \right) \left(\tilde{E}_x + i\tilde{E}_y \right) \right\}$$

$$\times \int_{-\infty}^{\infty} dP F'(P) \exp\left\{ i \left[\frac{\omega P}{c \gamma_z^2 \mathcal{E}_0} + C \right] (z' - z) \right\} , \qquad (5.7a)$$

$$\left(\nabla_\perp^2 + \frac{2i\omega}{c} \frac{\partial}{\partial z} \right) \left(\tilde{E}_x - i\tilde{E}_y \right) = 0 . \qquad (5.7b)$$

These equations describe the general case of electromagnetic wave amplification by the electron beam moving in the helical undulator. Equations (5.7a) and (5.7b) refer to the right- and left-helicity components of the wave, respectively. In the general case these components are not independent, but are connected by the boundary conditions on the waveguide walls.

Equations (5.7) are applicable to the case of an open electron beam, too (see Chap. 4). In the latter case the solutions for the right- and left-helicity waves are linearly independent. It follows from (5.7a) and (5.7b) that only those waves are amplified that have the same helicity as the undulator field itself. That is why in Chap. 4 we used (4.21) which is identical to (5.7a).

The self-consistent field equations (5.7) should be supplemented by the boundary conditions. We assume the waveguide walls to be perfectly conducting. The boundary conditions for this case force the vector of the electric field to be perpendicular to the waveguide wall:

$$(n_x E_y - n_y E_x)|_S = 0 , \qquad E_z|_S = 0 ,$$

where $\mathbf{n} = \mathbf{e}_x n_x + \mathbf{e}_y n_y$ is the unit vector perpendicular to the waveguide wall ($|\mathbf{n}| = 1$). Using Maxwell's equation,

$$\nabla \times \boldsymbol{H} = c^{-1} \partial \boldsymbol{E}/\partial t ,$$

we can represent the boundary condition $E_z|_S = 0$ as

$$(\nabla \times \boldsymbol{H})_z|_S = 0 .$$

The paraxial approximation assumes that the field amplitude does not change significantly along the z axis on the scale of the wavelength, and the field is nearly transversely polarized, i.e.

$$E_x \simeq H_y , \qquad E_y \simeq -H_y .$$

Thus, in the paraxial approximation, the boundary conditions on the waveguide wall take the form

$$(n_x E_y - n_y E_x)|_S = 0 , \qquad (\partial E_x/\partial x + \partial E_y/\partial y)|_S = 0 .$$

5.1.2 Integro-Differential Equation for the Beam Modulation

The self-consistent equations can also be reduced to a unique equation for the beam modulation. Let us illustrate this method for the case of a negligibly small energy spread in the beam. The initial distribution function is the delta function in this case, $F(P) = \delta(P - P_0)$, and the Vlasov equation can be reduced to

$$\left[\frac{d^2}{dz^2} + 2iC \frac{d}{dz} + \left(\frac{4\pi e}{c\gamma_z^2 \mathcal{E}_0} j_0(\boldsymbol{r}_\perp) - C^2 \right) \right] \tilde{j}_1(z, \boldsymbol{r}_\perp)$$
$$= -\frac{\omega}{c\gamma_z^2 \mathcal{E}_0} j_0(\boldsymbol{r}_\perp) U(z, \boldsymbol{r}_\perp) . \tag{5.8}$$

The complex amplitude of the effective potential, U, is connected with the components of the electromagnetic wave by relation (5.2). To close the problem, we should solve Maxwell's equations with the boundary conditions on the waveguide walls and express the fields in terms of the first harmonic of the beam current density. The electromagnetic field is expressed via the vector and the scalar potential as

$$\boldsymbol{H} = \nabla \times \boldsymbol{A} , \qquad \boldsymbol{E} = -\frac{1}{c}\frac{\partial \boldsymbol{A}}{\partial t} - \nabla \phi . \tag{5.9}$$

In the following we use the Coulomb gauge for the potentials,

$$\nabla \cdot \boldsymbol{A} = 0 . \tag{5.10}$$

Substitution of (5.9) into Maxwell's equations gives (for the Coulomb gauge):

$$\nabla^2 \boldsymbol{A} - \frac{1}{c^2}\frac{\partial^2 \boldsymbol{A}}{\partial t^2} = -\frac{4\pi}{c}\boldsymbol{j} + \frac{1}{c}\nabla\frac{\partial \phi}{\partial t} = -\frac{4\pi}{c}\boldsymbol{J} , \tag{5.11a}$$

$$\nabla^2 \phi = -4\pi \rho_e , \qquad \boldsymbol{j} = v\rho_e , \tag{5.11b}$$

where ρ_e is the charge density. It follows from (5.11b) and from the charge conservation law that

$$\nabla \cdot \boldsymbol{J} = 0 .$$

In the Coulomb gauge the scalar potential, ϕ, is the static Coulomb potential. The dynamical part of the field is associated with the vector potential, $\boldsymbol{A}$, only.

Let us consider a monochromatic external wave of frequency ω. Then $\boldsymbol{A}$, ϕ, and $\boldsymbol{j}$ may be written as

$$\boldsymbol{A} = \boldsymbol{A}_\omega e^{-i\omega t} + \text{C.C.} , \qquad \phi = \phi_\omega e^{-i\omega t} + \text{C.C.} , \qquad \boldsymbol{j} = \boldsymbol{j}_\omega e^{-i\omega t} + \text{C.C.}$$

Equation (5.11a) for $\boldsymbol{A}_\omega$ and $\boldsymbol{J}_\omega = \boldsymbol{j}_\omega + i\omega(4\pi)^{-1}\nabla\phi_\omega$ takes the form:

$$\nabla^2 \boldsymbol{A}_\omega + \frac{\omega^2}{c^2} \boldsymbol{A}_\omega = -\frac{4\pi}{c} \boldsymbol{J}_\omega . \tag{5.12}$$

We assume the waveguide walls to be perfectly conducting. The boundary conditions for this case force the vector of the electric field to be perpendicular to the waveguide wall. In addition to the Coulomb gauge condition (5.10), we impose the boundary condition for the scalar potential ϕ to be equal to a zero on the waveguide walls. The boundary conditions for the vector potential $\boldsymbol{A}$ are defined by the boundary conditions for the field:

$$\boldsymbol{n} \times \boldsymbol{A}_\omega|_S = 0 , \tag{5.13}$$

where $\boldsymbol{n}$ is the unit vector perpendicular to the waveguide wall ($|\boldsymbol{n}| = 1$).

Under these boundary conditions, the solution of the inhomogeneous Helmholtz equation (5.12) has the form:

$$A_\omega^\alpha(\boldsymbol{r}) = \sum_\beta \int G_\omega^{\alpha\beta}(\boldsymbol{r}, \boldsymbol{r}') J_\omega^\beta(\boldsymbol{r}') d\boldsymbol{r}' , \tag{5.14}$$

where $G_\omega^{\alpha\beta}(\boldsymbol{r}, \boldsymbol{r}')$ is the tensor Green's function of the waveguide and $\boldsymbol{r}$ and $\boldsymbol{r}'$ are the coordinates of the observation and the source point, respectively. It can be shown that the condition $\phi_\omega|_S = 0$ results in a zero value of the integral (see Appendix A.3):

$$\sum_\beta \int G_\omega^{\alpha\beta}(\boldsymbol{r}, \boldsymbol{r}')(\nabla\phi_\omega(\boldsymbol{r}'))^\beta d\boldsymbol{r}' = 0 . \tag{5.15}$$

Hence, (5.14) takes the form:

$$A_\omega^\alpha(\boldsymbol{r}) = \sum_\beta \int G_\omega^{\alpha\beta}(\boldsymbol{r}, \boldsymbol{r}') j_\omega^\beta(\boldsymbol{r}') d\boldsymbol{r}' . \tag{5.16}$$

Using (5.16), we write the following expression for the electric field of the radiated wave:

$$E_\omega^\alpha(\boldsymbol{r}) = i\frac{\omega}{c} \sum_\beta \int G_\omega^{\alpha\beta}(\boldsymbol{r}, \boldsymbol{r}') j_\omega^\beta(\boldsymbol{r}') d\boldsymbol{r}' , \tag{5.17}$$

where the tensor Green's function, $G_\omega^{\alpha\beta}$, in the paraxial approximation is given by:[1]

$$G_\omega = \frac{2\pi i}{\omega} \sum_\mu \exp\left\{i\left[\frac{\omega}{c} - \frac{c(k_\perp^{TE})_\mu^2}{2\omega}\right]|z-z'|\right\}$$

$$\times \left[e_x \frac{\partial \psi_\mu^{TE}(r_\perp)}{\partial y} - e_y \frac{\partial \psi_\mu^{TE}(r_\perp)}{\partial x}\right]$$

$$\otimes \left[e_x \frac{\partial \psi_\mu^{TE}(r'_\perp)}{\partial y'} - e_y \frac{\partial \psi_\mu^{TE}(r'_\perp)}{\partial x'}\right]$$

$$+ \frac{2\pi i}{\omega} \sum_\nu \exp\left\{i\left[\frac{\omega}{c} - \frac{c(k_\perp^{TM})_\nu^2}{2\omega}\right]|z-z'|\right\}$$

$$\times \left[e_x \frac{\partial \psi_\nu^{TM}(r_\perp)}{\partial x} + e_y \frac{\partial \psi_\nu^{TM}(r_\perp)}{\partial y}\right]$$

$$\otimes \left[e_x \frac{\partial \psi_\nu^{TM}(r'_\perp)}{\partial x'} + e_y \frac{\partial \psi_\nu^{TM}(r'_\perp)}{\partial y'}\right] , \quad (5.18)$$

Here the symbol $\otimes$ denotes the direct product of vectors. The waveguide functions, ψ^{TE} and ψ^{TM}, are the solutions of the Helmholtz equation:

$$\nabla_\perp^2 \psi + k_\perp^2 \psi = 0 , \quad (5.19)$$

with boundary conditions

$$\mathbf{n} \cdot \nabla \psi^{TE}|_S = 0, \quad \psi^{TM}|_S = 0 , \quad (5.20)$$

and normalization condition

$$\int |\nabla \psi|^2 d\mathbf{r}_\perp = 1 . \quad (5.21)$$

The βth component of the transverse beam current density, j_ω^β, is connected with the complex amplitude, $\tilde{j}_1$, as

$$j_\omega^\beta = \frac{1}{c} v^\beta(z) \tilde{j}_1(z, \mathbf{r}_\perp) \exp\left(i k_w z + i\omega \frac{z}{c}\right) . \quad (5.22)$$

Using the relation

$$\tilde{E}_{x,y} \exp\left(i\omega \frac{z}{c}\right) = E_\omega^{x,y} ,$$

and (5.2), (5.18), (5.17), and (5.22), we write the expression for the effective potential of the interaction between the particle and the radiated electromagnetic wave:

---

[1] Expression (5.18) contains only transverse components of $G_\omega^{\alpha\beta}$. In the case of an overmoded waveguide it provides sufficient accuracy for the calculation of the transverse components of the vector potential. The full expression for $G_\omega^{\alpha\beta}$ is presented in Appendix A.3. Taking into account the latter expression, one can check that the Coulomb gauge condition is fulfilled.

5.1 Self-Consistent Equations

$$U_i = -\frac{\pi i}{2c} e\theta_s^2 \int_0^z dz' \int d\mathbf{r}'_\perp \tilde{j}_1(z', \mathbf{r}'_\perp)$$

$$\times \left(\frac{\partial}{\partial x} + i\frac{\partial}{\partial y}\right)\left(\frac{\partial}{\partial x'} - i\frac{\partial}{\partial y'}\right)$$

$$\times \left\{\sum_\mu \psi_\mu^{\text{TE}}(\mathbf{r}_\perp)\psi_\mu^{\text{TE}}(\mathbf{r}'_\perp) \exp\left[-i\frac{c(k_\perp^{\text{TE}})_\mu^2}{2\omega}(z-z')\right]\right.$$

$$\left. + \sum_\nu \psi_\nu^{\text{TM}}(\mathbf{r}_\perp)\psi_\nu^{\text{TM}}(\mathbf{r}'_\perp) \exp\left[-i\frac{c(k_\perp^{\text{TM}})_\nu^2}{2\omega}(z-z')\right]\right\} . \quad (5.23)$$

The external electromagnetic wave can be expressed in terms of the eigenmodes of the passive waveguide:

$$\mathbf{e}_x(\tilde{E}_x)_{\text{ext}} + \mathbf{e}_y(\tilde{E}_y)_{\text{ext}}$$

$$= \sum_\mu C_\mu^{\text{TE}} \exp\left[-i\frac{c(k_\perp^{\text{TE}})_\mu^2 z}{2\omega}\right]$$

$$\times \left(\mathbf{e}_x \frac{\partial \psi_\mu^{\text{TE}}(\mathbf{r}_\perp)}{\partial y} - \mathbf{e}_y \frac{\partial \psi_\mu^{\text{TE}}(\mathbf{r}_\perp)}{\partial x}\right)$$

$$+ \sum_\nu C_\nu^{\text{TM}} \exp\left[-i\frac{c(k_\perp^{\text{TM}})_\nu^2 z}{2\omega}\right]$$

$$\times \left(\mathbf{e}_x \frac{\partial \psi_\nu^{\text{TM}}(\mathbf{r}_\perp)}{\partial x} + \mathbf{e}_y \frac{\partial \psi_\nu^{\text{TM}}(\mathbf{r}_\perp)}{\partial y}\right) , \quad (5.24)$$

where the coefficients C_μ^{TE} and C_ν^{TM} are given by

$$C_\mu^{\text{TE}} = \int d\mathbf{r}_\perp \left(\tilde{E}_x(0, \mathbf{r}_\perp)\frac{\partial \psi_\mu^{\text{TE}}}{\partial y} - \tilde{E}_y(0, \mathbf{r}_\perp)\frac{\partial \psi_\mu^{\text{TE}}}{\partial x}\right) ,$$

$$C_\nu^{\text{TM}} = \int d\mathbf{r}_\perp \left(\tilde{E}_x(0, \mathbf{r}_\perp)\frac{\partial \psi_\nu^{\text{TM}}}{\partial x} + \tilde{E}_y(0, \mathbf{r}_\perp)\frac{\partial \psi_\nu^{\text{TM}}}{\partial y}\right) . \quad (5.25)$$

Total power of the input radiation is expressed in terms of the coefficients C_μ^{TE} and C_ν^{TM} as[2]

$$W_{\text{ext}} = \frac{c}{2\pi}\left(\sum_\mu |C_\mu^{\text{TE}}|^2 + \sum_\nu |C_\nu^{\text{TM}}|^2\right) . \quad (5.26)$$

In the case of single-mode input radiation, we can let the value of the amplitude coefficient C_μ^{TE} (or C_ν^{TM}) be a real and positive constant, and write the expression for U_{ext} as

---

[2] This expression can be derived from (5.30) using the orthogonality condition for the eigenfunctions (5.32).

for TE-mode:

$$U_{\text{ext}}^{\text{TE}} = ie\theta_s \left(\frac{\pi W_{\text{ext}}}{2c}\right)^{1/2} \left(\frac{\partial \psi_\mu^{\text{TE}}(\boldsymbol{r}_\perp)}{\partial y} - i\frac{\partial \psi_\mu^{\text{TE}}(\boldsymbol{r}_\perp)}{\partial x}\right)$$
$$\times \exp\left[-i\frac{c(k_\perp^{\text{TE}})_\mu^2 z}{2\omega}\right], \qquad (5.27a)$$

for TM-mode:

$$U_{\text{ext}}^{\text{TM}} = ie\theta_s \left(\frac{\pi W_{\text{ext}}}{2c}\right)^{1/2} \left(\frac{\partial \psi_\nu^{\text{TM}}(\boldsymbol{r}_\perp)}{\partial x} + i\frac{\partial \psi_\nu^{\text{TM}}(\boldsymbol{r}_\perp)}{\partial y}\right)$$
$$\times \exp\left[-i\frac{c(k_\perp^{\text{TM}})_\nu^2 z}{2\omega}\right]. \qquad (5.27b)$$

Thus, we have obtained expressions for the effective potential of the particle interaction with the radiated wave, U_i, and with the external wave, U_{ext}. We are interested in the sum of these two contributions:

$$U = U_i + U_{\text{ext}} . \qquad (5.28)$$

Substituting (5.28), (5.23), and (5.27) into (5.8), we obtain the integro-differential equation for the first harmonic of the beam current density, $\tilde{j}_1(z, \boldsymbol{r}_\perp)$.

5.2 Power Balance

Let us study the balance of the radiated power and the power loss by the electron beam. The power flow density of the electromagnetic wave is given by Poynting's vector $\boldsymbol{\Pi}$:

$$\boldsymbol{\Pi} = \frac{c}{4\pi}(\boldsymbol{E} \times \boldsymbol{H}) .$$

The vectors of the electric and magnetic field of the monochromatic wave can be written as

$$\boldsymbol{E}(\boldsymbol{r},\,t) = \boldsymbol{E}_\omega(\boldsymbol{r})\exp(-i\omega t) + \boldsymbol{E}_\omega^*(\boldsymbol{r})\exp(i\omega t) ,$$
$$\boldsymbol{H}(\boldsymbol{r},\,t) = \boldsymbol{H}_\omega(\boldsymbol{r})\exp(-i\omega t) + \boldsymbol{H}_\omega^*(\boldsymbol{r})\exp(i\omega t) .$$

We write the vector product, $\boldsymbol{E} \times \boldsymbol{H}$, as

$$\boldsymbol{E} \times \boldsymbol{H} = [\boldsymbol{E}_\omega \times \boldsymbol{H}_\omega^* + \boldsymbol{E}_\omega^* \times \boldsymbol{H}_\omega]$$
$$+ [\boldsymbol{E}_\omega \times \boldsymbol{H}_\omega \exp(-2i\omega t) + \boldsymbol{E}_\omega^* \times \boldsymbol{H}_\omega^* \exp(2i\omega t)] . \qquad (5.29)$$

The time averaged value of this product is

$$\langle \boldsymbol{E} \times \boldsymbol{H} \rangle = \frac{1}{T}\int_0^T (\boldsymbol{E} \times \boldsymbol{H})\mathrm{d}t' = 2\,\mathrm{Re}(\boldsymbol{E}_\omega \times \boldsymbol{H}_\omega^*) .$$

5.2 Power Balance

Thus, the time averaged longitudinal component of Poynting's vector can be written in the form:
$$\langle \Pi_z \rangle = \frac{c}{2\pi} \operatorname{Re}\left[(\boldsymbol{E}_\omega)_x (\boldsymbol{H}_\omega^*)_y - (\boldsymbol{E}_\omega)_y (\boldsymbol{H}_\omega^*)_x \right] .$$
We stress again that we have use the paraxial approximation which assumes the waveguide to be overmoded. In this case the vectors of the electric and magnetic field are perpendicular to each other and have equal absolute values. Thus, we can write the following expression for the radiation power:
$$W(z) = \int \langle \Pi_z \rangle \mathrm{d}\boldsymbol{r}_\perp = \frac{c}{2\pi} \int |\boldsymbol{E}_\omega|^2 \mathrm{d}\boldsymbol{r}_\perp . \tag{5.30}$$
The field of the electromagnetic wave is a superposition of the fields of the external and the radiated waves:
$$\boldsymbol{E}_\omega = \boldsymbol{E}_\omega^{\mathrm{ext}} + \boldsymbol{E}_\omega^{\mathrm{i}} ,$$
so the total radiation power, W, consists of three summands:
$$W_{\mathrm{tot}} = W_1 + W_2 + W_3 .$$
The term W_1 refers to the radiated wave only,
$$W_1 = \frac{c}{2\pi} \int |\boldsymbol{E}_\omega^{\mathrm{i}}|^2 \mathrm{d}\boldsymbol{r}_\perp .$$
The term W_2 is connected with the external electromagnetic wave,
$$W_2 = W_{\mathrm{ext}} = \frac{c}{2\pi} \int |\boldsymbol{E}_\omega^{\mathrm{ext}}|^2 \mathrm{d}\boldsymbol{r}_\perp .$$
The term W_3 describes the interference between the external and the radiated waves,
$$W_3 = \frac{c}{2\pi} \int \mathrm{d}\boldsymbol{r}_\perp \left[(\boldsymbol{E}_\omega^*)^{\mathrm{i}} \cdot (\boldsymbol{E}_\omega)^{\mathrm{ext}} + \mathrm{C.C.} \right] .$$
Using (5.17), we can write an explicit expression for the term W_1:
$$W_1 = \frac{\omega^2}{2\pi c} \sum_{\alpha,\beta,\gamma} \int \mathrm{d}\boldsymbol{r}_\perp \int \mathrm{d}\boldsymbol{r}' G_\omega^{\gamma\beta}(\boldsymbol{r}, \boldsymbol{r}') j_\omega^\beta(\boldsymbol{r}')$$
$$\times \int \mathrm{d}\boldsymbol{r}'' (G_\omega^{\gamma\alpha}(\boldsymbol{r}, \boldsymbol{r}'') j_\omega^\alpha(\boldsymbol{r}''))^* . \tag{5.31}$$
Taking into account the orthogonality conditions for $\boldsymbol{\nabla}\psi_\mu^{\mathrm{TE}}$ and $\boldsymbol{\nabla}\psi_\nu^{\mathrm{TM}}$,
$$\int \mathrm{d}\boldsymbol{r}_\perp \boldsymbol{\nabla}\psi_\nu \cdot \boldsymbol{\nabla}\psi_{\nu'} = \delta_{\nu\nu'} , \tag{5.32}$$
and using (5.18), we obtain

$$W_1 = \frac{\pi\theta_s^2}{2c} \int d\mathbf{r}'_\perp \int d\mathbf{r}''_\perp \int_0^z dz' \int_0^z dz'' \tilde{j}_1(z', \mathbf{r}'_\perp) \tilde{j}_1^*(z'', \mathbf{r}''_\perp)$$

$$\times \left(\frac{\partial}{\partial x'} - i\frac{\partial}{\partial y'}\right)\left(\frac{\partial}{\partial x''} + i\frac{\partial}{\partial y''}\right) \left\{ \sum_\mu \psi_\mu^{\mathrm{TE}}(\mathbf{r}'_\perp) \psi_\mu^{\mathrm{TE}}(\mathbf{r}''_\perp) \right.$$

$$\times \exp\left[-i\frac{c(k_\perp^{\mathrm{TE}})_\mu^2}{2\omega}(z'' - z')\right]$$

$$\left. + \sum_\nu \psi_\nu^{\mathrm{TM}}(\mathbf{r}'_\perp) \psi_\nu^{\mathrm{TM}}(\mathbf{r}''_\perp) \exp\left[-i\frac{c(k_\perp^{\mathrm{TM}})_\nu^2}{2\omega}(z'' - z')\right] \right\}. \quad (5.33)$$

The product of the integrals over z' and z'' can be represented as in (4.27). Using (5.33) and (5.23), we can write the final expression for W_1:

$$W_1 = \frac{\pi\theta_s^2}{2c} \int d\mathbf{r}'_\perp \int d\mathbf{r}''_\perp \int_0^z dz' \int_0^{z'} dz'' \tilde{j}_1(z', \mathbf{r}'_\perp) \tilde{j}_1^*(z'', \mathbf{r}''_\perp)$$

$$\times \left(\frac{\partial}{\partial x'} - i\frac{\partial}{\partial y'}\right)\left(\frac{\partial}{\partial x''} + i\frac{\partial}{\partial y''}\right) \left\{ \sum_\mu \psi_\mu^{\mathrm{TE}}(\mathbf{r}'_\perp) \psi_\mu^{\mathrm{TE}}(\mathbf{r}''_\perp) \right.$$

$$\times \exp\left[-i\frac{c(k_\perp^{\mathrm{TE}})_\mu^2}{2\omega}(z'' - z')\right] +$$

$$\left. + \sum_\nu \psi_\nu^{\mathrm{TM}}(\mathbf{r}'_\perp) \psi_\nu^{\mathrm{TM}}(\mathbf{r}''_\perp) \exp\left[-i\frac{c(k_\perp^{\mathrm{TM}})_\nu^2}{2\omega}(z'' - z')\right] \right\} + \mathrm{C.C.}$$

$$= -\frac{i}{e} \int_0^z dz' \int d\mathbf{r}'_\perp U_i^*(z', \mathbf{r}'_\perp) \tilde{j}_1(z', \mathbf{r}'_\perp) + \mathrm{C.C.} \quad (5.34)$$

Let us consider the interference term W_3. Using (5.17), we rewrite the expression for W_3 in the following form:

$$W_3 = \frac{c}{2\pi} \int d\mathbf{r}_\perp \left[(\mathbf{E}_\omega)^{\mathrm{i}} \cdot (\mathbf{E}_\omega^*)^{\mathrm{ext}} + \mathrm{C.C.}\right]$$

$$= \frac{i\omega}{2\pi} \sum_{\alpha,\beta} \int d\mathbf{r}_\perp (E_\omega^*)_\alpha^{\mathrm{ext}} \int d\mathbf{r}' G_\omega^{\alpha\beta}(\mathbf{r}, \mathbf{r}') j_\omega^\beta(\mathbf{r}') + \mathrm{C.C.} \quad (5.35)$$

The field of the external electromagnetic wave, E_ω^{ext}, can be expanded in a series of eigenmodes of the passive waveguide (5.24):

$$\boldsymbol{E}_\omega^{\text{ext}} = \sum_\mu C_\mu^{\text{TE}} \exp\left[\mathrm{i}\frac{\omega z}{c} - \mathrm{i}\frac{c(k_\perp^{\text{TE}})_\mu^2 z}{2\omega}\right]$$

$$\times \left(\boldsymbol{e}_x \frac{\partial \psi_\mu^{\text{TE}}(\boldsymbol{r}_\perp)}{\partial y} - \boldsymbol{e}_y \frac{\partial \psi_\mu^{\text{TE}}(\boldsymbol{r}_\perp)}{\partial x}\right)$$

$$+ \sum_\nu C_\nu^{\text{TM}} \exp\left[\mathrm{i}\frac{\omega z}{c} - \mathrm{i}\frac{c(k_\perp^{\text{TM}})_\nu^2 z}{2\omega}\right]$$

$$\times \left(\boldsymbol{e}_x \frac{\partial \psi_\nu^{\text{TM}}(\boldsymbol{r}_\perp)}{\partial x} + \boldsymbol{e}_y \frac{\partial \psi_\nu^{\text{TM}}(\boldsymbol{r}_\perp)}{\partial y}\right) . \tag{5.36}$$

Substituting this expression into (5.35) and using (5.18) and (5.32), we obtain the following result:

$$W_3 = -\frac{\theta_s}{2} \int_0^z \mathrm{d}z' \int \mathrm{d}\boldsymbol{r}'_\perp \tilde{j}_1(z', \boldsymbol{r}'_\perp)$$

$$\times \left\{ \sum_\mu (C_\mu^{\text{TE}})^* \exp\left[\mathrm{i}\frac{c(k_\perp^{\text{TE}})_\mu^2 z'}{2\omega}\right] \left(\frac{\partial \psi_\mu^{\text{TE}}}{\partial y'} + \mathrm{i}\frac{\partial \psi_\mu^{\text{TE}}}{\partial x'}\right) \right.$$

$$+ \sum_\nu (C_\nu^{\text{TM}})^* \exp\left[\mathrm{i}\frac{c(k_\perp^{\text{TM}})_\nu^2 z'}{2\omega}\right]$$

$$\left. \times \left(\frac{\partial \psi_\nu^{\text{TM}}}{\partial x'} - \mathrm{i}\frac{\partial \psi_\nu^{\text{TM}}}{\partial y'}\right) \right\} + \text{C.C.} \tag{5.37}$$

Using (5.2) and (5.25), we rewrite (5.37) as

$$W_3 = -\frac{\mathrm{i}}{e} \int_0^z \mathrm{d}z' \int \mathrm{d}\boldsymbol{r}'_\perp U_{\text{ext}}^*(z', \boldsymbol{r}'_\perp) \tilde{j}_1(z', \boldsymbol{r}'_\perp) + \text{C.C.}$$

Thus, the radiated power is equal to

$$W_1 + W_3 = -\frac{\mathrm{i}}{e} \int_0^z \mathrm{d}z' \int \mathrm{d}\boldsymbol{r}'_\perp U^*(z', \boldsymbol{r}'_\perp) \tilde{j}_1(z', \boldsymbol{r}'_\perp) + \text{C.C.} , \tag{5.38}$$

where U is the complex amplitude of the effective potential,

$$U = U_\mathrm{i} + U_{\text{ext}} .$$

The power balance requires the radiated power to be equal to the difference between the electron beam power at the entrance and the exit of the undulator, i.e.

$$W_1 + W_3 = \frac{I_0}{e} (\langle \mathcal{E} \rangle_{\text{entr}} - \langle \mathcal{E} \rangle_{\text{exit}}) ,$$

where $\langle \mathcal{E} \rangle$ is the mean energy of the electrons and I_0 is the beam current. The rate of energy loss by a single electron is given by

$$\frac{d\mathcal{E}}{dz} = i\left(U(z,\,\boldsymbol{r}_\perp) + \frac{4\pi e}{\omega}\tilde{j}_1(z,\,\boldsymbol{r}_\perp)\right)\exp(i\psi) + \text{C.C.}$$

To obtain the mean power loss by the electron beam, we should multiply this value by the particle flux density,

$$-j_z(z,\,\psi,\,\boldsymbol{r}_\perp)/e\,,$$

perform an averaging over the phase ψ, and integrate over the beam cross-section and the undulator length. Finally, we obtain the following result:

$$\frac{I_0}{e}(\langle\mathcal{E}\rangle_{\text{entr}} - \langle\mathcal{E}\rangle_{\text{exit}})$$

$$= \frac{1}{2\pi e}\int_0^z dz'\int_0^{2\pi}d\psi\int d\boldsymbol{r}'_\perp j_z(z',\,\psi,\,\boldsymbol{r}'_\perp)\frac{d\mathcal{E}}{dz'}$$

$$= \frac{i}{e}\int_0^z dz'\int d\boldsymbol{r}'_\perp U(z',\,\boldsymbol{r}'_\perp)\tilde{j}_1^*(z',\,\boldsymbol{r}'_\perp) + \text{C.C.} \tag{5.39}$$

Comparison of the latter expression with (5.38) shows that power balance takes place.

5.3 Beam Radiation Modes in a Circular Waveguide

Equations (5.7) describe the general case of an FEL amplifier with helical undulator and homogeneous waveguide. In this section we study in detail the FEL amplifier with helical undulator, axisymmetric electron beam, and circular waveguide. A particular case of an overmoded waveguide is considered,

$$R^2\omega^2/c^2 \gg 1\,,$$

where R is the radius of the waveguide. It is assumed that the electron beam axis coincides with the axis of the waveguide. When considering axially symmetric systems, it is convenient to rewrite (5.7) using cylindrical coordinates $(r,\,\varphi,\,z)$. In this case the field amplitude components should be subjected to the following transformation:

$$\tilde{E}_x = \cos(\varphi)\tilde{E}_r - \sin(\varphi)\tilde{E}_\varphi = e^{-i\varphi}(\tilde{E}_r - i\tilde{E}_\varphi)/2 + e^{i\varphi}(\tilde{E}_r + i\tilde{E}_\varphi)/2\,,$$

$$\tilde{E}_y = \sin(\varphi)\tilde{E}_r + \cos(\varphi)\tilde{E}_\varphi = ie^{-i\varphi}(\tilde{E}_r - i\tilde{E}_\varphi)/2 - ie^{i\varphi}(\tilde{E}_r + i\tilde{E}_\varphi)/2\,.$$

The amplitude of the effective potential of interaction, U, takes the form:

$$U = -\frac{e\theta_s}{2i}(\tilde{E}_x + i\tilde{E}_y) = -\frac{e\theta_s}{2i}(\tilde{E}_r + i\tilde{E}_\varphi)e^{i\varphi}\,.$$

Substituting these relations into (5.7), we get:

5.3 Beam Radiation Modes in a Circular Waveguide

$$\left[\frac{\partial^2}{\partial r^2} + \frac{1}{r}\frac{\partial}{\partial r} + \frac{1}{r^2}\frac{\partial^2}{\partial \varphi^2} + \frac{2\mathrm{i}\omega}{c}\frac{\partial}{\partial z}\right]\left(\tilde{E}_r + \mathrm{i}\tilde{E}_\varphi\right)\mathrm{e}^{\mathrm{i}\varphi}$$

$$= \mathrm{i}j_0(\boldsymbol{r}_\perp)\int_0^z \mathrm{d}z'\left\{\frac{2\pi\omega\theta_s^2 e}{c^2}\left(\tilde{E}_r + \mathrm{i}\tilde{E}_\varphi\right)\mathrm{e}^{\mathrm{i}\varphi}\right.$$

$$+\frac{4\pi e}{\omega}\left[\frac{\partial^2}{\partial r^2} + \frac{1}{r}\frac{\partial}{\partial r} + \frac{1}{r^2}\frac{\partial^2}{\partial \varphi^2} + \frac{2\mathrm{i}\omega}{c}\frac{\partial}{\partial z'}\right]\left(\tilde{E}_r + \mathrm{i}\tilde{E}_\varphi\right)\mathrm{e}^{\mathrm{i}\varphi}\bigg\}$$

$$\times \int_{-\infty}^{\infty} \mathrm{d}P\,(\mathrm{d}F(P)/\mathrm{d}P)\exp\left\{\mathrm{i}\left[\frac{\omega}{c\gamma_z^2\mathcal{E}_0}P + C\right](z'-z)\right\}, \quad (5.40\mathrm{a})$$

$$\left[\frac{\partial^2}{\partial r^2} + \frac{1}{r}\frac{\partial}{\partial r} + \frac{1}{r^2}\frac{\partial^2}{\partial \varphi^2} + \frac{2\mathrm{i}\omega}{c}\frac{\partial}{\partial z}\right]\left(\tilde{E}_r - \mathrm{i}\tilde{E}_\varphi\right)\mathrm{e}^{-\mathrm{i}\varphi} = 0. \quad (5.40\mathrm{b})$$

To find the evolution of the radiation field, one should impose initial conditions at the undulator entrance and boundary conditions on the waveguide walls. The continuity conditions of $\tilde{E}_r$ and $\tilde{E}_\varphi$ and their derivatives should be fulfilled on the beam boundary.

Let us derive explicit expressions for the boundary conditions in a circular waveguide. We assume the waveguide walls to be perfectly conducting. Hence, the components E_φ and E_z must be equal to zero on the waveguide wall. Using Maxwell's equation,

$$\nabla \times \boldsymbol{H} = c^{-1}\partial \boldsymbol{E}/\partial t,$$

we find the relations between the electric and magnetic field components inside the waveguide:

$$r^{-1}\partial \tilde{H}_z/\partial \varphi - \mathrm{i}(\omega/c)\tilde{H}_\varphi - \partial \tilde{H}_\varphi/\partial z = -\mathrm{i}(\omega/c)\tilde{E}_r,$$

$$\mathrm{i}(\omega/c)\tilde{H}_r + \partial \tilde{H}_r/\partial z - \partial \tilde{H}_z/\partial r = -\mathrm{i}(\omega/c)\tilde{E}_\varphi,$$

$$r^{-1}\partial(r\tilde{H}_\varphi)/\partial r - r^{-1}\partial \tilde{H}_r/\partial \varphi = -\mathrm{i}(\omega/c)\tilde{E}_z.$$

For the electromagnetic wave propagating in the overmoded waveguide the vectors of the electric and magnetic field are equal in absolute value and perpendicular to each other, i.e.,

$$\tilde{E}_r \simeq \tilde{H}_\varphi, \qquad \tilde{E}_\varphi \simeq -\tilde{H}_r,$$

and the boundary conditions on the waveguide wall take the form:

$$\tilde{E}_\varphi|_{r=R} = 0, \qquad \partial(r\tilde{E}_r)/\partial r|_{r=R} = 0. \quad (5.41)$$

Equations (5.40) and boundary conditions (5.41) describe the linear mode of operation of the FEL amplifier with an axisymmetric electron beam, helical undulator and circular waveguide. There are two problems which may be studied in the linear approximation: the eigenvalue problem and the initial-value problem. We start with the solution of the eigenvalue problem. Analytical solutions are obtained for the cases of a stepped and parabolic profile

5.3.1 Stepped Profile of Electron Beam

In this section we study the case of an axisymmetric electron beam having a stepped profile of current density,

$$j_0(r) = \begin{cases} I_0/(\pi r_0^2) & \text{for } r < r_0, \\ 0 & \text{for } r > r_0, \end{cases}$$

where I_0 is the beam current and r_0 is the beam radius. To find the eigenvalue equation, one should consider the high-gain limit. Using the method of separation variables, we seek the solution of (5.40) in the form:

$$\tilde{E}_{r,\varphi}(z, r, \varphi) = F_{r,\varphi}^{(m)}(r) \exp(-im\varphi + \Lambda z), \tag{5.42}$$

where m is an integer. Inside the electron beam, for $r < r_0$, the complex amplitude $F_+^{(m)} = \left(F_r^{(m)} + iF_\varphi^{(m)}\right)$ satisfies the equation:

$$\left[\frac{d^2}{d\hat{r}^2} + \frac{1}{\hat{r}}\frac{d}{d\hat{r}} - \frac{(m-1)^2}{\hat{r}^2} + 2iB\hat{\Lambda} - \frac{2i\hat{D}}{1 - i\hat{D}\hat{\Lambda}_p^2}\right] F_+^{(m)} = 0, \tag{5.43a}$$

and outside the beam, for $r > r_0$, the equation

$$\left[\frac{d^2}{d\hat{r}^2} + \frac{1}{\hat{r}}\frac{d}{d\hat{r}} - \frac{(m-1)^2}{\hat{r}^2} + 2iB\hat{\Lambda}\right] F_+^{(m)} = 0. \tag{5.43b}$$

The complex amplitude $F_-^{(m)} = \left(F_r^{(m)} - iF_\varphi^{(m)}\right)$ satisfies the equation

$$\left[\frac{d^2}{d\hat{r}^2} + \frac{1}{\hat{r}}\frac{d}{d\hat{r}} - \frac{(m+1)^2}{\hat{r}^2} + 2iB\hat{\Lambda}\right] F_-^{(m)} = 0 \tag{5.43c}$$

inside and outside the electron beam. Here the following notation has been used:

$$\hat{r} = r/r_0, \quad \hat{\Lambda} = \Lambda/\Gamma, \quad \hat{C} = C/\Gamma, \quad \hat{\Lambda}_p^2 = \Lambda_p^2/\Gamma^2, \quad B = \Gamma r_0^2 \omega/c,$$

$$\Gamma = \left[I_0 \omega^2 \theta_s^2/(I_A c^2 \gamma_z^2 \gamma)\right]^{1/2}, \quad \Lambda_p = \left[4I_0/(I_A r_0^2 \gamma_z^2 \gamma)\right]^{1/2},$$

$$\hat{D} = \int_{-\infty}^{\infty} d\hat{P} \left(d\hat{F}(\hat{P})/d\hat{P}\right) \left[\hat{\Lambda} + i\hat{C} + i\hat{P}\right]^{-1},$$

where $\hat{F}(\hat{P})$ is the reduced distribution function. The distribution function of the Gaussian energy spread is

$$F(P) = (2\pi\langle(\Delta\mathcal{E})^2\rangle)^{-1/2} \exp\left[-P^2/\left(2\langle(\Delta\mathcal{E})^2\rangle\right)\right],$$

and the corresponding reduced distribution function $\hat{F}(\hat{P})$ is of the form

$$\hat{F}(\hat{P}) = \left[2\pi\hat{\Lambda}_T^2\right]^{-1/2} \exp\left[-\hat{P}^2/(2\hat{\Lambda}_T^2)\right],$$

where $\hat{\Lambda}_T^2 = \langle(\Delta\mathcal{E})^2\rangle/(\rho^2\mathcal{E}^2)$ is the energy spread parameter and $\rho = c\gamma_z^2\Gamma/\omega$ is the efficiency parameter. The function $\hat{D}$ is given by the expression:

$$\hat{D} = i\int_0^\infty \exp\left\{-\frac{\hat{\Lambda}_T^2\xi^2}{2} - (\hat{\Lambda} + i\hat{C})\right\}\xi d\xi \ .$$

The solution of (5.43) with the boundary conditions (5.41) allows us to find the eigenvalues and the eigenfunctions of the beam radiation modes in the circular waveguide. It is relevant to make some remarks prior to a detailed study of the problem. First, we should draw the attention of the reader to the fact that the reduced parameters used in the theory of the waveguide FEL are identical to those used in the theory of the FEL amplifier with an open electron beam (see Chap. 4 for more details). Imposing the boundary conditions on the waveguide walls will result in the appearance of an additional parameter of the theory, namely the waveguide diffraction parameter. Another consequence of the boundary conditions at the waveguide walls consists in the elimination of the double degeneration of the beam radiation modes which takes place in the case of an open electron beam (see Sect. 4.4 for more details). The double degeneration means that for each eigenvalue there exist two different eigenfunctions corresponding to each value of the azimuthal index n (see (4.118) and (4.121)). Let us analyze (5.43) in the limit of infinitely large radius of the waveguide, $R \to \infty$, corresponding to the open beam asymptote. The boundary conditions are simplified to

$$F_\pm^{(m)}(r) \to 0 \qquad \text{as} \quad r \to \infty \ ,$$

and we get from (5.43c) that

$$F_-^{(m)}(r) = 0 \qquad \text{for} \quad 0 < r < \infty \ .$$

According to (5.43a) and (5.43b), the function $F_+^{(m)}(r)$ depends on the value of the azimuthal index m as $|m - 1|$. Hence, the solutions of (5.43) are doubly degenerate in the limit of an open beam asymptote, as they should be. At finite value of the waveguide radius, one should apply boundary conditions on the waveguide walls for the functions $F_{r,\phi}^{(m)}(r)$. These boundary conditions mix the functions $F_-^{(m)}(r)$ and $F_+^{(m)}(r)$. As a result, $F_-^{(m)}(r)$ is no longer equal to zero. Remembering that the equations for $F_+^{(m)}(r)$ and $F_-^{(m)}(r)$ contain indexes $|m-1|$ and $|m+1|$, respectively, we come to the conclusion that the presence of the waveguide walls eliminates the double degeneration of the radiation modes in the FEL with circular waveguide and helical undulator.

Equations (5.43) are the Bessel equations. Solutions of these equations should be chosen in the form:

$$F_+^{(m)} = \begin{cases} C_1 J_{m-1}(\mu\hat{r}) & \text{for} \quad 0 < r < r_0 \\ C_2 I_{m-1}(g\hat{r}) + C_3 K_{m-1}(g\hat{r}) & \text{for} \quad r_0 < r < R \end{cases}$$

$$F_-^{(m)} = C_4 I_{m+1}(g\hat{r}) \qquad \text{for} \quad 0 < r < R \ , \qquad (5.44)$$

where
$$g^2 = -2\mathrm{i}B\hat{\Lambda}, \qquad \mu^2 = -2\mathrm{i}\hat{D}\left[1 - \mathrm{i}\hat{D}\hat{\Lambda}_\mathrm{p}^2\right]^{-1} - g^2 .$$

As a result, we obtain the following solutions for the complex amplitudes $F_{r,\varphi}^{(m)}$:

Region 1 ($r < r_0$):
$$F_r^{(m)} = \frac{1}{2}\left[C_1 J_{m-1}(\mu\hat{r}) + C_4 I_{m+1}(g\hat{r})\right],$$
$$F_\varphi^{(m)} = -\frac{\mathrm{i}}{2}\left[C_1 J_{m-1}(\mu\hat{r}) - C_4 I_{m+1}(g\hat{r})\right], \qquad (5.45\mathrm{a})$$

Region 2 ($r_0 < r < R$):
$$F_r^{(m)} = \frac{1}{2}\left[C_2 I_{m-1}(g\hat{r}) + C_3 K_{m-1}(g\hat{r}) + C_4 I_{m+1}(g\hat{r})\right],$$
$$F_\varphi^{(m)} = -\frac{\mathrm{i}}{2}\left[C_2 I_{m-1}(g\hat{r}) + C_3 K_{m-1}(g\hat{r}) - C_4 I_{m+1}(g\hat{r})\right] . \qquad (5.45\mathrm{b})$$

The boundary conditions on the waveguide walls give us the equations:
$$F_\varphi^{(m)}|_{r=R} = -\frac{\mathrm{i}}{2}\left[C_2 I_{m-1}(\Theta) + C_3 K_{m-1}(\Theta) - C_4 I_{m+1}(\Theta)\right]$$
$$= 0 ,$$
$$2\frac{\mathrm{d}}{\mathrm{d}r}\left(rF_r^{(m)}\right)|_{r=R} = C_2 \left[m I_{m-1}(\Theta) + \Theta I_m(\Theta)\right]$$
$$+ C_3 \left[m K_{m-1}(\Theta) - \Theta K_m(\Theta)\right]$$
$$+ C_4 \left[-m I_{m+1}(\Theta) + \Theta I_m(\Theta)\right]$$
$$= 0 , \qquad (5.46)$$

where $\Theta = g\sqrt{\Omega/B}$ and
$$\Omega = \Gamma R^2 \omega / c$$
is the waveguide diffraction parameter. The continuity conditions of $F_{\varphi,r}$ and $\mathrm{d}(F_{\varphi,r})/\mathrm{d}r$ at the beam boundary give us the equations:
$$C_1 J_{m-1}(\mu) = C_2 I_{m-1}(g) + C_3 K_{m-1}(g),$$
$$C_1 \left[(m-1) J_{m-1}(\mu) - \mu J_m(\mu)\right]$$
$$= C_2 \left[(m-1) I_{m-1}(g) + g I_m(g)\right]$$
$$+ C_3 \left[(m-1) K_{m-1}(g) - g K_m(g)\right] . \qquad (5.47)$$

Equations (5.46) and (5.47) constitute a system of four linear equations. To obtain nontrivial solutions for the coefficients $C_{1,2,3,4}$ we set its determinant equal to zero. As a result, we obtain the eigenvalue equation for the case of the stepped beam profile:

5.3 Beam Radiation Modes in a Circular Waveguide

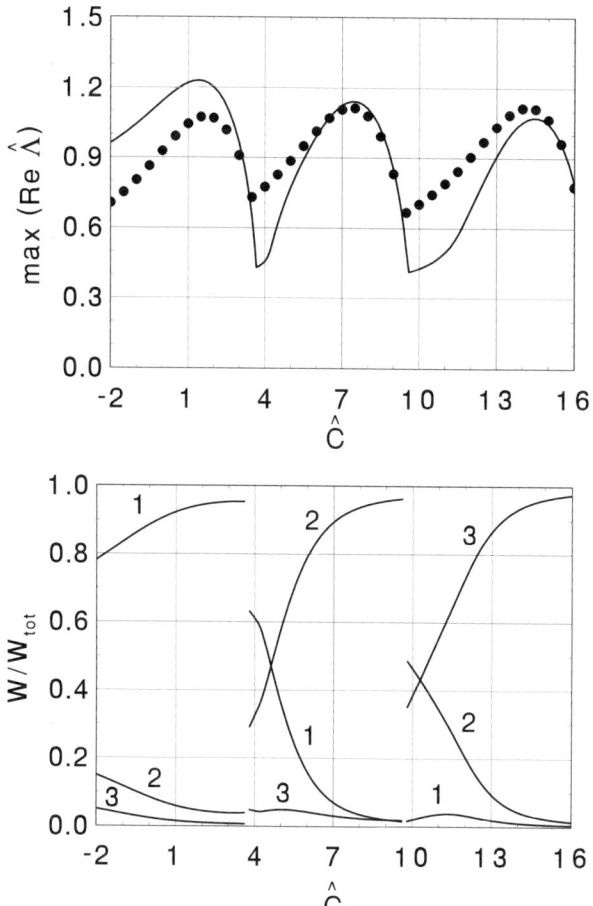

Fig. 5.1. Maximum reduced growth rate (upper plot) and partial contributions of the passive waveguide modes into full power of the beam radiation mode (lower plot) versus the detuning parameter at $m = 1$, $B = 0.1$, $\Omega = 1$, $\hat{\Lambda}_p^2 \to 0$, and $\hat{\Lambda}_T^2 = 0$. The curve in the upper plot is calculated with (5.48) and the circles with the single-mode approximation (5.63). Curves 1, 2, and 3 in the lower plot corresponds to the TE_{11}, TM_{11}, and TE_{12} mode, respectively

$$2\left[gK_m(g)J_{m-1}(\mu) - \mu J_m(\mu)K_{m-1}(g)\right]$$
$$= \left\{\frac{K_m(\Theta)}{I_m(\Theta)} + \frac{K'_m(\Theta)}{I'_m(\Theta)}\right\}\left[gI_m(g)J_{m-1}(\mu) + \mu J_m(\mu)I_{m-1}(g)\right]. \quad (5.48)$$

Here the prime denotes the derivative of the Bessel function over the argument. The corresponding field eigenmodes are given by the formulae:

Region 1 $(r < r_0)$:

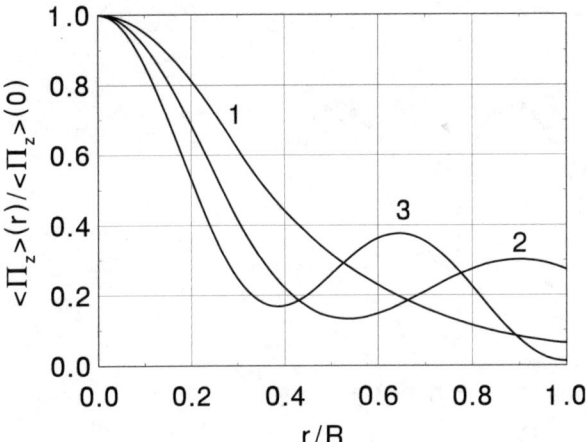

Fig. 5.2. Radiation power flow density of the beam radiation modes versus radius. Calculations were performed with (5.49) and (5.67). Curve 1: $\hat{C} = 1.4$. Curve 2: $\hat{C} = 7.4$. Curve 3: $\hat{C} = 14.4$. Here $m = 1$, $B = 0.1$, $\Omega = 1$, $\hat{\Lambda}_p^2 \to 0$, and $\hat{\Lambda}_T^2 = 0$

$$F_r^{(m)} = J_{m-1}(\mu\hat{r}) + \frac{1}{2}I_{m+1}(g\hat{r})\{\ \}_{(-)}[\ ]\ ,$$

$$F_\varphi^{(m)} = -\mathrm{i}J_{m-1}(\mu\hat{r}) + \frac{\mathrm{i}}{2}I_{m+1}(g\hat{r})\{\ \}_{(-)}[\ ]\ , \tag{5.49a}$$

Region 2 ($r_0 < r < R$):

$$F_r^{(m)} = \left(K_{m-1}(g\hat{r}) + \frac{1}{2}I_{m-1}(g\hat{r})\{\ \}_{(+)} + \frac{1}{2}I_{m+1}(g\hat{r})\{\ \}_{(-)}\right)[\ ]\ ,$$

$$F_\varphi^{(m)} = -\mathrm{i}\left(K_{m-1}(g\hat{r}) + \frac{1}{2}I_{m-1}(g\hat{r})\{\ \}_{(+)}\right.$$
$$\left.- \frac{1}{2}I_{m+1}(g\hat{r})\{\ \}_{(-)}\right)[\ ]\ , \tag{5.49b}$$

where the brackets $\{\ \}_{(\pm)}$ and $[\ ]$ denote

$$\{\ \}_{(\pm)} = \left\{\frac{K_m(\Theta)}{I_m(\Theta)} \pm \frac{K_m'(\Theta)}{I_m'(\Theta)}\right\}\ ,$$

$$[\ ] = [gI_m(g)J_{m-1}(\mu) + \mu J_m(\mu)I_{m-1}(g)]\ .$$

It should be noted that despite the fact that the factors g and μ are two-valued functions, (5.48) and (5.49) are invariant on replacing μ by $-\mu$ or g by $-g$. For instance, using the relations

$$J_m(-\xi) = (-1)^m J_m(\xi)\ , \qquad I_m(-\xi) = (-1)^m I_m(\xi)\ ,$$

we readily conclude that (5.48) is invariant at $\mu \to -\mu$. The same property holds when exchanging g for $-g$. The calculations are long in this case and we leave them to the reader.

5.3 Beam Radiation Modes in a Circular Waveguide

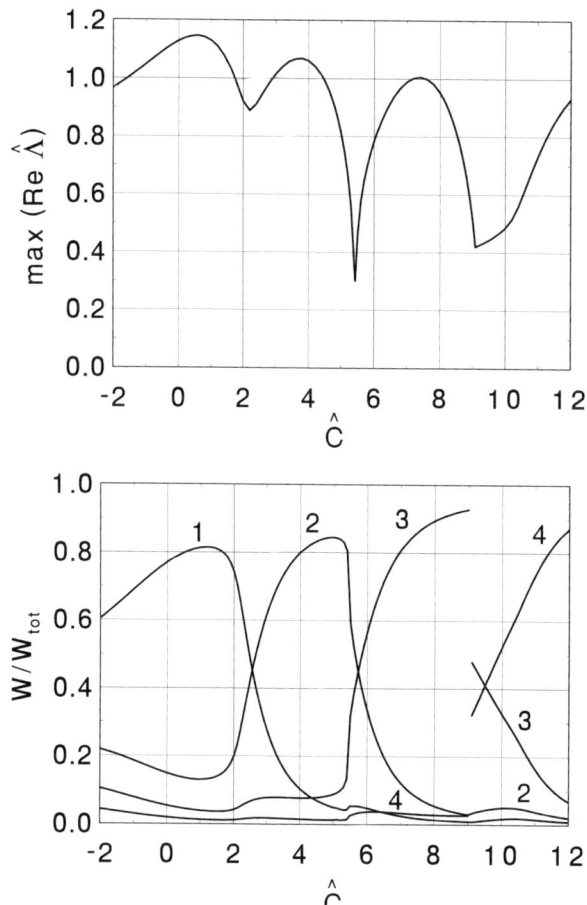

Fig. 5.3. Maximum reduced growth rate (upper plot) and partial contributions of the passive waveguide modes into full power of the beam radiation mode (lower plot) versus the detuning parameter at $m = 1$, $B = 0.1$, $\Omega = 2$, $\hat{\Lambda}_p^2 \to 0$, and $\hat{\Lambda}_T^2 = 0$. The upper plot is calculated with (5.48) and the circles with the single-mode approximation (5.63). Curves 1, 2, 3, and 4 in the lower plot corresponds to the TE_{11}, TM_{11}, TE_{12}, and TM_{12} mode, respectively

Let us analyze the nature of the beam radiation modes. The components of the electric field can be expressed in terms of the complex amplitudes $F_{r,\varphi}^{(m)}$:

$$E_{r,\varphi} = \tilde{E}_{r,\varphi} \exp\left[i\omega(z/c - t)\right] + \text{C.C.}$$
$$= |F_{r,\varphi}^{(m)}| \exp\left[\text{Re}\, \Lambda z\right] \cos\left[-m\varphi + \omega(z/c - t) + \text{Im}\, \Lambda z + \psi_{r,\varphi}\right] ,$$

where

$$F_{r,\varphi}(r) = |F_{r,\varphi}(r)| \exp\left[i\psi_{r,\varphi}(r)\right] .$$

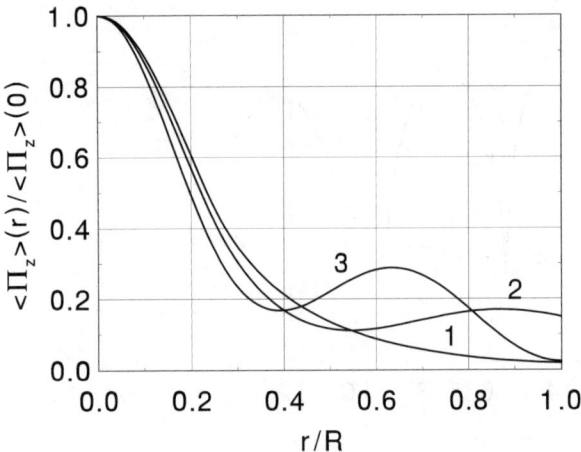

Fig. 5.4. Radiation power flow density of the beam radiation modes versus radius. Calculations have been performed with (5.49) and (5.67). Curve 1: $\hat{C} = 0.5$. Curve 2: $\hat{C} = 3.7$. Curve 3: $\hat{C} = 7.4$. Here $m = 1$, $B = 0.1$, $\Omega = 2$, $\hat{\Lambda}_p^2 \to 0$, and $\hat{\Lambda}_T^2 = 0$

It is seen from these relations that the field components have $|m|$ variations over the azimuthal angle φ, and the field vectors rotate in time around the waveguide axis at fixed value of the z coordinate. At positive values of m, the direction of the rotation is the same as the direction of the electron rotation in the undulator. At negative values of m the field vectors rotate in the opposite direction.

Let us consider the azimuthal distribution of the beam current density modulation. According to (5.3), its azimuthal dependence is the same as for $\mathrm{Re}(\tilde{E}_x + i\tilde{E}_y)$, and is proportional to $\cos\left[(m - 1)\varphi\right]$. This cosine density distribution rotates in time around the z axis at a fixed value of the z coordinate. At $m > 1$ the direction of rotation is the same as that of the electron rotation. At $m < 0$ it rotates in the opposite direction. For the radiation mode with $m = 1$ the beam modulation density does not depend on φ. When the FEL amplifier is tuned to the azimuthally symmetric radiation mode with $m = 0$, the beam modulation density changes as $\cos(\varphi)$ and rotates in the opposite direction to the electron rotation direction. Thus, we deal with the situation when the beam with the axisymmetric current density modulation generates nonsymmetric radiation modes and vice versa. Physically it is a consequence of the fact that the azimuthal symmetry of the system is violated by the helicity of the undulator.

It is relevant to compare the results for the FEL amplifier with the helical undulator and circular waveguide and the results for the open beam radiation modes presented in Chap. 4. In the case of the open electron beam, we obtained solutions for the eigenfunctions in the form of $\Phi_\pm \propto e^{\pm in\phi}$ with $n \geq 0$. Formally they are similar to those for the circular waveguide. Nevertheless,

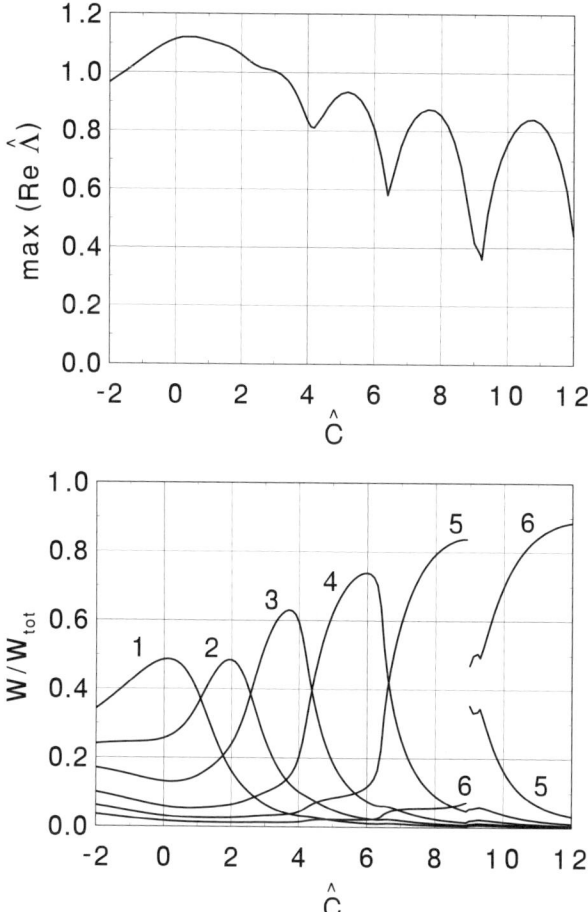

Fig. 5.5. Maximum reduced growth rate (upper plot) and partial contributions of the passive waveguide modes into full power of the beam radiation mode (lower plot) versus the detuning parameter at $m = 1$, $B = 0.1$, $\Omega = 5$, $\hat{\Lambda}_p^2 \to 0$, and $\hat{\Lambda}_T^2 = 0$. The upper plot is calculated with (5.48) and the circles with the single-mode approximation (5.63). Curves 1–6 in the lower plot corresponds to the TE_{11}, TM_{11}, TE_{12}, TM_{12}, TE_{13}, and TM_{13} mode, respectively

there is a principal difference connected with the double degeneration: the field of the beam mode can be represented in the form of "standing" (not rotating) solutions, proportional to $\sin(n\phi)$ and $\cos(n\phi)$. The presence of the waveguide walls leads to situation when a unique value of the eigenvalue, Λ, corresponds to each value of the azimuthal index, m. As a result, the eigenmodes of the active waveguide always have the appearance of rotating field configurations.

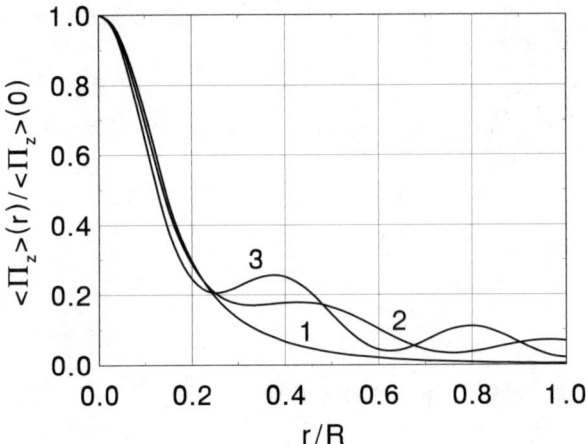

Fig. 5.6. Radiation power flow density of the beam radiation modes versus radius. Calculations have been performed with (5.49) and (5.67). Curve 1: $\hat{C} = 0.4$. Curve 2: $\hat{C} = 5.2$. Curve 3: $\hat{C} = 7.6$. Here $m = 1$, $B = 0.1$, $\Omega = 5$, $\hat{\Lambda}_\mathrm{p}^2 \to 0$, and $\hat{\Lambda}_\mathrm{T}^2 = 0$

It is instructive to calculate the expansion of the beam radiation modes (modes of an active waveguide) in the modes of a passive waveguide. Let us write the transverse field of the eigenmode in the form (see (5.24) and (5.25)):

$$\boldsymbol{e}_x \tilde{E}_x + \boldsymbol{e}_y \tilde{E}_y = \sum_\mu C_\mu^\mathrm{TE} \left(\boldsymbol{e}_x \frac{\partial \psi_\mu^\mathrm{TE}}{\partial y} - \boldsymbol{e}_y \frac{\partial \psi_\mu^\mathrm{TE}}{\partial x} \right)$$
$$+ \sum_\nu C_\nu^\mathrm{TM} \left(\boldsymbol{e}_x \frac{\partial \psi_\mu^\mathrm{TM}}{\partial x} + \boldsymbol{e}_y \frac{\partial \psi_\mu^\mathrm{TM}}{\partial y} \right), \quad (5.50)$$

$$C_\mu^\mathrm{TE} = \int \mathrm{d}\boldsymbol{r}_\perp \left(\tilde{E}_x \frac{\partial \psi_\mu^\mathrm{TE}}{\partial y} - \tilde{E}_y \frac{\partial \psi_\mu^\mathrm{TE}}{\partial x} \right), \quad (5.51)$$

$$C_\nu^\mathrm{TM} = \int \mathrm{d}\boldsymbol{r}_\perp \left(\tilde{E}_x \frac{\partial \psi_\nu^\mathrm{TM}}{\partial x} + \tilde{E}_y \frac{\partial \psi_\nu^\mathrm{TM}}{\partial y} \right). \quad (5.52)$$

The eigenfunctions of an empty waveguide, ψ^TE and ψ^TM, are the solutions of the Helmholtz equation (5.19) satisfying boundary conditions (5.20) and normalization condition (5.21). Explicit expressions for the eigenfunctions of a circular waveguide have the form (see Appendix A.4 for more details):

$$\psi_{mn}^\mathrm{TE} = A^\mathrm{TE} J_m(\mu_{mn} r/R) \begin{pmatrix} \sin(m\varphi) \\ \cos(m\varphi) \end{pmatrix},$$

$$\psi_{mn}^\mathrm{TM} = A^\mathrm{TM} J_m(\nu_{mn} r/R) \begin{pmatrix} \sin(m\varphi) \\ \cos(m\varphi) \end{pmatrix}, \quad (5.53)$$

where μ_{mn} is the nth root of the equation

$$J'_m(\mu) = 0 \ ,$$

and ν_{mn} is the nth root of the Bessel function,

$$J_m(\nu) = 0 \ .$$

Expressions for the normalizing coefficients, A^{TE} and A^{TM}, are presented in Appendix A.4.

Since we use the cylindrical coordinate system, it is convenient to calculate the expansion series for the field components $\tilde{E}_{r,\varphi}$ which are equal to

$$\tilde{E}_{r,\varphi} = F^{(m)}_{r,\varphi}(r)\exp(-im\varphi + \Lambda z) \ .$$

Using (5.51) and (5.52), we find that the expansion coefficients (TE or TM) corresponding to the $\sin(m\varphi)$ and $\cos(m\varphi)$ modes in (5.53) have equal modulus and differ by a factor of $(-i)$ only. This means that actually the expansion is performed in the rotating modes of an empty waveguide. Such a rotating mode is a linear superposition of sine-like and cosine-like modes in (5.53) having equal amplitudes, but shifted in phase by $\pi/2$. One of the modes oscillate in time as $\cos(\omega t)$, and the other one as $\sin(\omega t)$. A peculiar feature of the rotating mode is that the configuration of the electric field force lines rotates as a whole around the waveguide axis at a fixed value of the z coordinate. Denoting by C^{TE}_{mk} and C^{TM}_{mj} for cosine-like eigenfunctions in (5.53), we write the final expression for the expansion series for $m \neq 0$:

$$\tilde{E}_r = \sum_{k=1}^{\infty} \frac{\sqrt{2}C^{\text{TE}}_{mk}}{\sqrt{\pi(\mu^2_{mk}-m^2)}J_m(\mu_{mk})} \frac{1}{r}\frac{\partial}{\partial\varphi}[J_m(\mu_{mk}r/R)\exp(-im\varphi)]$$
$$+ \sum_{j=1}^{\infty} \frac{\sqrt{2}C^{\text{TM}}_{mj}}{\sqrt{\pi}\nu_{mj}J_{m-1}(\nu_{mj})} \frac{\partial}{\partial r}[J_m(\nu_{mj}r/R)\exp(-im\varphi)] \ , \quad (5.54)$$

$$\tilde{E}_\varphi = -\sum_{k=1}^{\infty} \frac{\sqrt{2}C^{\text{TE}}_{mk}}{\sqrt{\pi(\mu^2_{mk}-m^2)}J_m(\mu_{mk})} \frac{\partial}{\partial r}[J_m(\mu_{mk}r/R)\exp(-im\varphi)]$$
$$+ \sum_{j=1}^{\infty} \frac{\sqrt{2}C^{\text{TM}}_{mj}}{\sqrt{\pi}\nu_{mj}J_{m-1}(\nu_{mj})} \frac{1}{r}\frac{\partial}{\partial\varphi}[J_m(\nu_{mj}r/R)\exp(-im\varphi)] \ . \quad (5.55)$$

For an azimuthally symmetric mode, $m=0$, we have

$$\tilde{E}_r = \sum_{j=1}^{\infty} \frac{C^{\text{TM}}_{0j}}{\sqrt{\pi}R} \frac{J_1(\nu_{0j}r/R)}{J_1(\nu_{0j})} \ , \quad (5.56)$$

$$\tilde{E}_\varphi = -\sum_{k=1}^{\infty} \frac{C^{\text{TE}}_{0k}}{\sqrt{\pi}R} \frac{J_1(\mu_{0k}r/R)}{J_0(\mu_{0k})} \ . \quad (5.57)$$

Using (5.48), (5.49), and (5.54)–(5.57), and omitting an inessential common factor, we write the coefficients of expansion in the explicit form:

$$C_{mn}^{\text{TE}} = \frac{\mu_{mn}\left[\mu J_m(\mu) J_{m-1}(M_{mn}) - M_{mn} J_{m-1}(\mu) J_m(M_{mn})\right]}{\sqrt{\mu_{mn}^2 - m^2} J_m(\mu_{mn}) \left[\hat{\Lambda} + i\mu_{mn}^2/(2\Omega)\right](\mu^2 - M_{mn}^2)},$$

$$C_{mn}^{\text{TM}} = \frac{\left[\mu J_m(\mu) J_{m-1}(N_{mn}) - N_{mn} J_{m-1}(\mu) J_m(N_{mn})\right]}{J_{m-1}(\nu_{mn}) \left[\hat{\Lambda} + i\nu_{mn}^2/(2\Omega)\right](\mu^2 - N_{mn}^2)}, \qquad (5.58)$$

where $M_{mn} = \mu_{mn}\sqrt{B/\Omega}$ and $N_{mn} = \nu_{mn}\sqrt{B/\Omega}$. The relative contribution of the passive waveguide mode (TE_{mn} or TM_{mn}) to the total radiation power of the beam radiation mode, W_m, is given by the expression:

$$W_{mn}^{\text{TE,TM}}/W_m = |C_{mn}^{\text{TE,TM}}|^2 \left[\sum_i |C_{mi}^{\text{TE}}|^2 + \sum_j |C_{mj}^{\text{TM}}|^2\right]^{-1}. \qquad (5.59)$$

The reader may well be wondering why the coefficients (5.58) do not depend on the z coordinate. The answer lies in the self-reproducing nature of the radiation field. Let us compare the transverse distributions of the radiation field at different undulator lengths, z_1 and z_2. It follows from (5.42) and (5.49) that the transverse structure of the radiation field remains the same only the amplitude and the overall phase shift change, $\tilde{E}_{r,\varphi}(z_2, r, \varphi)/\tilde{E}_{r,\varphi}(z_1, r, \varphi) = \exp(\Lambda(z_2 - z_1))$. So, self-reproduction of the radiation field takes place. At each longitudinal coordinate, z, the radiation field can be expanded in a series of modes of the empty waveguide with coefficients given by (5.58). Noting that these coefficients do not depend on the z coordinate, one arrives at a paradox. Indeed, each mode of the passive waveguide has its own phase velocity. Using (5.58) at z_1, then calculating the fields of each mode of the passive waveguide at z_2, and summing them up, one does not obtain the field of the active waveguide mode at z_2. The paradox is resolved in a simple way. One should add to the result obtained the additional field radiated by the electron beam between the points z_1 and z_2. The final result is always the self-reproducing field configuration.

The maximal reduced growth rate, $\max(\text{Re}\,\hat{\Lambda})$, and the eigenfunction are universal functions of six parameters: azimuthal index, m, beam diffraction parameter, B, waveguide diffraction parameter, Ω, detuning parameter, $\hat{C}$, space charge parameter, $\hat{\Lambda}_{\text{p}}^2$, and energy spread parameter, $\hat{\Lambda}_{\text{T}}^2$. In this book we restrict consideration only to beam radiation modes having azimuthal index $m = 1$, 0, and 2 (see Figs. 5.1–5.10, 5.11–5.14, and 5.15–5.18, respectively). Analysis of the properties of these modes allows one to study all the physical effects typical for the FEL amplifier with helical undulator and circular waveguide.

The structure for presenting the material is similar to that used in the previous chapter. The only difference is that the set of dimensional parameters describing the waveguide FEL is supplemented by an additional parameter with respect to the set of parameters describing the case of an open electron beam (see Chap. 4). This parameter is the waveguide diffraction parameter, Ω. We realize that the equations and formulae describing the waveguide FEL

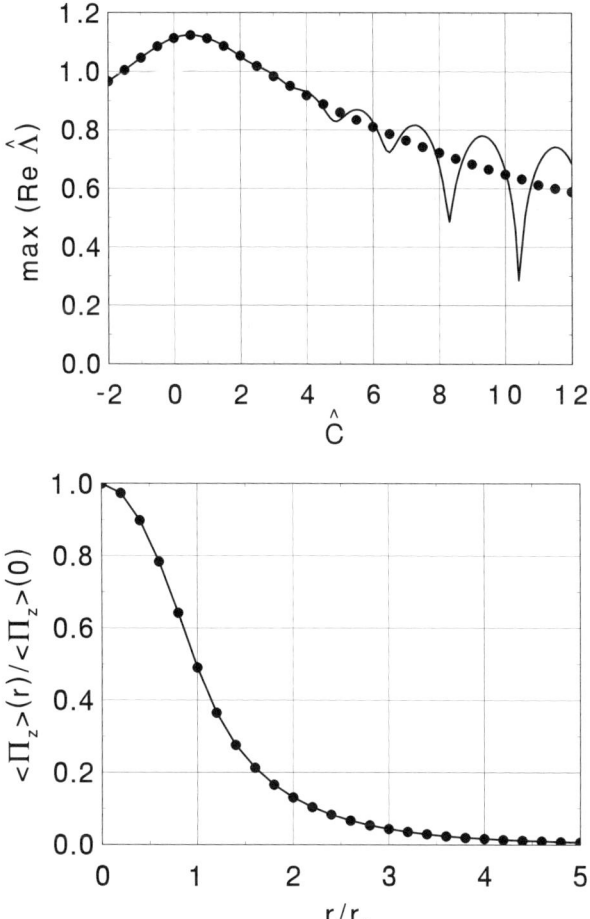

Fig. 5.7. Maximum reduced growth rate versus the detuning parameter (upper plot) and radiation power flow density of the beam radiation mode versus radius (lower plot). The curves are calculated with (5.48), (5.49), and (5.67). The circles are calculated with (5.66) and (5.68) for the fundamental symmetric TEM$_{00}$ mode of an open electron beam. Here $m = 1$, $B = 0.1$, $\Omega = 10$, $\hat{\Lambda}_{\rm P}^2 \to 0$, and $\hat{\Lambda}_{\rm T}^2 = 0$

are complicated. Also, numerical solution of these equations is not a simple problem. That is why we decided to present comprehensive graphical material illustrating different physical properties of the beam radiation modes of the waveguide FEL. These universal plots are prepared using similarity techniques, so they possess a high degree of generality. To give the reader a fuller notion of the object under study, we present all the main characteristics (maximal growth rate, beam radiation mode expansion in the modes of passive waveguide and the radiation power flow distribution) for two different

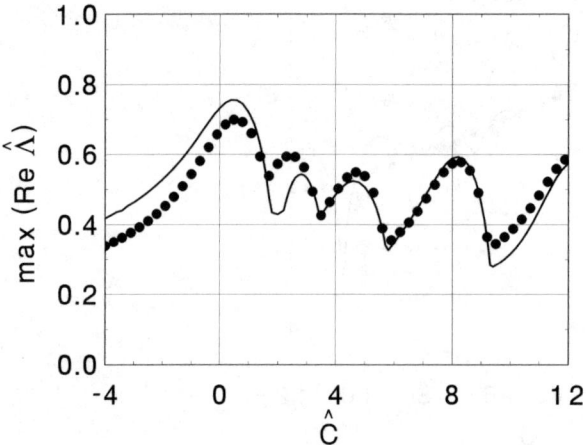

Fig. 5.8. Maximum reduced growth rate versus the detuning parameter. The curve is calculated with (5.48) and the circles with the single-mode approximation (5.63). Here $m = 1$, $B = 1$, $\Omega = 3$, $\hat{\Lambda}_p^2 \to 0$, and $\hat{\Lambda}_T^2 = 0$

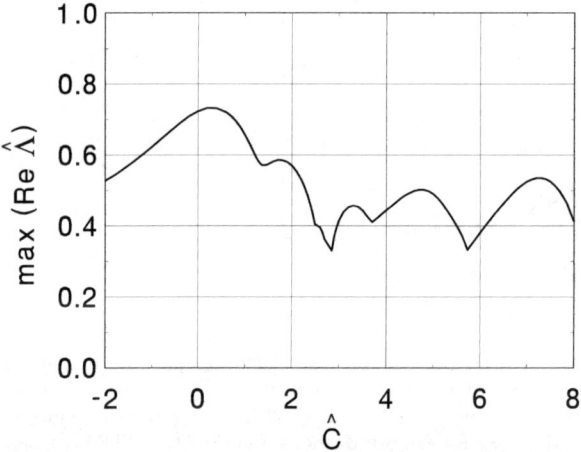

Fig. 5.9. Maximum reduced growth rate versus the detuning parameter. Calculations have been performed with (5.48). Here $m = 1$, $B = 1$, $\Omega = 5$, $\hat{\Lambda}_p^2 \to 0$, and $\hat{\Lambda}_T^2 = 0$

values of the beam diffraction parameter $B = 0.1$ and $B = 1$ and several values of the waveguide diffraction parameter Ω from $\Omega = 1$ up to $\Omega = 10$.

Let us study the dependence of the maximal growth rate on the parameters of the problem. Figures 5.1, 5.3, 5.5, 5.7, 5.8–5.11, 5.12, 5.13, 5.14, 5.15, 5.16, 5.17, and 5.18–5.21 illustrate the dependency of the maximal growth rate on the detuning parameter $\hat{C}$ for several values of diffraction parameters

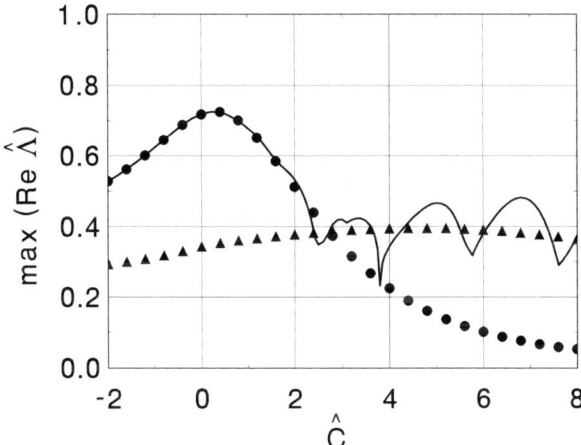

Fig. 5.10. Maximum reduced growth rate versus the detuning parameter. The curve is the solution of (5.48), the circles are the solutions of (5.66) for the fundamental TEM$_{00}$ mode, and the triangles are the solutions of (5.66) for the TEM$_{01}$ mode. Here $m = 1$, $B = 1$, $\Omega = 10$, $\hat{\Lambda}_{\rm p}^2 \to 0$, and $\hat{\Lambda}_{\rm T}^2 = 0$

B and Ω and three values of the azimuthal index $m = 0, 1$, and 2. It is clearly seen that for values $\Omega \simeq 1$ of the waveguide diffraction parameter these dependencies have the character of maxima series. The physical nature of this phenomenon is illustrated with the plots in Figs. 5.1, 5.3, and 5.5 showing the relative contribution of the passive waveguide modes into the power of the beam radiation mode. One can see from these plots that whenever the resonance condition of the beam with the corresponding passive waveguide mode is fulfilled, the contribution of the latter to the beam radiation mode dominates for values $\Omega \simeq 1$ of waveguide diffraction parameter. We will show below that this case is described rather well by a single-mode approximation. At higher values of the diffraction parameter Ω the width of the resonances becomes comparable with the distance between them, and several waveguide modes begin to contribute to the total power of the radiation mode. Finally, when the waveguide walls are placed far from the beam, i.e. at $\Omega \gg 1$, the dependency of the growth rate on the detuning becomes smooth and the open beam approximation becomes valid.

Let us now study in detail the asymptote of the eigenvalue equation (5.48) corresponding to the single-mode approximation. Analysis of (5.58) shows that when

$$\hat{\Lambda} \to \hat{\Lambda}_0 = -{\rm i}\mu_{mn}^2/(2\Omega) \;, \tag{5.60}$$

the contribution of the TE$_{mn}$ mode of the empty waveguide begins to dominate in the expansion of the beam radiation mode. Also, when

$$\hat{\Lambda} \to \hat{\Lambda}_0 = -{\rm i}\nu_{mn}^2/(2\Omega) \;, \tag{5.61}$$

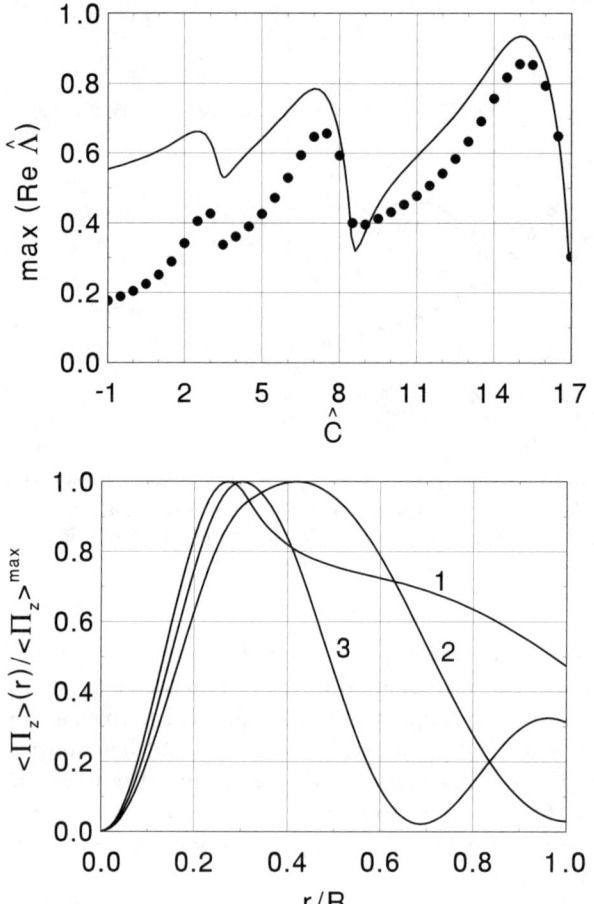

Fig. 5.11. Maximum reduced growth rate versus the detuning parameter (upper plot) and radiation power flow density versus radius (lower plot). The curve in the upper plot is calculated with (5.48) and the circles represent the single-mode approximation (5.63). The curves in the lower plot are calculated with (5.49) and (5.67). Curves 1, 2 and 3 correspond to the detuning parameter $\hat{C} = 2.4, 7$, and 15, respectively. Here $m = 0$, $B = 0.1$, $\Omega = 1$, $\hat{\Lambda}_P^2 \to 0$, and $\hat{\Lambda}_T^2 = 0$

the contribution of the TM$_{mn}$ mode dominates. Analysis of the self-consistent field equations (5.43) shows that the asymptotes (5.60) and (5.61) hold, in particular, at small values of the beam current, $I_0 \to 0$. In this case we have $B \to 0$, $\Omega/B = \text{const.}$, and $\Lambda/\Gamma \to \infty$. In the zeroth approximation we find that $\mu = \mathrm{i}g$. Then, taking into account the relations

$$K_{m-1}(\xi)I_m(\xi) + I_{m-1}(\xi)K_m(\xi) = \frac{1}{\xi}, \qquad J_m(\mathrm{i}\xi) = \mathrm{i}^m I_m(\xi),$$

we find that (5.48) transforms to:

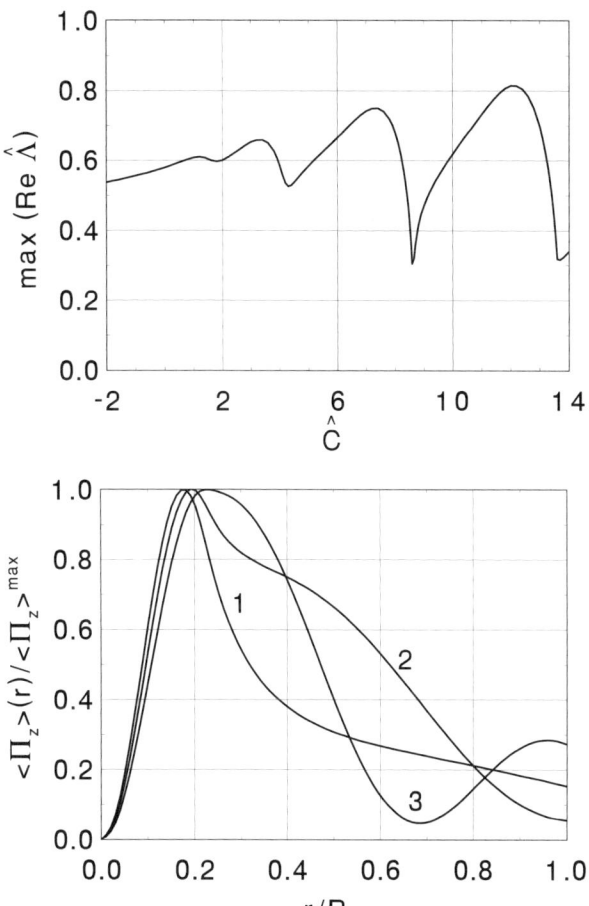

Fig. 5.12. Maximum reduced growth rate versus the detuning parameter (upper plot) and radiation power flow density versus radius (lower plot). The curve in the upper plot is calculated with (5.48). The curves in the lower plot are calculated with (5.49) and (5.67). Curves 1, 2, and 3 correspond to the detuning parameter $\hat{C} = 1.2$, 3.3, and 7.3, respectively. Here $m = 0$, $B = 0.1$, $\Omega = 2$, $\hat{\Lambda}_{\mathrm{p}}^2 \to 0$, and $\hat{\Lambda}_{\mathrm{T}}^2 = 0$

$$I_m(\Theta) I'_m(\Theta) = 0.$$

The zeroth approximation for the eigenvalue is given by (5.60) and (5.61). To find the first-order approximation for the eigenvalue, we write it as $\hat{\Lambda} = \hat{\Lambda}_0 + \delta\hat{\Lambda}$, where $\delta\hat{\Lambda}/\hat{\Lambda}_0$ is small. To the first order in $\delta\hat{\Lambda}/\hat{\Lambda}_0$, we obtain the eigenvalue equation of the single-mode approximation:

$$\left\{ V \left[\hat{\Lambda} - \hat{\Lambda}_0 \right]^{-1} - \hat{\Lambda}_{\mathrm{p}}^2 \right\} \hat{D} = \mathrm{i} \,, \tag{5.62}$$

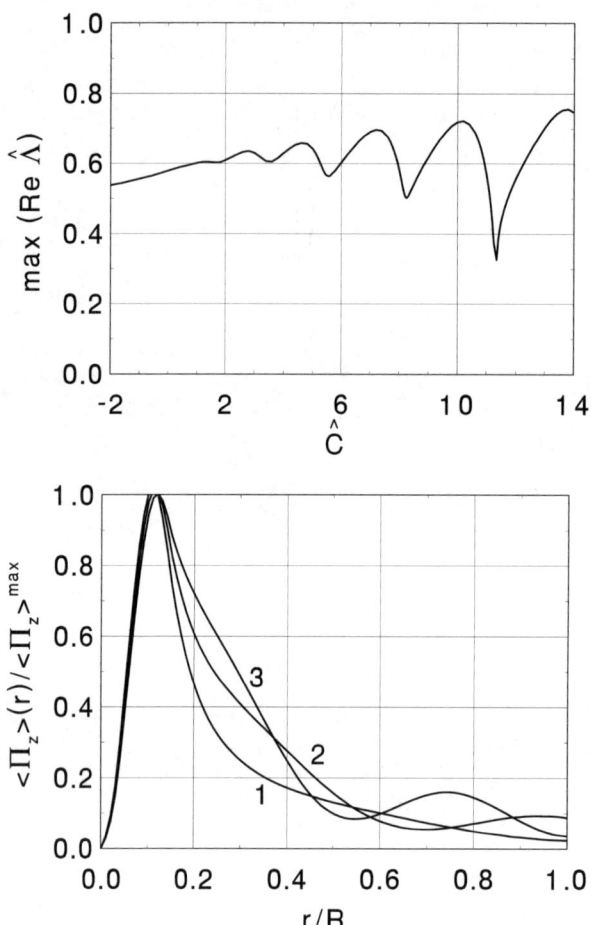

Fig. 5.13. Maximum reduced growth rate versus the detuning parameter (upper plot) and radiation power flow density versus radius (lower plot). The curve in the upper plot is calculated with (5.48). The circles represent the single-mode approximation (5.63). The curves in the lower plot are calculated with (5.49) and (5.67). Curves 1, 2, and 3 correspond to the detuning parameter $\hat{C} = 1.3$, 2.8, and 4.6, respectively. Here $m = 0$, $B = 0.1$, $\Omega = 5$, $\hat{\Lambda}_p^2 \to 0$, and $\hat{\Lambda}_T^2 = 0$

where the factor V is given by the expressions:

$$V^{\mathrm{TE}} = \frac{i\mu_{mn}^2}{2\Omega J_m^2(\mu_{mn})(\mu_{mn}^2 - m^2)} \left[J_{m-1}^2(M_{mn}) - J_{m-2}(M_{mn})J_m(M_{mn}) \right] ,$$

$$V^{\mathrm{TM}} = \frac{i}{2\Omega J_{m-1}^2(\nu_{mn})} \left[J_{m-1}^2(N_{mn}) - J_{m-2}(N_{mn})J_m(N_{mn}) \right] .$$

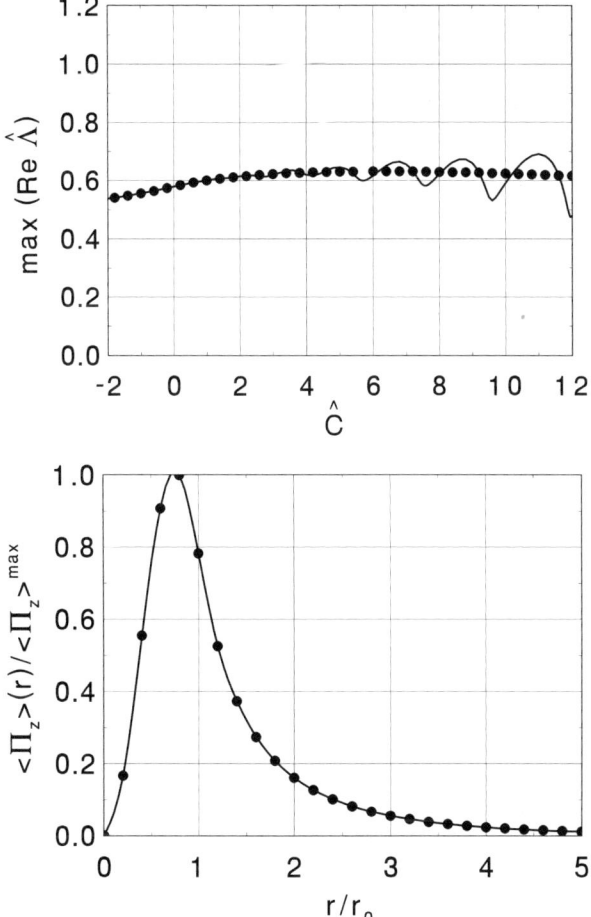

Fig. 5.14. Maximum reduced growth rate versus the detuning parameter (upper plot) and radiation power flow density of the beam radiation mode versus radius (lower plot). The curves are calculated with (5.48), (5.49), and (5.67). The circles are calculated with (5.66) and (5.68) for the TEM_{10} mode of an open electron beam. Here $m = 0$, $B = 0.1$, $\Omega = 10$, $\hat{\Lambda}_p^2 \to 0$, and $\hat{\Lambda}_T^2 = 0$

Here the superscripts TE and TM refers to the case of TE and TM modes, respectively. In the case of negligibly small energy spread, the eigenvalue equation (5.62) reduces to the cubic equation:

$$\left[\left(\hat{\Lambda} + i\hat{C}\right)^2 + \hat{\Lambda}_p^2\right]\left[\hat{\Lambda} - \hat{\Lambda}_0\right] = V . \tag{5.63}$$

Figures. 5.1, 5.8, 5.11, 5.15, and 5.20 illustrate the accuracy of the single-mode approximation. It is seen that the single-mode approximation provides good accuracy for $\Omega \simeq 1$. Also, the accuracy increases with increase of the

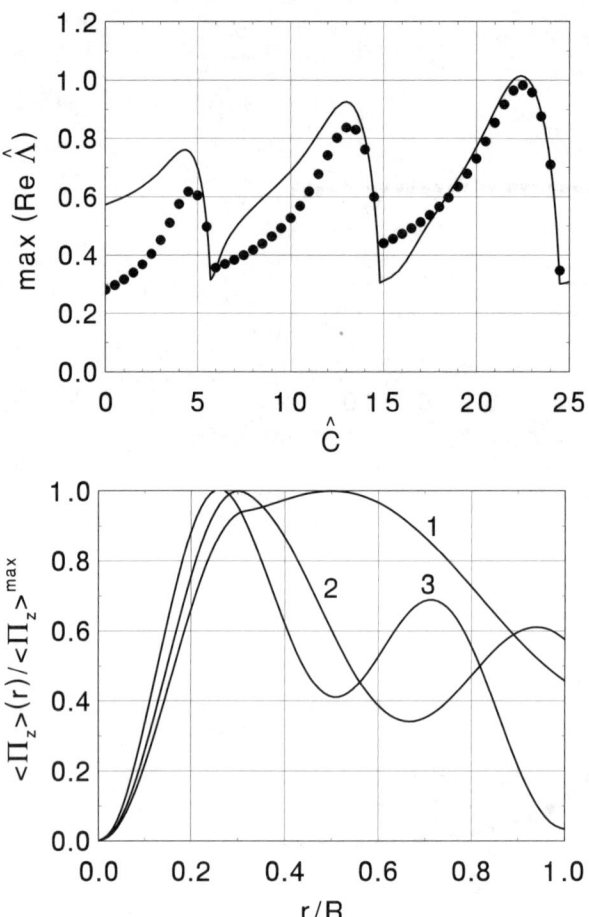

Fig. 5.15. Maximum reduced growth rate versus the detuning parameter (upper plot) and radiation power flow density versus radius (lower plot). The curve in the upper plot is calculated with (5.48) and the circles represent the single-mode approximation (5.63). The curves in the lower plot are calculated with (5.49) and (5.67). Curves 1, 2 and 3 correspond to the detuning parameter $\hat{C} = 4.4$, 13.0, and 22.4, respectively. Here $m = 2$, $B = 0.1$, $\Omega = 1$, $\hat{\Lambda}_P^2 \to 0$, and $\hat{\Lambda}_T^2 = 0$

mode number. This can be explained as follows. When deriving equations for the single-mode approximation, we required the value of $\delta\hat{\Lambda}/\hat{\Lambda}_0$ to be much less than unity. It follows from (5.60) and (5.61) that the larger the values of $\nu_{mn}^2/(2\Omega)$ (or $\mu_{mn}^2/(2\Omega)$), the better should be the accuracy of the single-mode approximation. These values are always large for the value $\Omega \simeq 1$ of the diffraction parameter.

5.3 Beam Radiation Modes in a Circular Waveguide

Another important physical asymptote corresponds to a large radius of the waveguide. In this case we should let $\Omega \to \infty$ in (5.48). Using the asymptotic expansion of the Bessel functions for large values of the argument ($\operatorname{Re} g > 0$):

$$I_m(\Theta), \; I'_m(\Theta) \to (2\pi\Theta)^{-1/2} \exp(\Theta),$$

$$K_m(\Theta), \; -K'_m(\Theta) \to (2\Theta/\pi)^{-1/2} \exp(-\Theta),$$

we find that

$$\left\{ \frac{K_m(\Theta)}{I_m(\Theta)} + \frac{K'_m(\Theta)}{I'_m(\Theta)} \right\} \to 0. \tag{5.64}$$

As a result, the eigenvalue equation (5.48) transforms to:[3]

$$\mu J_m(\mu) K_{m-1}(g) = g K_m(g) J_{m-1}(\mu). \tag{5.65}$$

Introducing the notation $n = |m - 1|$ and using the relations

$$(-1)^m J_{-m}(\zeta) = J_m(\zeta), \qquad K_{-m}(\zeta) = K_m(\zeta),$$

$$J_{m-1}(\zeta) = (2m/\zeta) J_m(\zeta) - J_{m+1}(\zeta),$$

$$K_{m-1}(\zeta) = -(2m/\zeta) K_m(\zeta) + K_{m+1}(\zeta),$$

we rewrite (5.65) as

$$\mu J_{n+1}(\mu) K_n(g) = g K_{n+1}(g) J_n(\mu). \tag{5.66}$$

The latter eigenvalue equation is identical to (4.121) obtained in Chap. 4 for an open axisymmetric electron beam. It follows from (5.48) and (5.66) that in the asymptote of the open beam all the eigenvalues become doubly degenerate, because two different values of the azimuthal index m correspond to one value of n. The only exception is the case of $n = 0$ which corresponds to the mode with azimuthal index $m = 1$. Investigation of the plots in Figs. 5.11–5.18. allows one to trace the degeneration for the modes with $m = 0$ and $m = 2$. It is seen from these plots that at large values of the waveguide diffraction parameter Ω (i.e. as $R \to \infty$) the growth rates of these modes

---

[3] We mentioned above that the eigenvalue equation (5.48) is invariant under the exchange $g \to -g$, or, when $\mu \to -\mu$. The same property holds for the limiting transition $\Omega \to \infty$, despite seeming to be nontrivial. Indeed, the expression in brackets in (5.64) tends to $2\pi i(-1)^{m+1}$ under the exchange of g by $-g$ and we obtain the equation

$$-g K_m(-g) J_{m-1}(\mu) - \mu J_m(\mu) K_{m-1}(-g)$$

$$= \pi i(-1)^{m+1} [-g I_m(-g) J_{m-1}(\mu) + \mu J_m(\mu) I_{m-1}(-g)]$$

instead of (5.65). Using the relation between the Bessel functions,

$$K_m(-g) - i\pi(-1)^{m+1} I_m(-g) = (-1)^m K_m(g),$$

this equation reduces to (5.65).

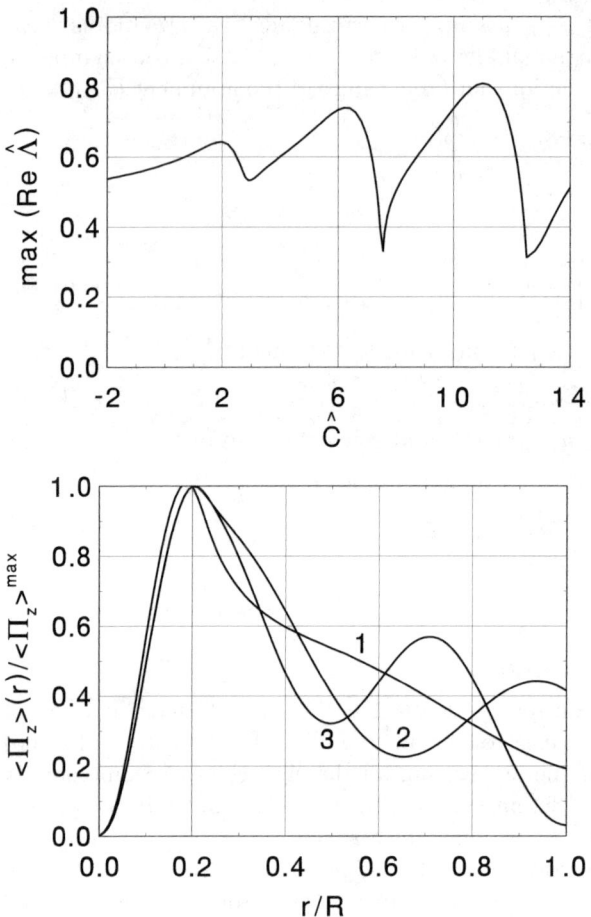

Fig. 5.16. Maximum reduced growth rate versus the detuning parameter (upper plot) and radiation power flow density versus radius (lower plot). The curve in the upper plot is calculated with (5.48). The curves in the lower plot are calculated with (5.49) and (5.67). Curves 1, 2, and 3 correspond to the detuning parameter $\hat{C} = 2$, 6.3, and 11, respectively. Here $m = 2$, $B = 0.1$, $\Omega = 2$, $\hat{\Lambda}_p^2 \to 0$, and $\hat{\Lambda}_T^2 = 0$

approach asymptotically the value of the growth rate of the TEM$_{10}$ mode of the open beam.

Let us discuss the region of applicability of the open beam asymptote. For small values of the detuning parameter, $\hat{C}$, the asymptotic results become valid at the value $\Omega \simeq 10$ of the waveguide diffraction parameter. When detuning increases, the difference between the rigorous and asymptotic results becomes significant. Thus, at large values of the detuning parameter, the open beam approximation becomes valid for a larger size of the waveguide.

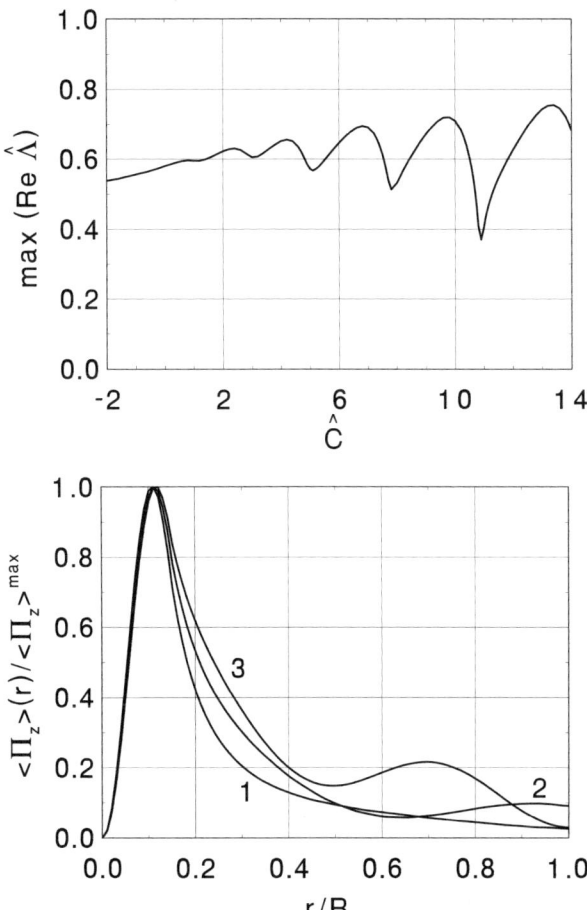

Fig. 5.17. Maximum reduced growth rate versus the detuning parameter (upper plot) and radiation power flow density versus radius (lower plot). The curve in the upper plot is calculated with (5.48). The curves in the lower plot are calculated with (5.49) and (5.67). Curves 1, 2, and 3 correspond to the detuning parameter $\hat{C} = 0.8$, 2.3, and 4.2, respectively. Here $m = 2$, $B = 0.1$, $\Omega = 5$, $\hat{\Lambda}_p^2 \to 0$, and $\hat{\Lambda}_T^2 = 0$

Figures 5.3 and 5.19 illustrate the influence of the energy spread in the beam on FEL amplifier operation. The main effect of the energy spread consists in the reduction of the maximal growth rate and the change in the resonant conditions for the local growth rate maxima.

With the help of the plots presented in Figs. 5.3 and 5.20 one can get an idea about the space charge influence on FEL amplifier operation. First of all, the space charge fields reduce the growth rate and change the resonant conditions. Another peculiarity consists in a sharper separation of the

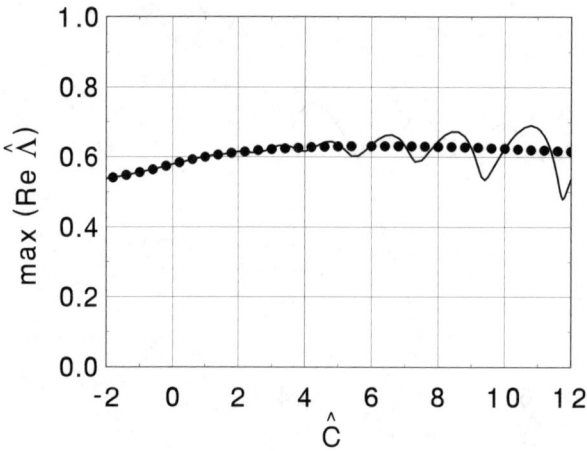

Fig. 5.18. Maximum reduced growth rate versus the detuning parameter. The curve is calculated with (5.48) and the circles are the solution of the asymptotic equation (5.66) for the TEM$_{10}$ mode. Here $m = 2$, $B = 0.1$, $\Omega = 10$, $\hat{\Lambda}_\mathrm{p}^2 \to 0$, and $\hat{\Lambda}_\mathrm{T}^2 = 0$

local maxima at values $\Omega \simeq 1$ of the waveguide diffraction parameter. The combined influence of the space charge fields and energy spread is illustrated in Fig. 5.21. It is seen that the presence of the energy spread leads to an overlapping of the local maxima.

In the high-gain limit the distribution of the mean flow density of the radiation power is given by the formula (in the paraxial approximation):

$$\langle \Pi_z \rangle = \frac{c}{4\pi T} \int_0^T |\boldsymbol{E}(z, r, \varphi, t)|^2 dt$$

$$= \mathrm{const.} \times \left(|F_r^{(m)}|^2 + |F_\varphi^{(m)}|^2 \right) \exp\left[2 \operatorname{Re} \Lambda z \right] , \quad (5.67)$$

where $\langle \Pi_z \rangle$ is the time-averaged longitudinal component of Poynting's vector, $\boldsymbol{E}$ is the electric field of the wave, and the eigenfunctions $F_{r,\varphi}^{(m)}$ are given by (5.49). In the case under study, the electric field $\boldsymbol{E}$ is a trigonometric function of the argument $m\varphi - \omega(z/c - t)$, so the time-averaged value of Π_z does not depend on the time t and the azimuthal angle φ, and it is a function of the radius r only. This feature follows naturally from the fact that the radiation modes of the axisymmetric electron beam in a circular waveguide and helical undulator are rotating. Figures 5.2, 5.4, 5.6, 5.7, 5.11, 5.12, 5.13, 5.14, 5.15, 5.16, and 5.17 present plots of the distributions of the radiation power flow density for several values of the diffraction parameters B and Ω and for three values of the azimuthal index $m = 0$, 1 and 2.

Let us study the behavior of the radiation mode field distribution in the asymptote of large waveguide radius, $R \to \infty$ ($\Omega \to \infty$). In the same way

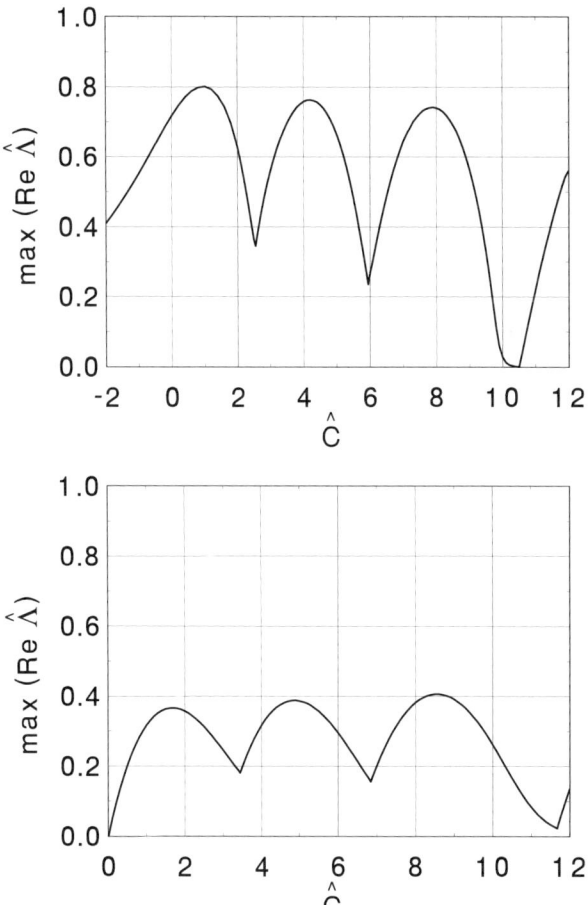

Fig. 5.19. Maximum reduced growth rate versus the detuning parameter for two values of the energy spread parameter, $\hat{\Lambda}_T^2 = 0.25$ (upper plot) and $\hat{\Lambda}_T^2 = 1$ (lower plot). Calculations have been performed with (5.48). Here $m = 1$, $B = 0.1$, $\Omega = 2$, and $\hat{\Lambda}_p^2 \to 0$.

as was done above, we substitute the asymptotic expansions of the Bessel functions into (5.49) and obtain

$$\tilde{E}_r + i\tilde{E}_\varphi \propto \exp(-im\varphi), \qquad \tilde{E}_r = i\tilde{E}_\varphi ,$$

$$\tilde{E}_x + i\tilde{E}_y \propto \exp[-i(m-1)\varphi + i\omega(z/c - t)], \qquad \tilde{E}_x = i\tilde{E}_y .$$

It is seen that in the limit of $\Omega \to \infty$, the field of the mode with azimuthal index m becomes circularly polarized and has $n = |m-1|$ variations in the angle φ. The eigenvalues of the beam radiation modes become doubly degenerate, because there are two linearly independent eigenfunctions $\Phi_1 \propto \cos(n\varphi)$ and $\Phi_2 \propto \sin(n\varphi)$ (or $\Phi_+ \propto \exp(in\varphi)$ and $\Phi_- \propto \exp(-in\varphi)$) for each value of n.

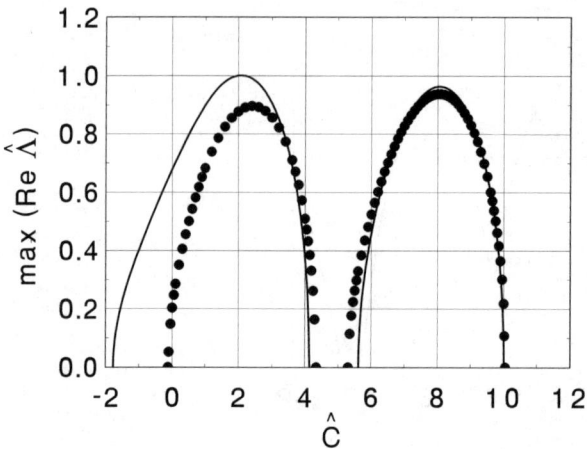

Fig. 5.20. Maximum reduced growth rate versus the detuning parameter. The curve is calculated with (5.48) and the circles are the solution of (5.63) (single-mode approximation). Here $m = 1$, $B = 0.1$, $\Omega = 2$, $\hat{\Lambda}_p^2 = 1$, and $\hat{\Lambda}_T^2 = 0$

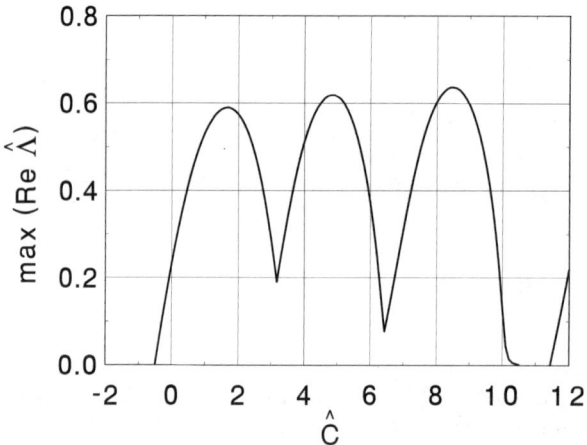

Fig. 5.21. Maximum reduced growth rate versus the detuning parameter. Calculations have been performed with (5.48). Here $m = 1$, $B = 0.1$, $\Omega = 2$, $\hat{\Lambda}_p^2 = 1$, and $\hat{\Lambda}_T^2 = 0.25$

As $\Omega \to \infty$, the radiation power flow density for both eigenfunctions Φ_+ and Φ_- is given by

$$\langle \Pi_z \rangle \propto \begin{cases} |J_n(\mu r/r_0)|^2 & \text{for} \quad r < r_0, \\ |J_n(\mu) K_n(gr/r_0)|^2 / |K_n(g)|^2 & \text{for} \quad r > r_0, \end{cases} \quad (5.68)$$

where μ and g are calculated from the eigenvalue equation (5.66).

The plots presented in Figs. 5.7 and 5.14 give one an idea about the validity region of the asymptote (5.68). Figure 5.7 presents the transverse distribution of the radiation power flow density of the mode with azimuthal index $m = 1$. Calculations have been performed with the rigorous formulae (5.48) and (5.49), and with the asymptotic formulae (5.66) and (5.68). The other plot in this figure corresponds to the calculations for the TEM$_{00}$ mode performed with the asymptotic formulae (5.66) and (5.68). It is seen that in the region $0.1 \lesssim B \lesssim 1$ of the beam diffraction parameter there is good agreement between the rigorous and asymptotic results when $\Omega = 10$. Similar dependencies for the radiation mode with $m = 0$ and the TE$_{10}$ mode of the open beam are presented in Fig. 5.14. We do not present separately the results of calculations for the mode with $m = 2$, since to the accuracy of the line thickness it is the same as that for $m = 0$ presented in Fig. 5.14.

5.3.2 Parabolic Profile

In the general case the beam current density of the axisymmetric electron beam is given by the expression:

$$j_0(r) = \begin{cases} I_0 S(r/r_0) \left[2\pi \int_0^{r_0} r S(r/r_0) \mathrm{d}r \right]^{-1} & \text{for } r < r_0 \,, \\ 0 & \text{for } r > r_0 \,, \end{cases} \quad (5.69)$$

where r_0 and I_0 are the radius and the current of the electron beam, respectively. $S(r/r_0)$ is a function describing the beam profile. To be specific, we assume that $S(0) = 1$.

In the high-gain limit we seek solutions of (5.40a) and (5.40b) in the form of

$$\tilde{E}_{r,\varphi}(z, r, \varphi) = F_{r,\varphi}^{(m)}(r) \exp\left(-\mathrm{i}m\varphi + \Lambda z\right) \,.$$

Introducing the complex amplitudes

$$F_{\pm}^{(m)} = \left(F_r^{(m)} \pm \mathrm{i} F_\varphi^{(m)} \right) \,,$$

we obtain the following reduced equations from (5.40):

$$\left[\frac{\mathrm{d}^2}{\mathrm{d}\hat{r}^2} + \frac{1}{\hat{r}} \frac{\mathrm{d}}{\mathrm{d}\hat{r}} - \frac{(m-1)^2}{\hat{r}^2} + 2\mathrm{i}B\hat{\Lambda} - \frac{2\mathrm{i}\hat{D}S(\hat{r})}{1 - \mathrm{i}\hat{D}S(\hat{r})\hat{\Lambda}_\mathrm{p}^2} \right] F_+^{(m)} = 0 \,, \quad (5.70\mathrm{a})$$

$$\left[\frac{\mathrm{d}^2}{\mathrm{d}\hat{r}^2} + \frac{1}{\hat{r}} \frac{\mathrm{d}}{\mathrm{d}\hat{r}} - \frac{(m+1)^2}{\hat{r}^2} + 2\mathrm{i}B\hat{\Lambda} \right] F_-^{(m)} = 0 \,, \quad (5.70\mathrm{b})$$

where the following notation has been introduced:

$$\hat{z} = \Gamma z, \quad \hat{r} = r/r_0, \quad \hat{\Lambda} = \Lambda/\Gamma, \quad \hat{C} = C/\Gamma \,,$$
$$\hat{\Lambda}_\mathrm{p}^2 = 4c^2(\theta_\mathrm{s} r_0 \omega)^{-2}, \quad \hat{\Lambda}_\mathrm{T}^2 = \langle(\Delta\mathcal{E})^2\rangle/(\rho^2 \mathcal{E}^2) \,,$$
$$\rho = c\gamma_z^2 \Gamma/\omega \,, \quad B = \Gamma r_0^2 \omega/c \,,$$

$$\Gamma = \left[I_0 \omega^2 \theta_s^2 \left(2I_A c^2 \gamma_z^2 \gamma \int_0^1 \xi S(\xi) d\xi \right)^{-1} \right]^{1/2},$$

$$\hat{D} = \mathrm{i} \int_0^\infty \exp\left[-\frac{\hat{\Lambda}_T^2 \xi^2}{2} - \left(\hat{\Lambda} + \mathrm{i}\hat{C} \right) \right] \xi d\xi .$$

To be specific, we assume the energy spread in the beam to be Gaussian.

As far as we know, a rigorous solution of (5.70a) can be found only for the bounded parabolic profile in the limit of a negligibly small space charge field, $\hat{\Lambda}_p^2 \to 0$. The beam current density for the bounded parabolic profile is given by

$$j_0(r) = \frac{I_0(1 - r^2/r_1^2)}{\pi r_0^2 [1 - r_0^2/(2r_1^2)]} \quad \text{for } r < r_0 ,$$
$$j_0(r) = 0 \quad \text{for } r > r_0 , \tag{5.71}$$

where $r_0 < r_1$. The profile function is

$$S(\hat{r}) = 1 - k_1^2 \hat{r}^2 \quad \text{for } \hat{r} < 1 ,$$
$$S(\hat{r}) = 0 \quad \text{for } \hat{r} > 1 ,$$

where $\hat{r} = r/r_0$ and $k_1 = r_0/r_1$. For $\hat{r} < 1$ we rewrite (5.70a) in the form:

$$\left[\frac{d^2}{d\hat{r}^2} + \frac{1}{\hat{r}} \frac{d}{d\hat{r}} - 2\mathrm{i}\hat{D} \left(1 - k_1^2 \hat{r}^2 \right) + 2\mathrm{i}B\hat{\Lambda} - \frac{(m-1)^2}{\hat{r}^2} \right] F_+^{(m)}(\hat{r}) = 0 ,$$

where $B = \Gamma r_0^2 \omega / c$ and

$$\Gamma = \left[I_0 \omega^2 \theta_s^2 \left(c^2 I_A \gamma_z^2 \gamma (1 - k_1^2/2) \right)^{-1} \right]^{1/2} .$$

Introducing the notation $\mu^2 = -2\mathrm{i}\hat{D} + 2\mathrm{i}B\hat{\Lambda}$ and $\delta^2 = -2\mathrm{i}\hat{D}k_1^2$, we reduce the latter equation to standard form:

$$\left[\frac{d^2}{d\hat{r}^2} + \frac{1}{\hat{r}} \frac{d}{d\hat{r}} + \mu^2 - \delta^2 \hat{r}^2 - \frac{(m-1)^2}{\hat{r}^2} \right] F_+^{(m)}(\hat{r}) = 0 . \tag{5.72}$$

The general solution of this equation has the form:

$$F_+^{(m)}(\hat{r}) = C_1 \hat{r}^n \exp(-\delta \hat{r}^2/2) {}_1F_1(\epsilon, n+1, \delta \hat{r}^2) \quad \text{for } \hat{r} < 1 ,$$

where $n = |m-1|$, ${}_1F_1$ is the confluent hypergeometric function, and

$$\epsilon = (n+1)/2 - \mu^2/(4\delta) .$$

The profile function $S(\hat{r})$ is equal to zero outside the beam, for $1 < \hat{r} < R/r_0$, so (5.70a) is the Bessel equation. The general solution of this equation is

$$F_+^{(m)}(\hat{r}) = C_2 I_n(g\hat{r}) + C_3 K_n(g\hat{r}) \quad \text{for } 1 < \hat{r} < R/r_0 ,$$

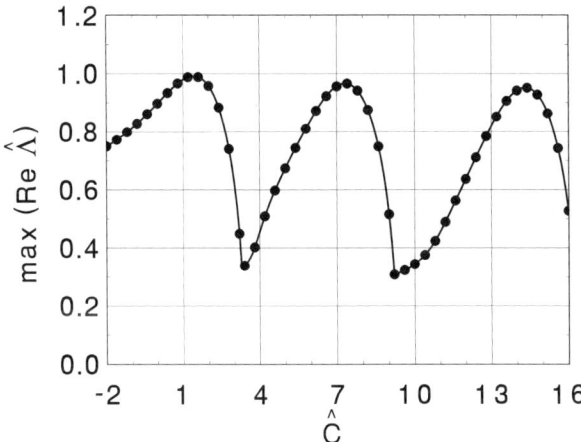

Fig. 5.22. Parabolic profile. Maximum reduced growth rate of the beam radiation modes with azimuthal index $m = 1$ versus the detuning parameter. The curve is calculated with (5.73) and the circles represent the results of calculations with the multilayer approximation method (see (5.77)–(5.82)). Here $B = 0.1$, $\Omega = 1, k_1 = 1$, $\hat{\Lambda}_{\mathrm{p}}^2 \to 0$, and $\hat{\Lambda}_{\mathrm{T}}^2 = 0$

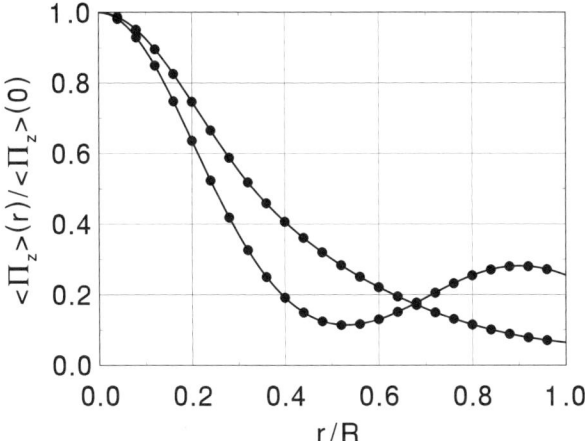

Fig. 5.23. Parabolic profile. Radiation power flow density of the beam radiation modes with azimuthal index $m = 1$ versus radius. The curves are calculated with (5.73) and (5.74) and the circles are calculated with the multilayer approximation method (see (5.77)–(5.82)). Curve 1: $\hat{C} = 1.4$. Curve 2: $\hat{C} = 7.4$. Here $B = 0.1$, $\Omega = 1, k_1 = 1$, $\hat{\Lambda}_{\mathrm{p}}^2 \to 0$, and $\hat{\Lambda}_{\mathrm{T}}^2 = 0$

where $g^2 = -2\mathrm{i}B\hat{\Lambda}$, $\mathrm{Re}\, g > 0$. The solution of (5.70b) for the complex amplitude $F_-^{(m)}$ is given by

$$F_-^{(m)}(\hat{r}) = C_4 I_{m+1}(g\hat{r})$$

in the whole region inside the waveguide, $0 < \hat{r} < R/r_0$. The boundary conditions on the waveguide walls, at $\hat{r} = R/r_0$, require $F_\varphi^{(m)}(\hat{r})$ and $d(\hat{r}F_r^{(m)}(\hat{r}))/d\hat{r}$ be equal to zero. Also, the eigenfunction and its derivative, $F_{\varphi,r}^{(m)}(\hat{r})$ and $d(F_{\varphi,r}^{(m)}(\hat{r}))/d\hat{r}$, must be continuous at the beam boundary, when $\hat{r} = 1$. The fulfillment of these conditions leads to a system of four linear equations for the coefficients $C_{1,2,3,4}$. The requirement of the compatibility of this system leads to the following eigenvalue equation:

$$\delta K_n(g)\left[-\frac{2\epsilon}{n+1}{}_1F_1(\epsilon+1,\ n+2,\ \delta) + {}_1F_1(\epsilon,\ n+1,\ \delta)\right]$$

$$-gK_{n+1}(g)\ {}_1F_1(\epsilon,\ n+1,\ \delta) + \frac{1}{2}\{\ \}_{(+)}[\ ] = 0\ . \tag{5.73}$$

The beam radiation modes are given by the expressions:

Region 1 ($r < r_0$):

$$F_r^{(m)} = \hat{r}^n \exp(-\delta\hat{r}^2/2)\ {}_1F_1(\epsilon,\ n+1,\ \delta\hat{r}^2) + \frac{1}{2}I_{m+1}(g\hat{r})\{\ \}_{(-)}[\ ],$$

$$F_\varphi^{(m)} = -i\hat{r}^n \exp(-\delta\hat{r}^2/2)\ {}_1F_1(\epsilon,\ n+1,\ \delta\hat{r}^2)$$

$$+\frac{i}{2}I_{m+1}(g\hat{r})\{\ \}_{(-)}[\ ]\ , \tag{5.74a}$$

Region 2 ($r_0 < r < R$):

$$F_r^{(m)} = \left\{K_n(g\hat{r}) + \frac{1}{2}I_n(g\hat{r})\{\ \}_{(+)} + \frac{1}{2}I_{m+1}(g\hat{r})\{\ \}_{(-)}\right\}[\ ]\exp(-\delta/2),$$

$$F_\varphi^{(m)} = -i\left\{K_n(g\hat{r}) + \frac{1}{2}I_n(g\hat{r})\{\ \}_{(+)}\right.$$

$$\left. - \frac{1}{2}I_{m+1}(g\hat{r})\{\ \}_{(-)}\right\}[\ ]\exp(-\delta/2)\ . \tag{5.74b}$$

Here the brackets $[\ ]$ and $\{\ \}_{(\pm)}$ denote:

$$[\ ] = \left[\delta I_n(g)\left(-\frac{2\epsilon}{n+1}{}_1F_1(\epsilon+1,\ n+2,\ \delta) + {}_1F_1(\epsilon,\ n+1,\ \delta)\right)\right.$$

$$\left. + gI_{n+1}(g){}_1F_1(\epsilon,\ n+1,\ \delta)\right], \tag{5.75a}$$

$$\{\ \}_{(\pm)} = \left\{\frac{K_m(\Theta)}{I_m(\Theta)} \pm \frac{K'_m(\Theta)}{I'_m(\Theta)}\right\}, \tag{5.75b}$$

where $\Theta = g\sqrt{\Omega/B}$ and $\Omega = \Gamma R^2 \omega/c$.

Figures 5.22 and 5.23 present plots for the growth rate and field distribution of the radiation modes of the electron beam with a bounded parabolic profile.

Let us study the asymptotic behavior of these solutions. The asymptote of the open electron beam corresponds to a large radius of the waveguide. When $R \to \infty$ (i.e. $\Omega \to \infty$), the term $\{\ \}_{\pm}$ tends quickly to zero and (5.73) transforms to

$$\delta K_n(g) \left[\frac{2\epsilon}{n+1} {}_1F_1(\epsilon+1,\ n+2,\ \delta) - {}_1F_1(\epsilon,\ n+1,\ \delta) \right]$$
$$+ g K_{n+1}(g)\ {}_1F_1(\epsilon,\ n+1,\ \delta) = 0\ , \tag{5.76}$$

which is identical to the eigenvalue equation (4.150) obtained in Chap. 4 for the open electron beam with bounded parabolic profile. It follows from (5.73) and (5.76) that double degeneration of the beam radiation modes takes place in the open beam limit, since two values of the azimuthal index m correspond to each value of n.

5.3.3 Arbitrary Gradient Profile

Analytical techniques allow us to obtain rigorous solutions of the eigenvalue problem for a limited number of beam profiles, namely for the stepped and for the parabolic profiles. In the case of an arbitrary gradient profile, the beam radiation modes can be calculated by means of numerical solution of (5.70). We divide the region $0 < \hat{r} < 1$ into K equal layers and assume the beam current density to be constant in each of them. The solution of (5.70a) in each layer has the form:

$$F_+^{(m)} = A_j J_n(\mu_j \hat{r}) + D_j N_n(\mu_j \hat{r})\ ,$$

where $(j-1)/K < \hat{r} < j/K$, A_j and D_j are complex constants, and J_n and N_n are Bessel functions of order n. The coefficients μ_j are given by the expression

$$\mu_j^2 = -2\mathrm{i}\hat{D}S_{j-1/2}\left(1 - \mathrm{i}\hat{A}_\mathrm{p}^2\hat{D}S_{j-1/2}\right)^{-1} - g^2, \qquad g^2 = -2\mathrm{i}B\hat{A}\ ,$$

where $S_{j-1/2} = S(\hat{r}_{j-1/2})$ and $\hat{r}_{j-1/2} = (j-1/2)/K$. To avoid the singularity of the solution at $\hat{r} = 0$, the constant D_1 in the first layer should be set equal to zero, $D_1 = 0$.

Outside the beam the solution of (5.70a) has the form:

$$F_+^{(m)}(\hat{r}) = A_{K+1} I_{m-1}(g\hat{r}) + D_{K+1} K_{m-1}(g\hat{r}) \qquad \text{for} \qquad 1 < \hat{r} < R/r_0\ .$$

The solution of (5.70b) for the complex amplitude $F_-^{(m)}$ inside and outside the beam is:

$$F_-^{(m)}(\hat{r}) = F_1 I_{m+1}(g\hat{r}) \qquad \text{for} \qquad 0 < \hat{r} < R/r_0\ .$$

The constants A_{j+1} and D_{j+1} are connected with the constants A_j and D_j by the continuity conditions of $F_+^{(m)}$ and $\mathrm{d}F_+^{(m)}/\mathrm{d}\hat{r}$ at the boundaries between the layers, and these may be written in the following matrix form:

$$\begin{pmatrix} A_{j+1} \\ D_{j+1} \end{pmatrix} = T_j \begin{pmatrix} A_j \\ D_j \end{pmatrix}, \qquad j = 1, 2, \ldots, K-1, \tag{5.77}$$

where the elements of the matrix T_j are given by the following expressions ($\hat{r}_j = j/K$):

$$\begin{aligned}
(T_j)_{11} &= \frac{\pi}{2}\hat{r}_j \left[\mu_j J_m(\mu_j \hat{r}_j) N_{m-1}(\mu_{j+1}\hat{r}_j) \right. \\
&\quad \left. - \mu_{j+1} J_{m-1}(\mu_j \hat{r}_j) N_m(\mu_{j+1}\hat{r}_j)\right], \\
(T_j)_{12} &= \frac{\pi}{2}\hat{r}_j \left[\mu_j N_m(\mu_j \hat{r}_j) N_{m-1}(\mu_{j+1}\hat{r}_j) \right. \\
&\quad \left. - \mu_{j+1} N_{m-1}(\mu_j \hat{r}_j) N_m(\mu_{j+1}\hat{r}_j)\right], \\
(T_j)_{21} &= -\frac{\pi}{2}\hat{r}_j \left[\mu_j J_m(\mu_j \hat{r}_j) J_{m-1}(\mu_{j+1}\hat{r}_j) \right. \\
&\quad \left. - \mu_{j+1} J_{m-1}(\mu_j \hat{r}_j) J_m(\mu_{j+1}\hat{r}_j)\right], \\
(T_j)_{22} &= -\frac{\pi}{2}\hat{r}_j \left[\mu_j N_m(\mu_j \hat{r}_j) J_{m-1}(\mu_{j+1}\hat{r}_j) \right. \\
&\quad \left. - \mu_{j+1} N_{m-1}(\mu_j \hat{r}_j) J_m(\mu_{j+1}\hat{r}_j)\right].
\end{aligned} \tag{5.78}$$

At the beam boundary, at $\hat{r} = 1$, the continuity conditions lead to the relations:

$$A_K J_{m-1}(\mu_K) + D_K N_{m-1}(\mu_K) = A_{K+1} I_{m-1}(g) + D_{K+1} K_{m-1}(g),$$
$$\mu_K A_K J_m(\mu_K) + \mu_K D_K N_m(\mu_K) = g A_{K+1} I_m(g) + g D_{K+1} K_m(g),$$

which may be written in the matrix form:

$$\begin{pmatrix} A_{K+1} \\ D_{K+1} \end{pmatrix} = T_K \begin{pmatrix} A_K \\ D_K \end{pmatrix}.$$

Finally, the constants A_{K+1} and D_{K+1} may be expressed in terms of the factor A_1 only:

$$\begin{pmatrix} A_{K+1} \\ D_{K+1} \end{pmatrix} = T_K \times T_{K-1} \times \cdots \times T_1 \begin{pmatrix} A_1 \\ 0 \end{pmatrix}. \tag{5.79}$$

The boundary conditions on the waveguide walls give us the equations:

$$A_{K+1} I_{m-1}(\Theta) + D_{K+1} K_{m-1}(\Theta) = F_1 I_{m+1}(\Theta),$$
$$A_{K+1} I_m(\Theta) + D_{K+1} K_m(\Theta) = F_1 I_m(\Theta). \tag{5.80}$$

where $\Theta = g\sqrt{\Omega/B}$. These equations can be rewritten in the matrix form

$$T_{K+1} \begin{pmatrix} A_{K+1} \\ D_{K+1} \end{pmatrix} = F_1 \begin{pmatrix} 1 \\ 1 \end{pmatrix}.$$

Now we express the coefficient F_1 in terms of A_1:

$$T_{K+1} \times T_K \times T_{K-1} \times \cdots \times T_1 \begin{pmatrix} A_1 \\ 0 \end{pmatrix} = T \begin{pmatrix} A_1 \\ 0 \end{pmatrix} = F_1 \begin{pmatrix} 1 \\ 1 \end{pmatrix}, \tag{5.81}$$

where the matrix T depends on the unknown quantity $\hat{\Lambda}$. The coefficient A_1 may be chosen arbitrarily, so without loss of generality we let $A_1 = 1$. Then, excluding F_1 from (5.81), we obtain the eigenvalue equation:

$$(T)_{11} = (T)_{21} \ . \tag{5.82}$$

This equation enables one to find the eigenvalues. The field eigenmodes may be found by solving (5.82) and using the matrix relations given above.

The multilayer approximation method can be tested by the rigorous solutions (5.73) and (5.74) for the parabolic profile. The results of the calculations are presented in Figs. 5.22 and 5.23. There is good agreement between the numerical and analytical results: the relative difference does not exceed 0.1% for the number of layers $K = 40$.

5.4 Initial-Value Problem

5.4.1 Analytical Solution

In this section we present the analytical solution of the initial-value problem for the particular case of a stepped electron beam profile. The vector components of the electric field of the amplified wave in cylindrical coordinates can be conveniently represented as

$$E_{r,\varphi}(z, r, \varphi, t) = \tilde{E}_{r,\varphi}(z, r, \varphi) \exp[i\omega(z/c - t)] + \text{C.C.}$$

Using the notation

$$\hat{r} = r/r_0 \ , \quad B = \Gamma r_0^2 \omega / c \ , \quad \hat{\Lambda}_{\text{p}}^2 = \Lambda_{\text{p}}^2 / \Gamma^2 \ ,$$

$$\Gamma = \left[I_0 \omega^2 \theta_{\text{s}}^2 / (I_A c^2 \gamma_z^2 \gamma)\right]^{1/2} \ , \quad \Lambda_{\text{p}} = \left[4 I_0 / (I_A r_0^2 \gamma_z^2 \gamma)\right]^{1/2} \ ,$$

equations (5.40) may be rewritten in the following dimensionless form:

Region 1 ($r < r_0$):

$$\left[\frac{\partial^2}{\partial \hat{r}^2} + \frac{1}{\hat{r}} \frac{\partial}{\partial \hat{r}} + \frac{1}{\hat{r}^2} \frac{\partial^2}{\partial \varphi^2} + 2\mathrm{i}B \frac{\partial}{\partial \hat{z}}\right] \left(\tilde{E}_r + \mathrm{i}\tilde{E}_\varphi\right) \mathrm{e}^{\mathrm{i}\varphi}$$

$$= \mathrm{i} \int_0^{\hat{z}} \mathrm{d}\hat{z}' \Bigg\{ \left(\tilde{E}_r + \mathrm{i}\tilde{E}_\varphi\right) \mathrm{e}^{\mathrm{i}\varphi}$$

$$+ \hat{\Lambda}_{\text{p}}^2 \left[\frac{\partial^2}{\partial \hat{r}^2} + \frac{1}{\hat{r}} \frac{\partial}{\partial \hat{r}} + \frac{1}{\hat{r}^2} \frac{\partial^2}{\partial \varphi^2} + 2\mathrm{i}B \frac{\partial}{\partial \hat{z}'}\right] \left(\tilde{E}_r + \mathrm{i}\tilde{E}_\varphi\right) \mathrm{e}^{\mathrm{i}\varphi} \Bigg\}$$

$$\times \int_{-\infty}^{\infty} \mathrm{d}\hat{P} \left(\mathrm{d}\hat{F}(\hat{P})/\mathrm{d}\hat{P}\right) \exp\left[\mathrm{i}\left(\hat{P} + \hat{C}\right)(\hat{z}' - \hat{z})\right] \ , \tag{5.83a}$$

$$\left[\frac{\partial^2}{\partial \hat{r}^2} + \frac{1}{r} \frac{\partial}{\partial \hat{r}} + \frac{1}{\hat{r}^2} \frac{\partial^2}{\partial \varphi^2} + 2\mathrm{i}B \frac{\partial}{\partial \hat{z}}\right] \left(\tilde{E}_r - \mathrm{i}\tilde{E}_\varphi\right) \mathrm{e}^{-\mathrm{i}\varphi} = 0 \ , \tag{5.83b}$$

Region 2 ($r_0 < r < R$):
$$\left[\frac{\partial^2}{\partial \hat{r}^2} + \frac{1}{r}\frac{\partial}{\partial \hat{r}} + \frac{1}{\hat{r}^2}\frac{\partial^2}{\partial \varphi^2} + 2\mathrm{i}B\frac{\partial}{\partial \hat{z}}\right]\left(\tilde{E}_r \pm \mathrm{i}\tilde{E}_\varphi\right) e^{\pm \mathrm{i}\varphi} = 0 . \qquad (5.83c)$$

We represent the complex field amplitudes $\tilde{E}_{r,\varphi}(z, r, \varphi)$ as a Fourier series in the angle φ:
$$\tilde{E}_{r,\varphi} = \sum_{m=-\infty}^{m=\infty} \tilde{E}_{r,\varphi}^{(m)}(z, r) \exp(-\mathrm{i}m\varphi) .$$

The solution for the Fourier coefficients $\tilde{E}_{r,\varphi}^{(m)}(z, r)$ can be found with the Laplace transform technique. The Laplace transforms of $\tilde{E}_{r,\varphi}^{(m)}(z, r)$,
$$\bar{E}_{r,\varphi}^{(m)}(p, \hat{r}) = \int_0^\infty \mathrm{d}\hat{z}\, \exp(-p\hat{z}) \tilde{E}_{r,\varphi}^{(m)}(\hat{z}, \hat{r}) , \qquad \mathrm{Re}\, p > 0 ,$$

satisfy the following equations:

Region 1 ($r < r_0$):
$$\left[\frac{\mathrm{d}^2}{\mathrm{d}\hat{r}^2} + \frac{1}{\hat{r}}\frac{\mathrm{d}}{\mathrm{d}\hat{r}} - \frac{(m-1)^2}{\hat{r}^2} + \bar{\mu}^2\right](\bar{E}_r^{(m)} + \mathrm{i}\bar{E}_\varphi^{(m)}) = f_+^{(m)}(\hat{r}) , \qquad (5.84a)$$

$$\left[\frac{\mathrm{d}^2}{\mathrm{d}\hat{r}^2} + \frac{1}{\hat{r}}\frac{\mathrm{d}}{\mathrm{d}\hat{r}} - \frac{(m+1)^2}{\hat{r}^2} - \bar{g}^2\right](\bar{E}_r^{(m)} - \mathrm{i}\bar{E}_\varphi^{(m)}) = f_-^{(m)}(\hat{r}) , \qquad (5.84b)$$

Region 2 ($r_0 < r < R$):
$$\left[\frac{\mathrm{d}^2}{\mathrm{d}\hat{r}^2} + \frac{1}{\hat{r}}\frac{\mathrm{d}}{\mathrm{d}\hat{r}} - \frac{(m\mp 1)^2}{\hat{r}^2} - \bar{g}^2\right](\bar{E}_r^{(m)} \pm \mathrm{i}\bar{E}_\varphi^{(m)}) = f_\pm^{(m)}(\hat{r}) . \qquad (5.84c)$$

Here the following notation is introduced:
$$f_\pm^{(m)}(\hat{r}) = 2\mathrm{i}B(\tilde{E}_r^{(m)} \pm \mathrm{i}\tilde{E}_\varphi^{(m)})|_{\hat{z}=0} ,$$

$$\bar{g}^2 = -2\mathrm{i}Bp , \qquad \bar{\mu}^2 = -2\mathrm{i}\hat{D}(p)\left[1 - \mathrm{i}\hat{D}(p)\hat{\Lambda}_\mathrm{p}^2\right]^{-1} - \bar{g}^2 ,$$

$$\hat{D}(p) = \int_{-\infty}^{\infty} \mathrm{d}\hat{P}\left(\mathrm{d}\hat{F}(\hat{P})/\mathrm{d}\hat{P}\right)\left[p + \mathrm{i}\hat{C} + \mathrm{i}\hat{P}\right]^{-1} .$$

Equations (5.84) are to be solved under the boundary conditions on the waveguide walls:
$$\bar{E}_\varphi^{(m)}|_{r=R} = 0 , \qquad \mathrm{d}(r\bar{E}_r^{(m)})/\mathrm{d}r|_{r=R} = 0 , \qquad (5.85a)$$

and the continuity condition at the beam boundary:
$$(\bar{E}_{r,\varphi}^{(m)})^{-1}(\mathrm{d}/\mathrm{d}r)\bar{E}_{r,\varphi}^{(m)}|_{r=r_0+0} = (\bar{E}_{r,\varphi}^{(m)})^{-1}(\mathrm{d}/\mathrm{d}r)\bar{E}_{r,\varphi}^{(m)}|_{r=r_0-0} . \qquad (5.85b)$$

The inhomogeneous differential equations (5.84) can be solved by means of the Green's functions method (see Sec. 4.3.3 for more details). Omitting intermediate calculations, we present the general solution for $\bar{E}_r^{(m)} + \mathrm{i}\bar{E}_\varphi^{(m)}$:

5.4 Initial-Value Problem

Region 1 ($r < r_0$):

$$\bar{E}_r^{(m)} + i\bar{E}_\varphi^{(m)} = C_1 J_{m-1}(\bar{\mu}r/r_0)$$

$$+ \frac{\pi}{2} N_{m-1}(\bar{\mu}r/r_0) \int_0^{r/r_0} J_{m-1}(\bar{\mu}\zeta) f_+^{(m)}(\zeta) \zeta d\zeta$$

$$+ \frac{\pi}{2} J_{m-1}(\bar{\mu}r/r_0) \int_{r/r_0}^1 N_{m-1}(\bar{\mu}\zeta) f_+^{(m)}(\zeta) \zeta d\zeta , \qquad (5.86)$$

Region 2 ($r_0 < r < R$):

$$\bar{E}_r^{(m)} + i\bar{E}_\varphi^{(m)} = C_2 I_{m-1}(\bar{g}r/r_0) + C_3 K_{m-1}(\bar{g}r/r_0)$$

$$- K_{m-1}(\bar{g}r/r_0) \int_1^{r/r_0} I_{m-1}(\bar{g}\zeta) f_+^{(m)}(\zeta) \zeta d\zeta$$

$$- I_{m-1}(\bar{g}r/r_0) \int_{r/r_0}^{R/r_0} K_{m-1}(\bar{g}\zeta) f_+^{(m)}(\zeta) \zeta d\zeta . \qquad (5.87)$$

The general solution of (5.84) for $\bar{E}_r^{(m)} - i\bar{E}_\varphi^{(m)}$ in the whole region inside the waveguide is given by

$$\bar{E}_r^{(m)} - i\bar{E}_\varphi^{(m)} = C_4 I_{m+1}(\bar{g}r/r_0)$$

$$- K_{m+1}(\bar{g}r/r_0) \int_0^{r/r_0} I_{m+1}(\bar{g}\zeta) f_-^{(m)}(\zeta) \zeta d\zeta$$

$$- I_{m+1}(\bar{g}r/r_0) \int_{r/r_0}^{R/r_0} K_{m+1}(\bar{g}\zeta) f_-^{(m)}(\zeta) \zeta d\zeta . \qquad (5.88)$$

Here (J_m, N_m) and (I_m, K_m) are Bessel functions and modified Bessel functions of the first and second kind of order m, respectively. The relations connecting the coefficients C_j are defined by conditions (5.85):

$$C_1 J_{m-1}(\bar{\mu}) + Q_1 N_{m-1}(\bar{\mu}) = (C_2 - Q_2) I_{m-1}(\bar{g}) + C_3 K_{m-1}(\bar{g}) ,$$
$$C_1 \bar{\mu} J_m(\bar{\mu}) + Q_1 \bar{\mu} N_m(\bar{\mu}) = (Q_2 - C_2) \bar{g} I_m(\bar{g}) + C_3 \bar{g} K_m(\bar{g}) ,$$
$$C_2 I_{m-1}(\bar{\Theta}) + (C_3 - Q_3) K_{m-1}(\bar{\Theta}) = C_4 I_{m+1}(\bar{\Theta}) - Q_4 K_{m+1}(\bar{\Theta}) ,$$
$$(C_2 + C_4) I_m(\bar{\Theta}) = (C_3 - Q_3 - Q_4) K_m(\bar{\Theta}) . \qquad (5.89)$$

To simplify these expressions, we use the following notation:

$$\bar{\Theta} = \bar{g}\sqrt{\Omega/B} , \quad \Omega = \Gamma R^2 \omega/c ,$$

$$Q_1 = \frac{\pi}{2}\int_0^1 J_{m-1}(\bar{\mu}\zeta)f_+^{(m)}(\zeta)\zeta\,\mathrm{d}\zeta \ ,$$

$$Q_2 = \int_1^{R/r_0} K_{m-1}(\bar{g}\zeta)f_+^{(m)}(\zeta)\zeta\,\mathrm{d}\zeta \ ,$$

$$Q_3 = \int_1^{R/r_0} I_{m-1}(\bar{g}\zeta)f_+^{(m)}(\zeta)\zeta\,\mathrm{d}\zeta \ ,$$

$$Q_4 = \int_0^{R/r_0} I_{m+1}(\bar{g}\zeta)f_-^{(m)}(\zeta)\zeta\,\mathrm{d}\zeta \ .$$

The solution of (5.89) for the coefficient C_1 is:

$$C_1 = \frac{1}{\Delta_3}\Big\{Q_1\left[\Delta_1\bar{\mu}N_m(\bar{\mu}) - \Delta_2 N_{m-1}(\bar{\mu})\right]$$
$$-Q_2 - \frac{1}{2}Q_3\{\ \}_{(+)} - \frac{1}{2}Q_4\{\ \}_{(-)}\Big\} \ ,$$

$$\Delta_1 = K_{m-1}(\bar{g}) + \frac{1}{2}I_{m-1}(\bar{g})\{\ \}_{(+)} \ ,$$

$$\Delta_2 = \bar{g}K_m(\bar{g}) - \frac{\bar{g}}{2}I_m(\bar{g})\{\ \}_{(+)} \ ,$$

$$\Delta_3 = \bar{g}K_m(\bar{g})J_{m-1}(\bar{\mu}) - \bar{\mu}J_m(\bar{\mu})K_{m-1}(\bar{g})$$
$$-\frac{1}{2}\{\ \}_{(+)}[\bar{g}I_m(\bar{g})J_{m-1}(\bar{\mu}) + \bar{\mu}J_m(\bar{\mu})I_{m-1}(\bar{g})] \ , \quad (5.90)$$

where the brackets denote

$$\{\ \}_{(\pm)} = \left\{\frac{K_m(\bar{\Theta})}{I_m(\bar{\Theta})} \pm \frac{K'_m(\bar{\Theta})}{I'_m(\bar{\Theta})}\right\} \ .$$

Using (5.89), we express the coefficients C_2, C_3 and C_4 via the value of the coefficient C_1:

$$C_2 = \frac{1}{2\Delta_2}\Big\{C_1\bar{\mu}J_m(\bar{\mu})\{\ \}_{(+)}$$
$$+[Q_1\bar{\mu}N_m(\bar{\mu}) - Q_2\bar{g}I_m(\bar{g}) - Q_3\bar{g}K_m(\bar{g})]\{\ \}_{(+)}$$
$$-Q_4\bar{g}K_m(\bar{g})\{\ \}_{(-)}\Big\} \ ,$$

$$C_3 = \{\ \}_{(+)}^{-1}\left[2C_2 + Q_3\{\ \}_{(+)} + Q_4\{\ \}_{(-)}\right] \ ,$$

$$C_4 = [I_{m+1}(\bar{\Theta})]^{-1}\left[C_2 I_{m-1}(\bar{\Theta}) + C_3 K_{m-1}(\bar{\Theta})\right.$$
$$\left. - Q_3 K_{m-1}(\bar{\Theta}) + Q_4 K_{m+1}(\bar{\Theta})\right] \ . \quad (5.91)$$

5.4 Initial-Value Problem   313

Substituting these coefficients into (5.86), (5.87), and (5.88), we express $\bar{E}_r^{(m)} \pm i\bar{E}_\varphi^{(m)}$ in terms of the coefficient C_1 only. The function $\bar{E}_r^{(m)} + i\bar{E}_\varphi^{(m)}$ inside the beam, for $r_0 < r < R$, is given by (5.86), and outside the beam, for $r_0 < r < R$, may be expressed as

$$\bar{E}_r^{(m)} + i\bar{E}_\varphi^{(m)} = \frac{1}{\Delta_2}\left[K_{m-1}(\bar{g}r/r_0) + \frac{1}{2}I_{m-1}(\bar{g}r/r_0)\{\ \}_{(+)}\right]$$

$$\times \left[C_1\bar{\mu}J_m(\bar{\mu}) + Q_1\bar{\mu}N_m(\bar{\mu}) - Q_2\bar{g}I_m(\bar{g})\right.$$

$$\left. -Q_3\bar{g}K_m(\bar{g}) - Q_4\bar{g}K_m(\bar{g})\{\ \}_{(-)}\{\ \}_{(+)}^{-1}\right]$$

$$+K_{m-1}(\bar{g}r/r_0)\left[Q_3 + Q_4\{\ \}_{(-)}\{\ \}_{(+)}^{-1}\right]$$

$$-K_{m-1}(\bar{g}r/r_0)\int_1^{r/r_0} I_{m-1}(\bar{g}\zeta)f_+^{(m)}(\zeta)\zeta\mathrm{d}\zeta$$

$$-I_{m-1}(\bar{g}r/r_0)\int_{r/r_0}^{R/r_0} K_{m-1}(\bar{g}\zeta)f_+^{(m)}(\zeta)\zeta\mathrm{d}\zeta\ . \qquad (5.92)$$

The function $\bar{E}_r^{(m)} - i\bar{E}_\varphi^{(m)}$ is given by

$$\bar{E}_r^{(m)} - i\bar{E}_\varphi^{(m)} = \frac{I_{m+1}(\bar{g}r/r_0)}{I_{m+1}(\bar{\Theta})}\left\{\frac{1}{\Delta_2}\left[K_{m-1}(\bar{\Theta}) + \frac{1}{2}I_{m-1}(\bar{\Theta})\right.\right.$$

$$\times \{\ \}_{(+)}\right]\left[C_1\bar{\mu}J_m(\bar{\mu}) + Q_1\bar{\mu}N_m(\bar{\mu}) - Q_2\bar{g}I_m(\bar{g})\right.$$

$$\left. -Q_3\bar{g}K_m(\bar{g}) - Q_4\bar{g}K_m(\bar{g})\{\ \}_{(-)}\{\ \}_{(+)}^{-1}\right]$$

$$+Q_4\left[K_{m-1}(\bar{\Theta})\{\ \}_{(-)}\{\ \}_{(+)}^{-1} + K_{m+1}(\bar{\Theta})\right]\Big\}$$

$$-K_{m+1}(\bar{g}r/r_0)\int_0^{r/r_0} I_{m+1}(\bar{g}\zeta)f_-^{(m)}(\zeta)\zeta\mathrm{d}\zeta$$

$$-I_{m+1}(\bar{g}r/r_0)\int_{r/r_0}^{R/r_0} K_{m+1}(\bar{g}\zeta)f_-^{(m)}(\zeta)\zeta\mathrm{d}\zeta \qquad (5.93)$$

in the whole region inside the waveguide. To find $\tilde{E}_{r,\varphi}^{(m)}(\hat{z},\ \hat{r})$, we use the inverse Laplace transformation:

$$\tilde{E}_{r,\varphi}^{(m)}(\hat{z},\ \hat{r}) = \frac{1}{2\pi i}\int_{\gamma'-i\infty}^{\gamma'+i\infty} \bar{E}_{r,\varphi}^{(m)}(\lambda,\ \hat{r})\exp(\lambda\hat{z})\mathrm{d}\lambda\ . \qquad (5.94)$$

314   5. Waveguide FELs

Here we limit our consideration to the important practical case of the high-gain regime. In this case only one beam radiation mode survives, namely the one with a maximal field growth rate. Thus, expression (5.94) for the field amplitude becomes that for a single residue of the integrand taken at the pole corresponding to the largest increasing root of the eigenvalue equation

$$\Delta_3(\lambda_1) = 0, \quad \operatorname{Re} \lambda_1 > 0. \tag{5.95}$$

The final result of the asymptotic solution for $\bar{E}_r^{(m)} + i\bar{E}_\varphi^{(m)}$ has the form:

Region 1 ($r < r_0$):

$$\tilde{E}_+^{(m)} = \tilde{E}_r^{(m)} + i\tilde{E}_\varphi^{(m)} = u_1 \exp(\lambda_1 \hat{z}) J_{m-1}(\mu_1 r/r_0), \tag{5.96}$$

Region 2 ($r_0 < r < R$):

$$\tilde{E}_+^{(m)} = u_1 \exp(\lambda_1 \hat{z}) \left[K_{m-1}(g_1 r/r_0) + \frac{1}{2} I_{m-1}(g_1 r/r_0)\{\ \}_{(+)} \right]$$

$$\times \mu_1 J_m(\mu_1) \left[g_1 K_m(g_1) - \frac{1}{2} g_1 I_m(g_1)\{\ \}_{(+)} \right]^{-1}. \tag{5.97}$$

The asymptotic solution for $\bar{E}_r^{(m)} - i\bar{E}_\varphi^{(m)}$ in the whole region inside the waveguide has the form:

$$\tilde{E}_-^{(m)} = \tilde{E}_r^{(m)} - i\tilde{E}_\varphi^{(m)}$$
$$= u_1 \exp(\lambda_1 \hat{z}) I_{m+1}(g_1 r/r_0) \mu_1 J_m(\mu_1) (I_{m+1}(\Theta_1))^{-1}$$
$$\times \left[K_{m-1}(\Theta_1) + \frac{1}{2} I_{m-1}(\Theta_1)\{\ \}_{(+)} \right]$$
$$\times \left[g_1 K_m(g_1) - \frac{1}{2} g_1 I_m(g_1)\{\ \}_{(+)} \right]^{-1}. \tag{5.98}$$

The amplitude coefficient u_1 is given by the expression:

$$u_1 = \left(\frac{d\Delta_3}{d\lambda}\right)^{-1}_{\lambda=\lambda_1} \left\{ \left\{ \mu_1 N_m(\mu_1) \left[K_{m-1}(g_1) + \frac{1}{2} I_{m-1}(g_1)\{\ \}_{(+)} \right] \right. \right.$$
$$\left. - N_{m-1}(\mu_1) \left[g_1 K_m(g_1) - \frac{1}{2} g_1 I_m(g_1)\{\ \}_{(+)} \right] \right\}$$
$$\times \frac{\pi}{2} \int_0^1 J_{m-1}(\mu_1 \zeta) f_+^{(m)}(\zeta) \zeta d\zeta - \int_1^{R/r_0} K_{m-1}(g_1 \zeta) f_+^{(m)}(\zeta) \zeta d\zeta$$
$$- \frac{1}{2} \{\ \}_{(+)} \int_1^{R/r_0} I_{m-1}(g_1 \zeta) f_+^{(m)}(\zeta) \zeta d\zeta$$
$$\left. - \frac{1}{2} \{\ \}_{(-)} \int_0^{R/r_0} I_{m+1}(g_1 \zeta) f_-^{(m)}(\zeta) \zeta d\zeta \right\}. \tag{5.99}$$

To simplify (5.96)–(5.99), we use the following notation:
$$\mu_1 = \bar{\mu}|_{\lambda=\lambda_1}, \quad g_1 = \bar{g}|_{\lambda=\lambda_1}, \quad \Theta_1 = \bar{\Theta}|_{\lambda=\lambda_1},$$
$$\{\,\}_{(\pm)} = \left\{ \frac{K_m(\Theta_1)}{I_m(\Theta_1)} \pm \frac{K'_m(\Theta_1)}{I'_m(\Theta_1)} \right\}.$$

Taking into account that $\Delta_3 = 0$ at $\lambda = \lambda_1$, we write the following expression for the derivative:

$$\left(\frac{\mathrm{d}\Delta_3}{\mathrm{d}\lambda}\right)_{\lambda=\lambda_1} = P_1 \left(\frac{\mathrm{d}\bar{g}}{\mathrm{d}\lambda}\right)_{\lambda=\lambda_1} + P_2 \left(\frac{\mathrm{d}\bar{\mu}}{\mathrm{d}\lambda}\right)_{\lambda=\lambda_1}$$
$$-\frac{1}{2}\{\,\}_{(+)} \left[P_3 \left(\frac{\mathrm{d}\bar{g}}{\mathrm{d}\lambda}\right)_{\lambda=\lambda_1} + P_4 \left(\frac{\mathrm{d}\bar{\mu}}{\mathrm{d}\lambda}\right)_{\lambda=\lambda_1} \right]$$
$$-\frac{P_5}{2g_1}\left(\frac{\mathrm{d}\bar{g}}{\mathrm{d}\lambda}\right)_{\lambda=\lambda_1} \left[\frac{\Theta_1^2 + m^2}{\Theta_1^2 [I'_m(\Theta_1)]^2} - \frac{1}{[I_m(\Theta_1)]^2} \right], \quad (5.100)$$

where

$$P_1 = \mu_1 J_m(\mu_1) \left[K_m(g_1) - \frac{2(m-1)}{g_1} K_{m-1}(g_1) \right]$$
$$\qquad - g_1 K_{m-1}(g_1) J_{m-1}(\mu_1),$$
$$P_2 = g_1 K_m(g_1) \left[\frac{2(m-1)}{\mu_1} J_{m-1}(\mu_1) - J_m(\mu_1) \right]$$
$$\qquad - \mu_1 J_{m-1}(\mu_1) K_{m-1}(g_1),$$
$$P_3 = \mu_1 J_m(\mu_1) \left[\frac{2(m-1)}{g_1} I_{m-1}(g_1) + I_m(g_1) \right] + g_1 I_{m-1}(g_1) J_{m-1}(\mu_1),$$
$$P_4 = g_1 I_m(g_1) \left[\frac{2(m-1)}{\mu_1} J_{m-1}(\mu_1) - J_m(\mu_1) \right] + \mu_1 J_{m-1}(\mu_1) I_{m-1}(g_1),$$
$$P_5 = g_1 I_m(g_1) J_{m-1}(\mu_1) + \mu_1 J_m(\mu_1) I_{m-1}(g_1). \quad (5.101)$$

Remembering the definition of $\bar{g}^2 = -2\mathrm{i}\lambda B$, we can write:
$$\left(\frac{\mathrm{d}\bar{g}}{\mathrm{d}\lambda}\right)_{\lambda=\lambda_1} = -\frac{\mathrm{i}B}{g_1}.$$

For a Gaussian energy spread of particles in the beam, the parameter $\bar{\mu}$ is given by
$$\bar{\mu}^2 = -\frac{2\mathrm{i}\hat{D}(\lambda)}{1 - \mathrm{i}\hat{\Lambda}_\mathrm{p}^2 \hat{D}(\lambda)} - \bar{g}^2,$$

where the function $\hat{D}(\lambda)$ is
$$\hat{D}(\lambda) = \mathrm{i} \int_0^\infty \exp\left[-\frac{\hat{\Lambda}_\mathrm{T}^2 \xi^2}{2} - \lambda\xi - \mathrm{i}\hat{C}\xi \right] \xi \mathrm{d}\xi.$$

The derivative of $\bar{\mu}$ with respect to λ at the point $\lambda = \lambda_1$ has the form:

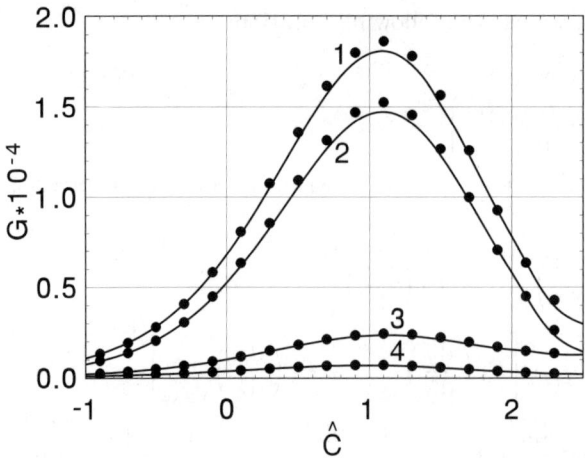

Fig. 5.24. Gain G (curve 1) and partial contributions of individual clockwise-rotating modes of a passive waveguide (curve 2: TE$_{11}$ mode, curve 3: TM$_{11}$ mode, curve 4: TE$_{12}$ mode) versus detuning parameter $\hat{C}$. The input signal is a clockwise-rotating TE$_{11}$ mode. Here $B = 0.1$, $\Omega = 2$, $\hat{\Lambda}_p^2 \to 0$, $\hat{\Lambda}_T^2 = 0$, and $\hat{z} = 5.5$

$$\left(\frac{d\bar{\mu}}{d\lambda}\right)_{\lambda=\lambda_1} = \frac{1}{i\mu_1(1-i\hat{\Lambda}_p^2\hat{D}_1)^2}\left(\frac{d\hat{D}}{d\lambda}\right)_{\lambda=\lambda_1} + \frac{iB}{\mu_1},$$

where $\hat{D}_1 = \hat{D}(\lambda)|_{\lambda=\lambda_1}$. Taking into account the definition of the function $\hat{D}$, we get

$$\left(\frac{d\hat{D}}{d\lambda}\right)_{\lambda=\lambda_1} = \left(\hat{D}_1 - \frac{i}{\hat{\Lambda}_T^2}\right)\left(\frac{1}{\lambda_1+i\hat{C}} + \frac{\lambda_1+i\hat{C}}{\hat{\Lambda}_T^2}\right) + \frac{\lambda_1+i\hat{C}}{\hat{\Lambda}_T^4}.$$

In the limit of negligibly small energy spread, $\hat{\Lambda}_T^2 \to 0$, the function $\hat{D}$ and its derivative with respect to λ at $\lambda = \lambda_1$ takes the following simple form:

$$\hat{D}_1 = \frac{i}{(\lambda_1+i\hat{C})^2}, \quad \left(\frac{d\hat{D}}{d\lambda}\right)_{\lambda=\lambda_1} = -\frac{2i}{(\lambda_1+i\hat{C})^3}.$$

The power gain, G, of the radiation wave having azimuthal index m is given by the formula:

$$G = \int_0^R \left[\left|\tilde{E}_+^{(m)}(z,r)\right|^2 + \left|\tilde{E}_-^{(m)}(z,r)\right|^2\right]rdr$$

$$\times \left\{\int_0^R \left[\left|\tilde{E}_+^{(m)}(0,r)\right|^2 + \left|\tilde{E}_-^{(m)}(0,r)\right|^2\right]rdr\right\}^{-1}. \quad (5.102)$$

In the high-gain limit, the power gain may be represented in the form

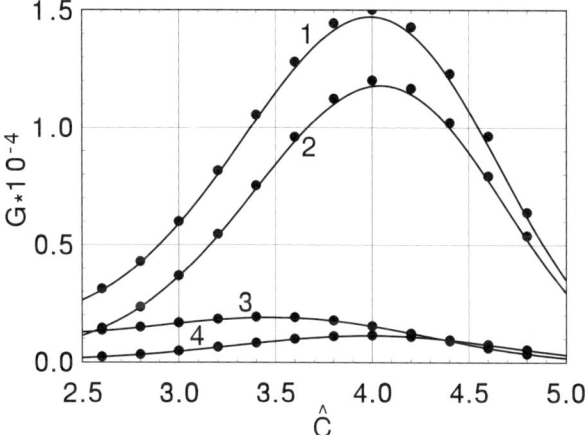

Fig. 5.25. Gain G (curve 1) and partial contributions of individual clockwise-rotating modes of a passive waveguide (curve 2: TM_{11} mode, curve 3: TE_{11} mode, curve 4: TE_{12} mode) versus detuning parameter $\hat{C}$. The input signal is a clockwise-rotating TM_{11} mode. Here $B = 0.1$, $\Omega = 2$, $\hat{\Lambda}_P^2 = 0$, $\hat{\Lambda}_T^2 = 0$, and $\hat{z} = 5.5$

$$G = A \exp[2 \operatorname{Re} \lambda_1 \hat{z}], \tag{5.103}$$

where A and $\operatorname{Re} \lambda_1$ can be calculated using (5.95)–(5.102). The relative contribution of the passive waveguide mode to the full power of the active waveguide mode can be represented as

$$G_{mn}^{TE,TM} = G|C_{mn}^{TE,TM}|^2 \left[\sum_i |C_{mi}^{TE}|^2 + \sum_j |C_{mj}^{TM}|^2 \right]^{-1}, \tag{5.104}$$

where the gain, G, and the expansion coefficients, C_{mk}, are given by (5.102) and (5.58), respectively.

Figures 5.24–5.26 show the gain G and the partial contributions of the passive waveguide modes for several values of the parameters. The curves in these figures are calculated with the analytic formulae (5.58) and (5.95)–(5.104). The circles are the results of numerical calculations with (5.131)–(5.133).

Let us find the asymptotic behavior of these solutions. The asymptote of the open beam corresponds to a large radius of the waveguide, $R \to \infty$. The waveguide diffraction parameter, Ω, tends to infinity, too. Using the asymptotic expressions for the Bessel functions at large values of the argument, we find that

$$\{\ \}_{(\pm)} = \left\{ \frac{K_m(\Theta_1)}{I_m(\Theta_1)} \pm \frac{K'_m(\Theta_1)}{I'_m(\Theta_1)} \right\} \to 0.$$

It follows from (5.99) that in the limit $\Omega \to \infty$, the amplitude coefficient, u_1, is given by

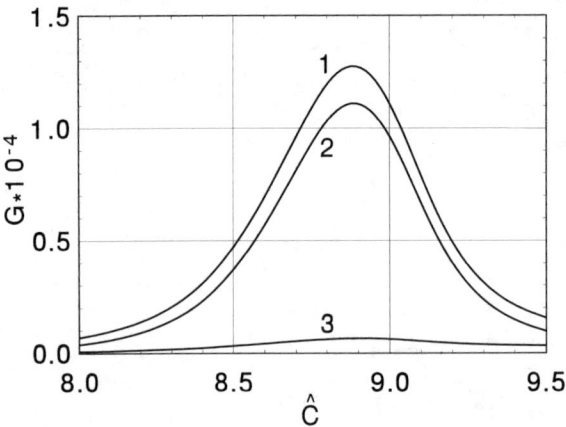

Fig. 5.26. Gain G (curve 1) and partial contributions of individual counterclockwise-rotating modes of a passive waveguide (curve 2: TM_{14} mode, curve 3: TE_{15} mode) versus detuning parameter $\hat{C}$. The input signal is a counterclockwise-rotating TM_{14} mode. Here $B = 0.1$, $\Omega = 10$, $\hat{\Lambda}_p^2 = 0$, $\hat{\Lambda}_T^2 = 0$, and $\hat{z} = 17$

$$u_1 = [d/d\lambda(\bar{g}K_m(\bar{g})J_{m-1}(\bar{\mu}) - \bar{\mu}J_m(\bar{\mu})K_{m-1}(\bar{g}))]_{\lambda=\lambda_1}^{-1}$$

$$\times \Bigg\{ \{\mu_1 N_m(\mu_1)K_{m-1}(g_1) - N_{m-1}(\mu_1)g_1 K_m(g_1)\}$$

$$\times \frac{\pi}{2} \int_0^1 J_{m-1}(\mu_1\zeta) f_+^{(m)}(\zeta) \zeta d\zeta$$

$$- \int_1^\infty K_{m-1}(g_1\zeta) f_+^{(m)}(\zeta) \zeta d\zeta \Bigg\} . \tag{5.105}$$

Using the following relations between the Bessel functions,

$$J_m(i\zeta) = i^m I_m(\zeta) , \quad N_m(i\zeta) = i^{m+1} I_m(\zeta) - \frac{2}{\pi i^m} K_m(\zeta) ,$$

$$I_m'(\zeta)K_m(\zeta) - I_m(\zeta)K_m'(\zeta) = \frac{1}{\zeta} ,$$

we can rewrite the expression for the amplitude coefficient as

$$u_1 = [d/d\lambda \left(\bar{\mu}J_{n+1}(\bar{\mu})K_n(\bar{g}) - \bar{g}K_{n+1}(\bar{g})J_n(\bar{\mu})\right)]_{\lambda=\lambda_1}^{-1}$$

$$\times \Bigg\{ \frac{K_n(g_1)}{J_n(\mu_1)} \int_0^1 J_n(\mu_1\zeta) \hat{f}^{(n)}(\zeta) \zeta d\zeta$$

$$+ \int_1^\infty K_n(\mu_1\zeta) \hat{f}^{(n)}(\zeta) \zeta d\zeta \Bigg\} , \tag{5.106}$$

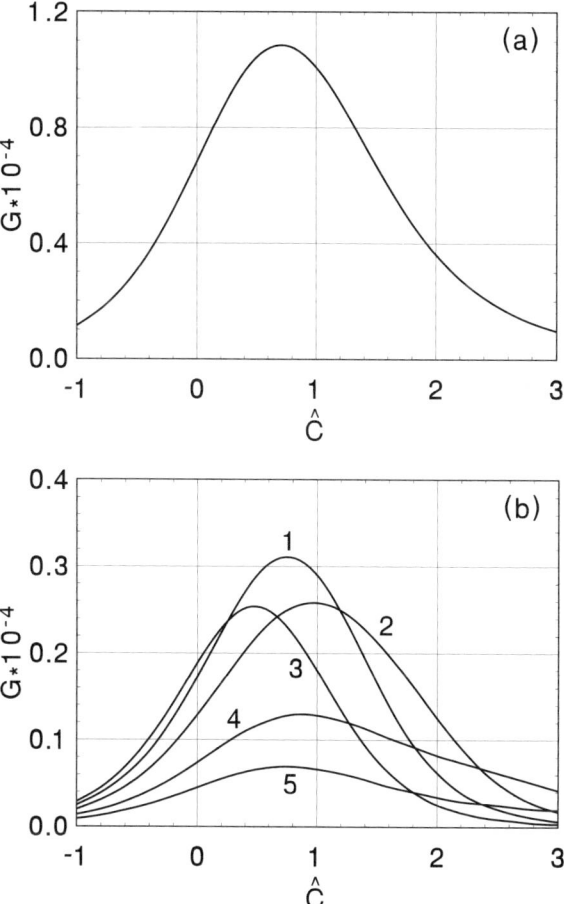

Fig. 5.27. Graph (a): gain G, graph (b): partial contributions of individual clockwise-rotating modes of a passive waveguide (curve 1: TM_{11} mode, curve 2: TE_{12} mode, curve 3: TE_{11} mode, curve 4: TM_{12} mode, curve 5: TE_{13} mode) versus detuning parameter $\hat{C}$. The input signal is a clockwise-rotating TE_{11} mode. Here $B = 0.1$, $\Omega = 10$, $\hat{\Lambda}_P^2 = 0$, $\hat{\Lambda}_T^2 = 0$, and $\hat{z} = 6$

where $n = m - 1$ and
$$\lim_{\Omega \to \infty} f_+^{(m)}(r) = \hat{f}^{(n)}(r) \ .$$

When solving the initial-value problem for the open electron beam (see Sec. 4.4.3), we found that $\hat{f}^{(n)}(r)$ and the components of the circularly polarized wave at the undulator entrance are related by

$$(E_x + iE_y)_{z=0} = \sum_{n=-\infty}^{\infty} \tilde{E}_{\text{ext}}^{(n)}(r) \exp\left[-i(n\varphi + \omega t)\right] \ ,$$

$$\hat{f}^{(n)}(r) = 2iB\tilde{E}^{(n)}_{\text{ext}}(r) \, .$$

Using (5.96)–(5.98), we find in the limit $\Omega \to \infty$ that the field components of the wave with azimuthal index m are related by

$$\tilde{E}_r + i\tilde{E}_\varphi \propto \exp(-im\varphi), \qquad \tilde{E}_r = i\tilde{E}_\varphi \, .$$

Transformation of the field components from cylindrical to Cartesian coordinates leads to:

$$E_x + iE_y \propto \exp\left[-i(m-1)\varphi + i\omega(z/c - t)\right], \qquad E_x = iE_y \, .$$

Thus, we find that in the limit of an open electron beam, $\Omega \to \infty$, the function

$$\hat{f}_+^{(m)}(r) = 2iB\left(\tilde{E}_r^{(m)} + i\tilde{E}_\varphi^{(m)}\right)|_{z=0}$$

is equal to the function $\hat{f}^{(n)}(r)$ introduced in Chap. 4. Thus, we can conclude that the solution of the initial-value problem provides correct results in the asymptote of an open electron beam.

5.4.2 Effective Potential for a Circular Waveguide

We illustrated above that analytical techniques provide powerful tools for the solution of the self-consistent equations (5.7) for the field amplitude. The reader should remember that the self-consistent field equations were solved under boundary conditions (5.41) on the waveguide walls written down in the paraxial approximation. There is another approach to the description of the amplification process which is more suitable for numerical simulations. At the beginning of this chapter we derived (5.8). The effective potential, U, entering the right-hand side of this equation, is expressed in terms of the first harmonic of the beam current density (see (5.23)). Thus, we obtain the integro-differential equation for the first harmonic of the beam current density. It is worth mentioning that within the framework of this approach, we start from the exact expression for the Green's function, and only then simplify it using the paraxial approximation. It is important that both methods lead to the same results.

The general expression for the complex amplitude of the effective potential of the interaction between the particle and the electromagnetic wave, $U(z, \boldsymbol{r}_\perp)$, is given by (5.23). In the case of a circular waveguide, the functions of the passive waveguide, ψ^{TE} and ψ^{TM}, entering (5.23), are given by (see Appendix A.4):

$$\psi_{\text{TE}} = A_{mk}^{\text{TE}} J_m(\mu_{mk} r/R) \begin{pmatrix} \sin(m\varphi) \\ \cos(m\varphi) \end{pmatrix} ,$$

$$\psi_{\text{TM}} = A_{mk}^{\text{TM}} J_m(\nu_{mk} r/R) \begin{pmatrix} \sin(m\varphi) \\ \cos(m\varphi) \end{pmatrix} , \qquad (5.107)$$

where

$$A_{mk}^{\text{TE}} = \frac{\sqrt{2}}{\sqrt{\pi(\mu_{mk}^2 - m^2)}J_m(\mu_{mk})} \quad \text{for} \quad m > 0,$$

$$A_{0k}^{\text{TE}} = \frac{1}{\sqrt{\pi}\mu_{0k}J_0(\mu_{0k})},$$

$$A_{mk}^{\text{TM}} = \frac{\sqrt{2}}{\sqrt{\pi}\nu_{mk}J_{m-1}(\nu_{mk})} \quad \text{for} \quad m > 0,$$

$$A_{0k}^{\text{TM}} = \frac{1}{\sqrt{\pi}\nu_{0k}J_1(\nu_{0k})},$$

and μ_{mk} and ν_{mk} are the kth roots of the equations

$$J'_m(\mu) = 0,$$
$$J_m(\nu) = 0.$$

Substituting (5.107) into (5.23), we obtain an expression for the effective potential of the interaction of the particle with the radiated electromagnetic wave:

$$U_{\text{i}} = -\frac{\text{i}}{2c}e\theta_{\text{s}}^2 \int_0^z dz' \int_0^{2\pi} d\varphi' \int_0^{r_0} dr' r' \tilde{j}_1(z', r', \varphi')$$

$$\times \mathcal{D}\mathcal{D}' \Bigg\{ \sum_{k=1}^{\infty} \frac{1}{\mu_{0k}^2 J_0^2(\mu_{0k})} J_0(\mu_{0k}r/R)$$

$$\times J_0(\mu_{0k}r'/R) \exp\left[-\frac{\text{i}\mu_{0k}^2(z-z')c}{2\omega R^2}\right]$$

$$+ \sum_{m=1}^{\infty} \sum_{k=1}^{\infty} \frac{2}{(\mu_{mk}^2 - m^2)J_m^2(\mu_{mk})} J_m(\mu_{mk}r/R)J_m(\mu_{mk}r'/R)$$

$$\times [\sin(m\varphi)\sin(m\varphi') + \cos(m\varphi)\cos(m\varphi')] \exp\left[-\frac{\text{i}\mu_{mk}^2(z-z')c}{2\omega R^2}\right]$$

$$+ \sum_{k=1}^{\infty} \frac{1}{\nu_{0k}^2 J_1^2(\nu_{0k})} J_0(\nu_{0k}r/R)J_0(\nu_{0k}r'/R) \exp\left[-\frac{\text{i}\nu_{0k}^2(z-z')c}{2\omega R^2}\right]$$

$$+ \sum_{m=1}^{\infty} \sum_{k=1}^{\infty} \frac{2}{\nu_{mk}^2 J_{m-1}^2(\nu_{mk})} J_m(\nu_{mk}r/R)J_m(\nu_{mk}r'/R)$$

$$\times [\sin(m\varphi)\sin(m\varphi') + \cos(m\varphi)\cos(m\varphi')]$$

$$\times \exp\left[-\frac{\text{i}\nu_{mk}^2(z-z')c}{2\omega R^2}\right] \Bigg\}, \tag{5.108}$$

where $\mathcal{D}\mathcal{D}'$ denotes

$$\mathcal{D}\mathcal{D}' = \left(\frac{\partial}{\partial x} + \text{i}\frac{\partial}{\partial y}\right)\left(\frac{\partial}{\partial x'} - \text{i}\frac{\partial}{\partial y'}\right).$$

The Cartesian coordinates (x, y) relate to the cylindrical coordinates (r, φ) as

$$r = \sqrt{x^2 + y^2}, \quad \varphi = \arctan(y/x) .$$

The expression for the operator $\mathcal{DD}'$ in cylindrical coordinates takes the form:

$$\mathcal{DD}' = e^{i(\varphi-\varphi')} \left(\frac{\partial}{\partial r} + \frac{i}{r}\frac{\partial}{\partial \varphi} \right) \left(\frac{\partial}{\partial r'} - \frac{i}{r'}\frac{\partial}{\partial \varphi'} \right) .$$

Let us apply the operator $\mathcal{DD}'$ to the function

$$J_m(\mu_{mk}r/R) J_m(\mu_{mk}r'/R) \cos[m(\varphi-\varphi')] .$$

Using the relations

$$2J'_m(\zeta) = J_{m-1}(\zeta) - J_{m+1}(\zeta) ,$$

$$\frac{2m}{\zeta} J_m(\zeta) = J_{m-1}(\zeta) + J_{m+1}(\zeta) ,$$

we obtain

$$\mathcal{DD}' J_m(\mu_{mk}r/R) J_m(\mu_{mk}r'/R) \cos[m(\varphi-\varphi')]$$
$$= \frac{1}{2} e^{i(\varphi-\varphi')} \left\{ \frac{\mu_{mk}^2}{R^2} J_{m-1}(\mu_{mk}r/R) J_{m-1}(\mu_{mk}r'/R) e^{-im(\varphi-\varphi')} \right.$$
$$+ \left[\frac{\mu_{mk}^2}{R^2} J_{m-1}(\mu_{mk}r/R) J_{m-1}(\mu_{mk}r'/R) \right.$$
$$- \frac{2m\mu_{mk}}{Rr} J_m(\mu_{mk}r/R) J_{m-1}(\mu_{mk}r'/R)$$
$$- \frac{2m\mu_{mk}}{Rr'} J_{m-1}(\mu_{mk}r/R) J_m(\mu_{mk}r'/R)$$
$$+ \left. \left. \frac{4m^2}{rr'} J_m(\mu_{mk}r/R) J_m(\mu_{mk}r'/R) \right] e^{im(\varphi-\varphi')} \right\} . \quad (5.109)$$

For $m > 0$ this expression takes the form:

$$\mathcal{DD}' J_m(\mu_{mk}r/R) J_m(\mu_{mk}r'/R) \cos[m(\varphi-\varphi')]$$
$$= \frac{\mu_{mk}^2}{2R^2} e^{i(\varphi-\varphi')} \left\{ J_{m+1}(\mu_{mk}r/R) J_{m+1}(\mu_{mk}r'/R) e^{im(\varphi-\varphi')} \right.$$
$$+ \left. J_{m-1}(\mu_{mk}r/R) J_{m-1}(\mu_{mk}r'/R) e^{-im(\varphi-\varphi')} \right\} , \quad (5.110)$$

and for $m = 0$ we have:

$$\mathcal{DD}' J_0(\mu_{0k}r/R) J_0(\mu_{0k}r'/R)$$
$$= \frac{\mu_{0k}^2}{R^2} J_1(\mu_{0k}r/R) J_1(\mu_{0k}r'/R) e^{i(\varphi-\varphi')} . \quad (5.111)$$

Now we can write down an explicit expression for the effective potential of the interaction, U_i:

$$U_i = -\frac{i}{2c} e\theta_s^2 \int_0^z dz' \int_0^{2\pi} d\varphi' \int_0^{r_0} dr' r' \tilde{j}_1(z', r', \varphi')$$

$$\times \left\{ \sum_{m=-\infty}^{\infty} \sum_{k=1}^{\infty} \frac{\mu_{mk}^2}{R^2(\mu_{mk}^2 - m^2) J_m^2(\mu_{mk})} J_{m-1}(\mu_{mk} r/R) \right.$$

$$\times J_{m-1}(\mu_{mk} r'/R) \exp\left[-i(m-1)(\varphi - \varphi')\right] \exp\left[-\frac{i\mu_{mk}^2(z-z')c}{2\omega R^2}\right]$$

$$+ \sum_{m=-\infty}^{\infty} \sum_{k=1}^{\infty} \frac{1}{R^2 J_{m-1}^2(\nu_{mk})} J_{m-1}(\nu_{mk} r/R) J_{m-1}(\nu_{mk} r'/R)$$

$$\left. \times \exp\left[-i(m-1)(\varphi - \varphi')\right] \exp\left[-\frac{i\nu_{mk}^2(z-z')c}{2\omega R^2}\right] \right\} . \quad (5.112)$$

Let us show that in the limit of an open electron beam this expression transforms to the corresponding expression obtained in the previous chapter. The Bessel functions $J_m(\mu_{mk})$ and $J_{m-1}(\nu_{mk})$ can be expanded asymptotically for $k \gg 1$:

$$J_m(\zeta) = \frac{\sqrt{2}}{\sqrt{\pi\zeta}} \cos\left(\zeta - \frac{\pi m}{2} - \frac{\pi}{4}\right) .$$

Also, we obtain that

$$\mu_{mk} \simeq (k + m/2 - 3/4)\pi , \quad \nu_{mk} \simeq (k + m/2 - 1/4)\pi ,$$

$$|\cos(\mu_{mk} - m\pi/2 - \pi/4)| \simeq 1 , |\cos(\nu_{mk} - (m-1)\pi/2 - \pi/4)| \simeq 1 ,$$

for $k \gg 1$. Using these relations, we find

$$J_m^2(\mu_{mk}) \simeq \frac{2}{\pi\mu_{mk}} , \quad J_{m-1}^2(\nu_{mk}) \simeq \frac{2}{\pi\nu_{mk}} .$$

Then we replace the summation over discrete values of μ_{mk} and ν_{mk} by integration:

$$U_i = -\frac{i}{2c} e\theta_s^2 \int_0^z dz' \int_0^{2\pi} d\varphi' \int_0^{r_0} dr' r' \tilde{j}_1(z', r', \varphi')$$

$$\times \left\{ \sum_{m=-\infty}^{\infty} \int_0^{\infty} d\zeta \, \zeta J_{m-1}(\zeta r) J_{m-1}(\zeta r') \right.$$

$$\left. \times \exp\left[-i(m-1)(\varphi - \varphi')\right] \exp\left[-\frac{i\zeta^2(z-z')c}{2\omega}\right] \right\} . \quad (5.113)$$

After integration we obtain:

324   5. Waveguide FELs

$$U_i = -\frac{e\theta_s^2 \omega}{2c^2} \int_0^z dz' \int_0^{2\pi} d\varphi' \int_0^{r_0} dr' r' \frac{\tilde{j}_1(z', r', \varphi')}{z - z'}$$

$$\times \left\{ \sum_{m=-\infty}^{\infty} i^{-(m-1)} \exp\left[-i(m-1)(\varphi - \varphi')\right] J_{m-1}\left[\frac{\omega r r'}{c(z - z')}\right] \right\}$$

$$\times \exp\left\{\frac{i\omega[r^2 + (r')^2]}{2(z - z')c}\right\} . \tag{5.114}$$

Taking into account that

$$J_0(\zeta) + 2\sum_{m=1}^{\infty} i^{-m} \cos(m\xi) J_m(\zeta) = \exp[-i\zeta \cos(\xi)] , \tag{5.115}$$

and introducing the variables $\xi = \varphi - \varphi'$ and $\zeta = \omega r r'[c(z-z')]^{-1}$, we transform (5.114) to

$$U_i = -\frac{e\theta_s^2 \omega}{2c^2} \int_0^z dz' \int d\mathbf{r}'_\perp \frac{\tilde{j}_1(z', \mathbf{r}'_\perp)}{z - z'} \exp\left[\frac{i\omega|\mathbf{r}_\perp - \mathbf{r}'_\perp|^2}{2c(z - z')}\right] . \tag{5.116}$$

The expression (5.116) is identical to that obtained in Chap. 4 by means of the solution of the inhomogeneous paraxial wave equation (see (4.22) and (5.2)).

To simplify the following study, we consider the specific case of azimuthal symmetry of the beam modulation, i.e. the value of $\tilde{j}_1$ does not depend on the angle φ. Under this assumption the complex amplitude of the effective potential (5.112) takes the form:

$$U_i = -i\frac{e}{c}\pi\theta_s^2 \int_0^z dz' \int_0^{r_0} dr' \, r' \tilde{j}_1(z', r') \left\{ \sum_{n=1}^{\infty} \frac{\mu_{1n}^2}{R^2(\mu_{1n}^2 - 1) J_1^2(\mu_{1n})} \right.$$

$$\times J_0(\mu_{1n} r/R) J_0(\mu_{1n} r'/R) \exp\left[-\frac{i\mu_{1n}^2 c(z - z')}{2\omega R^2}\right]$$

$$+ \sum_{n=1}^{\infty} \frac{1}{R^2 J_0^2(\nu_{1n})} J_0(\nu_{1n} r/R) J_0(\nu_{1n} r'/R)$$

$$\left. \times \exp\left[-\frac{i\nu_{1n}^2 c(z - z')}{2\omega R^2}\right] \right\} . \tag{5.117}$$

In the limit of a negligibly small energy spread in the electron beam, we obtain the following equation for $\tilde{j}_1(z, r)$:

$$\frac{d^2 \tilde{j}_1}{dz^2} + 2iC\frac{d\tilde{j}_1}{dz} + \left(\Lambda_p^2 - C^2\right)\tilde{j}_1 = -\frac{\omega j_0}{c\gamma_z^2 \mathcal{E}_0}\left(U_i + U_{\text{ext}}\right) , \tag{5.118}$$

where $j_0 = I/(\pi r_0^2)$ and U_i is given by (5.117).

The integro-differential equation obtained for the first harmonic of the beam current density is equivalent to the integro-differential equation for the field. In particular, this can be illustrated with the derivation of the eigenvalue equation. In the high-gain limit, we seek the solution of (5.118) in the form:

$$\tilde{j}_1(z, r) = a(r)\exp(\Lambda z) \ .$$

Thus, the equation for $a(r)$ is

$$\left[\left(\hat{\Lambda}+\mathrm{i}\hat{C}\right)^2 + \hat{\Lambda}_\mathrm{p}^2\right]a(\hat{r}) = \frac{\mathrm{i}}{\Omega}\int_0^\infty \mathrm{d}\zeta \exp\left(-\hat{\Lambda}\zeta\right)\int_0^1 \mathrm{d}\hat{r}'\ \hat{r}'$$

$$\times a(\hat{r}')\Bigg\{\sum_{n=1}^\infty \frac{\mu_{1n}^2}{(\mu_{1n}^2-1)J_1^2(\mu_{1n})}$$

$$\times J_0(\mu_{1n}\chi\hat{r})J_0(\mu_{1n}\chi\hat{r}')\exp\left[-\frac{\mathrm{i}\mu_{1n}^2\zeta}{2\Omega}\right]$$

$$+\sum_{n=1}^\infty \frac{1}{J_0^2(\nu_{1n})}J_0(\nu_{1n}\chi\hat{r})J_0(\nu_{1n}\chi\hat{r}')$$

$$\times \exp\left[-\frac{\mathrm{i}\nu_{1n}^2\zeta}{2\Omega}\right]\Bigg\}\ , \qquad (5.119)$$

where $\chi = \sqrt{B/\Omega}$. The sum in the right-hand side of this equation is calculated as follows. It can be written as a contour integral in the complex plane λ, assuming that each term of the sum is the residue at the point

$$\lambda = \lambda_k = -\frac{\mathrm{i}\mu_{1k}^2}{2\Omega}\ , \qquad \lambda = \lambda_k = -\frac{\mathrm{i}\nu_{1k}^2}{2\Omega}\ . \qquad (5.120)$$

The integration is performed along the contour consisting of a line parallel to the imaginary axis and lying in the right-hand half of the complex λ plane. The contour is closed by a semicircle of infinite radius in the left-hand side of the complex plane. The integration along this semicircle does not contribute to the integral, since we assume that the conditions of Jordan's lemma are fulfilled for the integrand. Thus, (5.119) can be rewritten as

$$\left[\left(\hat{\Lambda}+\mathrm{i}\hat{C}\right)^2 + \hat{\Lambda}_\mathrm{p}^2\right]a(\hat{r}) = \frac{1}{2\pi\mathrm{i}}\int_0^1 \mathrm{d}\hat{r}'\ \hat{r}'\ a(\hat{r}')\int_0^\infty \mathrm{d}\zeta \exp(-\hat{\Lambda}\zeta)$$

$$\times \int_{-\mathrm{i}\infty+\gamma'}^{\mathrm{i}\infty+\gamma'} \mathrm{d}\lambda\ \exp(\lambda\zeta)f(\lambda,\ \hat{r},\ \hat{r}')\ , \qquad (5.121)$$

where $f(\lambda,\ \hat{r},\ \hat{r}')$ is a superposition of the Bessel function of the first and the second kind. Application of the inverse Laplace transformation leads to

$$\left[\left(\hat{\Lambda}+i\hat{C}\right)^2 + \hat{\Lambda}_p^2\right] a(\hat{r}) = \int_0^1 d\hat{r}' \, \hat{r}' \, a(\hat{r}') f(\hat{\Lambda}, \hat{r}, \hat{r}') \,.$$

Thus, we need to find function $f(\lambda, \hat{r}, \hat{r}')$, satisfying the conditions of Jordan's lemma and analytic in the whole complex λ plane except for the poles (5.120). These requirements uniquely define the function $f(\lambda, \hat{r}, \hat{r}')$. It is shown in Appendix A.5 that

$$\int_0^\infty d\zeta \exp(-\hat{\Lambda}\zeta) \left\{ \sum_{n=1}^\infty \frac{\mu_{1n}^2}{(\mu_{1n}^2 - 1)J_1^2(\mu_{1n})} \right.$$
$$\left. \times J_0(\mu_{1n}r/R)J_0(\mu_{1n}r'/R) \exp\left[-\frac{i\mu_{1n}^2\zeta}{2\Omega}\right] \right\}$$
$$= i\Omega I_0(g\hat{r})I_0(g\hat{r}') \frac{K_1'\left(g\sqrt{\Omega/B}\right)}{I_1'\left(g\sqrt{\Omega/B}\right)}$$
$$+ i\Omega \begin{pmatrix} K_0(g\hat{r})I_0(g\hat{r}') & \text{for} & \hat{r} > \hat{r}' \\ I_0(g\hat{r})K_0(g\hat{r}') & \text{for} & \hat{r}' > \hat{r} \end{pmatrix}, \quad (5.122)$$

$$\int_0^\infty d\zeta \, \exp(-\hat{\Lambda}\zeta) \left\{ \sum_{n=1}^\infty \frac{1}{J_0^2(\nu_{1n})} J_0(\nu_{1n}r/R)J_0(\nu_{1n}r'/R) \exp\left[-\frac{i\nu_{1n}^2\zeta}{2\Omega}\right] \right\}$$
$$= i\Omega I_0(g\hat{r})I_0(g\hat{r}') \frac{K_1\left(g\sqrt{\Omega/B}\right)}{I_1\left(g\sqrt{\Omega/B}\right)}$$
$$+ i\Omega \begin{pmatrix} K_0(g\hat{r})I_0(g\hat{r}') & \text{for} & \hat{r} > \hat{r}' \\ I_0(g\hat{r})K_0(g\hat{r}') & \text{for} & \hat{r}' > \hat{r} \end{pmatrix}, \quad (5.123)$$

where $g^2 = -2iB\hat{\Lambda}$ and $B/\Omega = r_0^2/R^2$. Using these relations, we rewrite (5.119) in the form:

$$\left[\left(\hat{\Lambda}+i\hat{C}\right)^2 + \hat{\Lambda}_p^2\right] a(\hat{r}) = -I_0(g\hat{r}) \left[\frac{K_1(\Theta)}{I_1(\Theta)} + \frac{K_1'(\Theta)}{I_1'(\Theta)}\right]$$
$$\times \int_0^1 d\zeta \, \zeta a(\zeta) I_0(g\zeta)$$
$$+ 2K_0(g\hat{r}) \int_0^{\hat{r}} d\zeta \, \zeta a(\zeta) I_0(g\zeta)$$
$$+ 2I_0(g\hat{r}) \int_{\hat{r}}^1 d\zeta \, \zeta a(\zeta) K_0(g\zeta) \,. \quad (5.124)$$

Using this equation, we calculate the first and second derivatives of $a(\hat{r})$, and find that $a(\hat{r})$ satisfies the Bessel equation:

$$\hat{r}^2 \frac{d^2 a(\hat{r})}{d\hat{r}^2} + \hat{r} \frac{da(\hat{r})}{d\hat{r}} + \mu^2 \hat{r}^2 a(\hat{r}) = 0 ,$$

where

$$\mu^2 = 2 \left[\left(\hat{\Lambda} + i\hat{C} \right)^2 + \hat{\Lambda}_p^2 \right]^{-1} - g^2 .$$

To avoid the singularity of the solution at $\hat{r} = 0$, the solution for $a(\hat{r})$ should be chosen in the form:

$$a(\hat{r}) = a_0 J_0(\mu \hat{r}) .$$

This solution should be substituted into the integral equation (5.124). Taking into account that

$$\int_0^\zeta \zeta' K_0(g\zeta') J_0(\mu \zeta') d\zeta' = \frac{1}{\mu^2 + g^2} \left[1 + \mu\zeta J_1(\mu\zeta) K_0(g\zeta) \right.$$

$$\left. - g\zeta J_0(\mu\zeta) K_1(g\zeta) \right] , \qquad (5.125)$$

$$\int_0^\zeta \zeta' I_0(g\zeta') J_0(\mu \zeta') d\zeta' = \frac{1}{\mu^2 + g^2} \left[1 + \mu\zeta J_1(\mu\zeta) I_0(g\zeta) \right.$$

$$\left. + g\zeta J_0(\mu\zeta) I_1(g\zeta) \right] , \qquad (5.126)$$

and using the relation

$$I_0(\zeta) K_1(\zeta) + K_0(\zeta) I_1(\zeta) = \frac{1}{\zeta} ,$$

we obtain the eigenvalue equation

$$[\mu J_1(\mu) K_0(g) - g J_0(\mu) K_1(g)] + \frac{1}{2} \left[\frac{K_1(\Theta)}{I_1(\Theta)} + \frac{K_1'(\Theta)}{I_1'(\Theta)} \right]$$

$$\times [\mu J_1(\mu) I_0(g) + g J_0(\mu) I_1(g)] = 0 . \qquad (5.127)$$

Analysis of this equation shows that it describes the particular case of an azimuthal index $m = 1$ of the eigenvalue equation (5.48), obtained from the differential equations (5.43) with boundary conditions (5.41). Thus, we can state that the solution of both the differential equation for the field amplitude and the integral equation for the beam density modulation, leads to the same result. In both cases the equations were solved using the paraxial approximation, but at different stages of the solution. In the case of the field equations, we used the paraxial wave equation and the approximate boundary conditions (5.41). In the case of the equation for the beam current density modulation, we obtained an exact expression for the Green's function and only then simplified it using the paraxial approximation.

5.4.3 Numerical Solution

In this section we present an algorithm for the numerical solution of the initial-value problem. The initial conditions correspond to the case when an unmodulated axisymmetric electron beam and the electromagnetic wave are fed to the undulator entrance. When performing numerical simulations, it is more natural to use the equations for the first harmonic of the beam current density. In the case of negligibly small energy spread the evolution of the first harmonic of the beam current density, $\tilde{j}_1$, is described by (5.118). The value of the effective potential connected with the radiated wave, U_i, should be calculated with (5.112). To simplify the formulae, we consider only the azimuthally symmetric case when the first harmonic of the beam current density, $\tilde{j}_1$, does not depend on the angle φ. To provide azimuthal symmetry of the electron beam modulation in the amplification process, the external wave should be a rotating mode of the empty waveguide (TE_{1k} or TM_{1k}). For instance, the TE_{1k} rotating mode is a superposition of the modes

$$\psi_{1k}^{\text{TE}} \propto J_1(\mu_{1k} r/R) \begin{pmatrix} \sin(\varphi) \\ \cos(\varphi) \end{pmatrix}$$

with equal amplitudes, but having a relative phase shift equal to $\pi/2$. One of the modes should oscillate in time as $\cos(\omega t)$, and the other as $\sin(\omega t)$. The expression for the effective potential connected with the input wave, U_{ext}, takes the form in this case:

$$U_{\text{ext}} = -\frac{e\theta_s}{2\text{i}}(\tilde{E}_x + \text{i}\tilde{E}_y) = -\frac{e\theta_s}{2\text{i}}(\tilde{E}_r + \text{i}\tilde{E}_\varphi)e^{\text{i}\varphi}, \quad (5.128)$$

where $\tilde{E}_r$, $\tilde{E}_\varphi$ are given by (5.54) and (5.55):

$$\tilde{E}_r = \sum_{k=1}^{\infty} \frac{\sqrt{2} C_{1k}^{\text{TE}}}{\sqrt{\pi(\mu_{1k}^2 - 1)} J_1(\mu_{1k})} \frac{1}{r} \frac{\partial}{\partial \varphi} [J_1(\mu_{1k} r/R) \exp(-\text{i}\varphi)]$$

$$+ \sum_{j=1}^{\infty} \frac{\sqrt{2} C_{1j}^{\text{TM}}}{\sqrt{\pi} \nu_{1j} J_1(\nu_{1j})} \frac{\partial}{\partial r} [J_1(\nu_{1j} r/R) \exp(-\text{i}\varphi)], \quad (5.129)$$

$$\tilde{E}_\varphi = -\sum_{k=1}^{\infty} \frac{\sqrt{2} C_{1k}^{\text{TE}}}{\sqrt{\pi(\mu_{1k}^2 - 1)} J_1(\mu_{1k})} \frac{\partial}{\partial r} [J_1(\mu_{1k} r/R) \exp(-\text{i}\varphi)]$$

$$+ \sum_{j=1}^{\infty} \frac{\sqrt{2} C_{1j}^{\text{TM}}}{\sqrt{\pi} \nu_{1j} J_1(\nu_{1j})} \frac{1}{r} \frac{\partial}{\partial \varphi} [J_1(\nu_{1j} r/R) \exp(-\text{i}\varphi)]. \quad (5.130)$$

The amplitude coefficients C_{1k}^{TE}, C_{1j}^{TM} can be expressed in terms of the input signal power. Let the input wave be a rotating TE_{1k} mode having power W_{ext}. The coefficient C_{1k}^{TE} is equal to $C_{1k}^{\text{TE}} = \sqrt{\pi W_{\text{ext}}/c}$ in this case and the effective potential connected with the external wave is

$$U_{\text{ext}}^{\text{TE}} = \frac{e\theta_s \mu_{1k}}{2R} \frac{\sqrt{2 W_{\text{ext}}/c}}{\sqrt{(\mu_{1k}^2 - 1)} J_1(\mu_{1k})} J_0(\mu_{1k} r/R) \exp\left(-\frac{\text{i} \mu_{1k}^2 z c}{2\omega R^2}\right).$$

It is convenient to rewrite in dimensionless form the equation for the first harmonic of the beam current density using the reduced parameters defined above: the detuning parameter $\hat{C}$, the space charge parameter $\hat{\Lambda}_p^2$, and the diffraction parameters B and Ω. We introduce the normalized complex amplitude $\hat{a}_1 = \pi r_0^2 \tilde{j}_1/I_0$ and rewrite (5.118) as

$$\left[\frac{d^2}{d\hat{z}^2} + 2i\hat{C}\frac{d}{d\hat{z}} + (\hat{\Lambda}_p^2 - \hat{C}^2)\right]\hat{a}_1$$

$$= -\frac{\hat{C}_{1k}^{TE}}{2}J_0(\mu_{1k}\chi\hat{r})\exp\left(-\frac{i\mu_{1k}^2 \hat{z}}{2\Omega}\right)$$

$$+\frac{i}{\Omega}\int_0^{\hat{z}} d\hat{z}' \int_0^1 d\hat{r}'\hat{r}'\hat{a}_1(\hat{z}', \hat{r}')K(\hat{z} - \hat{z}', \hat{r}, \hat{r}'), \qquad (5.131)$$

where the transverse coordinate is normalized to the beam radius, $\hat{r} = r/r_0$, $\chi = r_0/R = \sqrt{B/\Omega}$,

$$\hat{C}_{1k}^{TE} = \frac{\sqrt{2\hat{W}_{ext}}}{\sqrt{\Omega(\mu_{1k}^2 - 1)}J_1(\mu_{1k})}\mu_{1k},$$

$$\hat{W}_{ext} = W_{ext}/W_0, \quad W_0 = I_0 \mathcal{E}_0 \rho/e.$$

The kernel of the integrand in the right-hand side of (5.131) is given by

$$K(\hat{z} - \hat{z}', \hat{r}, \hat{r}')$$

$$= \sum_{k=1}^{\infty} \frac{\mu_{1k}^2}{(\mu_{1k}^2 - 1)J_1^2(\mu_{1k})} J_0(\mu_{1k}\chi\hat{r})J_0(\mu_{1k}\chi\hat{r}') \exp\left[-\frac{i\mu_{1k}^2(\hat{z} - \hat{z}')}{2\Omega}\right]$$

$$+ \sum_{k=1}^{\infty} \frac{1}{J_0^2(\nu_{1k})} J_0(\nu_{1k}\chi\hat{r})J_0(\nu_{1k}\chi\hat{r}') \exp\left[-\frac{i\nu_{1k}^2(\hat{z} - \hat{z}')}{2\Omega}\right]. \qquad (5.132)$$

The power gain is calculated with the following formula:

$$G = 1 + \frac{2}{\Omega \hat{W}_{ext}} \int_0^{\hat{z}} d\hat{z}' \int_0^{\hat{z}'} d\hat{z}'' \int_0^1 d\hat{r}'\hat{r}'$$

$$\times \int_0^1 d\hat{r}''\hat{r}''\hat{a}_1(\hat{z}', \hat{r}')\hat{a}_1^*(\hat{z}'', \hat{r}'')K(\hat{z}' - \hat{z}'', \hat{r}', \hat{r}'')$$

$$+i\frac{\hat{C}_{1k}^{TE}}{\hat{W}_{ext}} \int_0^{\hat{z}} d\hat{z}' \int_0^1 d\hat{r}'\hat{r}'$$

$$\times \hat{a}_1^*(\hat{z}', \hat{r}')J_0(\mu_{1k}\chi\hat{r}')\exp\left[-\frac{i\mu_{1k}^2\hat{z}'}{2\Omega}\right] + \text{C.C.} \qquad (5.133)$$

The numerical algorithm for the solution of the integro-differential equation (5.131) can be organized as follows. The electron beam is divided into M layers assuming the beam modulation to be constant within each layer. Then M ordinary differential equations can be integrated by means of well-known techniques (e.g. by the Runge Kutta technique). The integral in the right-hand side of (5.131) should be recalculated numerically at each integration step. There exists the possibility of simplifying the calculation of this integral taking into account the specific dependence of the kernel on $\hat{z} - \hat{z}'$. In principle, its calculation can be reduced to the integration of a system of ordinary differential equations. Numerical integration over the transverse coordinate is performed only at the final step of integration in this case. The reader can easily check this by taking the derivative of this integral with respect to $\hat{z}$. The recommended technique allows one to increase significantly the accuracy and the speed of calculations.

It is our experience that the accuracy of calculations can be controlled by an appropriate choice of the simulation parameters (integration step, number of radial divisions, number of modes for calculations of the radiation fields, etc.) in order to achieve the desired accuracy. Also, the extension of the algorithm to an arbitrary gradient profile of the electron beam and arbitrary initial conditions is straightforward.

In the previous section we obtained an analytical solution of the initial-value problem valid in the high-gain linear regime. It is interesting to check the accuracy of this asymptotic solution with the results of numerical simulations. In Figs. 5.24 and 5.25 we present the results of the calculation of the power gain. Partial contributions of the passive waveguide modes are plotted, too. The curves are calculated with the analytical formulae (5.96)–(5.104). The circles are the results of numerical simulation code. It is seen that at a power gain of about 40 dB the asymptotic formulae for the gain provide good accuracy. Nevertheless, one can see that there is small systematic excess of the simulation results over analytical results. The reason is that the numerical simulation code calculates total gain, while the analytical formulae describe only the contribution to the total power of one beam radiation mode having maximal gain.

5.5 Nonlinear Mode of Operation

In this section we present a brief description of a simple algorithm for numerical simulations of the FEL amplifier with an axisymmetric electron beam, circular waveguide, and helical undulator. Basic approximations of the physical model are the same as those used in Chap. 4. We use the three-dimensional treatment of the radiation field, while the motion of particles is considered to be one-dimensional. The radiation fields are calculated using a Green's function written down in the paraxial approximation. Space charge fields and the energy spread of the electrons in the beam can also be taken into account. We

5.5 Nonlinear Mode of Operation

restrict the considerations to the specific case of the initial conditions when an unmodulated electron beam and rotating TE$_{1k}$ (or TM$_{1k}$) mode are fed to the undulator entrance. Under these initial conditions the azimuthal symmetry of the electron beam modulation exists in the amplification process.

The electron motion is described in the energy-phase variables with the phase $\psi = \int k_w(z)dz - \omega(t - z/c)$ as canonical coordinate and energy $\mathcal{E}$ as canonical momentum. At small deviations of the electron energy from the nominal value $\mathcal{E}_0$, the canonical equations of motion are:

$$\frac{dP}{dz} = [iU\exp(i\psi) + \text{C.C.}] - eE_z,$$

$$\frac{d\psi}{dz} = \frac{\omega}{c\gamma_z^2 \mathcal{E}_0} P + C, \qquad (5.134)$$

where $P = \mathcal{E} - \mathcal{E}_0$ is the deviation of the electron energy from the nominal value and C is the detuning. The complex amplitude of the effective potential of interaction, $U = U_i + U_{\text{ext}}$, is given by (5.23) and (5.27). The space charge field is calculated using the approximation of a large transverse size of the electron beam, $r_b^2/\gamma_z^2 \gg c^2/\omega^2$. In this case the longitudinal component of the electric field is given by the expression:

$$E_z = 4\pi\omega^{-1} \sum_{n=1}^{\infty} n^{-1} \left[\tilde{j}_n(z, \boldsymbol{r}_\perp) e^{in\psi} + \text{C.C.}\right],$$

where $\tilde{j}_n$ is the complex amplitude of the nth harmonic of the longitudinal component of the beam current density. Also, under the condition of $r_b^2/\gamma_z^2 \gg c^2/\omega^2$, we can neglect the reduction of the space charge field in the waveguide.

The dimensionless equations of the particle motion are as follows:

$$\frac{d\hat{P}}{d\hat{z}} = -2\,\text{Im}(e^{i\psi}\hat{U} + \hat{\Lambda}_p^2 \hat{U}_c), \qquad (5.135)$$

$$\frac{d\psi}{d\hat{z}} = \hat{P} + \hat{C}, \qquad (5.136)$$

where $\hat{z} = \Gamma z$, $\hat{P} = P/(\rho\mathcal{E}_0)$ is the reduced energy deviation, $\rho = c\gamma_z^2\Gamma/\omega$ is the efficiency parameter, $\hat{C} = C/\Gamma$ is the detuning parameter and $\hat{\Lambda}_p^2 = 4c^2/(\omega^2 r_0^2 \theta_s^2)$ is the space charge parameter. The effective potential of the interaction with the electromagnetic wave is given by the expression:

$$\hat{U} = \frac{1}{2}\hat{C}_{1k}^{\text{TE}} J_0(\mu_{1k}\chi\hat{r}) \exp\left[-\frac{i\mu_{1k}^2 \hat{z}}{2\Omega}\right]$$

$$-\frac{i}{\Omega}\int_0^{\hat{z}} d\hat{z}' \int_0^1 d\hat{r}'\hat{r}'\hat{a}_1(\hat{z}',\hat{r}')\,K(\hat{z}-\hat{z}',\hat{r},\hat{r}'), \qquad (5.137)$$

where the kernel $K(\hat{z} - \hat{z}', \hat{r}, \hat{r}')$ is given by (5.132), $\chi = \sqrt{B/\Omega}$, and $\hat{r} = r/r_0$. The factor $\hat{C}_{1k}^{\text{TE}}$, corresponding to the amplitude of the external wave, is defined in the same way as in (5.131).

The expression for the effective potential of the space charge field, $\hat{U}_c$, is of the form (4.163):

$$\hat{U}_c = \sum_{n=1}^{\infty} e^{in\psi} \frac{\hat{a}_n(\hat{z},\hat{r})}{n} . \tag{5.138}$$

The complex amplitudes ($n = 1, 2, ...$)

$$\hat{a}_n = |\hat{a}_n| \exp(i\psi_n) ,$$

entering (5.137) and (5.138), are calculated with the local macroparticle ensemble as follows:

$$|\hat{a}_n| = \frac{1}{N} \left[\left[\sum_{k=1}^{N} \cos(n\psi_{(k)}) \right]^2 + \left[\sum_{k=1}^{N} \sin(n\psi_{(k)}) \right]^2 \right]^{1/2} ,$$

$$\psi_n = -\arctg \left[\frac{\sum_{k=1}^{N} \sin(n\psi_{(k)})}{\sum_{k=1}^{N} \cos(n\psi_{(k)})} \right] . \tag{5.139}$$

It was shown in Sec. 2.2.5 that the Fourier series (5.138) of the space charge harmonics can be reduced to the sum of a trigonometric series, and the latter results in a simple algebraic function.

The power gain, G, is calculated using (5.133). An additional method for the calculation of the gain is based on the calculation of the mean energy loss by the electron beam:

$$G = 1 + \Delta W / W_{\text{ext}} = 1 - \langle \hat{P} \rangle / \hat{W}_{\text{ext}} ,$$

where $\langle \hat{P} \rangle$ is the mean reduced energy deviation and

$$\hat{W}_{\text{ext}} = W_{\text{ext}}/W_0 , \quad W_0 = \rho I_0 \mathcal{E}_0 / e .$$

The reduced efficiency is defined in the same way as in Chap. 4:

$$\hat{\eta} = \eta/\rho = -\langle \hat{P} \rangle .$$

The simulation is performed with the macroparticle method. The macroparticle ensemble is prepared as follows: the electron beam is divided into M layers over the radius and in each layer we distribute uniformly N macroparticles over the phase ψ from 0 to 2π. The initial energy spread is simulated with the additional distribution of the particles according to the Gaussian law (see Sect. 2.2.6 for more details). The system of $2 \times N \times M$ ordinary differential equations (5.135) and (5.136) is integrated with the Runge Kutta technique.

The saturation mechanism in the waveguide FEL is the same as that described in Chaps. 2 and 4. To give the reader an idea about the nonlinear mode of operation of the waveguide FEL, we present in Figs. 5.28–5.34 the results of numerical simulations. The region of physical parameters covers values of the beam diffraction parameter from $B = 0.1$ to $B = 1$, and values of the waveguide diffraction parameter from $\Omega \simeq 1$ to $\Omega \simeq 100$. The plots in

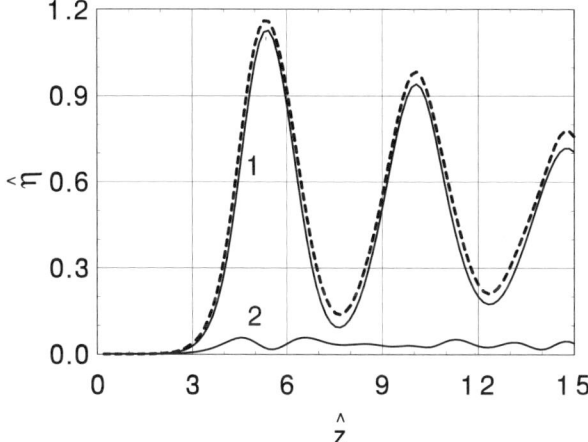

Fig. 5.28. The normalized total efficiency (dashed curve) and partial contributions of the TE$_{11}$ mode (curve 1) and TM$_{11}$ mode (curve 2) versus $\hat{z}$. Here $B = 0.1$, $\Omega = 2$, $\hat{C} = 0.5$, $\hat{\Lambda}_p^2 \to 0$, and $\hat{\Lambda}_T^2 = 0$. The normalized input power is $\hat{W}_{\text{ext}} = 10^{-3}$

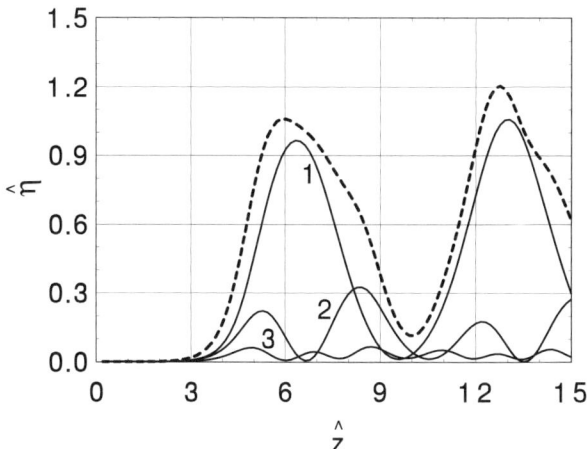

Fig. 5.29. The normalized total efficiency (dashed curve) and partial contributions of the TE$_{11}$ mode (curve 1), TM$_{11}$ mode (curve 2) and TE$_{12}$ mode (curve 3) versus $\hat{z}$. Here $B = 0.1$, $\Omega = 5$, $\hat{C} = 0.4$, $\hat{\Lambda}_p^2 \to 0$, and $\hat{\Lambda}_T^2 = 0$. The normalized input power is $\hat{W}_{\text{ext}} = 10^{-3}$

Figs. 5.28–5.33 present the dependencies on the undulator length of the FEL efficiency and of the partial contribution of the passive waveguide modes for different values of Ω and two values of B, 0.1, and 1.

First, we refer the reader to Chap. 2 dealing with the one-dimensional theory of the FEL amplifier. It was mentioned in the concluding remarks to Chap. 2 that we might expect the operation of the waveguide FEL to be

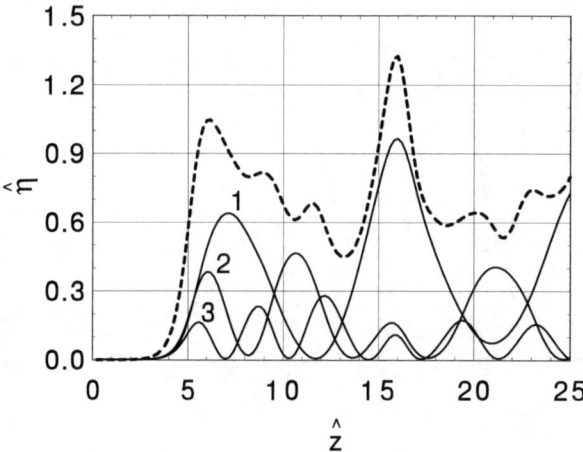

Fig. 5.30. The normalized total efficiency (dashed curve) and partial contributions of the TE_{11} mode (curve 1), TM_{11} mode (curve 2) and TE_{12} mode (curve 3) versus $\hat{z}$. Here $B = 0.1$, $\Omega = 10$, $\hat{C} = 0.4$, $\hat{\Lambda}_p^2 \to 0$, and $\hat{\Lambda}_T^2 = 0$. The normalized input power is $\hat{W}_{\text{ext}} = 10^{-3}$

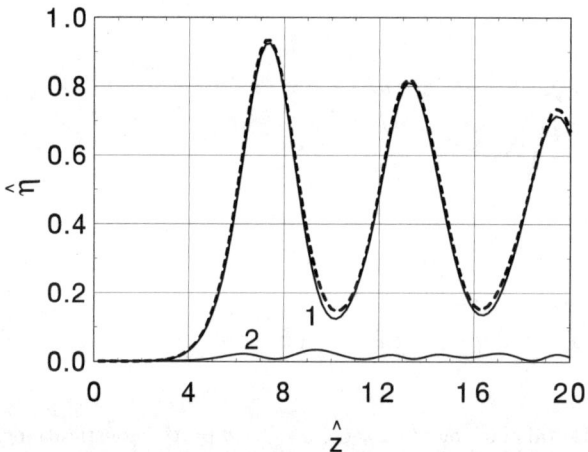

Fig. 5.31. The normalized total efficiency (dashed curve) and partial contributions of the TE_{11} mode (curve 1) and TM_{11} mode (curve 2) versus $\hat{z}$. Here $B = 1$, $\Omega = 3$, $\hat{C} = 0.5$, $\hat{\Lambda}_p^2 \to 0$, and $\hat{\Lambda}_T^2 = 0$. The normalized input power is $\hat{W}_{\text{ext}} = 10^{-3}$

similar to that predicted by the one-dimensional model. Qualitatively this situation should take place when the structure of the beam radiation mode is close to that of an empty waveguide. Now we are able to obtain a quantitative description. When considering the linear mode of operation (see Sect. 5.3), we find that for the value of the waveguide parameter $\Omega \simeq 1$, the beam radiation

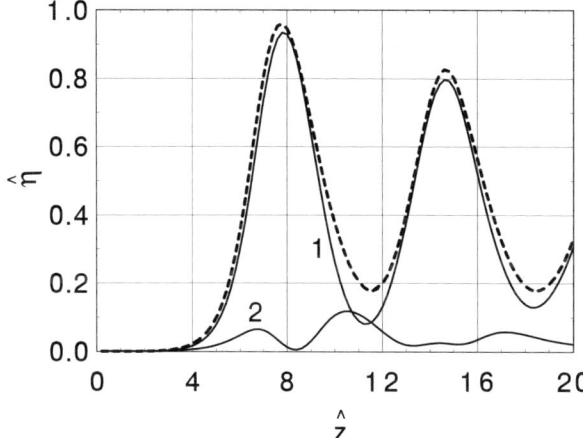

Fig. 5.32. The normalized total efficiency (dashed curve) and partial contributions of the TE_{11} mode (curve 1) and TM_{11} mode (curve 2) versus $\hat{z}$. Here $B = 1$, $\Omega = 5$, $\hat{C} = 0.3$, $\hat{\Lambda}_p^2 \to 0$, and $\hat{\Lambda}_T^2 = 0$. The normalized input power is $\hat{W}_{\text{ext}} = 10^{-3}$

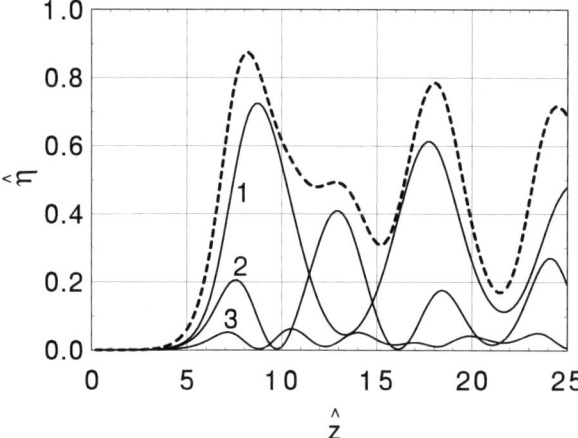

Fig. 5.33. The normalized total efficiency (dashed curve) and partial contributions of the TE_{11} mode (curve 1), TM_{11} mode (curve 2) and TE_{12} mode (curve 3) versus $\hat{z}$. Here $B = 1$, $\Omega = 10$, $\hat{C} = 0.25$, $\hat{\Lambda}_p^2 \to 0$, and $\hat{\Lambda}_T^2 = 0$. The normalized input power is $\hat{W}_{\text{ext}} = 10^{-3}$

mode is quite close to the mode of an empty waveguide. Also, the eigenvalue equation becomes similar to that of the one-dimensional theory. In particular, when the energy spread is negligibly small, the eigenvalue equation is cubic (5.63). Analysis of the latter equation and (2.32) gives the reader an idea of how the parameters of the problem should be used in the equations of the one-dimensional theory. Analysis of the plots presented in Figs. 5.28 and 5.31

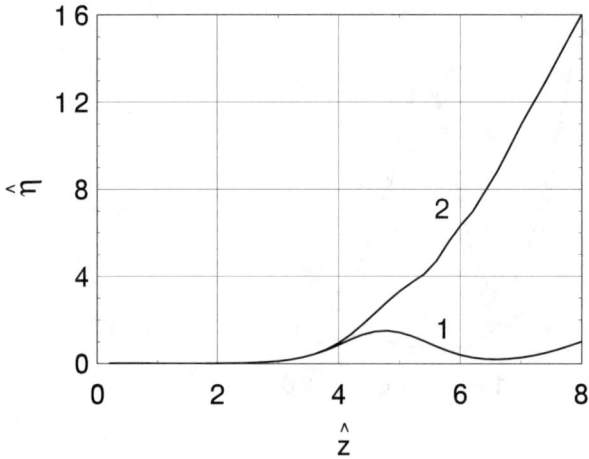

Fig. 5.34. The normalized total efficiency versus $\hat{z}$: curve 1 without tapering ($\hat{C} = 1.4$); curve 2 with tapering by the law $\hat{C} = 1.4 + 1.25(\hat{z} - \hat{z}_i) + 0.8(\hat{z} - \hat{z}_i)^2$, where $\hat{z}_i = 3.3$. Here $B = 0.1$, $\Omega = 1$, $\hat{\Lambda}_p^2 \to 0$, and $\hat{\Lambda}_T^2 = 0$. The normalized input power is $\hat{W}_{\text{ext}} = 10^{-3}$

shows that the nonlinear mode of operation of the waveguide FEL at values of the waveguide diffraction parameter $\Omega \simeq 1$ seems to be similar to that obtained in the framework of the one-dimensional theory (see Fig. 2.15). Also, we might expect that an optimal law of the undulator tapering for the waveguide FEL with $\Omega \simeq 1$ should be similar to the one-dimensional case: namely, the undulator parameters should be tapered quadratically with undulator length. This statement is confirmed by the plot presented in Fig. 5.34. The parameters for the numerical example are: $\Omega = 1$, $B = 0.1$, and $\hat{W}_{\text{ext}} = 10^{-3}$. The parameters of the undulator tapering have been optimized in order to achieve maximal output power at the undulator exit. The result of the optimization is that the law of undulator tapering should be quadratic:

$$\hat{C} = \begin{cases} 1.4 & \text{for } \hat{z} < 3.3 \\ 1.4 + 1.25(\hat{z} - 3.3) + 0.8(\hat{z} - 3.3)^2 & \text{for } \hat{z} > 3.3 \ . \end{cases} \quad (5.140)$$

The output power grows quadratically with undulator length, as in the one-dimensional theory. Summarizing the discussion on the waveguide FEL with the value of $\Omega \simeq 1$, we can state that it can be simulated in the framework of the one-dimensional model at an appropriate redetermination of the problem parameters. The lower the ratio B/Ω, the better the expected agreement, since the variation of the field eigenmode across the beam becomes smaller.

Analysis of the plots presented in Figs. 5.28–5.33 shows that the larger the value of Ω, the larger the number of passive waveguide modes that begin to contribute to the total radiation power. When the amplifier operates in

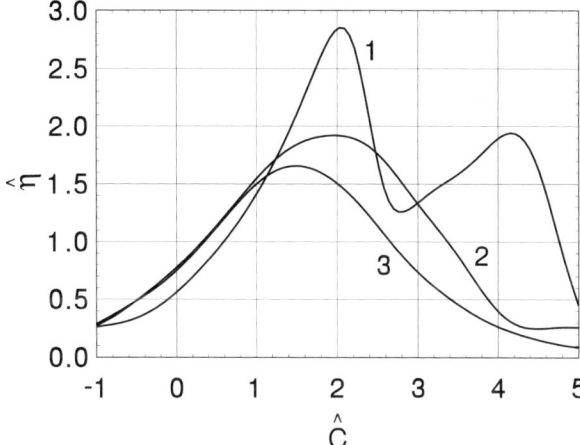

Fig. 5.35. The normalized efficiency versus detuning parameter. Here $B = 0.1$, $\hat{\Lambda}_\mathrm{p}^2 \to 0$, $\hat{\Lambda}_\mathrm{T}^2 = 0$. The normalized input power is $\hat{W}_\mathrm{ext} = 10^{-3}$ and the reduced undulator length is $\hat{z} = 6$. Curve 1: $\Omega = 2$. Curve 2: $\Omega = 5$. Curve 3: $\Omega = 10$

the linear regime, the relative contribution of each passive waveguide mode remains fixed, while the amplitudes of all the modes grow exponentially with undulator length with the gain equal for all the modes. The situation changes drastically in the nonlinear regime as is seen from the plots in Figs. 5.28–5.33. The partial contribution of one particular mode is not fixed, but oscillates with undulator length. For all these numerical examples the FEL amplifier is tuned to the maximum of the gain, and the TE$_{11}$ mode of the passive waveguide dominates in the linear regime (see, e.g. Figs. 5.1 and 5.3). The relative contribution of this mode is even increased at the saturation point. Nevertheless, in the deep nonlinear regime we find that other modes can dominate. An example is presented in Fig. 5.33. The TE$_{11}$ mode dominates at the saturation point, while the TM$_{11}$ mode dominates in the second maximum. When the parameters of the FEL amplifier fall into this intermediate region, its output characteristics can be simulated only within the framework of the waveguide FEL theory.

We should also like to mention another interesting feature which may occur in the waveguide FEL. In Fig. 5.35 we present the dependence of the output efficiency versus the detuning at fixed undulator length. For large values of Ω the curves show conventional behavior (see, e.g. Figs. 2.17 and 4.64 calculated in the framework of the one-dimensional theory and three-dimensional theory (for an open beam), respectively). But for values of $\Omega \simeq 1$, the dependence of the output power on the detuning seems to be complicated. This phenomenon can be explained if one looks at Fig. 5.3 showing the dependence of the maximal gain on the detuning. This dependence has a nonmonotonic and breaking behavior. The parts of these curves to the left and to the right

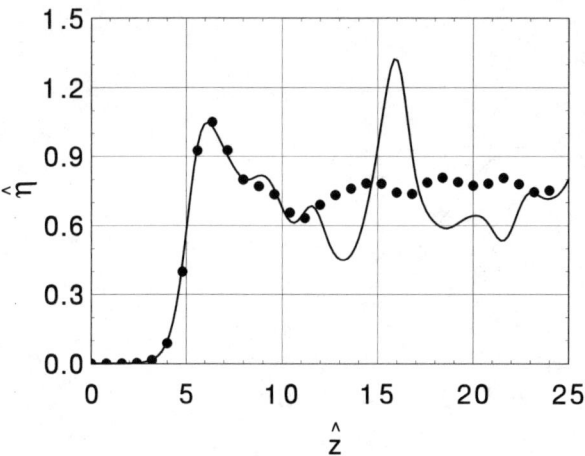

Fig. 5.36. The normalized total efficiency (curve) versus $\hat{z}$. Here $B = 0.1$, $\Omega = 10$, $\hat{C} = 0.4$, $\hat{A}_p^2 \to 0$, and $\hat{A}_T^2 = 0$. The normalized input power is $\hat{W}_{\text{ext}} = 10^{-3}$. The circles are the results of calculations with the nonlinear simulation code for an open beam

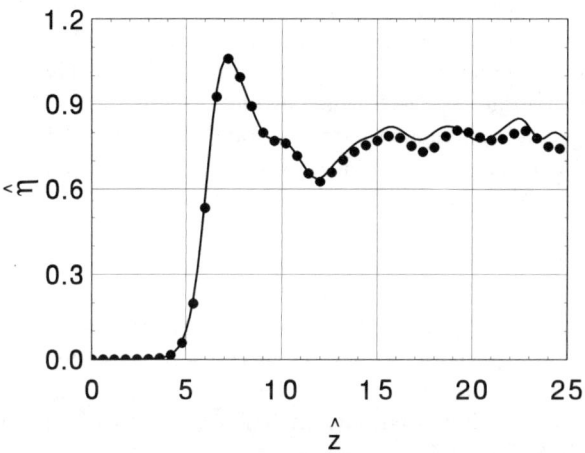

Fig. 5.37. The normalized total efficiency (curve) versus $\hat{z}$. Here $B = 0.1$, $\Omega = 100$, $\hat{C} = 0.4$, $\hat{A}_p^2 \to 0$, and $\hat{A}_T^2 = 0$. The normalized input power is $\hat{W}_{\text{ext}} = 10^{-3}$. The circles are the results of calculations with the nonlinear simulation code for an open beam

of the break near $\hat{C} \simeq 2$ correspond to different beam radiation modes. As a result, the two maxima in curve 1 also correspond to different beam radiation modes. The TE_{11} mode of the passive waveguide dominates in the first maximum, while the TM_{11} mode dominates in the second.

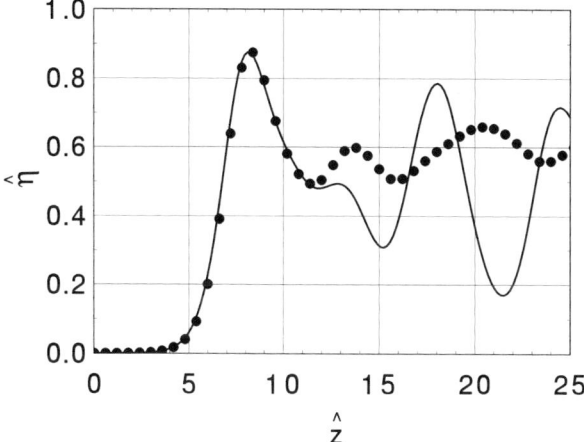

Fig. 5.38. The normalized total efficiency (curve) versus $\hat{z}$. Here $B = 1$, $\Omega = 10$, $\hat{C} = 0.25$, $\hat{\Lambda}_p^2 \to 0$, and $\hat{\Lambda}_T^2 = 0$. The normalized input power is $\hat{W}_{\text{ext}} = 10^{-3}$. The circles are the results of calculations with the nonlinear simulation code for an open beam

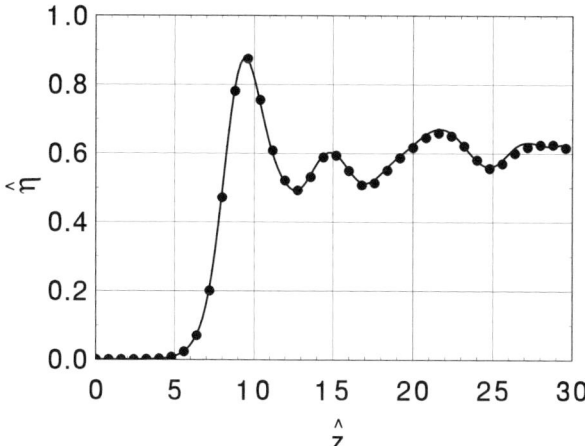

Fig. 5.39. The normalized total efficiency (curve) versus $\hat{z}$. Here $B = 1$, $\Omega = 100$, $\hat{C} = 0.25$, $\hat{\Lambda}_p^2 \to 0$, and $\hat{\Lambda}_T^2 = 0$. The normalized input power is $\hat{W}_{\text{ext}} = 10^{-3}$. The circles are the results of calculations with the nonlinear simulation code for an open beam

Another interesting problem is to define the region of applicability of the results of the nonlinear theory obtained in the framework of the model of the open electron beam (see Chap. 4). Analysis of the high-gain linear mode of operation shows that when $\Omega \simeq 10$ and the detuning is not too large, there is almost perfect agreement of the gain calculated with and without taking into account the waveguide walls (see Figs. 5.7 and 5.14). Also, the field distributions match perfectly (see Figs. 5.7 and 5.14). Using the plots presented in Figs. 5.36–5.39 we can give a quantitative answer to the question about the region of applicability of the model of the open electron beam. If one is interested only in calculating the characteristics at saturation, the open beam asymptote gives reliable results for $\Omega \gtrsim 10$. On the other hand, agreement for the deep nonlinear regime occurs at significantly larger values of $\Omega \simeq 100$. This is explained in a simple way. When the FEL amplifier operates in the nonlinear regime, the field expands significantly beyond the electron beam due to diffraction effects. The larger the length of the nonlinear stage, the wider the radiation expansion is. As a result, the transverse waveguide size should be large in order to avoid the influence of the waveguide walls on the FEL process.

5.6 Rectangular Waveguide

In this section we present some results for a rectangular waveguide. In the case of a circular waveguide and an axisymmetric electron beam we obtained analytical results for the high-gain linear regime. This was mainly connected with the possibility of separating the variables in the self-consistent equations. This is not the case for the rectangular waveguide, where only numerical methods are applicable. Here we describe an approach which can be used for constructing a numerical simulation algorithm for the FEL amplifier with a rectangular waveguide.

We consider a rectangular waveguide with the walls parallel to the (x, y) axes of the Cartesian coordinate system. The waveguide function, $\psi(\boldsymbol{r}_\perp)$, is the solution of the Helmholtz equation:

$$\frac{\partial^2 \psi}{\partial x^2} + \frac{\partial^2 \psi}{\partial y^2} + k_\perp^2 \psi = 0 , \tag{5.141}$$

satisfying the boundary conditions,

$$\frac{\partial \psi^{\text{TE}}}{\partial x} = 0 , \quad \psi^{\text{TM}} = 0 \quad \text{at} \quad x = 0, l_x ,$$

$$\frac{\partial \psi^{\text{TE}}}{\partial y} = 0 , \quad \psi^{\text{TM}} = 0 \quad \text{at} \quad y = 0, l_y .$$

We impose the following normalization condition for the waveguide functions:

$$\int_0^{l_x} dx \int_0^{l_y} dy \left[\left(\frac{\partial \psi}{\partial x}\right)^2 + \left(\frac{\partial \psi}{\partial y}\right)^2 \right] = 1 .$$

The solutions of (5.141), satisfying the above conditions, are

5.6 Rectangular Waveguide

$$\psi^{\text{TE}} = \frac{2}{(k_\perp)_{mn}\sqrt{l_x l_y}} \cos(m\pi x/l_x)\cos(n\pi y/l_y) \quad \text{for} \quad m, n > 0 ,$$

$$\psi^{\text{TE}} = \frac{\sqrt{2}}{(k_\perp)_{mn}\sqrt{l_x l_y}} \cos(m\pi x/l_x) \quad \text{for} \quad m > 0, n = 0 ,$$

$$\psi^{\text{TE}} = \frac{\sqrt{2}}{(k_\perp)_{mn}\sqrt{l_x l_y}} \cos(n\pi y/l_y) \quad \text{for} \quad m = 0, n > 0 ,$$

$$\psi^{\text{TM}} = \frac{2}{(k_\perp)_{mn}\sqrt{l_x l_y}} \sin(m\pi x/l_x)\sin(n\pi y/l_y) \quad \text{for} \quad m, n > 0 .$$

The eigenvalues (for both TE and TM modes) are:

$$(k_\perp)_{mn} = \sqrt{(\pi m/l_x)^2 + (\pi n/l_y)^2} .$$

Let us consider an FEL amplifier with helical undulator and rectangular waveguide. Substituting the expressions for the waveguide functions into (5.23), we find an expression for the effective potential of the interaction of the particle with the radiated wave:

$$\begin{aligned}
U_i = &-\mathrm{i}\frac{2\pi e\theta_s^2}{cl_x l_y} \int_0^z \mathrm{d}z' \int_0^{l_x} \mathrm{d}x' \int_0^{l_y} \mathrm{d}y' \, \tilde{j}_1(z', x', y') \\
&\times \mathcal{DD}' \Bigg\{ \sum_{m=1}^\infty \sum_{n=1}^\infty \frac{1}{(k_\perp)_{mn}^2} \exp\left[-\mathrm{i}\frac{(k_\perp)_{mn}^2(z-z')c}{2\omega}\right] \\
&\times [\cos(m\pi x/l_x)\cos(m\pi x'/l_x)\cos(n\pi y/l_y)\cos(n\pi y'/l_y) \\
&+ \sin(m\pi x/l_x)\sin(m\pi x'/l_x)\sin(n\pi y/l_y)\sin(n\pi y'/l_y)] \\
&+ \sum_{m=1}^\infty \frac{1}{2(k_\perp)_{m0}^2} \exp\left[-\mathrm{i}\frac{(k_\perp^2)_{m0}(z-z')c}{2\omega}\right] \\
&\times \cos(m\pi x/l_x)\cos(m\pi x'/l_x) \\
&+ \sum_{n=1}^\infty \frac{1}{2(k_\perp^2)_{0n}} \exp\left[-\mathrm{i}\frac{(k_\perp)_{0n}^2(z-z')c}{2\omega}\right] \\
&\times \sin(n\pi y/l_y)\sin(n\pi y'/l_y) \Bigg\} ,
\end{aligned} \quad (5.142)$$

where $\mathcal{DD}'$ denotes

$$\mathcal{DD}' = \left(\frac{\partial}{\partial x} + \mathrm{i}\frac{\partial}{\partial y}\right)\left(\frac{\partial}{\partial x'} - \mathrm{i}\frac{\partial}{\partial y'}\right) .$$

Taking derivatives, we rewrite (5.142) as

$$U_{\rm i} = -{\rm i}\frac{\pi e\theta_{\rm s}^2}{c l_x l_y}\int_0^z dz' \int_0^{l_x} dx' \int_0^{l_y} dy'\, \tilde{j}_1(z', x', y')$$

$$\times \left\{ \sum_{m=1}^\infty \sum_{n=1}^\infty \exp\left[-{\rm i}\frac{(k_\perp)_{mn}^2 (z-z')c}{2\omega}\right]\right.$$

$$\times \{\cos[m\pi(x-x')/l_x]\cos[n\pi(y-y')/l_y]$$
$$- \cos[m\pi(x+x')/l_x]\cos[n\pi(y+y')/l_y]\}$$

$$+ \frac{1}{2}\sum_{m=1}^\infty \exp\left[-{\rm i}\frac{(k_\perp^2)_{m0}(z-z')c}{2\omega}\right]$$

$$\times \{\cos[m\pi(x-x')/l_x] - \cos[m\pi(x+x')/l_x]\}$$

$$+ \frac{1}{2}\sum_{n=1}^\infty \exp\left[-{\rm i}\frac{(k_\perp)_{0n}^2 (z-z')c}{2\omega}\right]$$

$$\left. \times \{\cos[n\pi(y-y')/l_y] + \cos[n\pi(y+y')/l_y]\}\right\}. \qquad (5.143)$$

Let us illustrate the practical application of these formulae. For instance, when performing simulations of the linear regime, we can use the equation for the first harmonic of the beam current density. In the case of a negligibly small energy spread, we have:

$$\frac{d^2\tilde{j}_1}{dz^2} + 2{\rm i}C\frac{d\tilde{j}_1}{dz} + \left(\frac{4\pi e j_0(\boldsymbol{r}_\perp)}{\gamma_z^2 \mathcal{E}_0} - C^2\right)\tilde{j}_1 = -\frac{\omega j_0(\boldsymbol{r}_\perp)}{c\gamma_z^2 \mathcal{E}_0}(U_{\rm i} + U_{\rm ext})\,. \quad (5.144)$$

Here $j_0(\boldsymbol{r}_\perp)$ is the beam current density at the undulator entrance. The effective potential of the external electromagnetic wave, $U_{\rm ext}$, should be calculated using (5.2), (5.24), and (5.25). The power, radiated by the electron beam, is given by (5.39):

$$W = -\frac{I}{e}<P> = \frac{\rm i}{e}\int_0^z dz' \int d\boldsymbol{r}_\perp U(z',\boldsymbol{r}_\perp)\tilde{j}_1^*(z',\boldsymbol{r}_\perp) + {\rm C.C.}$$

This formula can also be used for the calculation of the partial contributions of the passive waveguide modes to the total radiation power, since $U_{\rm i}$ is expanded in a series of empty waveguide modes (see (5.143)).

5.7 Wall Resistance Effects

All the results obtained in this chapter refer to the case of perfectly conducting waveguide walls having conductivity $\sigma \to \infty$. In reality the conductivity always has a finite value. Let us consider a medium having a conductivity and permittivity of σ and ϵ, respectively. Then the electromagnetic field in this medium satisfies Maxwell's equations:

$$\nabla \times \boldsymbol{H} = \frac{4\pi\sigma}{c}\boldsymbol{E} + \frac{\epsilon}{c}\frac{\partial \boldsymbol{E}}{\partial t},$$

$$\nabla \times \boldsymbol{E} = -\frac{1}{c}\frac{\partial \boldsymbol{H}}{\partial t},$$

$$\nabla \cdot \boldsymbol{E} = 0, \quad \nabla \cdot \boldsymbol{H} = 0.$$

For a monochromatic wave, these equations can be rewritten as:

$$\nabla \times \boldsymbol{H}_\omega = -\mathrm{i}\frac{\omega}{c}\epsilon'\boldsymbol{E}_\omega, \quad \nabla \times \boldsymbol{E}_\omega = \mathrm{i}\frac{\omega}{c}\boldsymbol{H}_\omega,$$

$$\nabla \cdot \boldsymbol{E}_\omega = 0, \quad \nabla \cdot \boldsymbol{H}_\omega = 0,$$

where

$$\epsilon' = \epsilon + \mathrm{i}\frac{4\pi\sigma}{\omega}.$$

As a result, we obtain a unique equation for $\boldsymbol{E}_\omega$:

$$\nabla^2 \boldsymbol{E}_\omega + (n')^2 \frac{\omega^2}{c^2}\boldsymbol{E}_\omega = 0,$$

where

$$n' = \sqrt{\epsilon + \mathrm{i}\frac{4\pi\sigma}{\omega}}$$

is the refractive index of the medium. The simplest solution of the latter equation is the plane wave,

$$\boldsymbol{E}_\omega \exp(-\mathrm{i}\omega t) + \mathrm{C.C.} = \boldsymbol{E}_0 \exp[\mathrm{i}(\boldsymbol{k} \cdot \boldsymbol{r} - \omega t)] + \mathrm{C.C.},$$

where $k_x^2 + k_y^2 + k_z^2 = (n')^2\omega^2/c^2$ and $\boldsymbol{E}_0$ is an arbitrary vector. Using Maxwell's equations, we find the expression for $\boldsymbol{H}_\omega$:

$$\boldsymbol{H}_\omega = \frac{c}{\omega}(\boldsymbol{k} \times \boldsymbol{E}_\omega). \tag{5.145}$$

Let us consider a plane electromagnetic wave entering a plane metallic surface. The angle of incidence of the wave is θ and the angle of wave propagation in the metal is θ_1. The tangential component of the wave vector should be continuous on the boundary, which gives the relation:[4]

$$n'\sin(\theta_1) = \sin(\theta). \tag{5.146}$$

The refractive index of a metal is mainly defined by the conductivity, σ, and has a complex value and large modulus:

---

[4] For complex values of the refractive index θ_1 cannot be interpreted as a geometrical angle. Its value is defined by the relations $\sin\theta_1 = k_\mathrm{t}/k$ and $\cos\theta_1 = k_n/k$, where $k = n'\omega/c$ and k_t and k_n are, respectively, the tangential and normal components of the wave vector $\boldsymbol{k}$ in the metal.

$$n' = \sqrt{\epsilon + i\frac{4\pi\sigma}{\omega}} \simeq \sqrt{i\frac{4\pi\sigma}{\omega}} \ .$$

Taking into account that $|n'| \gg 1$, we obtain from (5.146) that

$$\cos(\theta_1) = \sqrt{1 - \frac{\sin^2(\theta)}{(n')^2}} \simeq 1 \ , \tag{5.147}$$

which means that the direction of the wave propagating in a metal is almost perpendicular to the surface and is almost independent of the value of the incidence angle.

It follows from Maxwell's equations that the tangential components of E_t and H_t are continuous on the surface. It follows from (5.147) that the wave components in the metal are E_t and H_t. Using (5.145), we find that E_t and H_t inside the metal are connected by the relation:

$$E_t/H_t = 1/n' \ . \tag{5.148}$$

We can consider the continuity condition (5.148) for E_t and H_t on the metal surface as the boundary condition. Despite the fact that we considered above only the case of a plane wave, the boundary condition (5.148) is valid for an arbitrary wave. Indeed, any wave can be represented as a linear superposition of plane waves with different values of the angles of incidence, θ. Since the boundary condition is valid for each plane wave, it is valid also for their linear superposition. The boundary condition (5.148) can be extended to the case of an arbitrary shape of the metallic surface when the typical value of the curvature radius is much larger than the wavelength.

So, we have presented an example of introducing an approximate boundary conditions. This approach is popular, since it allows one to significantly simplify the solution of the electrodynamic problem. The approximate boundary conditions (5.148) were derived by Leontovich and are named after him. They are valid on the surface of a material having a large value of the modulus of the refractive index. Leontovich's boundary conditions can be written in the vector form:

$$(\boldsymbol{n} \times \boldsymbol{E}_\omega)|_S = \frac{1}{n'} \boldsymbol{n} \times (\boldsymbol{n} \times \boldsymbol{H}_\omega)|_S \ , \tag{5.149}$$

where $\boldsymbol{n}$ is the unit vector perpendicular to the surface, and directed inside the waveguide.

Let us consider the problem of the excitation of a waveguide having finite conductivity. The vector and scalar potential are connected with the field components as follows:

$$\boldsymbol{H} = \boldsymbol{\nabla} \times \boldsymbol{A} \ , \quad \boldsymbol{E} = -\frac{1}{c}\frac{\partial \boldsymbol{A}}{\partial t} - \boldsymbol{\nabla}\phi \ .$$

Taking into account the Coulomb gauge for the vector potential,

$$\boldsymbol{\nabla} \cdot \boldsymbol{A} = 0 \ ,$$

we find from Maxwell's equations that the vector potential satisfied the equation:

$$\nabla^2 \boldsymbol{A} - \frac{1}{c^2}\frac{\partial^2 \boldsymbol{A}}{\partial t^2} = -\frac{4\pi}{c}\boldsymbol{j} + \frac{1}{c}\frac{\partial}{\partial t}\nabla\phi = -\frac{4\pi}{c}\boldsymbol{J}\ .$$

When the waveguide is excited at frequency w, we have:

$$\nabla^2 \boldsymbol{A}_\omega + \frac{\omega^2}{c^2}\boldsymbol{A}_\omega = -\frac{4\pi}{c}\boldsymbol{J}_\omega\ ,$$

where $\boldsymbol{J}_\omega = \boldsymbol{j}_\omega + i\omega(4\pi)^{-1}\nabla\phi_\omega$. The boundary condition for the scalar potential on the waveguide walls is[5] $\phi_\omega = 0$. The boundary conditions for the vector potential, $\boldsymbol{A}_\omega$, must provide the boundary conditions (5.149) for the field:

$$(\boldsymbol{n}\times\boldsymbol{A}_\omega)|_S = -\frac{ic}{\omega n'}\boldsymbol{n}\times(\boldsymbol{n}\times(\nabla\times\boldsymbol{A}_\omega))|_S\ . \tag{5.150}$$

The problem of the excitation can be solved in the same way as for a perfectly conducting waveguide (see Appendix A.3). We start with the problem of the excitation of a cavity consisting of a piece of homogeneous waveguide of length l_z and two plane side walls perpendicular to the z axis. Then we perform the limiting transition $l_z \to \infty$ and obtain the result for the waveguide. Thus, we need to solve the equations:

$$\nabla^2 \boldsymbol{A}_\omega + \frac{\omega^2}{c^2}\boldsymbol{A}_\omega = -\frac{4\pi}{c}\boldsymbol{j}_\omega - i\frac{\omega}{c}\nabla\phi_\omega\ , \tag{5.151}$$

$$\nabla\cdot\boldsymbol{A}_\omega = 0\ , \tag{5.152}$$

with the boundary conditions on the waveguide walls:

$$(\boldsymbol{n}\times\boldsymbol{A}_\omega)|_S = -\frac{ic}{\omega n'}\boldsymbol{n}\times(\boldsymbol{n}\times(\nabla\times\boldsymbol{A}_\omega))|_S\ ,\qquad \phi_\omega|_S = 0, \tag{5.153}$$

where $n' = n'(\omega)$ is the refractive index of the waveguide material. First, we need to solve the eigenvalue problem for an unloaded cavity:

$$\nabla^2 \boldsymbol{F}_\nu + k_\nu^2 \boldsymbol{F}_\nu = 0\ , \tag{5.154}$$

$$\nabla\cdot\boldsymbol{F}_\nu = 0\ , \tag{5.155}$$

$$(\boldsymbol{n}\times\boldsymbol{F}_\nu)|_S = -\frac{ic}{\omega n'}\boldsymbol{n}\times(\boldsymbol{n}\times(\nabla\times\boldsymbol{F}_\nu))|_S\ , \tag{5.156}$$

where $\boldsymbol{F}_\nu$ and k_ν are the eigenfunctions and the eigenvalues, respectively. The eigenfunctions are orthogonal, i.e.

---

[5] Maxwell's equations and the Coulomb gauge, $\nabla\cdot\boldsymbol{A} = 0$, are invariant under a gradient transformation,

$$\phi \to \phi + \frac{1}{c}\frac{\partial\psi}{\partial t}\ ,\qquad \boldsymbol{A}\to\boldsymbol{A} - \nabla\psi\ ,$$

where ψ is an arbitrary function satisfying the condition $\nabla^2\psi = 0$. It is always possible to choose the function ψ in order to set $\phi|_S = 0$.

$$\int d\mathbf{r}\, \mathbf{F}_{\nu_1} \cdot \mathbf{F}_{\nu_2} = 0 \quad \text{at} \quad \nu_1 \neq \nu_2 , \tag{5.157}$$

where the integration is performed over the cavity volume. This statement can be proven using the considerations presented in Appendix A.3. Using the relation

$$\nabla \times (\nabla \times \mathbf{F}_\nu) = \nabla(\nabla \cdot \mathbf{F}_\nu) - \nabla^2 \mathbf{F}_\nu$$

we can rewrite the equation for $\mathbf{F}_\nu$ as

$$\nabla \times (\nabla \times \mathbf{F}_\nu) - k_\nu^2 \mathbf{F}_\nu = 0 . \tag{5.158}$$

We write this equation for two different eigenfunctions, $\mathbf{F}_{\nu_1}$ and $\mathbf{F}_{\nu_2}$. Then we calculate two expressions. The first is the scalar product of the equation for $\mathbf{F}_{\nu_1}$ with $\mathbf{F}_{\nu_2}$, and the second is the scalar product of the equation for $\mathbf{F}_{\nu_2}$ with $\mathbf{F}_{\nu_1}$. The second expression is subtracted from the first, and the result is integrated over the cavity volume. Using the Gauss divergence theorem, we go over to integration over the cavity surface:

$$(k_{\nu_1}^2 - k_{\nu_2}^2) \int d\mathbf{r}\, \mathbf{F}_{\nu_1} \cdot \mathbf{F}_{\nu_2} = \int d\sigma\, \{ (\mathbf{F}_{\nu_1} \times (\nabla \times \mathbf{F}_{\nu_2})) \cdot \mathbf{n}$$
$$- (\mathbf{F}_{\nu_2} \times (\nabla \times \mathbf{F}_{\nu_1})) \cdot \mathbf{n} \} . \tag{5.159}$$

Since $\mathbf{F}_{\nu_1}$ and $\mathbf{F}_{\nu_2}$ satisfy Leontovich's boundary conditions on the waveguide wall, the integral over the cavity surface is equal to zero. Thus,

$$(k_{\nu_1}^2 - k_{\nu_2}^2) \int d\mathbf{r}\, \mathbf{F}_{\nu_1} \cdot \mathbf{F}_{\nu_2} = 0 .$$

Without loss of generality, we normalize the eigenfunction of the passive cavity as

$$\int d\mathbf{r}\, \mathbf{F}_\nu \cdot \mathbf{F}_\nu = 1 .$$

It is relevant to mention that the orthogonality and the normalization conditions are formulated without complex conjugation. Substituting

$$\mathbf{A}_\omega = \sum_\nu C_\nu \mathbf{F}_\nu(\mathbf{r})$$

into (5.12), and taking into account (5.154), we find:

$$\sum_\nu C_\nu \left(\frac{\omega^2}{c^2} - k_\nu^2 \right) \mathbf{F}_\nu = -\frac{4\pi}{c} \mathbf{J}_\omega .$$

The scalar product of the latter expression with $\mathbf{F}_\nu(\mathbf{r})$ is integrated over the resonator volume. Taking into account the orthogonality and normalization conditions for the eigenfunctions, we find:

$$C_\nu = -\frac{4\pi}{c} \int d\mathbf{r}\, \frac{\mathbf{J}_\omega \cdot \mathbf{F}_\nu}{\omega^2/c^2 - k_\nu^2} .$$

Finally, we obtain the following expression for the Green's function:

$$G_\omega^{\alpha\beta}(\boldsymbol{r},\ \boldsymbol{r}') = -\frac{4\pi}{c}\sum_\nu \frac{F_\nu^\alpha(\boldsymbol{r})F_\nu^\beta(\boldsymbol{r}')}{\omega^2/c^2 - k_\nu^2}\ , \tag{5.160}$$

which is identical to (A.3.10) obtained for the case of perfectly conducting waveguide walls. The only difference is that the eigenvalues become complex values for the case of waveguide walls with finite conductivity, i.e.

$$k_\nu = k_\nu' + \mathrm{i}k_\nu''\ .$$

The next step is to perform the limiting transition from a cavity to an infinite waveguide (see Appendix A.3). Omitting the intermediate calculations, we present the final result for the Green's function of the overmoded waveguide with finite conductivity of the walls:

$$\begin{aligned}
G_\omega = {}& \frac{2\pi\mathrm{i}}{\omega}\sum_\mu \exp\left\{\mathrm{i}\left[\frac{\omega}{c} - \frac{c(k_\perp^{\mathrm{TE}})_\mu^2}{2\omega}\right]|z-z'|\right\} \\
& \times \left[\boldsymbol{e}_x\frac{\partial\psi_\mu^{\mathrm{TE}}(\boldsymbol{r}_\perp)}{\partial y} - \boldsymbol{e}_y\frac{\partial\psi_\mu^{\mathrm{TE}}(\boldsymbol{r}_\perp)}{\partial x}\right] \\
& \otimes \left[\boldsymbol{e}_x\frac{\partial\psi_\mu^{\mathrm{TE}}(\boldsymbol{r}'_\perp)}{\partial y'} - \boldsymbol{e}_y\frac{\partial\psi_\mu^{\mathrm{TE}}(\boldsymbol{r}'_\perp)}{\partial x'}\right] \\
& + \frac{2\pi\mathrm{i}}{\omega}\sum_\nu \exp\left\{\mathrm{i}\left[\frac{\omega}{c} - \frac{c(k_\perp^{\mathrm{TM}})_\nu^2}{2\omega}\right]|z-z'|\right\} \\
& \times \left[\boldsymbol{e}_x\frac{\partial\psi_\nu^{\mathrm{TM}}(\boldsymbol{r}_\perp)}{\partial x} + \boldsymbol{e}_y\frac{\partial\psi_\nu^{\mathrm{TM}}(\boldsymbol{r}_\perp)}{\partial y}\right] \\
& \otimes \left[\boldsymbol{e}_x\frac{\partial\psi_\nu^{\mathrm{TM}}(\boldsymbol{r}'_\perp)}{\partial x'} + \boldsymbol{e}_y\frac{\partial\psi_\nu^{\mathrm{TM}}(\boldsymbol{r}'_\perp)}{\partial y'}\right]\ ,
\end{aligned} \tag{5.161}$$

where the symbol $\otimes$ denotes the direct product of vectors. The scalar functions ψ^{TE} and ψ^{TM} are the solutions of the Helmholtz equation:

$$\nabla_\perp^2 \psi + k_\perp^2 \psi = 0\ ,$$

and are normalized as

$$\int \boldsymbol{\nabla}\psi_\nu \cdot \boldsymbol{\nabla}\psi_\nu \mathrm{d}\boldsymbol{r}_\perp = 1\ . \tag{5.162}$$

The boundary conditions for ψ^{TE} and ψ^{TM}, written in the paraxial approximation, are as follows:

$$\left[\boldsymbol{n}\cdot\boldsymbol{\nabla}\psi^{\mathrm{TE}} + (\boldsymbol{e}_z\times\boldsymbol{n})\cdot\boldsymbol{\nabla}\psi^{\mathrm{TM}}\right]\big|_S = -\frac{\mathrm{i}c}{\omega n'}(k_\perp^{\mathrm{TE}})^2\psi^{\mathrm{TE}}\big|_S\ ,$$

$$(k_\perp^{\mathrm{TM}})^2\psi^{\mathrm{TM}}\big|_S = \frac{\mathrm{i}\omega}{cn'}\left[\boldsymbol{n}\cdot\boldsymbol{\nabla}\psi^{\mathrm{TM}} - (\boldsymbol{e}_z\times\boldsymbol{n})\cdot\boldsymbol{\nabla}\psi^{\mathrm{TE}}\right]\big|_S\ .$$

When $\omega/(|n'k_\perp|c) \ll 1$, these boundary conditions can be written in the form:

$$\mathbf{n} \cdot \nabla \psi^{\mathrm{TE}}|_{\mathrm{S}} = -\frac{ic}{\omega n'}(k_\perp^{\mathrm{TE}})^2 \psi^{\mathrm{TE}}|_{\mathrm{S}} + \frac{i\omega}{n'c(k_\perp^{\mathrm{TE}})^2}\left[(\mathbf{e}_z \times \mathbf{n}) \cdot \nabla\right]^2 \psi^{\mathrm{TE}}|_{\mathrm{S}} ,$$

$$(k_\perp^{\mathrm{TM}})^2 \psi^{\mathrm{TM}}|_{\mathrm{S}} = \frac{i\omega}{cn'} \mathbf{n} \cdot \nabla \psi^{\mathrm{TM}}|_{\mathrm{S}} .$$

Let us consider the case of a circular waveguide. To find the eigenfunctions, we use the perturbation method. The eigenfunction of a perfectly conducting waveguide (5.107) is chosen as a zero-order approximation. Then we find the first-order perturbations for small values of the parameter $\omega/(|n'k_\perp|c)$:[6]

$$\psi_{mn}^{\mathrm{TE}} = A^{\mathrm{TE}} J_m[(\mu_{mn} + \delta\mu_{mn})r/R]\binom{\sin(m\varphi)}{\cos(m\varphi)} ,$$

$$\psi_{mn}^{\mathrm{TM}} = A^{\mathrm{TM}} J_m[(\nu_{mn} + \delta\nu_{mn})r/R]\binom{\sin(m\varphi)}{\cos(m\varphi)} ,$$

where $J'_m(\mu_{mn}) = 0$ and $J_m(\nu_{mn}) = 0$. To find the values of $\delta\mu_{mn}$ and $\delta\nu_{mn}$, we use the boundary conditions for ψ^{TE} and ψ^{TM}, and obtain:

$$\frac{\delta\mu_{mn}}{\mu_{mn}} = -\frac{ic}{n'\omega R}\frac{[\mu_{mn}^2 + \omega^2 m^2 R^2/(c^2\mu_{mn}^2)]}{\mu_{mn}^2 - m^2} ,$$

$$\frac{\delta\nu_{mn}}{\nu_{mn}} = -\frac{i\omega R}{cn'\nu_{mn}^2} .$$

The functions ψ^{TE} and ψ^{TM} are solutions of the Helmholtz equation and satisfy Leontovich's boundary conditions with an accuracy of $o(\omega/(|n'k_\perp|c))$. Thus, the orthogonality condition for the eigenfunctions of a nonideal waveguide is satisfied with the same accuracy. The normalizing coefficients A^{TE} and A^{TM} are calculated using the normalization condition (5.162).

The next problem is to describe the operation of the FEL amplifier with a nonideal waveguide. Detailed analysis of the problem falls outside the scope of this book. Here we present only an example which will help the reader to estimate the power of the effect of the finite conductivity of the waveguide walls. Let us consider the FEL amplifier with a value of the waveguide diffraction parameter, say $\Omega \simeq 1$. It was shown in Sect. 5.3 that at an appropriate tuning the FEL amplifier, the beam radiation mode is close to one of the modes of the passive waveguide. To be specific, we consider the case when the beam radiation mode is close to the TE_{1n} mode of the passive waveguide. To simplify the study, we consider the case of a negligibly small energy

---

[6] It should be stressed that the parameter for the perturbation theory is $\omega/(|n'k_\perp|c)$. The requirement $|n'| \gg 1$ is necessary, but not sufficient for the overmoded waveguide.

spread. The TE$_{1n}$ mode is considered as the zero-order approximation for the eigenmode in the integral equation:[7]

$$\left[\left(\hat{\Lambda}+\mathrm{i}\hat{C}\right)^2+\hat{\Lambda}_\mathrm{p}^2\right]a(\hat{r})$$
$$=\frac{\mathrm{i}}{\Omega}\int_0^\infty \mathrm{d}\zeta\exp\left(-\hat{\Lambda}\zeta\right)\exp\left[-\frac{\mathrm{i}(\mu_{1n}+\delta\mu_{1n})^2\zeta}{2\Omega}\right]$$
$$\times\int_0^1 \mathrm{d}\hat{r}'\,\hat{r}'a(\hat{r}')\frac{\mu_{1n}^2}{(\mu_{1n}^2-1)J_1^2(\mu_{1n})}$$
$$\times J_0\left(\mu_{1n}\sqrt{B/\Omega}\hat{r}\right)J_0\left(\mu_{1n}\sqrt{B/\Omega}\hat{r}'\right)\ . \tag{5.163}$$

We multiply both sides of the latter equation by

$$\hat{r}J_0\left(\mu_{1n}\sqrt{B/\Omega}\hat{r}\right),$$

integrate over $\hat{r}$ from 0 to 1, and finally obtain the eigenvalue equation:

$$\left[\left(\hat{\Lambda}+\mathrm{i}\hat{C}\right)^2+\hat{\Lambda}_\mathrm{p}^2\right]\left[\hat{\Lambda}+\frac{\mathrm{i}\mu_{1n}^2}{2\Omega}+\frac{\mathrm{i}\mu_{1n}\delta\mu_{1n}}{\Omega}\right]$$
$$=\frac{\mathrm{i}}{2\Omega}\frac{\mu_{1n}^2}{(\mu_{1n}^2-1)J_1^2(\mu_{1n})}\left[J_0^2\left(\mu_{1n}\sqrt{B/\Omega}\right)+J_1^2\left(\mu_{1n}\sqrt{B/\Omega}\right)\right]\ . \tag{5.164}$$

Earlier we obtained a similar eigenvalue equation for the case of a perfectly conducting waveguide (see (5.63)). Comparison of (5.164) and (5.63) shows that the nonideality of the waveguide leads to a decrease of the field growth rate which is given by the real part of the eigenvalue $\hat{\Lambda}$. This immediately follows from (5.164) if one takes into account that $\delta\mu_{1n}$ is a complex value.

The single-mode approximation for the eigenvalue equation for an arbitrary mode (TE$_{mn}$ or TM$_{mn}$) can be obtained in the same way as (5.164). Even though the single-mode analysis is an approximation, it allows one to find a simple estimate of the effect of the nonideality of the waveguide walls. In conclusion to this section we should like to stress that the approach described above for the problem of waveguide excitation is valid for a metallic or dielectric overmoded waveguide. The only requirement is that of a large value of the refractive index of the waveguide walls which allows one to use Leontovich's boundary conditions.

---

[7] The accuracy of this approximation is given by the value of Ω/μ_{1n}^2 which is assumed to be small for the single-mode approximation. Thus, we can neglect the perturbation of the eigenfunction of a nonideal waveguide.

5.8 Concluding Remarks

The analysis of the waveguide FEL refers to the most complicated problems in FEL physics. To appreciate the origin of the problem, we invite the reader to remember the theory of the FEL amplifier with an open electron beam. We demonstrated in Chap. 4 that the solution of the inhomogeneous paraxial wave equation can be expressed in terms of a scalar Green's function relating the complex amplitudes $\tilde{E}$ and $\tilde{j}_1$. The self-consistent equations, written in reduced form, are identical for both planar and helical undulator configurations. In the presence of the waveguide walls the relation between the current sources and electromagnetic filed is no longer so simple. The analysis is still straightforward, however, and is fulfilled by the introduction of the tensor Green's function $G^{\alpha\beta}$ relating the components E_α and j_β. The result is that the self-consistent equations become more complicated and their structure depends on the type of undulator and the geometry of the waveguide. This is a subject which is difficult to treat in a general way, and particular situations must be considered individually.

In this chapter we confined our study to a system with a circular waveguide, helical undulator, and axisymmetric electron beam. Although this problem relates to rather specific applications, it illustrates the main features of the waveguide FEL. This applies especially to the effect of self-reproduction of the radiation field in the active waveguide. The wide range of parameters required to adequately describe the waveguide FEL makes a general discussion difficult. Even a simple situation, such as the high-gain linear regime, leads to a universal function depending on six dimensionless parameters,

$$\operatorname{Re}\hat{\Lambda} = \mathcal{D}(m,\, \hat{C},\, B,\, \Omega,\, \hat{\Lambda}_{\mathrm{p}}^2,\, \hat{\Lambda}_{\mathrm{T}}^2)\,.$$

When illustrating the operation of the waveguide FEL, we confined our attention to the situations where space charge and energy spread effects are negligible, $\hat{\Lambda}_{\mathrm{p}}^2 \to 0$ and $\hat{\Lambda}_{\mathrm{T}}^2 \to 0$. However, the reader can simply extend the analysis with these effects taken into account.

It is straightforward to repeat the analysis of the previous sections for a system consisting of a circular waveguide, planar undulator, and axisymmetric electron beam. Discussion of these somewhat idealized examples does help in understanding the phenomena which occur in more complicated practical systems. Some of them have much in common with the systems already discussed. However, one should take into account the differences which can be listed as follows. A very frequent situation is that the millimeter-range FEL amplifier is driven by quasi-relativistic electron beams. Also, the ratio $\lambda/\lambda_{\mathrm{w}}$ is about 0.1–0.3. In this case one should discard the paraxial and ultrarelativistic approximations. Also, it frequently happens that the focusing of the electron beam is provided by a longitudinal magnetic field H_z. The amplification process can be treated in a straightforward manner for sufficiently small values of H_z. For large values of the longitudinal field, say at

$eH_z/(\gamma m_e c^2) \simeq k_{\rm w}$, the problem of making the solution self-consistent is of significant complexity and should be studied separately.

5.9 Suggested Bibliography

The most comprehensive study of the waveguide FEL is presented in the book by Freund and Antonsen [5.1]. The problem of the waveguide FEL with an axial magnetic field is intensively discussed by the authors. Also, the reader can find there an extended list of references to original papers relevant to the problem. The optical guiding aspects of the theory of the waveguide FEL are treated in [5.2]. Paper [5.3] is devoted to the theory of the FEL amplifier with axisymmetric electron beam, overmoded circular waveguide and helical undulator. Rigorous results were obtained for the electron beam with stepped and parabolic profiles. Some rigorous results of the theory of the FEL amplifier with a sheet electron beam, overmoded planar waveguide, and planar undulator were obtained in [5.4]. Coupled mode theory applied to the waveguide FEL forms the subject of [5.5]. Application of the Green's function method to a planar waveguide is described in [5.6]. A general treatment of the Green's function method may be found in the text by Morse and Feshbach [5.7]. An extended theoretical treatment of Leontovich's boundary conditions appears in the monograph by Brekovskikh [5.8].

6. FEL Amplifier Start-up from Shot Noise

Till now we have considered the electron beam as a continuous medium when describing the theory of the FEL amplifier. To some extent this is an idealization, since in reality the beam current is produced by a large number of moving electrons. If we consider the microstructure of the electron current, we find that electrons enter the undulator randomly in time and space. When the electron passes the undulator, it interacts with the electromagnetic field produced by other electrons and also emits radiation. So, we can expect that the FEL amplifier should possess intrinsic noise properties. Taking into account this complicated picture, we can formulate several problems. The first one is the region of applicability of the steady-state theory of the FEL amplifier: namely, to what extent can we neglect the noise properties of the FEL amplifier? The second one is what kind of physical effects are hidden in the steady-state theory. Finally, we study how these physical effects can be used in practical devices.

The answer to the last question is the simplest one. FEL amplifier start-up from shot noise is capable of producing coherent, tunable FEL radiation spanning a very short wavelength, down to a fraction of an Angstrom. Here it is relevant to note that the problem of the X-ray laser has not yet been solved, despite significant effort during last 40 years. Nowadays we have the situation that the wavelength range below 100 nm is not covered by quantum lasers. The main problem is that radiation absorption in the active medium increases drastically when approaching the VUV band. Progress in the development of short-wavelength lasers is rather moderate and one can hardly expect a breakthrough. If we consider the FEL amplifier, we find that there are no principal physical limitations on the way towards short wavelengths. Indeed, the active medium in the FEL amplifier is the electron beam moving in the undulator. Since the amplification process develops in vacuum during one pass of the electron beam through the undulator, the problem of the absorption of the radiation does not exist at all. On the other hand, the physical principles of the FEL amplifier operating in the VUV and X-ray wavelength bands are the same as those in the optical range. The requirements are that the parameters of the electron beam and the undulator become more severe when going down to shorter wavelength. Nevertheless, these requirements evolve slowly with the operating wavelength. During the last few decades there has been extremely rapid progress in the development of accelerators, and it became possible to produce electron beams which can be used as driving beams for short-wavelength FELs. As a result, X-ray FEL projects have been initiated recently at several laboratories around the world. The peak and average brilliance of VUV and X-ray FELs will exceed by many orders of magnitude the corresponding values available in third-generation synchrotron light sources. Successful construction of X-ray FELs will significantly influence a multitude of scientific and applied research.

Here it is relevant to remember that FELs form a separate class of vacuum-tube devices. The analysis of the noise properties of traditional vacuum-tube amplifiers has always been an important problem. This has been mainly connected with the practical need for reducing the intrinsic noise of the amplifier. The result of the experience obtained during these investigations shows that there always exists the fundamental effect of shot noise originating from the random emission (in time and space) of the electrons from the cathode. When we analyze this effect for the parameters of traditional microwave amplifiers, we find that it is complicated. In particular, suppression of the shot noise in some frequency band can occur due to space charge effects. Besides the shot noise effect, there are a number of different sources of noise which influence the operation of traditional vacuum-tube amplifiers.

As for the FEL amplifier operating in the visible or shorter wavelength range, its noise properties are defined only by the shot noise. To prove this statement, we refer to the perfect agreement between the predicted and measured properties of the incoherent synchrotron and undulator radiation. It is well known that the synchrotron radiation itself has its origin in the density fluctuations in the electron beam. The theoretical predications are based on the assumption that these fluctuations are due to the shot noise in the electron beam. If we accept this assumption, we find that the energy, radiated by the electron beam, should be equal to the number of particles in the beam times the energy radiated by a single electron. Since the properties of the synchrotron radiation have been measured to a high accuracy, we can state that suppression of the shot noise does not occur in the short wavelength band (otherwise, the measured energy of incoherent radiation would be less than that predicted). Therefore, this is a reliable assumption about the random arrival of the electrons at the undulator entrance.

An FEL amplifier which starts up from shot noise is frequently known as a self-amplified spontaneous emission (SASE) FEL. Following tradition, we also use this terminology. However, it is worth mentioning that such an essentially quantum terminology does not reflect the actual physics of the process. The amplification process in the SASE FEL has its origin in the density fluctuations in the electron beam. The latter effect is completely classical.

The shot noise in the electron beam causes fluctuations of the beam density which are random in time and space. As a result, the radiation, produced by such a beam, has random amplitudes and phases in time and space. These kinds of radiation fields can be described in terms of statistical optics, branch of optics that has been developed intensively during the last few decades and there exists a lot of experience and a theoretical basis for the description of fluctuating electromagnetic fields.

In the framework of statistical optics the radiation field is characterized by notions such as time and space coherence. Let us illustrate these notions. To be specific, we consider the radiation pulse of nearly monochromatic radiation having a duration and bandwidth equal to T and $\Delta\omega$, respectively. If $T \gg 1/\Delta\omega$, the radiation is partially coherent in time. The time coherence is of about $\tau_c \simeq 1/\Delta\omega$. The physical sense of this notion is as follows. Let us separate the radiation pulse by a plate. These two pulses pass different path lengths and are then combined together (in the same way as in the Michelson interferometer). If the difference between the path lengths is less than $c\tau_c$, we see the interference pattern at the end for a shot-to-shot averaging (averaging over a large number of radiation pulses). Using simple physical language, we can say that the radiation field is correlated within the time of coherence.

The notion of space coherence can be explained in the same way. Let us direct the radiation onto a screen with two pinholes and look at the interference pattern in the far diffraction zone when changing the distance between the pinholes. When the pinholes are located close to each other, we see a clear diffraction pattern. As the distance between the holes increases to some value, $D > \Delta r_c$, we come to the situation when we do not see an interference pattern after averaging over the ensemble of pulses. Qualitatively, the value of $\pi(\Delta r_c)^2$ is referred as the area of coherence of the radiation pulse.

The coherence volume is defined as the product of the coherence area, $\pi(\Delta r_c)^2$, with the coherence length, $c\tau_c$. If one uses the notions of quantum mechanics, this volume corresponds to one cell in the phase space of the photons. The number of photons in the coherence volume is also referred to as the degeneracy parameter, δ_w. Physically this parameter means the average number of photons which can interfere, or the number of photons in one quantum state (one mode). Analysis of VUV and X-ray SASE FELs shows that a feature of these devices is that typical values of δ_w will be about 10^{10}–10^{14}. Since the degeneracy parameter is large, we can state that the classical approach is adequate for a description of the statistical properties of the radiation from the SASE FEL.

The general analysis of start-up from shot noise is a rather complicated problem. We start the investigations in the framework of the one-dimensional theory in order to describe in the clearest way the effects connected with longitudinal coherence. The results of the one-dimensional theory can be applied to the high-gain FEL amplifier when the transverse coherence of the radiation is settled (this will be demonstrated later when considering 3-D start-up from noise). The shot noise in the electron beam is a Gaussian random process. The FEL amplifier, operating in the linear regime, can be considered as a linear filter which does not change the statistics of the signal. As a result, we can define general statistical properties of the output radiation without any calculations. For instance, in the case of the SASE FEL the real and imaginary parts of the slowly varying complex amplitudes of the electric field of the electromagnetic wave have a Gaussian distribution, the instantaneous radiation power fluctuates in accordance with the negative exponential distribution, and the energy in the radiation pulse fluctuates in accordance with the gamma distribution. We can also state that the spectral density of the radiation energy and the first-order time correlation function should form a Fourier transform pair (this is the so-called Wiener Khintchine theorem). Also, the higher-order correlation functions (time and spectral) should be expressed in terms of the first-order correlation functions. These properties are well known in statistical optics as properties of completely chaotic polarized radiation.

When describing the physical principles, it is always important to find ways of analytical description. From this point of view the SASE FEL is a rather complicated object. Indeed, in the general case we deal with a nonstationary random process. Analytical study of such a process is very complicated, since the electron bunch combines the features of the input signal and the active medium with time-dependent parameters. It is important to find a model which provides the possibility of an analytical description without loss of essential information about the features of the nonstationary process. Approximations satisfying these conditions are the use of a long rectangular electron bunch and application of the steady-state spectral Green's function. In the framework of this model it becomes possible to describe analytically all the statistical properties of the radiation from a SASE FEL: time and spectral correlation functions, parameters of the distribution of the radiation power after the monochromator, etc. Finding analytical solutions is always fruitful for understanding FEL physics and testing numerical simulation codes.

A complete description of the SASE FEL can be performed only with time-dependent numerical simulation codes. Application of the numerical calculations allows one to describe the general case of SASE FEL operation, including the case of an arbitrary axial profile of the electron bunch, the effects of finite pulse duration and nonlinear effects. Since construction of the time-dependent codes is a nontrivial problem, we devote significant attention to the description of all the steps of the algorithm. The methodology of statistical simulations is also described in detail. Analytical results for the high-gain linear regime serve as a primary standard for testing the codes.

Special attention is devoted to the analysis of the nonlinear mode of SASE FEL operation in the framework of the one-dimensional model. In particular, it is shown that the statistics of the radiation from the SASE FEL operating at saturation changes significantly with respect to the linear mode of operation (e.g., with respect to Gaussian statistics). Using similarity techniques, we present universal plots and formulae for calculations of the characteristics of the SASE FEL radiation.

The actual physical picture of start-up from noise should take into account that the fluctuations of the current density in the electron beam are uncorrelated in time and space. Thus, a large number of transverse radiation modes are excited when the electron beam enters the undulator. These radiation modes have different growth rates. For a sufficiently long undulator the fundamental mode, which has maximal gain, should survive. Information on transverse coherence formation can be obtained with fully three-dimensional, time-dependent simulation codes. In this chapter we devote significant attention to the description of the simulation algorithm. Using the results of numerical simulations, we illustrate the main features of three-dimensional start-up from noise. Analysis of the data obtained shows that the statistical properties of the radiation from the SASE FEL, operating in the linear regime, can be described with Gaussian statistics.

At the end of the chapter we apply the theory to the calculation of a practical device. The experimental data of the University of California at Los Angeles and the Los Alamos National Laboratory high-gain SASE FEL are analyzed in detail.

6.1 Shot Noise in the Electron Beam

Fluctuations of the electron beam current density serve as the input signal in the SASE FEL. These fluctuations always exist in the electron beam due to the effect of shot noise. When the electron beam enters the undulator, the presence of the beam modulation at frequencies close to the resonance frequency of the FEL amplifier initiates the process of amplification of coherent radiation. In this section we study the statistical properties of the shot noise in the electron beam. It should be noted that the process under study is nonstationary with finite pulse duration, so in what follows the averaging symbol $\langle \ldots \rangle$ means the ensemble average over bunches.

Let us consider the microscopic picture of the electron beam current at the entrance to the undulator. The electron beam current is made up of moving electrons randomly arriving at the entrance to the undulator:

$$I(t) = (-e) \sum_{k=1}^{N} \delta(t - t_k) \, ,$$

where $\delta(\ldots)$ is the delta function, $(-e)$ is the charge of the electron, N is the number of electrons in a bunch, and t_k is the random arrival time of the electron at the undulator entrance. The electron bunch profile is described by the profile function $F(t)$. The beam current averaged over an ensemble of bunches can be written in the form:

$$\langle I(t) \rangle = (-e) N F(t) \, . \tag{6.1}$$

For instance, for an electron beam with Gaussian distribution of the current along the beam, the profile function $F(t)$ is

$$F(t) = \frac{1}{\sqrt{2\pi}\sigma_T} \exp\left(-\frac{t^2}{2\sigma_T^2}\right) . \tag{6.2}$$

The probability of arrival of an electron during the time interval $t, t + dt$ is equal to $F(t)dt$.

The electron beam current, $I(t)$, and its Fourier transform, $\bar{I}(\omega)$, are connected by

$$\bar{I}(\omega) = \int_{-\infty}^{\infty} e^{i\omega t} I(t) dt = (-e) \sum_{k=1}^{N} e^{i\omega t_k} ,$$

$$I(t) = \frac{1}{2\pi} \int_{-\infty}^{\infty} \bar{I}(\omega) e^{-i\omega t} d\omega = (-e) \sum_{k=1}^{N} \delta(t - t_k) . \tag{6.3}$$

It follows from (6.3) that the Fourier transform of the input current, $\bar{I}(\omega)$, is the sum of a large number of complex phasors with random phases $\varphi_k = \omega t_k$. When the electron pulse duration σ_T is long, $\omega\sigma_T \gg 1$, the phases φ_k can be regarded as uniformly distributed on the interval $(0, 2\pi)$. In this case we can use the central limit theorem and conclude that the real part and the imaginary part of $\bar{I}(\omega)$ are distributed in accordance with the Gaussian law. The probability density distribution of $|\bar{I}(\omega)|^2$ is given by the negative exponential distribution:

$$p(|\bar{I}(\omega)|^2) = \frac{1}{\langle|\bar{I}(\omega)|^2\rangle} \exp\left(-\frac{|\bar{I}(\omega)|^2}{\langle|\bar{I}(\omega)|^2\rangle}\right) . \tag{6.4}$$

First-Order Spectral Correlation. Let us calculate the first-order correlation of the complex Fourier harmonics $\bar{I}(\omega)$ and $\bar{I}(\omega')$:

$$\langle \bar{I}(\omega) \bar{I}^*(\omega') \rangle = e^2 \left\langle \sum_{k=1}^{N} \sum_{n=1}^{N} \exp(i\omega t_k - i\omega' t_n) \right\rangle .$$

Expanding this relation, we can write:

$$\langle \bar{I}(\omega) \bar{I}^*(\omega') \rangle$$
$$= e^2 \left\langle \sum_{k=1}^{N} \exp[i(\omega - \omega')t_k] \right\rangle + e^2 \left\langle \sum_{k \neq n} \exp(i\omega t_k - i\omega' t_n) \right\rangle$$
$$= e^2 \sum_{k=1}^{N} \langle \exp[i(\omega - \omega')t_k] \rangle + e^2 \sum_{k \neq n} \langle \exp(i\omega t_k) \rangle \langle \exp(-i\omega' t_n) \rangle . \tag{6.5}$$

Taking into account (6.1) and (6.3), we find that $\langle \exp(i\omega t_k) \rangle$ is equal to the Fourier transform of the bunch profile function $F(t)$:

$$\langle \exp(i\omega t_k)\rangle = \int_{-\infty}^{\infty} F(t_k)e^{i\omega t_k}dt_k = \bar{F}(\omega) \ . \tag{6.6}$$

The Fourier transform of the Gaussian profile function (6.2) has the form:

$$\bar{F}(\omega) = \exp\left(-\frac{\omega^2 \sigma_T^2}{2}\right) \ .$$

Substituting (6.6) into (6.5), we obtain:

$$\langle \bar{I}(\omega)\bar{I}^*(\omega')\rangle = e^2 N \bar{F}(\omega - \omega') + e^2 N(N-1)\bar{F}(\omega)\bar{F}^*(\omega') \ . \tag{6.7}$$

When

$$N|\bar{F}(\omega)|^2 \ll 1 \ , \tag{6.8}$$

we can write the following expression for the first-order spectral correlation:

$$\langle \bar{I}(\omega)\bar{I}^*(\omega')\rangle = e^2 N \bar{F}(\omega - \omega') \ . \tag{6.9}$$

When $\omega = \omega'$, (6.7) gives the value of the spectral density of the process, $\langle |\bar{I}(\omega)|^2\rangle$. When (6.8) is fulfilled, we can neglect the contribution of coherent bunching (the second term in the right-hand side of (6.7)) to the spectral density and consider only the shot noise contribution, $e^2 N$. This is described by (6.9).

For the specific cases of a Gaussian and rectangular profile of the electron bunch, the first-order correlation of the complex Fourier harmonics, $\bar{I}(\omega)$ and $\bar{I}(\omega')$, has the form:

Gaussian profile:

$$\langle \bar{I}(\omega)\bar{I}^*(\omega')\rangle = e^2 N \exp\left[-\frac{(\omega-\omega')^2 \sigma_T^2}{2}\right] \ , \tag{6.10a}$$

Rectangular profile with pulse duration T:

$$\langle \bar{I}(\omega)\bar{I}^*(\omega')\rangle = e^2 N \left[\frac{(\omega-\omega')T}{2}\right]^{-1} \sin\left[\frac{(\omega-\omega')T}{2}\right] \ . \tag{6.10b}$$

Let us discuss the region of validity of the approximation (6.8). The physical meaning of (6.8) is that the frequency ω has to be large enough, $\omega \sigma_T \gg 1$. To be specific, we consider a numerical example for a Gaussian bunch profile. The value of $|\bar{F}(\omega)|^2$ is equal to $\exp(-100)$ when $\sigma_T \omega = 10$. As a rule, the number of particles in the bunch, N, is less than or about 10^{11}, so condition (6.8) is always fulfilled in practice. If we consider the case of a rectangular bunch profile, we find that the region of applicability of condition (6.8) is less than that for the case of a Gaussian bunch of the same duration. This is due to the fact that the bunch form factor, $|\bar{F}(\omega)|^2$, decreases more slowly with an increase in frequency. On the other hand, in a realistic situation there is no sharp boundary of the bunch and the beam current falls to zero during some time interval $\Delta\sigma_T \ll T$. When the beam current at the edge falls in accordance with a Gaussian law, $\Delta\sigma_T$ must obey the following conditions:

$$\frac{\Delta\sigma_\mathrm{T}}{T} \ll 1, \qquad \Delta\sigma_\mathrm{T}\omega \gg 1, \qquad \frac{N}{(\Delta\sigma_\mathrm{T}\omega)^4(\omega T)^2} \ll 1 \ .$$

In this case (6.10b) is valid within the boundaries:

$$|(\omega - \omega')| \ll \frac{1}{\Delta\sigma_\mathrm{T}} \ .$$

Second-Order Spectral Correlation. Let us calculate the second-order correlation of the complex Fourier harmonics $\bar{I}(\omega)$ and $\bar{I}(\omega')$:

$$\langle |\bar{I}(\omega)|^2 |\bar{I}(\omega')|^2 \rangle$$

$$= e^4 \left\langle \sum_{n=1}^{N}\sum_{m=1}^{N}\sum_{p=1}^{N}\sum_{q=1}^{N} \exp\left[\mathrm{i}\omega(t_n - t_m) + \mathrm{i}\omega'(t_p - t_q)\right] \right\rangle \ .$$

The N^4 terms in this sum can be set in 15 different classes. When condition (6.8) is fulfilled, only two of them are of importance, corresponding to ($n = m$, $p = q$, $n \neq p$) and ($n = q$, $m = p$, $n \neq m$). Thus, we can write:

$$\langle |\bar{I}(\omega)|^2 |\bar{I}(\omega')|^2 \rangle = \langle |\bar{I}(\omega)|^2 \rangle \langle |\bar{I}(\omega')|^2 \rangle + |\langle \bar{I}(\omega)\bar{I}^*(\omega') \rangle|^2 \ . \qquad (6.11)$$

Substituting (6.9) into (6.11), we obtain:

$$\langle |\bar{I}(\omega)|^2 |\bar{I}(\omega')|^2 \rangle = e^4 N^2 (1 + |\bar{F}(\omega - \omega')|^2) \ . \qquad (6.12)$$

6.2 One-Dimensional Theory of SASE FEL

6.2.1 Analytical Description of the Linear Regime

Green's Function. In the previous section we described the properties of the input shot noise signal in the frequency domain. The next step is the derivation of the spectral Green's function connecting the Fourier amplitudes of the output field and the Fourier amplitudes of the input noise signal. We consider a rectangular electron bunch of sufficient duration T, such as

$$\rho\omega_0 T \gg 1 \ , \qquad (6.13)$$

where ω_0 is the resonance frequency and ρ is the efficiency parameter. The definitions of the parameters used in this section are identical to those used in Chap. 2. The physical interpretation of approximation (6.13) is that the electron bunch is much longer than the slippage of the radiation with respect to the electrons per gain length. It allows us to neglect edge effects and to use the steady-state Green's function for any frequency within the FEL bandwidth. It is assumed that only the fluctuations of the beam current density define the value of the input signal. This means that we can neglect the effect of longitudinal velocity fluctuations connected with the finite energy spread in the beam. One can show that the ratio of the noise signal due to the velocity fluctuations to the shot noise signal is of the order of the normalized energy spread parameter $\hat{\Lambda}_\mathrm{T}^2$. In practice this parameter is always small otherwise it destroys FEL amplifier operation. We also assume that the transversely coherent fraction of the input shot noise signal is defined by the value of the beam current:

$$\frac{\bar{j}(\omega)}{j_0} = \frac{\bar{I}(\omega)}{I_0} ,\tag{6.14}$$

where $\bar{j}(\omega)$ is the Fourier harmonic of the beam current density in the one-dimensional model, $I_0 = eN/T$ is the average beam current, $j_0 = I_0/S$, and S is the transverse area of the electron beam.

Let us recall some results of the steady-state, one-dimensional theory of the FEL amplifier (see Chap. 2). The one-dimensional model describes the amplification of the plane electromagnetic wave,[1]

$$E_y = \tilde{E}_y(\omega, z) \exp\left[i\omega(z/c - t)\right] + \text{C.C.} ,$$

by the electron beam in the undulator. In Chap. 2 we considered the special case of initial conditions when the electron beam, modulated at frequency ω, is fed to the undulator entrance. The output radiation also has the same frequency, ω. In the high-gain linear regime the complex amplitude, $\tilde{E}_y(\omega, z)$, grows exponentially with undulator length:

$$\tilde{E}_y(\omega, z) \propto \exp(\Lambda z) .$$

In this chapter we study the case when the initial modulation of the electron beam is defined by the shot noise and has a white spectrum. Under the accepted limitations the results of the steady-state theory can be extended to this complicated case. Indeed, we can decompose the input signal into Fourier harmonics. Since in the linear regime all the harmonics are amplified independently, we can use the results of the steady-state theory for each harmonic and calculate the corresponding Fourier harmonics of the output radiation field. The electric field of the electromagnetic wave in the time domain, $E_y(z, t)$, and its Fourier transform, $\bar{E}(\omega, z)$, are connected by

$$E_y(z, t) = \frac{1}{2\pi} \int_{-\infty}^{\infty} \bar{E}(\omega, z) e^{-i\omega t} d\omega .\tag{6.15}$$

The Fourier harmonic of the electromagnetic field at the undulator exit and the Fourier harmonic of the input current are connected by the relation:

$$\bar{E}(\omega, z) = H_A(\omega, z) \bar{I}(\omega), \qquad \omega > 0 ,\tag{6.16}$$

where H_A is the spectral Green's function of the FEL amplifier. When $\omega < 0$ the Fourier harmonic is defined by the relation $\bar{E}^*(\omega, z) = \bar{E}(-\omega, z)$. We consider the case of a negligibly small space charge field and energy spread in the electron beam, $\hat{\Lambda}_p^2 \to 0$ and $\hat{\Lambda}_T^2 \to 0$. In the high-gain linear regime, the

---

[1] To be specific, we consider in this chapter the case of a planar undulator and linearly polarized electromagnetic wave. All the results can also be used for the case of a helical undulator and circularly polarized radiation by corresponding redetermination of the reduced parameters.

contribution of the term corresponding to the growing root of the eigenvalue equation,

$$\hat{\Lambda}(\hat{\Lambda} + i\hat{C})^2 = i,$$

dominates in the Green's function (see Section 2.1.4). The growing root can be expanded near exact resonance as

$$\operatorname{Re}\hat{\Lambda} = \frac{\sqrt{3}}{2}\left(1 - \frac{\hat{C}^2}{9}\right), \quad \operatorname{Im}\hat{\Lambda} = \frac{1}{2}\left(1 - \frac{4\hat{C}}{3}\right), \qquad (6.17)$$

where $\hat{\Lambda} = \Lambda/\Gamma$ is the normalized eigenvalue, $\hat{C} = C/\Gamma = (\omega_0 - \omega)/(2\rho\omega_0)$ is the detuning parameter and ω_0 is the resonance frequency. Finally, the Green's function is written in the following form:

$$H_A(\omega) = \frac{2}{3}\exp\left(i\frac{\omega}{c}z\right)\exp\left[\frac{\sqrt{3}}{2}\left(1 - \frac{\hat{C}^2}{9}\right)\hat{z}\right.$$
$$\left. + \frac{i}{2}\left(1 - \frac{4\hat{C}}{3}\right)\hat{z}\right]\frac{E_0}{I_0}, \qquad (6.18)$$

where E_0 is the normalizing factor for the electric field of the wave. All the normalizing factors and reduced parameters for the case of a planar undulator are defined in Sect. 2.2.7.

Prior to obtaining a rigorous solution of the problem, it is relevant to present a qualitative physical picture. First, we should remember that the initial modulation of the electron beam is defined by the shot noise and has a white spectrum. Second, we obtain from (6.18) that the high-gain FEL amplifier cuts and amplifies only a narrow frequency band of the initial spectrum. We can expect that the typical width of the radiation spectrum envelope, $\Delta\omega$, should be of the order of $\rho\omega_0\Delta\hat{C} \simeq \rho\omega_0$. Also, a single-shot spectrum of the radiation pulse having duration T should contain spikes with a typical width of about $1/T$. Thus, the number of spikes in the spectrum should be about $\rho\omega_0 T$. As for the time structure of the radiation pulse, we should expect that the radiation pulse should consist of about $\rho\omega_0 T$ spikes (wavepackets) with a typical duration of about $1/(\rho\omega_0)$ given by the inverse width of the spectrum envelope. Analysis of the plots in Fig. 6.1 an 6.2 shows that these simple physical estimates agree well with the actual properties of the radiation.

Let us undertake a quantitative investigation of the problem. We start with the calculation of the average radiation power at the undulator exit. Using the expression for Poynting's vector and Parseval's theorem, we calculate the radiation energy in one radiation pulse:

$$\mathcal{E} = \frac{cS}{4\pi}\int_0^T E_y^2(t)\mathrm{d}t = \frac{cS}{4\pi^2}\int_0^\infty |\bar{E}(\omega)|^2\mathrm{d}\omega, \qquad (6.19)$$

where S is the transverse area of the electron beam. The radiation energy, averaged over an ensemble, is given by the expression:

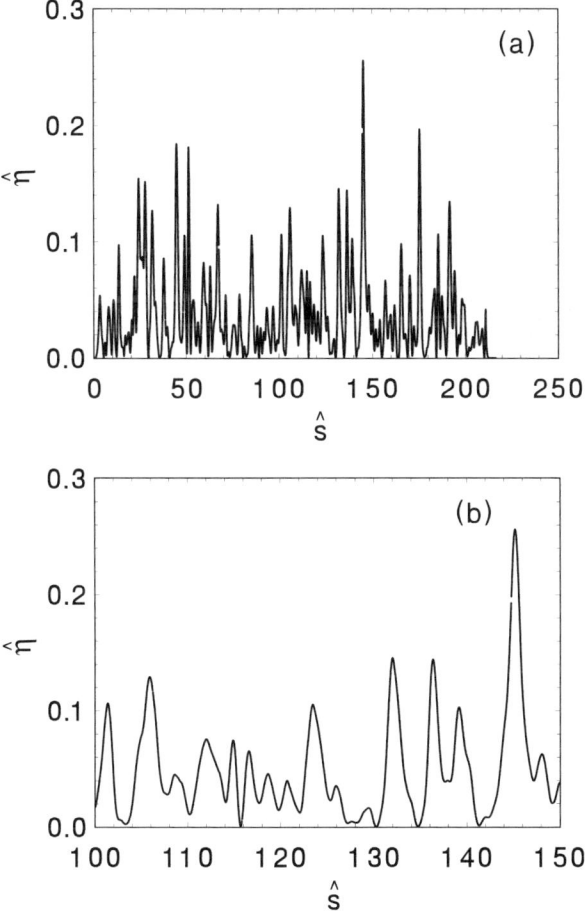

Fig. 6.1. Normalized power in one radiation pulse versus $\hat{s} = \rho\omega_0(z/\bar{v}_z - t)$ for the reduced length $\hat{z} = 10$ of the FEL amplifier. Graph (**a**) is plotted over the full length of the radiation pulse and graph (**b**) presents an enlarged fraction of graph (**a**). The calculations have been performed with the linear simulation code

$$\langle \mathcal{E} \rangle = \frac{cS}{4\pi^2} \int_0^\infty \langle |\bar{E}(\omega)|^2 \rangle d\omega, \tag{6.20}$$

where $\langle |\bar{E}(\omega)|^2 \rangle$ is calculated using (6.16) and (6.18):

$$\langle |\bar{E}(\omega)|^2 \rangle = A \exp\left[-\frac{(\omega - \omega_0)^2}{2\sigma_A^2}\right] \langle |\bar{I}(\omega)|^2 \rangle, \qquad \omega > 0. \tag{6.21}$$

Here the following notation is introduced:

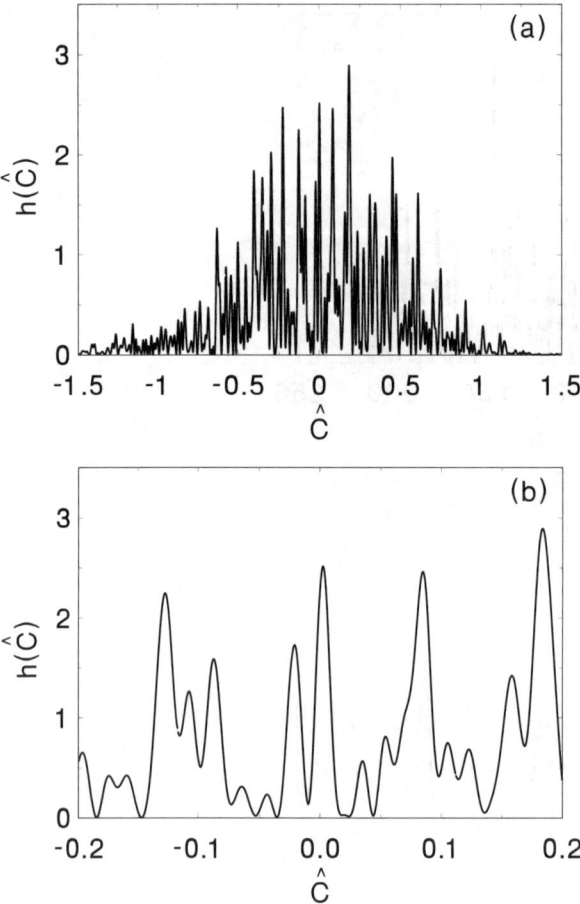

Fig. 6.2. Normalized energy spectrum of one radiation pulse for the reduced length $\hat{z} = 10$ of the FEL amplifier. Graph (**a**) is plotted over the full spectrum width and graph (**b**) presents an enlarged fraction of graph (**a**). The calculations have been performed with the linear simulation code

$$A = \frac{4}{9}\left(\frac{E_0}{I_0}\right)^2 \exp(\sqrt{3}\hat{z}), \qquad \sigma_A = 3\sqrt{\frac{2}{\sqrt{3}}\frac{\rho\omega_0}{\sqrt{\hat{z}}}}. \tag{6.22}$$

We integrate (6.20) and divide the result by the pulse duration, T. This gives us the average radiation power:

$$\langle W_{\text{out}} \rangle = \rho W_{\text{b}} \frac{\sqrt{4\pi\rho}}{3\sqrt{\sqrt{3}\hat{z}N_\lambda}} \exp(\sqrt{3}\hat{z}), \tag{6.23}$$

where $W_{\text{b}} = \gamma m_e c^2 I_0/e$ is the electron beam power and $N_\lambda = 2\pi I_0/(e\omega_0)$ is the number of electrons per radiation wavelength. Introducing the notion of the number of cooperating electrons, $N_c = N_\lambda/(2\pi\rho)$, we rewrite (6.23) in the following normalized form:

$$\langle\hat{\eta}\rangle = \frac{\langle W_{\text{out}}\rangle}{\rho W_b} = \frac{\exp(\sqrt{3}\hat{z})}{3\sqrt{\sqrt{3}\pi\hat{z}}N_c} . \tag{6.24}$$

One can see that the normalized output power (or, the normalized FEL efficiency) is a function of the normalized undulator length and of the number of cooperating electrons.

The radiation of the SASE FEL consists of a large number of independent wavepackets. An important characteristic of the wavepacket is the group velocity,

$$\frac{1}{v_g} = \frac{dk}{d\omega} = \frac{1}{c} + \frac{d\,\text{Im}\,\Lambda}{d\omega} , \tag{6.25}$$

where $k = \omega/c + \text{Im}\,\Lambda$ is the wavenumber. Expanding the imaginary part of the eigenvalue (6.17) near exact resonance, we can write:

$$\frac{d\,\text{Im}\,\Lambda}{d\omega} = -\frac{1}{2c\gamma_\ell^2}\frac{d\,\text{Im}\,\hat{\Lambda}}{d\hat{C}} = \frac{1}{3c\gamma_\ell^2} . \tag{6.26}$$

It follows from (6.25) and (6.26) that

$$v_g = c\left(1 - \frac{1}{3\gamma_\ell^2}\right) . \tag{6.27}$$

It is interesting to compare the slippage rate of the radiation wavepackets with respect to the electrons and the usual kinematic slippage rate, $c - \bar{v}_z = c/(2\gamma_\ell^2)$:

$$\frac{v_g - \bar{v}_z}{c - \bar{v}_z} = \frac{1}{3} . \tag{6.28}$$

We can conclude that the relative slippage of the wavepackets (spikes) with respect to the electron beam is three time less than the kinematic slippage. This effect takes place due to the dispersive properties of the active medium — the electron beam in the undulator.

Analysis of the Radiation Properties in the Frequency Domain. In the previous section we studied the properties of the Fourier harmonics of the shot noise in the frequency domain. The Fourier harmonics of the output radiation field are connected with the Fourier harmonics of the input shot noise by (6.16), where the Green's function is given by (6.18). It follows from (6.30) that the statistical properties of the Fourier amplitudes $\bar{E}(\omega)$ are defined by the statistical properties of the Fourier amplitudes of the input current, $\bar{I}(\omega)$. In particular, it follows from (6.4) that $|\bar{E}(\omega)|^2$ is distributed in accordance with the negative exponential probability function:

$$p(|\bar{E}(\omega)|^2) = \frac{1}{\langle|\bar{E}(\omega)|^2\rangle}\exp\left(-\frac{|\bar{E}(\omega)|^2}{\langle|\bar{E}(\omega)|^2\rangle}\right) . \tag{6.29}$$

For many practical applications of the SASE FEL radiation a monochromator has to be installed at the FEL amplifier exit. We denote the transmission function of the monochromator as $H_\mathrm{m}(\omega)$. Thus, (6.16) for the Fourier harmonic of the radiation field can be generalized in the following way:

$$\bar{E}(\omega) = H_\mathrm{m}(\omega) H_A(\omega) \bar{I}(\omega) \ , \qquad \omega > 0 \ . \tag{6.30}$$

For a narrow-band monochromator (with resolution better than the typical width of the spike in the spectrum), the energy in the radiation pulse, $\mathcal{E}$, after the monochromator is proportional to $|\bar{E}(\omega)|^2$. So, this energy fluctuates from pulse to pulse in accordance with the negative exponential probability distribution (6.29). It is worth mentioning that such a distribution is a feature of completely chaotic polarized radiation. In Fig. 6.5 below we present the results of numerical simulations of the probability distribution of the radiation energy after the monochromator. The upper plot corresponds to the case of a narrow-band monochromator. It is seen that within statistical accuracy the numerical results fit perfectly with the general result (6.29) predicted by the central limit theorem.

First-Order Spectral Correlations. The first-order spectral correlation function is defined as

$$g_1(\omega, \omega') = \frac{\langle \bar{E}(\omega) \bar{E}^*(\omega') \rangle}{\left[\langle |\bar{E}(\omega)|^2 \rangle \langle |\bar{E}(\omega')|^2 \rangle \right]^{1/2}} \ . \tag{6.31}$$

Substituting (6.30) and (6.9) into (6.31), we obtain an analytical expression for the first-order spectral correlation function of the FEL amplifier with a rectangular electron bunch:

$$|g_1(\omega, \omega')| = |\bar{F}(\omega - \omega')| = \left[\frac{(\omega - \omega')T}{2} \right]^{-1} \sin\left[\frac{(\omega - \omega')T}{2} \right] \ . \tag{6.32}$$

We define the spectral interval of coherence as

$$\Delta \omega_\mathrm{c} = \int_{-\infty}^{\infty} |g_1(\omega - \omega')|^2 \mathrm{d}(\omega - \omega') \ . \tag{6.33}$$

The value of the spectral coherence for the case of a rectangular bunch is given by

$$\Delta \omega_\mathrm{c} = \frac{2\pi}{T} \ . \tag{6.34}$$

Second-Order Spectral Correlations. The second-order spectral correlation function is defined as

$$g_2(\omega, \omega') = \frac{\langle |\bar{E}(\omega)|^2 |\bar{E}(\omega')|^2 \rangle}{\langle |\bar{E}(\omega)|^2 \rangle \langle |\bar{E}(\omega')|^2 \rangle} \ . \tag{6.35}$$

Using (6.30), (6.12) and (6.35), we find that the first and second order correlation functions are connected by the relation:

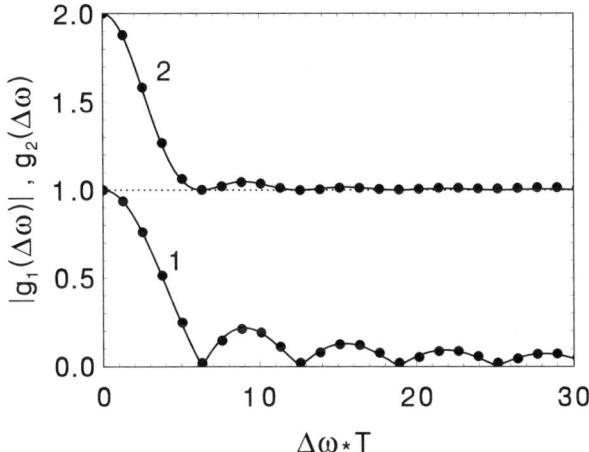

Fig. 6.3. Spectral correlation functions for the reduced length $\hat{z} = 10$ of the FEL amplifier. The solid curves are the results of calculations with analytical formulae (6.32) and (6.37). The circles are the result of calculations with linear simulation code over 5×10^4 independent statistical events. Curve 1 refers to the modulus of the first-order correlation function and curve 2 refers to the second-order correlation function

$$g_2(\omega, \omega') = 1 + |g_1(\omega, \omega')|^2 \;, \tag{6.36}$$

which is also a general property of completely chaotic polarized radiation. The explicit expression for the second-order spectral correlation function of the FEL amplifier with rectangular electron bunch has the form:

$$g_2(\omega, \omega') = 1 + \left[\frac{(\omega - \omega')T}{2}\right]^{-2} \sin^2\left[\frac{(\omega - \omega')T}{2}\right] \;. \tag{6.37}$$

Plots of the first and second order correlation functions are presented in Fig. 6.3.

Fluctuations of the Radiation Energy at the Exit of the Monochromator. Let us consider a typical experimental situation when the monochromator is installed at the exit of the FEL amplifier. The important practical problem concerns the fluctuations of the radiation energy, $\mathcal{E}$, after the monochromator. Using (6.20) and (6.30), we calculate the average energy measured after the monochromator:

$$\langle \mathcal{E} \rangle = \frac{cS}{4\pi^2} \int_0^\infty \langle |\bar{E}(\omega)|^2 \rangle d\omega = \frac{ce^2SN}{4\pi^2} \int_0^\infty |H_{\mathrm{m}}(\omega)|^2 |H_{\mathrm{A}}(\omega)|^2 d\omega \;. \tag{6.38}$$

It is seen that the average energy is a function of the frequency profile of the monochromator and of the gain profile of the FEL amplifier.

The normalized dispersion of the energy distribution is calculated as follows:

368   6. FEL Amplifier Start-up from Shot Noise

$$\sigma_{\mathcal{E}}^2 = \frac{\langle (\mathcal{E} - \langle \mathcal{E} \rangle)^2 \rangle}{\langle \mathcal{E} \rangle^2} = \frac{\int_0^\infty d\omega \int_0^\infty d\omega' \langle |\bar{E}(\omega)|^2 |\bar{E}(\omega')|^2 \rangle}{\int_0^\infty d\omega \langle |\bar{E}(\omega)|^2 \rangle \int_0^\infty d\omega' \langle |\bar{E}(\omega')|^2 \rangle} - 1 \ . \tag{6.39}$$

Using definition (6.35) of the second-order correlation function and (6.36), we reduce this expression to:

$$\sigma_{\mathcal{E}}^2 = \frac{\int_0^\infty d\omega \int_0^\infty d\omega' \langle |\bar{E}(\omega)|^2 \rangle \langle |\bar{E}(\omega')|^2 \rangle |g_1(\omega', \omega')|^2}{\int_0^\infty d\omega \langle |\bar{E}(\omega)|^2 \rangle \int_0^\infty d\omega' \langle |\bar{E}(\omega')|^2 \rangle} \ . \tag{6.40}$$

Analysis of this expression shows that the energy dispersion after the monochromator is a function of the frequency profile of the monochromator, of the gain profile of the FEL amplifier, and of the electron bunch form factor.

Let the monochromator transmission function, $H_m(\omega)$, be symmetric with respect to the FEL resonance frequency, ω_0. We consider two cases of the monochromator line profile, a Gaussian and a rectangular profile. The rectangular line of the monochromator is given by ($\omega > 0$):

$$|H_m(\omega)|^2 = 1 \quad \text{for} \quad |\omega - \omega_0| < \frac{\Delta \omega_m}{2} \ ,$$

$$|H_m(\omega)|^2 = 0 \quad \text{for} \quad |\omega - \omega_0| > \frac{\Delta \omega_m}{2} \ , \tag{6.41}$$

and the Gaussian line of the monochromator is defined as

$$|H_m(\omega)|^2 = \exp\left[-\frac{(\omega - \omega_0)^2}{2\sigma_m^2}\right] \ . \tag{6.42}$$

Substituting (6.30), (6.21), (6.32), and (6.41) into (6.40) we obtain an expression for the radiation energy dispersion after the monochromator with a rectangular line profile:

$$\sigma_{\mathcal{E}}^2 = \left[\frac{\pi}{2} \hat{\sigma}_A^2 \mathrm{erf}^2\left(\frac{\Delta \hat{\omega}_m}{2\sqrt{2} \hat{\sigma}_A}\right)\right]^{-1}$$

$$\times \left\{ \sqrt{\pi} \hat{\sigma}_A \mathrm{erf}\left(\frac{\Delta \hat{\omega}_m}{2 \hat{\sigma}_A}\right) \left[\mathrm{Si}(\Delta \hat{\omega}_m) - \frac{1 - \cos(\Delta \hat{\omega}_m)}{\Delta \hat{\omega}_m}\right] \right.$$

$$\left. + \exp\left(-\frac{\Delta \hat{\omega}_m^2}{4 \hat{\sigma}_A^2}\right) [\mathrm{Ci}(\Delta \hat{\omega}_m) - \ln(\Delta \hat{\omega}_m) - \gamma_E] \right\} \ , \tag{6.43}$$

where $\Delta \hat{\omega}_m = \Delta \omega_m T$, $\hat{\sigma}_A = \sigma_A T$, $\gamma_E = 0.577...$ is Euler's constant, erf(...) is the error function, and Ci(...) and Si(...) are the integral cosine and sine, respectively.

The corresponding expression for the Gaussian monochromator line has the form:

$$\sigma_{\mathcal{E}}^2 = \frac{\sqrt{\pi}}{\hat{\sigma}^2} \int_0^{\hat{\sigma}} \mathrm{erf}(x)\,\mathrm{d}x \,, \tag{6.44}$$

where

$$\hat{\sigma} = \frac{\hat{\sigma}_\mathrm{A}\hat{\sigma}_\mathrm{m}}{\sqrt{\hat{\sigma}_\mathrm{A}^2 + \hat{\sigma}_\mathrm{m}^2}}\,, \qquad \hat{\sigma}_\mathrm{m} = \sigma_\mathrm{m} T \,.$$

Let us study the asymptotic behavior of (6.43) and (6.44). When the monochromator linewidth is much less than the interval of spectral coherence (6.34), the normalized dispersion tends to unity:

$$\sigma_{\mathcal{E}}^2 \simeq 1 \qquad \text{for} \quad (\Delta\omega_\mathrm{m} T \,,\ \sigma_\mathrm{m} T) \ll 1 \,. \tag{6.45}$$

When the monochromator linewidth is much larger than the interval of spectral coherence and much less than the FEL amplifier bandwidth, the dispersion is inversely proportional to the monochromator linewidth,

$$\begin{aligned}\sigma_{\mathcal{E}}^2 &\simeq 2\pi/(\Delta\omega_\mathrm{m} T) & \text{for} \quad 1 \ll \Delta\omega_\mathrm{m} T \ll \sigma_\mathrm{A} T \,, \\ \sigma_{\mathcal{E}}^2 &\simeq \sqrt{\pi}/(\sigma_\mathrm{m} T) & \text{for} \quad 1 \ll \sigma_\mathrm{m} T \ll \sigma_\mathrm{A} T \,,\end{aligned} \tag{6.46}$$

for the rectangular and the Gaussian monochromator line, respectively. When the monochromator linewidth is much larger than the bandwidth of the FEL amplifier, the energy fluctuations are defined by the bandwidth of the FEL amplifier:

$$\sigma_{\mathcal{E}}^2 \simeq \frac{\sqrt{\pi}}{\sigma_\mathrm{A} T} \qquad \text{at} \quad (\Delta\omega_\mathrm{m}\,,\ \sigma_\mathrm{m}) \gg \sigma_\mathrm{A} \,. \tag{6.47}$$

Now we complicate the investigation and calculate the probability distribution of the radiation energy after the monochromator, $p(\mathcal{E})$. Using the well-known results obtained in the framework of statistical optics, we can state that the distribution of the radiation energy after the monochromator is described rather well by the gamma probability density function:

$$p(\mathcal{E}) = \frac{M^M}{\Gamma(M)} \left(\frac{\mathcal{E}}{\langle\mathcal{E}\rangle}\right)^{M-1} \frac{1}{\langle\mathcal{E}\rangle} \exp\left(-M\frac{\mathcal{E}}{\langle\mathcal{E}\rangle}\right) \,, \tag{6.48}$$

where $\Gamma(M)$ is the gamma function with argument M, and

$$M = \frac{1}{\sigma_{\mathcal{E}}^2} \,. \tag{6.49}$$

This distribution provides correct values for the mean value of $\mathcal{E}$ and for the dispersion $\sigma_{\mathcal{E}}^2 = 1/M$:

$$\int_0^\infty \mathcal{E} p(\mathcal{E})\,\mathrm{d}\mathcal{E} = \langle\mathcal{E}\rangle, \qquad \int_0^\infty \frac{(\mathcal{E} - \langle\mathcal{E}\rangle)^2}{\langle\mathcal{E}\rangle^2} p(\mathcal{E})\,\mathrm{d}\mathcal{E} = \frac{1}{M} \,.$$

The parameter M can be interpreted as the average number of degrees of freedom (or modes) in the radiation pulse. It follows from (6.40) that this

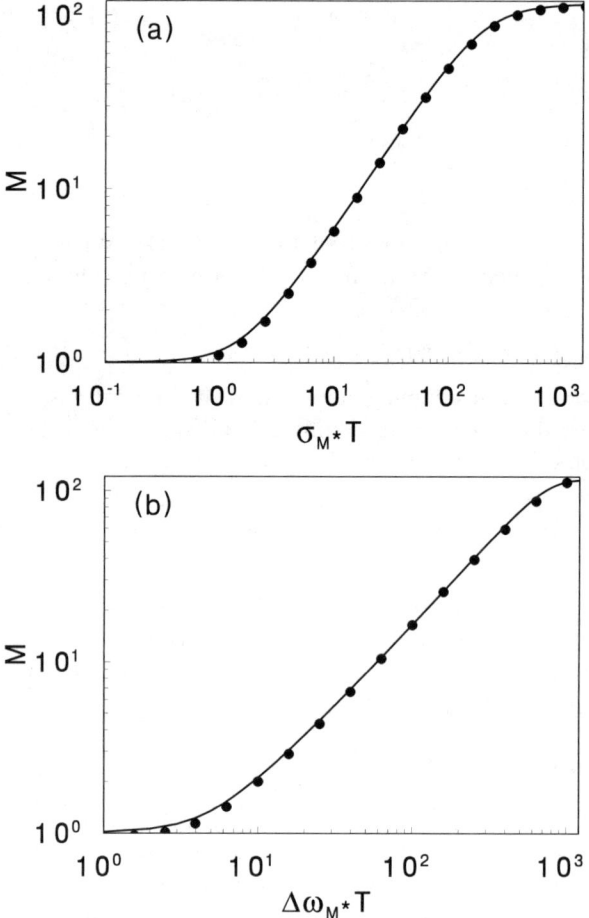

Fig. 6.4. The dependence of the parameter M on the value of the width of the monochromator with Gaussian profile (graph (**a**)) and with rectangular profile (graph (**b**)). The FEL amplifier operates in the linear regime at the reduced length $\hat{z} = 10$ of the FEL amplifier. Solid curves are the result of calculations with analytical formulae (6.44) and (6.43). Circles are the results of calculations with linear simulation code over 2400 shots

parameter cannot be less than unity. When M tends to the unity, (6.48) tends to the negative exponential distribution (6.29). For large values of M the distribution (6.48) tends to a Gaussian distribution. Figure 6.4 presents the plots for the parameter M as a function of the monochromator linewidth. Figure 6.5 shows the distribution of the energy after the monochromator for different values of the linewidth (which correspond to different values of the parameter M).

6.2 One-Dimensional Theory of SASE FEL 371

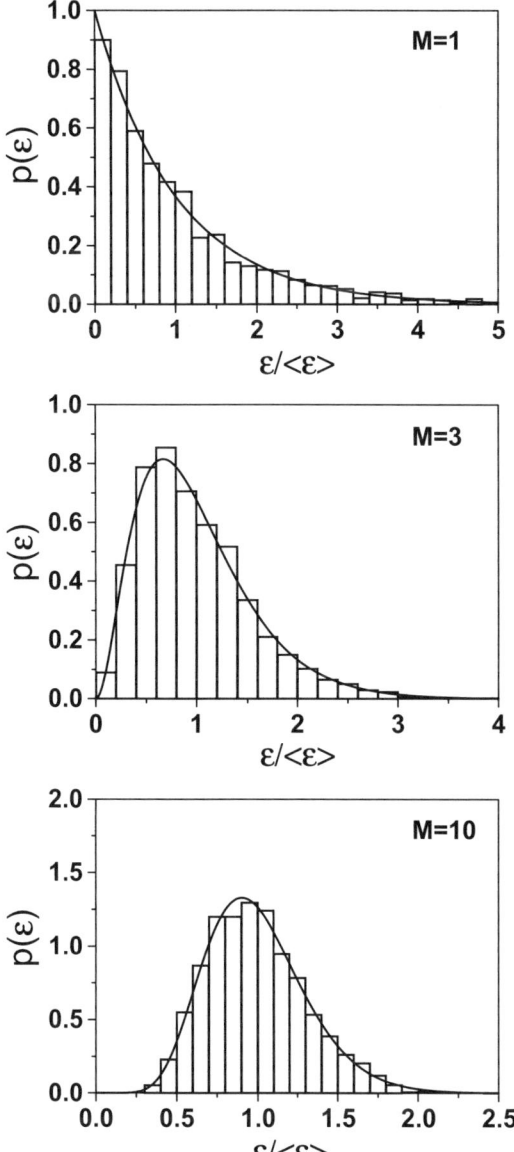

Fig. 6.5. Histograms of the probability density distribution, $p(\mathcal{E})$, of the radiation energy after the monochromator for different values of the width of the monochromator with Gaussian profile. Calculations have been performed with linear simulation code over 2400 shots (see Fig. 6.2). $\langle \mathcal{E} \rangle$ denotes the average energy. The solid curves represent the gamma distribution (6.48) where the values of M have been calculated with (6.44) and (6.49) (see also Fig. 6.4)

Analysis of the Radiation Properties in the Time Domain. An expression for the electric field of the electromagnetic wave as a function of time t and coordinate z can be obtained using the Fourier transform of (6.16). It is convenient to isolate explicitly the slowly varying complex amplitude:

$$E_y(z,t) = \tilde{E}(z,t)e^{i\omega_0(z/c-t)} + \text{C.C.} , \qquad (6.50)$$

where

$$\tilde{E}(z,t)e^{i\omega_0(z/c-t)} = \frac{1}{2\pi}\int_0^\infty \bar{E}(\omega)e^{-i\omega t}d\omega .$$

Using (6.3), (6.30), (6.16), (6.18), and (6.50), we calculate the complex amplitude $\tilde{E}(t)$ of the electric field at a distance $\hat{z} = \Gamma z$ from the undulator entrance:

$$\frac{\tilde{E}(t)}{E_0} = \frac{2\sqrt{2}\rho\omega_0 T}{\sqrt{\sqrt{3\pi}\hat{z}}} \exp\left[(\sqrt{3}+i)\frac{\hat{z}}{2}\right]$$

$$\times \frac{1}{N}\sum_{k=1}^{N} \exp(i\omega_0 t_k)\exp\left[-\sigma_A^2(t-t_z-t_k)^2\right] , \qquad (6.51)$$

where

$$t_z = \frac{z}{c}\left(1 + \frac{1}{3\gamma_\ell^2}\right) ,$$

and σ_A is defined by (6.22). Here and below we omit the z coordinate in the list of arguments assuming that $\tilde{E}(t)$ is calculated at the given z coordinate along the undulator. Since

$$\sigma_A \ll \omega_0 , \qquad (6.52)$$

we can approximately let the amplitude and the phase of each random phasor contributing to the sum in (6.51) be independent of each other, and the phases be uniformly distributed on the interval $(0, 2\pi)$. Thus, the distribution of the instantaneous radiation power, $W \propto |\tilde{E}|^2$, is the negative exponential distribution:

$$p(|\tilde{E}(t)|^2) = \frac{1}{\langle|\tilde{E}(t)|^2\rangle} \exp\left(-\frac{|\tilde{E}(t)|^2}{\langle|\tilde{E}(t)|^2\rangle}\right) . \qquad (6.53)$$

In Fig. 6.6 we present the results of numerical simulations of the fluctuations of the instantaneous radiation power. It is seen that this distribution is negative exponential (6.53).

First-Order Time Correlations. The correlation between the radiation fields at times t and t' has the form:

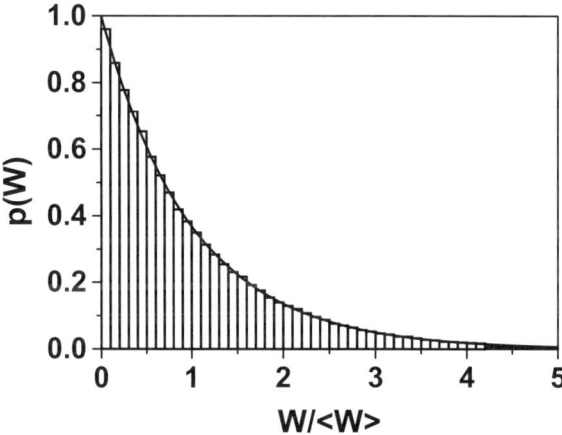

Fig. 6.6. Histogram of the probability density distribution, $p(W)$, of the instantaneous output power for the reduced length $\hat{z} = 10$ of the FEL amplifier. Calculations have been performed with linear simulation code over 2×10^5 independent statistical events (see Fig. 6.1). $\langle W \rangle$ denotes the average power. The solid curve represents the negative exponential distribution $p(W) = \exp(-W/\langle W \rangle)$

$$\langle \tilde{E}(t)\tilde{E}^*(t') \rangle \exp[-i\omega_0(t-t')]$$

$$= \frac{1}{4\pi^2} \int_0^\infty d\omega \int_0^\infty d\omega' \exp(-i\omega t + i\omega' t') \langle \bar{E}(\omega)\bar{E}^*(\omega') \rangle$$

$$= \frac{1}{4\pi^2} \int_0^\infty d\omega \int_0^\infty d\omega' \exp(-i\omega t + i\omega' t')$$

$$\times \left[\langle |\bar{E}(\omega)|^2 \rangle \langle |\bar{E}(\omega')|^2 \rangle \right]^{1/2} \bar{F}(\omega - \omega') \ . \tag{6.54}$$

Here (6.31) and (6.32) have been used when rewriting the integral. In the present consideration we use approximation (6.13) which assumes the interval of spectral coherence, $\Delta\omega_c = 2\pi/T$, to be much less than the FEL amplifier bandwidth, $\sigma_A \simeq \rho\omega_0$. Thus, we can simplify the integral in (6.54) in the following way. We replace the expression in square brackets by $\langle |\bar{E}(\omega)|^2 \rangle$, and after integration over ω and $\Delta\omega = (\omega - \omega')$ we obtain:

$$\langle \tilde{E}(t)\tilde{E}^*(t') \rangle \exp[-i\omega_0(t-t')]$$

$$\simeq \frac{1}{4\pi^2} \int_{-\infty}^\infty d(\Delta\omega) \exp(-i\Delta\omega t) \bar{F}(\Delta\omega) \int_0^\infty d\omega \exp[-i\omega(t-t')] \langle |\bar{E}(\omega)|^2 \rangle$$

$$= \frac{F(t)}{2\pi} \int_0^\infty d\omega \exp[-i\omega(t-t')] \langle |\bar{E}(\omega)|^2 \rangle \ , \tag{6.55}$$

where $F(t)$ is the radiation pulse profile. In this book we consider narrow-band signals and define the first-order time correlation function as follows:

$$g_1(t-t') = \frac{\langle \tilde{E}(t)\tilde{E}^*(t') \rangle}{\left[\langle |\tilde{E}(t)|^2 \rangle \langle |\tilde{E}(t')|^2 \rangle \right]^{1/2}} . \tag{6.56}$$

Using (6.55), we can write for the long rectangular pulse (see (6.13)):

$$g_1(t-t') = \frac{\int\limits_0^\infty d\omega \langle |\bar{E}(\omega)|^2 \rangle \exp\left[-i(\omega - \omega_0)(t - t')\right]}{\int\limits_0^\infty d\omega \langle |\bar{E}(\omega)|^2 \rangle} .$$

Since we deal with a narrow-band signal, the latter expression may be rewritten as follows

$$g_1(t-t') = \frac{\int\limits_{-\infty}^\infty d(\Delta\omega) \langle |\bar{E}(\Delta\omega)|^2 \rangle \exp\left[-i(\Delta\omega)(t - t')\right]}{\int\limits_{-\infty}^\infty d(\Delta\omega) \langle |\bar{E}(\Delta\omega)|^2 \rangle} . \tag{6.57}$$

where $\Delta\omega = (\omega - \omega_0)$. Therefore, the slowly varying correlation function and the normalized spectrum of the narrow-band signal are a Fourier transform pair. Remembering relation (6.16), we rewrite (6.57) in the following way:

$$g_1(t-t') = \frac{\int\limits_{-\infty}^\infty d(\Delta\omega)|H_A(\Delta\omega)|^2 \exp\left[-i\Delta\omega(t-t')\right]}{\int\limits_{-\infty}^\infty d(\Delta\omega)|H_A(\Delta\omega)|^2} . \tag{6.58}$$

It is seen from this expression that the first-order correlation function possesses the property $g_1(t-t') = g_1^*(t'-t)$. When the FEL gain curve is symmetrical with respect to the resonance frequency, ω_0, the function g_1 is real. In the high-gain linear regime, the FEL gain function $|H_A(\Delta\omega)|^2$ is symmetric (see (6.18)), and we obtain the following expression for the first-order time correlation function:

$$g_1(\tau) = \exp\left(-\frac{9\rho^2\omega_0^2\tau^2}{\sqrt{3}\hat{z}}\right) , \tag{6.59}$$

where $\tau = (t - t')$. It is convenient to write the expression for g_1 in terms of dimensionless variables:

$$g_1(\hat{\tau}) = \exp\left(-\frac{9\hat{\tau}^2}{\sqrt{3}\hat{z}}\right) ,$$

where $\hat{\tau} = \rho\omega_0\tau$. Following the approach of Mandel, we define the coherence time, τ_c, as

$$\tau_c = \int\limits_{-\infty}^\infty |g_1(\tau)|^2 d\tau . \tag{6.60}$$

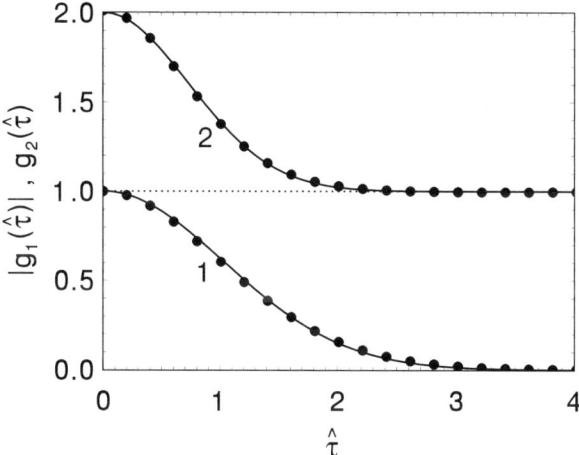

Fig. 6.7. Temporal correlation functions for the reduced length $\hat{z} = 10$ of the FEL amplifier. Solid curves are the results of calculations with analytical formulae (6.59) and (6.67). Circles are the result of calculations with linear simulation code over 5×10^4 independent statistical events. Curve 1 refers to the modulus of the first-order correlation function and curve 2 refers to the second-order correlation function

When the SASE FEL operates in the high-gain linear regime, the explicit expression for the coherence time is

$$\tau_c = \frac{\sqrt{\pi}}{\sigma_A} = \sqrt{\frac{\sqrt{3\pi \hat{z}}}{18}} \frac{1}{\rho \omega_0} . \tag{6.61}$$

Second-Order Time Correlations. The correlation between the radiation intensities at times t and t' is defined as

$$\langle |\tilde{E}(t)|^2 |\tilde{E}(t')|^2 \rangle$$
$$= \frac{1}{16\pi^4} \int_0^\infty d\omega_1 \int_0^\infty d\omega_2 \int_0^\infty d\omega_3 \int_0^\infty d\omega_4 \exp\left[-i(\omega_1 - \omega_2)t - i(\omega_3 - \omega_4)t'\right]$$
$$\times \langle \bar{E}(\omega_1) \bar{E}(\omega_3) \bar{E}^*(\omega_2) \bar{E}^*(\omega_4) \rangle . \tag{6.62}$$

Taking into account (6.16) and (6.8), we simplify the correlation in the integrand of (6.62) in the following way:

$$\langle \bar{E}(\omega_1) \bar{E}(\omega_3) \bar{E}^*(\omega_2) \bar{E}^*(\omega_4) \rangle$$
$$= \langle \bar{E}(\omega_1) \bar{E}^*(\omega_2) \rangle \langle \bar{E}(\omega_3) \bar{E}^*(\omega_4) \rangle$$
$$+ \langle \bar{E}(\omega_1) \bar{E}^*(\omega_4) \rangle \langle \bar{E}(\omega_3) \bar{E}^*(\omega_2) \rangle , \tag{6.63}$$

and present the integral (6.62) as the sum of two terms. To calculate the integral, one should take into account (6.31), (6.32), (6.55), and (6.13), which leads to

$$\langle |\tilde{E}(t)|^2 |\tilde{E}(t')|^2 \rangle = \langle |\tilde{E}(t)|^2 \rangle \langle |\tilde{E}(t')|^2 \rangle + |\langle \tilde{E}(t) \tilde{E}^*(t') \rangle|^2 \ . \tag{6.64}$$

The second-order time correlation function is defined as follows:

$$g_2(t - t') = \frac{\langle |\tilde{E}(t)|^2 |\tilde{E}(t')|^2 \rangle}{\langle |\tilde{E}(t)|^2 \rangle \langle |\tilde{E}(t')|^2 \rangle} \ . \tag{6.65}$$

It follows from (6.64) and the definitions of $g_1(t - t')$ and $g_2(t - t')$ that

$$g_2(t - t') = 1 + |g_1(t - t')|^2 \ . \tag{6.66}$$

For the SASE FEL operating in the high-gain linear regime, the explicit expression for the second-order time correlation function is (see (6.59) and (6.66)):

$$g_2(\tau) = 1 + \exp\left(-\frac{18\rho^2 \omega_0^2 \tau^2}{\sqrt{3}\hat{z}}\right) \ . \tag{6.67}$$

Analysis of the obtained relations, (6.53) and (6.66), shows that the radiation of the SASE FEL operating in the high-gain linear regime possesses all the features of completely chaotic polarized radiation (see also (6.29) and (6.36) obtained in the frequency domain). In Fig. 6.7 we present the plots of the first and second order time correlation functions calculated analytically and numerically.

Fluctuations of the Energy in the Radiation Pulse. The next problem is the description of the fluctuations of the radiation energy $\mathcal{E}$ detected during a finite time interval δT:

$$\mathcal{E} = \int_{t}^{t+\delta T} W(t) \mathrm{d}t \ .$$

It can be shown that such a distribution is described rather well by the gamma probability density function (6.48) with parameter M equal to (see Fig. 6.8b):

$$M^{-1} = \sigma_{\mathcal{E}}^2 = \frac{\sqrt{\pi}}{\delta T \sigma_{\mathrm{A}}} \mathrm{erf}(\delta T \sigma_{\mathrm{A}}) - \frac{1 - \exp\left[-(\delta T \sigma_{\mathrm{A}})^2\right]}{(\delta T \sigma_{\mathrm{A}})^2} \ . \tag{6.68}$$

Figure 6.8a presents the probability distribution of the energy for the specific value of $M = 4.9$. It is seen that this fits the gamma distribution well. When δT is less than the coherence time τ_{c} (6.61), the parameter M tends to unity and the gamma distribution tends to the negative exponential distribution. In the opposite case, when $\delta T \gg \tau_{\mathrm{c}}$, we can write:

$$M^{-1} = \frac{\sqrt{\pi}}{\delta T \sigma_{\mathrm{A}}} = \frac{\tau_{\mathrm{c}}}{\delta T} \ , \tag{6.69}$$

and the gamma distribution tends to a Gaussian distribution. At $\delta T = T$ we obtain (6.47).

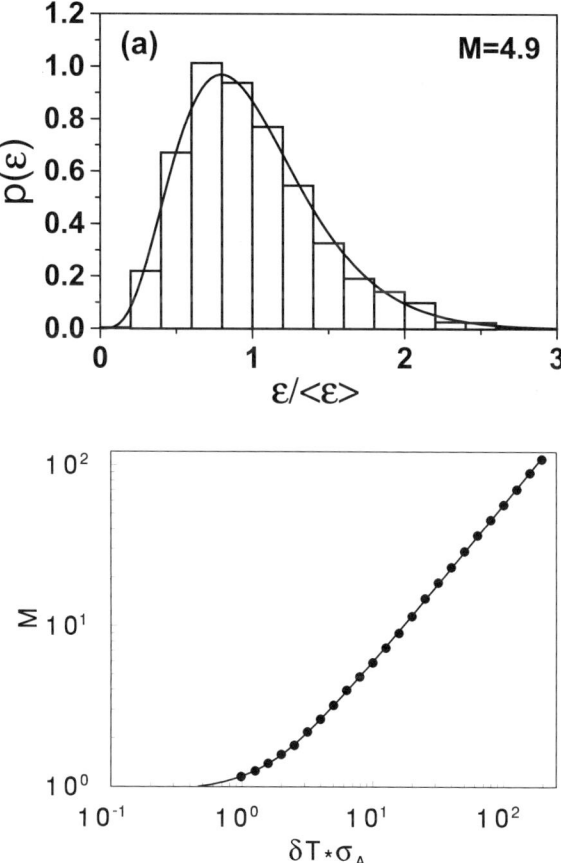

Fig. 6.8. Histogram of the probability density distribution, $p(\mathcal{E})$, of the radiation energy $\mathcal{E}$ detected during time δT (upper plot). $\langle \mathcal{E} \rangle$ denotes the average energy. The solid curve represents the gamma distribution (6.48). The dependence of the parameter M on the measurement time interval δT is presented in the lower plot. The solid curve is the result of calculations with the analytical formula (6.68). Circles are the result of calculations with linear simulation code. Calculations have been performed over 2400 shots (see Fig. 6.1)

Arbitrary Gain. The simple analytical results obtained above describe the high-gain linear regime of SASE FEL operation. In this section we generalize the investigations for the case of arbitrary gain. All the other physical approximations are the same as in the previous sections. We use the general solution of the initial-value problem for the case when a modulated electron beam is fed to the undulator entrance. The Fourier harmonic of the electric field at the undulator exit and the Fourier harmonic of the beam current modulation are connected by (6.16), where the Green's function is (see Section 2.1.4):

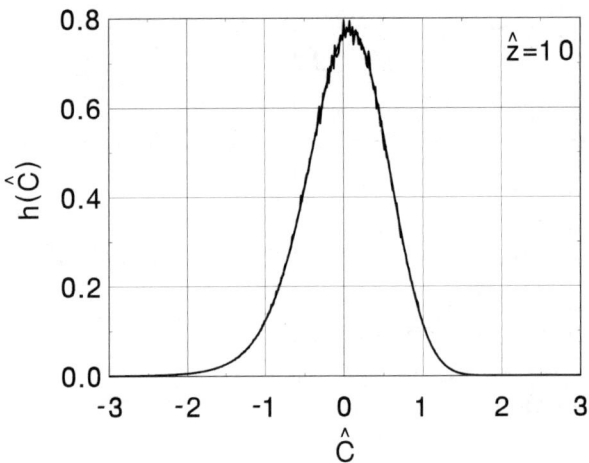

Fig. 6.9. Normalized spectrum of the radiation from a SASE FEL for $\hat{z} = 10$ calculated by averaging over 3000 shots (linear simulation code). The smooth curve is calculated analytically with (6.74) and (6.70)

$$H_A(\omega) = -2\left(M_{12} - i\hat{C}M_{13}\right)\frac{E_0}{I_0}\exp\left(i\frac{\omega}{c}z\right), \tag{6.70}$$

where

$$\begin{aligned} M_{12} &= -(\lambda_2 + \lambda_3)B_1 - (\lambda_1 + \lambda_3)B_2 - (\lambda_1 + \lambda_2)B_3, \\ M_{13} &= B_1 + B_2 + B_3, \end{aligned} \tag{6.71}$$

with

$$\begin{aligned} B_1 &= \frac{\exp(\lambda_1 \hat{z})}{(\lambda_1 - \lambda_2)(\lambda_1 - \lambda_3)}, \\ B_2 &= \frac{\exp(\lambda_2 \hat{z})}{(\lambda_2 - \lambda_1)(\lambda_2 - \lambda_3)}, \\ B_3 &= \frac{\exp(\lambda_3 \hat{z})}{(\lambda_3 - \lambda_1)(\lambda_3 - \lambda_2)}. \end{aligned} \tag{6.72}$$

Here λ are the roots of the eigenvalue equation:

$$\lambda\left(\lambda + i\hat{C}\right)^2 = i. \tag{6.73}$$

Analytical solution for the Green's function $|H_A(\omega)|^2$ gives us an envelope of the spectral power density. According to (6.58), the first-order time correlation function is the Fourier transform of the spectral power density. As a result, we are able to calculate the time correlation functions for arbitrary gain. In particular, this is important for a thorough testing of the numerical simulation codes.

In Fig. 6.9 we present the normalized spectrum envelope

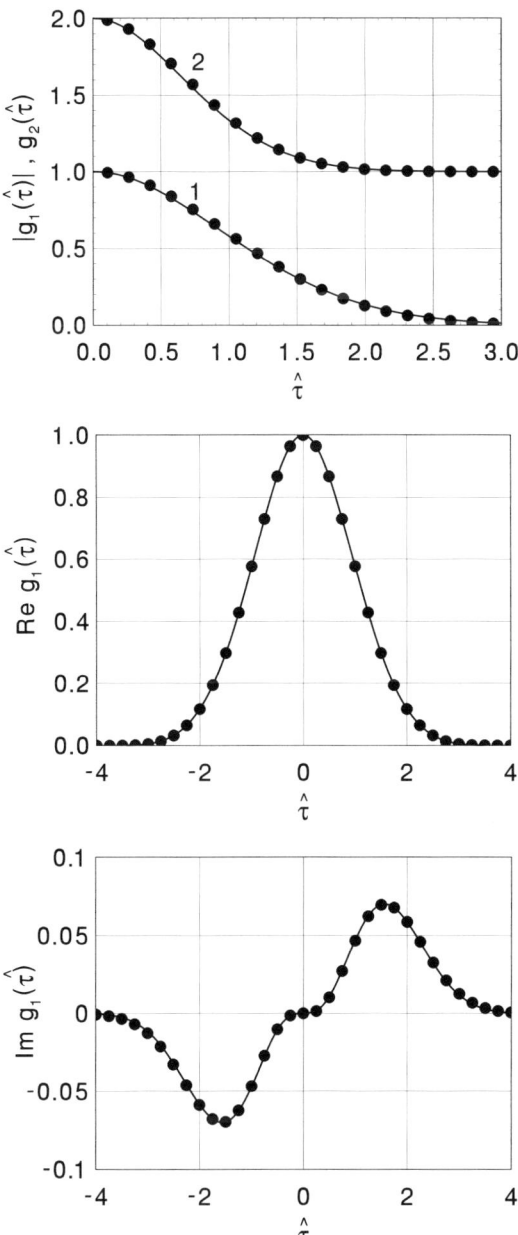

Fig. 6.10. The first and second order time correlation functions, $g_1(\tau)$ and $g_2(\tau)$, of the radiation pulse for $\hat{z} = 10$. The curves are calculated using (6.58) and (6.70), and the circles are the results of numerical simulations with linear code

$$h(\hat{C}) = \frac{|H_A(\hat{C})|^2}{\int_{-\infty}^{\infty} d\hat{C} |H_A(\hat{C})|^2} \tag{6.74}$$

of the radiation from the SASE FEL operating in the linear regime (the smooth curve). We also present in this figure the radiation spectrum, calculated with the linear simulation code and averaged over many shots. Figure 6.10 presents plots of the time correlation functions. The curves are calculated analytically using (6.58), (6.70), and (6.66). The circles are the results of simulations with the linear code. We have stressed above that in the general case the first-order time correlation function, $g_1(\hat{\tau})$, has an even real part and an odd imaginary part. Since the spectrum envelope is not exactly symmetric at the chosen undulator length $\hat{z} = 10$ (see Fig. 6.9), there is a nonzero imaginary part of $g_1(\hat{\tau})$. Its value gives us an idea about the region of validity of the high-gain asymptote described in the previous section.

6.2.2 Numerical Simulation Algorithm

In this section we describe a one-dimensional, time-dependent algorithm for the simulation of the FEL amplifier. In particular, it can be used for simulations of the FEL amplifier starting from the shot noise. To simplify the formulae, we consider here a rectangular electron bunch and neglect the effects of the energy spread and of the space charge. The main emphasis is put on the description of time-dependent effects and the initial conditions of the shot noise. All formulae are written for the case of a planar undulator. Generalization of the equations for the case of a helical undulator is straightforward (see Chap. 2).

Basic Equations. Let us start with the solution of the electrodynamic problem. The wave equation for the electromagnetic wave, $\boldsymbol{E} = \boldsymbol{e}_y E_y(z,t)$, amplified by the electron beam in the planar undulator, has the form:

$$\frac{\partial^2 E_y}{\partial z^2} - \frac{1}{c^2}\frac{\partial^2 E_y}{\partial t^2} = \frac{4\pi}{c^2}\frac{\partial j_y}{\partial t} \, . \tag{6.75}$$

The transverse current density may be written in the form

$$j_y = v_y(z) j_z(z,t)/c \, . \tag{6.76}$$

The velocity components of the electron moving in the field of the planar undulator are given by the expressions:

$$v_x = 0 \, ,$$
$$v_y = -c\theta_\ell \sin(k_w z) \, ,$$
$$v_z = v - \frac{c\theta_\ell^2}{2}\sin^2(k_w z) \, . \tag{6.77}$$

Here v is the total velocity of the electron. Substituting (6.76) and (6.77) in (6.75), one can obtain that the radiation of the electron beam in the undulator is in resonance with the electron motion when

$$k_\mathrm{w} + \frac{\omega_0}{c} - \frac{\omega_0}{\bar{v}_z} = 0 , \qquad (6.78)$$

where $\bar{v}_z = v - c\theta_\ell^2/4$ is the velocity of the electron along the z axis, averaged over the undulator period. The resonance properties of the radiation of the electron beam in the undulator should be understood as follows. Since we study the start-up from shot noise, we assume the input current to have a homogeneous spectral distribution. We also omit from the considerations the spectral components connected with the finite pulse duration (see the discussion in Sect. 6.1). At a sufficiently large undulator length, $k_\mathrm{w} z \gg 1$, the spectrum of the radiation is concentrated within the narrow band,

$$\Delta\omega/\omega_0 \lesssim (k_\mathrm{w} z)^{-1} ,$$

near the resonance frequency ω_0. Therefore, the electric field of the wave can be presented as

$$E_y(z,t) = \tilde{E}(z,t) e^{i\omega_0(z/c-t)} + \mathrm{C.C.} , \qquad (6.79)$$

where $\tilde{E}$ is the slowly varying complex amplitude:

$$|\partial \tilde{E}/\partial t| \ll \omega_0 |\tilde{E}| , \qquad |\partial \tilde{E}/\partial z| \ll k_\mathrm{w} |\tilde{E}| .$$

It is convenient to write down an expression for the current density in the following form:

$$j_z(z,t) = -j_0 + \tilde{j}_\mathrm{a}(z,t) e^{i\psi'} + \mathrm{C.C.} , \qquad (6.80)$$

where

$$\psi' = k_\mathrm{w} z + \omega_0(z/c - t) - Q \sin(2k_\mathrm{w} z) , \qquad Q = \frac{\omega_0 \theta_\ell^2}{8 k_\mathrm{w} c} ,$$

and $-j_0$ is the ensemble average current density. To be specific, we consider the model of a long rectangular bunch.

Substituting (6.79) and (6.80) into (6.75), we obtain a relation between the complex amplitudes $\tilde{E}$ and $\tilde{j}_\mathrm{a}$. Since $\tilde{E}$ is a slowly varying amplitude, the second derivatives in the left-hand side of (6.75) may be neglected. Although $\tilde{j}_\mathrm{a}$ in general is not a slowly varying function (as we have already mentioned, at the undulator entrance the current density has a white spectrum), the output field will be defined by the narrow-band fraction of the input current density fluctuations. Therefore, we can also neglect the derivative $\partial \tilde{j}_\mathrm{a}/\partial t$. As a result, we obtain

$$\left(\frac{\partial}{\partial z} + \frac{1}{c} \frac{\partial}{\partial t} \right) \tilde{E}(z,t) = i \frac{\pi \theta_\ell A_\mathrm{JJ}}{c} \tilde{j}_\mathrm{a}(z,t) . \qquad (6.81)$$

In the steady-state regime (no time dependence) this equation coincides with (2.73). The solution of this equation has the form:

$$\tilde{E}(z,t) = i\frac{\pi\theta_\ell A_{JJ}}{c} \int_0^z \tilde{j}_a\left(z', t - \frac{z-z'}{c}\right) dz' . \tag{6.82}$$

To perform the numerical simulations of the SASE FEL, we should go over to discrete quantities. Suppose, we have an electron bunch of length l_b. We divide it into $N_b = l_b/\lambda$ boxes, where $\lambda = 2\pi c/\omega_0$. When the number of particles in the bunch is equal to N, the number of particles per box is equal to $N_\lambda = N/N_b$ (we neglect fluctuations of this number with the relative standard deviation $1/\sqrt{N_\lambda}$, since they are always small). Let us consider the coordinate $s = z - \bar{v}_z t$ and define the position of the bunch tail as $s = 0$ and that of the bunch head as $s = l_b$. Then we assign a number to each box in accordance with the formula $j = [s/\lambda]$, where $[\ldots]$ denotes the integer part of a number. At any fixed point z along the undulator the arrival times of boxes differ by $\Delta t = t_j - t_{j+1} = \lambda/\bar{v}_z$. The position of each particle within the bunch is defined by the box number j and by the phase $0 \leq \psi' \leq 2\pi$. Indeed, for a particle moving with nominal energy $\mathcal{E}_0$ (which corresponds to the nominal velocity $\bar{v}_z$) the phase ψ' is constant:

$$\psi' = k_w z + \omega_0 \frac{z}{c} - \omega_0 \int_0^z \frac{dz'}{v_z(z')} - Q\sin(2k_w z)$$
$$= k_w z + \omega_0 \frac{z}{c} - \omega_0 \frac{z}{\bar{v}_z} + \text{const.}$$
$$= \text{const.}$$

The phase of each particle may be written in the form:

$$\psi' = 2\pi\left\{\frac{s}{\lambda} - \left[\frac{s}{\lambda}\right]\right\} .$$

Let us now introduce the notion of bunching in the jth box:

$$\hat{a}_1^{(j)} = \frac{1}{N_\lambda} \sum_{k=1}^{N_\lambda} e^{-i\psi_k'^{(j)}} , \tag{6.83}$$

where $\psi_k'^{(j)}$ is the phase of the kth particle inside the jth box. The complex amplitudes $\tilde{j}_a$ and $\hat{a}_1^{(j)}$ are connected by the relation:

$$\tilde{j}_a(z, t_j)/j_0 = -\hat{a}_1^{(j)}(z) .$$

The minus sign appears in this expression because electrons have charge $(-e)$ and move in the positive direction of the z axis.

When performing numerical simulations, the electron beam is simulated by the number of macroparticles per wavelength, N_m, which is usually much less than the actual number, N_λ. Thus, the bunching in each box is calculated as

$$\hat{a}_1^{(j)} = \frac{1}{N_m} \sum_{k=1}^{N_m} e^{-i\psi_k'^{(j)}} . \tag{6.84}$$

The next point is the solution of the dynamical problem. As we have already mentioned, for a particle moving with nominal velocity $\bar{v}_z$ (which corresponds to the nominal energy $\mathcal{E}_0$), the phase ψ' is constant along z. Therefore, $\mathrm{d}\psi'/\mathrm{d}z \neq 0$ only in the case when the particle energy deviates from the nominal value: $P = \mathcal{E} - \mathcal{E}_0 \neq 0$. Then the equations of motion averaged over an undulator period have the form (see Sect. 2.2.7):

$$\frac{\mathrm{d}P}{\mathrm{d}z} = -\frac{\mathrm{i}}{2}A_{\mathrm{JJ}}e\theta_\ell \tilde{E}(z,t)\mathrm{e}^{\mathrm{i}\psi'} + \mathrm{C.C.} \,,$$

$$\frac{\mathrm{d}\psi'}{\mathrm{d}z} = \frac{\omega P}{c\gamma_\ell^2 \mathcal{E}_0} \,.$$

In the same way as was done in Sect. 2.2.7, we perform the normalization procedure and rewrite the latter equations and expression (6.82) in the reduced form:

$$\frac{\mathrm{d}\hat{P}_k^{(j)}}{\mathrm{d}\hat{z}} = -\frac{\mathrm{i}}{2}\hat{E}^{(j)}(\hat{z})\mathrm{e}^{\mathrm{i}\psi_k^{'(j)}} + \mathrm{C.C.} \,,$$

$$\frac{\mathrm{d}\psi_k^{'(j)}}{\mathrm{d}\hat{z}} = \hat{P}_k^{(j)} \,, \tag{6.85}$$

$$\hat{E}^{(j)}(\hat{z}) = -2\mathrm{i}\Delta\hat{z}\sum_{m=1}^{n}\hat{a}_1^{(j-m)}(\hat{z} - m\Delta\hat{z}) \,, \tag{6.86}$$

where $\hat{E}^{(j)}(\hat{z}) = \tilde{E}(\hat{z}, t_j)/E_0$ is the normalized field in the jth box, $\hat{P}_k^{(j)} = P_k^{(j)}/(\rho\mathcal{E}_0)$ is the normalized energy deviation from the nominal value of the kth particles in the jth box, $\hat{z} = \Gamma z$, $\Delta\hat{z} = \Gamma\lambda_\mathrm{w}$, and $m = [\hat{z}/\Delta\hat{z}] = [z/\lambda_\mathrm{w}]$. When writing down (6.86) we have used a discrete representation of (6.82) within the accuracy of the ultrarelativistic approximation $(c - \bar{v}_z \ll c)$.

Simultaneous solution of (6.86) and (6.85) allows one to calculate the evolution in time and space of the electromagnetic field and of the particle motion under given initial conditions at the undulator entrance. The beam bunching is calculated according to (6.84). The simulation proceeds in the following way. At each integration step over the $\hat{z}$ coordinate the normalized field amplitude is calculated in each box using (6.86) and (6.84). Then equations (6.85) for the particle motion are integrated within each box. At the next step of integration, at $\hat{z} + \Delta\hat{z}$, the bunching and the field are recalculated, and the procedure is repeated. For simulations of the linear mode of operation of the FEL amplifier, it is more convenient solve the equation for the beam bunching instead of (6.85):

$$\frac{\mathrm{d}^2\hat{a}_1^{(j)}}{\mathrm{d}\hat{z}^2} = -\frac{1}{2}\hat{E}^{(j)}(\hat{z}) \,. \tag{6.87}$$

Initial Conditions. To find the evolution of the beam bunching $\hat{a}_1^{(j)}$ and of the electromagnetic field, we should define the initial conditions at $\hat{z} = 0$. At the entrance of the undulator there is no radiation field and the fluctuations of the beam current density caused by shot noise in the electron beam play the role of the input signal.

Linear Simulation Code. In principle, there exists a straightforward method for a rigorous simulation of the initial conditions for the linear code. Indeed, the number of particles in the beam is finite, and we can calculate the beam bunching in each box using (6.83) and the actual number of particles in the beam. One can easily check that this procedure takes a reasonable amount of CPU time on a modern computer. Nevertheless, the initial conditions can be prepared in a more elegant way, providing the same results without extra consumption of CPU time. The idea is pretty simple. As a rule, the number of particles per box, N_λ, is large. For instance, typical values of N_λ are about 10^4–10^6 for X-ray and VUV FELs. When N_λ is large, the bunching in each box is the sum of a large number of random phasors with fixed amplitudes and phases uniformly distributed in $(0, 2\pi)$. Using the central limit theorem, we can state that the squared modulus of the amplitudes, $|\hat{a}_1|^2$, are distributed according to the negative exponential distribution:

$$p(|\hat{a}_1|^2) = \frac{1}{\langle|\hat{a}_1|^2\rangle} \exp\left(-\frac{|\hat{a}_1|^2}{\langle|\hat{a}_1|^2\rangle}\right) , \qquad (6.88)$$

where $\langle|\hat{a}_1|^2\rangle = 1/N_\lambda$. So, the procedure for preparing the initial conditions becomes as follows. We use a negative exponential random generator with mean value equal to $1/N_\lambda$ which produces the values of $|\hat{a}_1|^2$ for each box. Then we extract the square root to find the values of $|\hat{a}_1|$. The phases of $\hat{a}_1$ are produced by a random number generator for a uniform distribution from 0 to 2π. The initial condition for the derivative of the bunching is set equal to zero, since the effect of the velocity fluctuations at the undulator entrance can be neglected as discussed above.

Nonlinear Simulation Code. The following technique can be used for the preparation of the initial conditions for the nonlinear simulation code. We choose the number of macroparticles in each box to be equal to $N_m \ll N_\lambda$. The particles are distributed evenly over the phase ψ', so the bunching is equal to zero. Then we calculate the amplitudes and phases of the bunching in the boxes in the same way as described above, i.e. using a random number generator for the negative exponential and uniform distributions. The next step is to change the phases of a few macroparticles within the box in order to reproduce the prescribed values of the amplitude and phase of the bunching.

Data Analysis. Calculations for the SASE FEL require us to perform a large number of statistically independent simulation runs. The result of each run contains parameters of the output radiation (field and phase) stored in the boxes over the full length of the radiation pulse. At the next stage of the numerical experiment the data arrays are handled with postprocessor codes to extract different statistical properties of the SASE FEL radiation. The first and second order time correlation functions are calculated using (6.56) and (6.65). The first and second order spectral correlation functions are calculated using (6.31) and (6.35). The spectrum of the radiation pulse is calculated by means of the Fourier transformation of the temporal structure. The energy spectra from individual shots can be summed in order to obtain the averaged spectrum envelope. Also, the envelope of the radiation spectrum can be reconstructed from the first-order time correlation function,

$$|H(\Delta\omega)|^2 \propto \int_{-\infty}^{\infty} d\tau g_1(\tau) \exp(i\Delta\omega\tau) \ . \qquad (6.89)$$

Using normalized variables, we get

$$|H(\hat{C})|^2 \propto \int_{-\infty}^{\infty} d\hat{\tau} g_1(\hat{\tau}) \exp(-2i\hat{C}\hat{\tau}) \ . \qquad (6.90)$$

Then the normalized spectrum envelope is calculated as

$$h(\hat{C}) = \frac{|H(\hat{C})|^2}{\int_{-\infty}^{\infty} d\hat{C} |H(\hat{C})|^2} \ . \qquad (6.91)$$

6.2.3 Numerical Simulations of the Main Characteristics of a SASE FEL

In this section we present the results of numerical studies of the operation of the SASE FEL in the linear and nonlinear regimes. We will use similarity techniques when presenting the results of the numerical simulation. In the framework of the accepted model, the input parameters of the system are the number of cooperating electrons $N_c = N_\lambda/(2\pi\rho)$ and the normalized bunch length $\hat{l}_b = \rho\omega_0 l_b/c$. Most of the statistical characteristics of the SASE FEL process are functions of N_c only in the fixed z coordinate. A typical range of the values of N_c is 10^6–10^9 for the SASE FELs of wavelength range from X-ray up to infrared. The numerical results, presented in this section, are calculated for the value $N_c = 3 \times 10^7$ which is typical for a VUV FEL. It is worth mentioning that the dependence of the output parameters of the SASE FEL on the value of N_c is rather weak, in fact logarithmic. Therefore, the obtained results are pretty general and can be used for the estimation of the parameters of actual devices with sufficient accuracy.

Temporal Characteristics. We start with the analysis of the temporal structure of the radiation within one sample pulse at different positions along the undulator. It is convenient to present the normalized power $\hat{\eta} = W/(\rho W_b)$ as a function of the normalized variable $\hat{s} = \rho\omega_0(z/\bar{v}_z - t)$. For this choice of the time dependence, any given value of $\hat{s}$ corresponds to a certain point in the electron bunch moving with longitudinal velocity $\bar{v}_z$. The value $\hat{s} = 0$ corresponds to the coordinate of the tail of the rectangular bunch. The head of the pulse is located in the positive direction of $\hat{s}$. Figure. 6.11 presents a

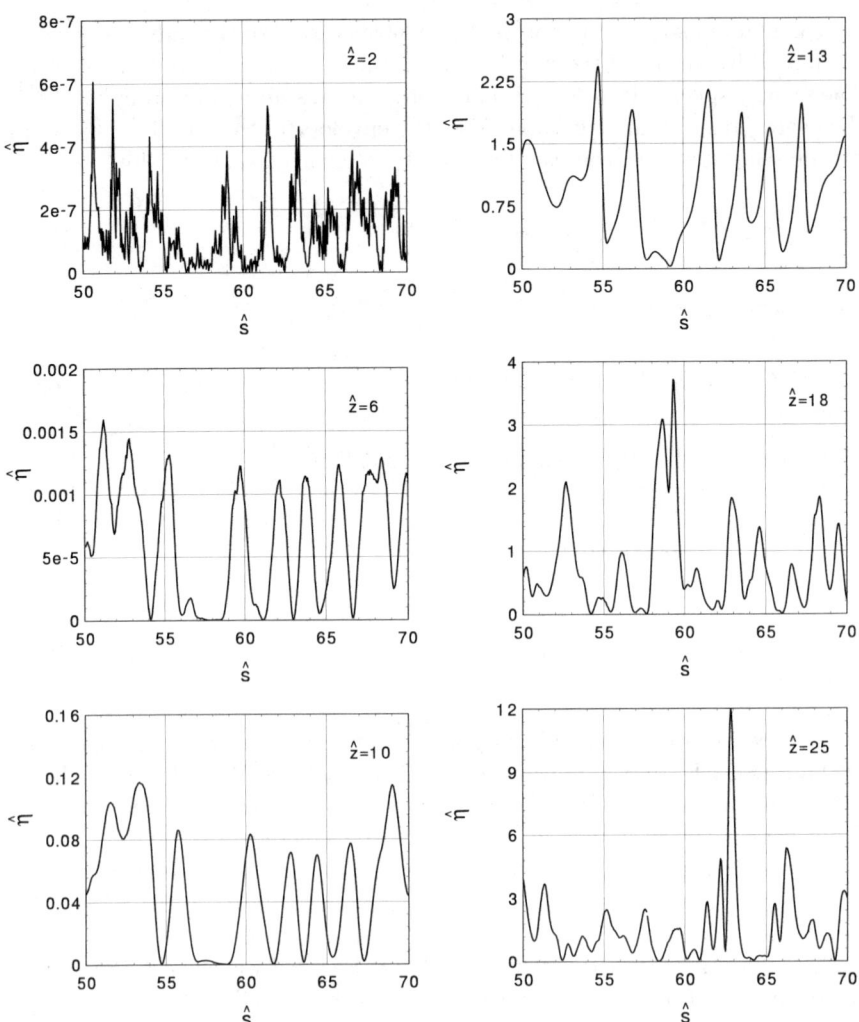

Fig. 6.11. Normalized power in the radiation pulse versus $\hat{s} = \rho\omega_0(z/\bar{v}_z - t)$ at different lengths of the FEL amplifier. Here $N_c = 3 \times 10^7$. The calculations have been performed with the nonlinear simulation code

typical time structure of the radiation from a SASE FEL operating in the linear and nonlinear regimes. The power is plotted within a certain window inside the electron bunch. The two-dimensional (on $\hat{z} - \hat{s}$ plane) intensity picture of the SASE FEL process for the same window in the bunch is presented in Fig. 6.12. The black color corresponds to maximal intensity. The lower plot in this figure is an enlarged fraction of the upper one.

It is seen from these figures that there are two quite different modes of SASE FEL operation. The first one is the linear mode of operation which

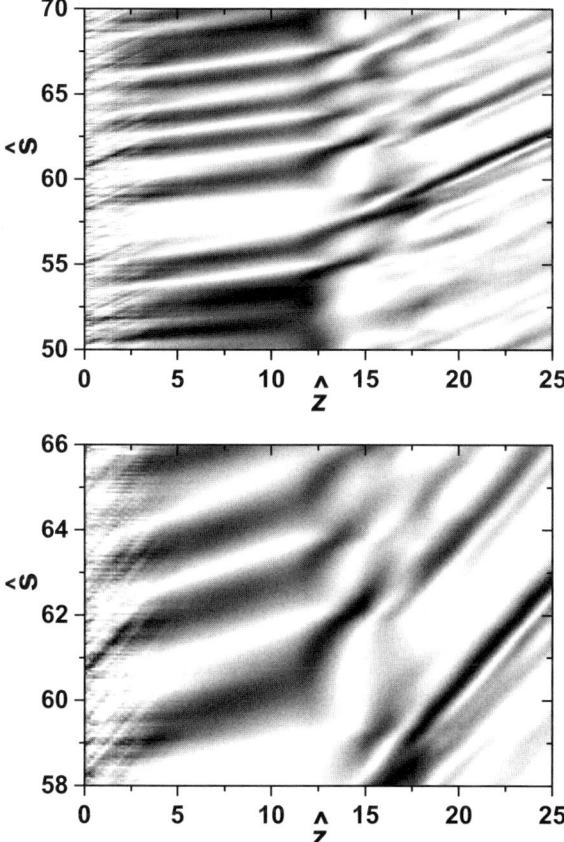

Fig. 6.12. Normalized power in the radiation pulse as a function of $\hat{s} = \rho\omega_0(z/\bar{v}_z - t)$ and $\hat{z}$. The lower graph is an enlarged fraction of the upper one. The calculations have been performed with the nonlinear simulation code

takes place up to $\hat{z} \simeq 11$. The effect of saturation clearly occurs at $\hat{z} \simeq 13$, and when the undulator length is increased further, we obtain a typical picture of the nonlinear mode of operation. We saw above that when the FEL amplifier operates in the linear mode of operation, the relative slippage of the wavepackets (spikes) with respect to the electron beam should be three time less than the kinematic slippage. Using Fig. 6.12, we can calculate the slippage rate of the spikes with respect to the electron beam. Indeed, it is easy to show that

$$\frac{v_g - \bar{v}_z}{c - \bar{v}_z} = 2\frac{d\hat{s}}{d\hat{z}},$$

i.e. one needs to find double the slope of a line with equal color intensity. In the high-gain linear regime this is about 1/3 which is in agreement with the

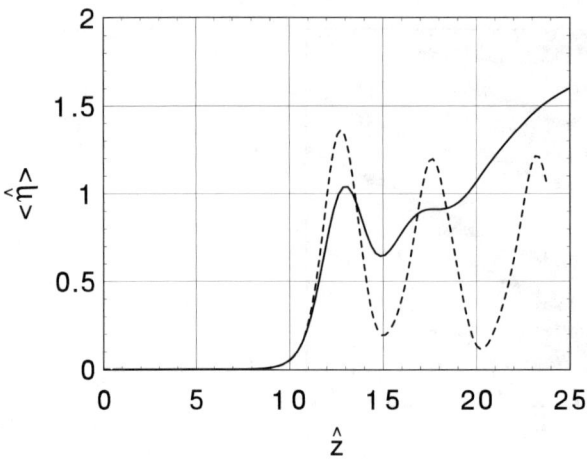

Fig. 6.13. Normalized averaged power of a SASE FEL as a function of normalized undulator length. Here $N_c = 3 \times 10^7$. The dashed line is calculated with steady-state code where the input power is equal to the effective power of the shot noise (6.95)

theoretically predicted value (see (6.28)). In the nonlinear regime $2d\hat{s}/d\hat{z} \simeq 1$, and the group velocity is close to the velocity of light. Also, one can see from Fig. 6.12 that at the beginning of the amplification process, the wavepackets move with the velocity of light.

In Fig. 6.13 we present the dependence of the normalized averaged power on the normalized longitudinal coordinate. Saturation for the given value $N_c = 3 \times 10^7$ is achieved at $\hat{z}_{sat} \simeq 13$ and the normalized efficiency $\hat{\eta}$ at the saturation point is about unity. In the general case the saturation point can be found from the relation

$$\hat{z}_{sat} \simeq 3 + \frac{1}{\sqrt{3}} \ln N_c \ . \tag{6.92}$$

The normalized efficiency at saturation is almost independent of the value of N_c.

In the high-gain linear regime the dependence of the reduced efficiency on the undulator length, $\hat{\eta}(\hat{z})$, is given by the analytical formula (6.24). It is convenient to introduce the notion of the effective power of the shot noise which can be used for numerical simulation of the SASE FEL with the steady-state codes. The output power of the steady-state, high-gain FEL amplifier grows exponentially in the high-gain linear regime (see Chap. 2):

$$W_{out} = \frac{1}{9} W_{in} \exp(\sqrt{3}\hat{z}) \ , \tag{6.93}$$

where W_{in} is the input power. Comparing (6.24) and (6.93), we can write the following formula for the effective power of the shot noise, W_{sh}, of the SASE FEL:

$$W_{\text{sh}} = W_{\text{in}} = \frac{3^{3/4}\rho W_{\text{b}}}{N_{\text{c}}\sqrt{\pi\hat{z}}} \ . \tag{6.94}$$

This formula should be treated as follows. Suppose, we perform numerical simulations with the steady-state simulation code of the FEL amplifier having undulator of length $\hat{z}$. If we take the value of the input radiation power equal to (6.94), the value of the output power will be equal to the averaged value of the output power of the SASE FEL. To provide correct simulation of the saturation length, we can approximate the effective input power as follows. We substitute (6.92) into (6.94) and obtain with logarithmic accuracy:

$$\frac{W_{\text{sh}}}{\rho W_{\text{b}}} \simeq \frac{3}{N_{\text{c}}\sqrt{\pi \ln N_{\text{c}}}} \ . \tag{6.95}$$

This expression gives a good estimate of the effective power of the shot noise. This can be seen from Fig. 6.13, where the dashed line is calculated with the steady-state code. Let us define the gain G of the SASE FEL at the saturation point as the ratio of the saturation power to the effective power of the shot noise. The gain at saturation is a function only of the number of cooperating electrons, N_{c}:

$$G \simeq \frac{1}{3} N_{\text{c}} \sqrt{\pi \ln N_{\text{c}}} \ .$$

One can also obtain from Fig. 6.13 that there is a difference between the SASE FEL and the steady-state FEL amplifier in the nonlinear stage of amplification. In the case of the SASE FEL, the radiation power continues to grow after the saturation point due to spectrum broadening. The latter effect will be illustrated below.

The next step in our investigation is the behavior of the probability density distribution of the instantaneous power. In the linear regime, the instantaneous radiation power fluctuates in accordance with the negative exponential distribution (see (6.53) and Fig. 6.6). This is a general feature of chaotic polarized radiation. In the nonlinear regime, near the saturation point, the distribution changes significantly (see Fig. 6.14). In the deep nonlinear regime (see the histogram for $\hat{z} = 24$), the probability density distribution is pretty close to, but does not exactly coincide with, the negative exponential distribution.

In Fig. 6.15 we show the normalized rms deviation of the instantaneous radiation power, $\sigma_{\text{w}} = \langle (W - \langle W \rangle)^2 \rangle^{1/2}/\langle W \rangle$, as a function of the undulator length. In the linear regime of SASE FEL operation, the value of the deviation is equal to unity. This case has been analyzed in detail in Sect. 6.2.1 devoted to the linear theory. In the nonlinear mode of operation, the deviation of the power fluctuations differs significantly from unity and exhibits complicated behavior.

In Figs. 6.16–6.18 we show the evolution of the time correlation functions of first and second order. At each normalized position along the undulator, $\hat{z}$, they are plotted versus the normalized variable

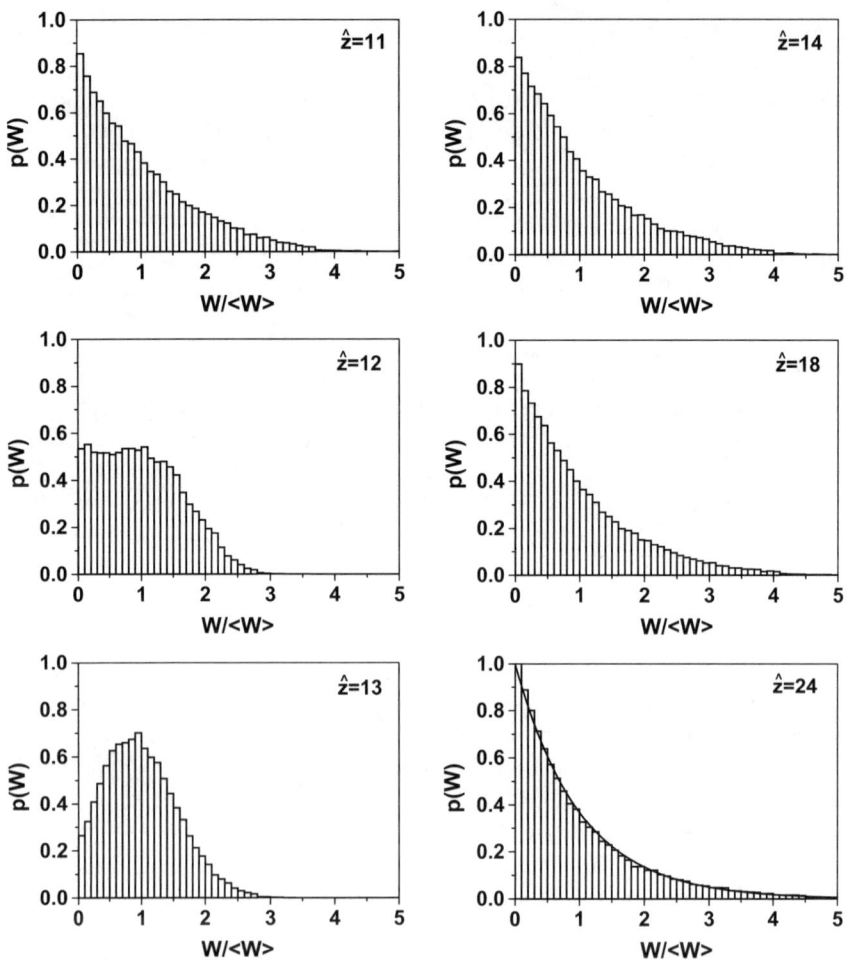

Fig. 6.14. Histograms of the probability density distribution, $p(W)$, of the instantaneous output power at different reduced lengths of the FEL amplifier. Calculations have been performed with nonlinear simulation code over 10^4 independent statistical events. Here $\langle W \rangle$ denotes the average power

$$\hat{\tau} = \rho\omega_0(t-t') = \hat{s}' - \hat{s} \ .$$

Figure 6.16 corresponds to the linear stage of SASE FEL operation. In this case we deal with a Gaussian random process and the relation (6.66) between the correlation functions holds for $g_2(t-t') = 1 + |g_1(t-t')|^2$. For a short undulator length, the correlation functions are those of the undulator radiation (see the upper plot in Fig. 6.16). The first-order time correlation function, g_1, linearly decreases and takes a zero value when $c\tau$ is equal to the total slippage. This is explained by the fact that at the beginning of the

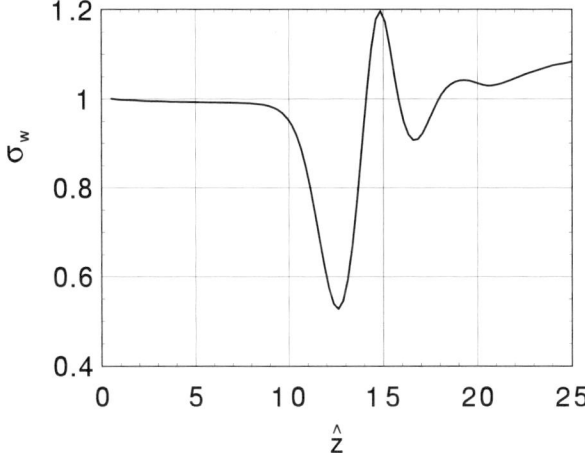

Fig. 6.15. Normalized rms deviation of the fluctuations of the instantaneous radiation power as s function of the normalized undulator length

undulator the electron beam radiates an electromagnetic field, but this field is too small to influence the particle motion (there is no amplification). When $\hat{z} = 3$ the shapes of the correlation functions deviate visibly from those of the undulator radiation, which indicates that the FEL amplification process develops. When $\hat{z} = 6$ the correlation functions are pretty close to those given by the high-gain asymptotic formulae (6.59) and (6.67).

Figures 6.17 and 6.18 present plots of the time correlation functions corresponding to the nonlinear mode of SASE FEL operation. Figure 6.19 presents the real and imaginary parts of the first-order correlation function at the saturation point. The nontrivial behavior of the correlation functions reflects the complicated nonlinear evolution of the SASE FEL process. One can see that (6.66) is no longer valid. The second-order correlation function of zero argument, $g_2(0)$, takes values smaller or larger than two, but always larger than unity.[2] It is a well-known result of statistical optics that the cases of $g_2(0) = 1$ and $g_2(0) = 2$ correspond to stabilized single-mode laser radiation and to completely chaotic radiation from a thermal source, respectively. The values of $g_2(0)$ between 1 and 2 belong to some intermediate situation. In classical optics, a radiation source with $g_2(0) < 1$ cannot exist but the case of $g_2(0) > 2$ is possible. As one can see from Figs. 6.17 and 6.18, the latter phenomenon (known as superbunching) may occur in the SASE FEL operating in the nonlinear regime.

In Fig. 6.20 we present the dependence on the undulator length of the normalized coherence time $\hat{\tau}_c = \rho\omega_0\tau_c$, where τ_c is defined in (6.60). In the linear regime $\hat{\tau}_c$ depends only on $\hat{z}$ and is described analytically. In particular,

---

[2] There is a simple relation between $g_2(0)$ and the normalized rms power deviation: $g_2(0) = 1 + \sigma_w^2$.

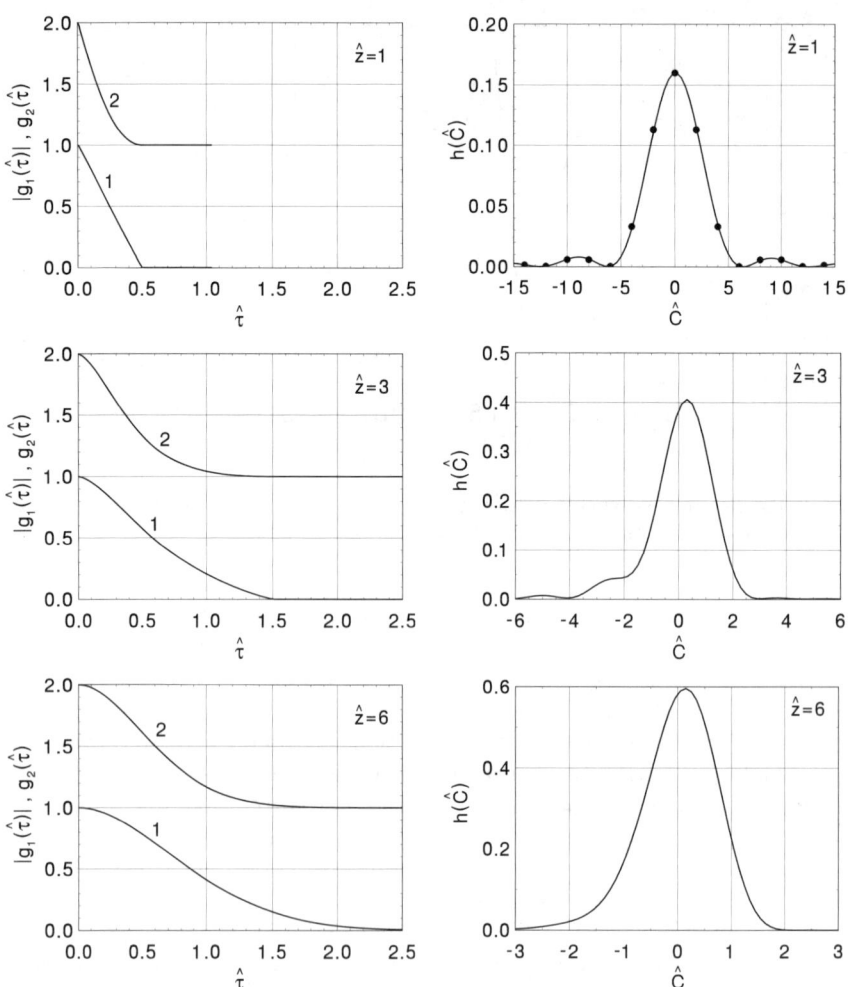

Fig. 6.16. Left column: first and second order time correlation functions, $|g_1(\hat{\tau})|$ (curve 1) and $g_2(\hat{\tau})$ (curve 2), of the radiation pulse from a SASE FEL operating in the linear regime. Right column: normalized spectrum of the radiation. Circles refer to the spectrum of the undulator radiation

in the high-gain linear regime, (6.61) is valid. The coherence time achieves its maximal value near the saturation point and then decreases drastically. The maximal value of $\hat{\tau}_c$ depends on the saturation length and, therefore, on the value of the parameter N_c. We can approximate this maximal value in the same way as the effective power of the shot noise. We substitute (6.92) into (6.61) and obtain with logarithmic accuracy:

$$(\hat{\tau}_c)_{\max} \simeq \sqrt{\frac{\pi \ln N_c}{18}} \ .$$

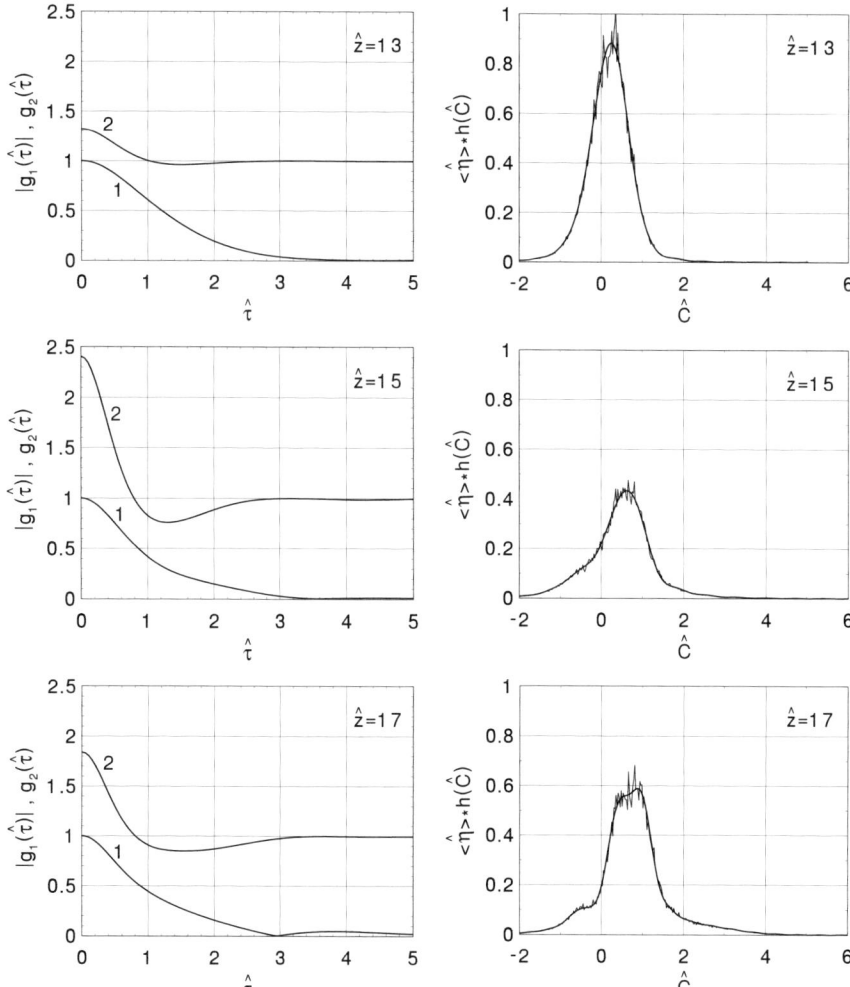

Fig. 6.17. Left column: first and second order time correlation functions, $|g_1(\hat{\tau})|$ (curve 1) and $g_2(\hat{\tau})$ (curve 2), of the radiation pulse from a SASE FEL operating in the nonlinear regime. Here $\hat{z} = 13$ is the saturation point. Right column: normalized spectrum, multiplied by the normalized efficiency. The smooth curve is reconstructed from the first order time correlation function. Calculations have been performed with nonlinear simulation code

Spectral Characteristics. The typical spectral density of the radiation energy for one radiation pulse is presented in Fig. 6.2. When comparing radiation spectra, it is convenient to use the normalized spectral density, $h(\hat{C})$, defined as

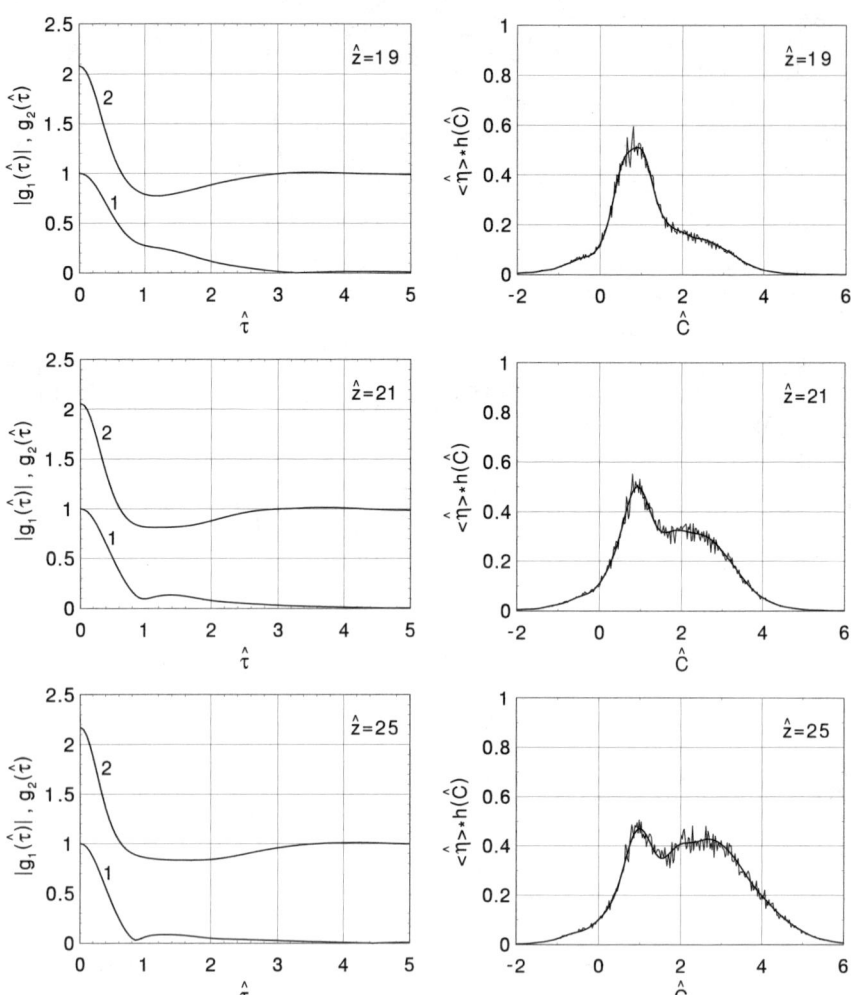

Fig. 6.18. Left column: first and second order time correlation functions, $|g_1(\hat{\tau})|$ (curve 1) and $g_2(\hat{\tau})$ (curve 2), of the radiation pulse from a SASE FEL operating in the nonlinear regime. Right column: normalized spectrum, multiplied by the normalized efficiency. The smooth curve is reconstructed from the first-order time correlation function. Calculations have been performed with nonlinear simulation code

$$\int_{-\infty}^{\infty} d\hat{C} h(\hat{C}) = 1 \ .$$

The frequency deviation, $\Delta\omega$, from the nominal value of ω_0 can be recalculated as $\Delta\omega = -2\rho\omega_0 \hat{C}$. Since we consider the model of a long rectangular

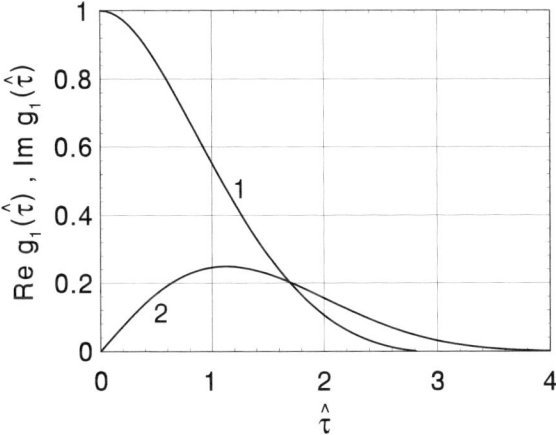

Fig. 6.19. Real (curve 1) and imaginary (curve 2) parts of the first-order time correlation function. The SASE FEL operates at the saturation point, $\hat{z} = 13$

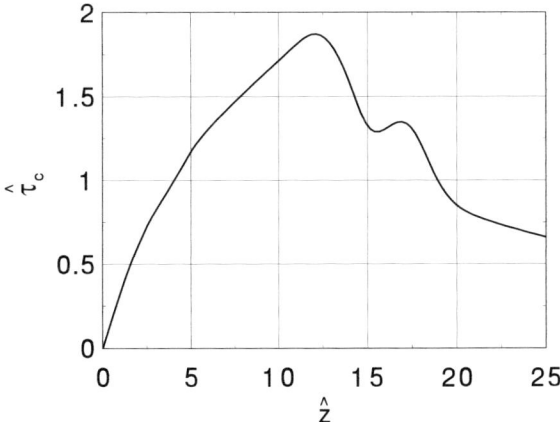

Fig. 6.20. Normalized coherence time of a SASE FEL as a function of the normalized undulator length

bunch, the function $h(\hat{C})$ can be treated as the normalized spectral density of both the radiation energy and the power.

In the linear regime, $h(\hat{C})$ can be calculated by means of two independent methods. The first method is analytical (see (6.70) and (6.74)). The spectrum envelope can also be calculated by averaging the results of numerical simulation code. Both methods give identical results (see Fig. 6.9). In Fig. 6.16 we show the spectrum envelope of the SASE FEL operating in the linear regime. At the beginning of the undulator the spectrum envelope is simply that of the undulator radiation,

6. FEL Amplifier Start-up from Shot Noise

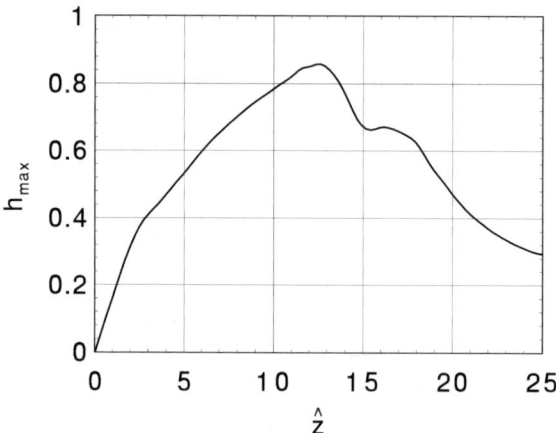

Fig. 6.21. Maximal value of the function $h(\hat{C})$ versus normalized undulator length

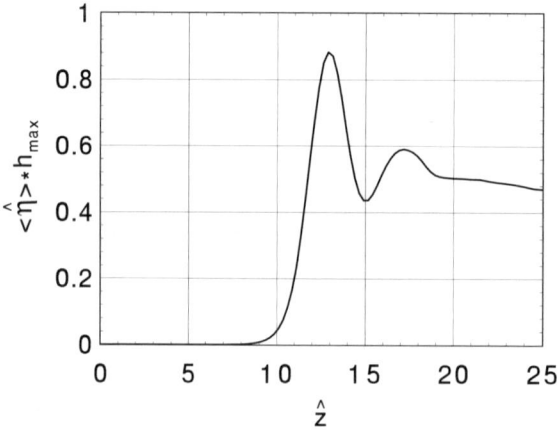

Fig. 6.22. Maximal value of the function $h(\hat{C})$, multiplied by the normalized efficiency, versus normalized undulator length

$$h(\hat{C}) = \frac{\hat{z}}{2\pi} \frac{\sin^2(\hat{C}\hat{z}/2)}{(\hat{C}\hat{z}/2)^2}.$$

One can see that this formula works well at $\hat{z} = 1$. Then, at $\hat{z} = 3$, the spectrum envelope is visibly modified due to the amplification, and at $\hat{z} = 6$ it becomes close enough to a Gaussian line for the high-gain asymptote.

It is relevant to discuss the results on the spectrum calculations for the initial stage of amplification. One should remember that here we study start-up from noise in the framework of the one-dimensional approximation which assumes that the output radiation has full transverse coherence. This assumption is justified for the high-gain regime, while it may be questionable

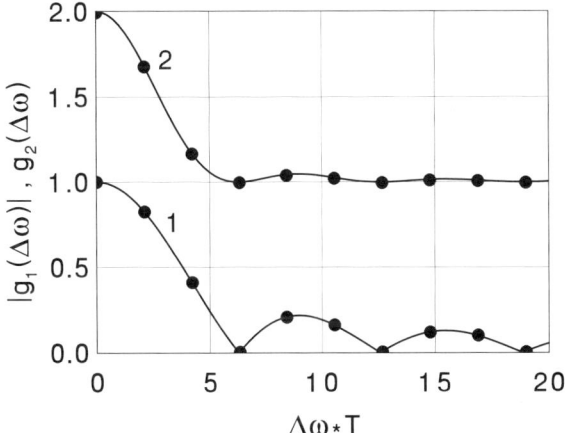

Fig. 6.23. Spectral correlation functions for the reduced length $\hat{z} = 13$ (saturation point). Solid curves are the results of calculations with analytical formulae (6.32) and (6.37). Circles are the result of calculations with nonlinear simulation code. Curve 1 refers to the modulus of the first-order correlation function and curve 2 refers to the second-order correlation function

for the low-gain regime. Thus, one can formulate the question: what is the physical interpretation of the plots in Fig. 6.16? This question is simply answered if we consider the radiation pattern in the far diffraction zone. The spectra presented in Fig. 6.16 should be treated as those corresponding to the radiation observed near the zero angle within $\Delta\theta \lesssim (\gamma_z \sqrt{k_w l_g})^{-1}$, where l_g is the gain length. The larger the gain, the larger part of the radiation power concentrates within this coherent angle $\Delta\theta \lesssim (\gamma_z \sqrt{k_w l_g})^{-1}$.

Figures 6.17 and 6.18 present the spectrum of a SASE FEL operating in the nonlinear regime. Here we plot the normalized spectrum $h(\hat{C})$, multiplied by the normalized FEL efficiency $\langle\hat{\eta}\rangle$. The averaged power spectral density can be calculated as

$$\frac{d\langle W \rangle}{d\omega} = \frac{W_b}{2\omega_0} \langle\hat{\eta}\rangle h(\hat{C}) \ . \tag{6.96}$$

The broken lines in Figs. 6.17 and 6.18 are the results of the direct averaging of the spectra for 200 pulses. The smooth curves are calculated using (6.90) and (6.91). It is our experience that the latter technique provides higher accuracy for spectra reconstruction. If we trace the spectrum evolution from the saturation point to the deep nonlinear regime, we find that the spectrum width and shape change significantly. The spectrum gets broader due to the effect of sideband instability (it already discussed in Chap. 3). The total power increases, while the maximal value of the spectral density decreases.

For each value of the undulator length, $\hat{z}$, there exists a maximal value of the normalized power spectral density, $h_{\max}$. The plot of $h_{\max}$ is presented in Fig. 6.21. An important practical characteristic of the SASE FEL is the maximal value of the power spectral density (users of VUV and X-ray radiation often exploit the notion of photon flux: the photon rate per frequency interval $\Delta\omega/\omega_0 = 10^{-3}$). One can easily express the maximal photon flux in terms of the maximal normalized power spectral density, $\langle\hat{\eta}\rangle h_{\max}$. The plot of the function $\langle\hat{\eta}\rangle h_{\max}$ is presented in Fig. 6.22. The absolute maximum of $\langle\hat{\eta}\rangle h_{\max}$ is achieved at the saturation point and is close to unity. Its value depends weakly on the value of the parameter N_c.

Finally, we present in Fig. 6.23 the spectral correlation functions at the saturation point. It is seen that they are the same as those in the linear mode of operation (see Fig. 6.3), and are well described by analytical formulae (6.32) and (6.37).

Arbitrary Axial Bunch Profile and Short-Pulse Effects. The analytical results of Sect. 6.2.1 refer to the specific model of a long rectangular bunch only. This model has proven to be very fruitful, providing the possibility of obtaining simple analytical expressions for the main statistical characteristics of the radiation from a SASE FEL operating in the high-gain linear regime. There is no doubt that they are useful for quick estimate and deeper understanding of FEL physics. On the other hand, the electron bunch in an actual device always has a gradient axial profile of the current. Also, there can be practical situations when the bunch length is comparable with the slippage length. In this section we briefly discuss these problems.

To be specific, we consider an electron beam with a Gaussian axial profile of the current density:

$$S(\hat{s}) = \frac{j(\hat{s})}{j_{\max}} = \exp\left(-\frac{\hat{s}^2}{2\hat{\sigma}_b^2}\right) ,$$

where $\hat{\sigma}_b = \rho\omega_0\sigma_b/c$ and σ_b is the rms bunch length. Here and below, the normalization is performed with respect to the maximal current density, $j_{\max}$. The rms bunch length is assumed to be large, $\omega_0\sigma_b/c \gg 1$ or, in normalized form:

$$\hat{\sigma}_b \gg \rho .$$

In practice, condition (6.8) is fulfilled under this assumption, and we can neglect the contribution of the coherent seed to the input signal of the FEL amplifier starting from the shot noise. Since ρ is always much less than unity, we can investigate short-pulse effects, when the bunch is comparable to (or even much shorter than) the typical slippage distance $c/(\rho\omega_0)$.

To simulate the SASE FEL driven by the electron bunch with a gradient profile $S(\hat{s})$, we should modify the equations presented in Sect. 6.2.2. Equation (6.86) for the complex field amplitude should be replaced by

$$\hat{E}^{(j)}(\hat{z}) = -2\mathrm{i}\Delta\hat{z}\sum_{m=1}^{n} S^{(j-m)}\hat{a}_1^{(j-m)}(\hat{z} - m\Delta\hat{z}) , \qquad (6.97)$$

where $S^{(j)}$ is the value of the function $S(\hat{s})$ in the jth box. Also, the initial conditions should be calculated taking into account that the number of par-

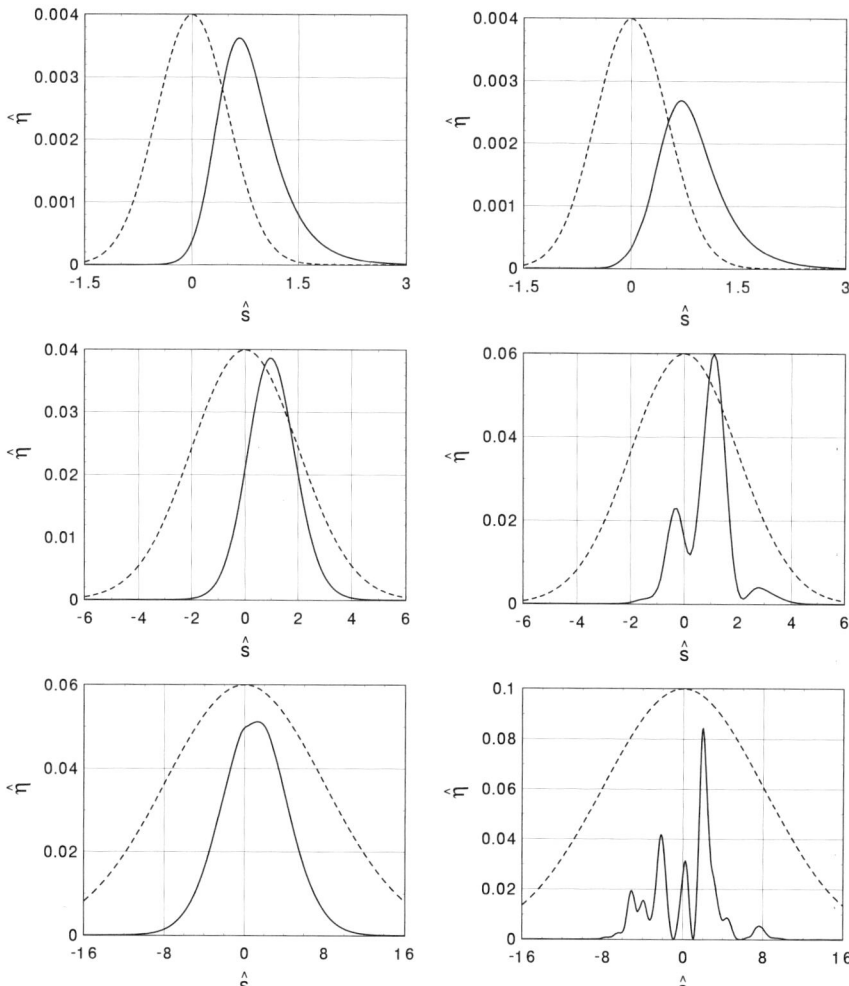

Fig. 6.24. Averaged (left column) and typical single-shot (right column) axial distribution over $\hat{s} = \rho\omega_0(z/\bar{v}_z - t)$) of the normalized radiation power from the SASE FEL for different rms bunch lengths of $\hat{\sigma}_{\mathrm{b}} = 0.5$, 2, and 8 (upper, middle, and lower plots, respectively). Dashed lines represent the axial profile of the beam current. The normalized length of the undulator is $\hat{z} = 10$. Simulations have been performed with linear simulation code

ticles in the jth box, $N_\lambda^{(j)}$, depends on the box number and is defined by the local current. Under the accepted limitations on the axial gradient of the current, we neglect the gradient within the box. Therefore, we still consider the phases of the initial bunching in each box to be uniformly distributed on $(0, 2\pi)$ and the squared modulus of amplitudes to be subjected to the nega-

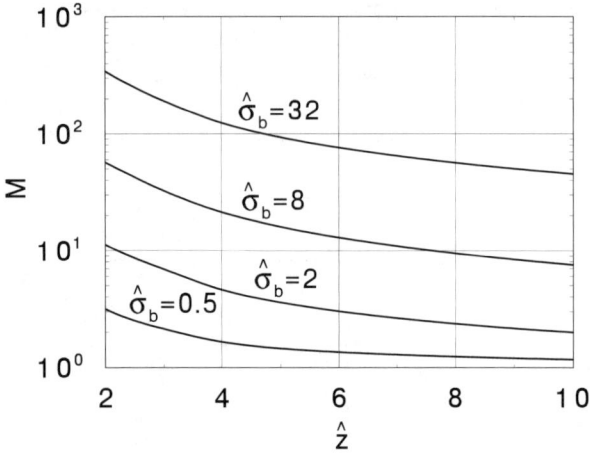

Fig. 6.25. Number of modes in the radiation pulse for Gaussian electron bunches of different lengths versus the undulator length. The SASE FEL operates in the linear regime

tive exponential distribution. Under these initial conditions the input signal is just the shot noise.

Even without performing numerical simulations, we can describe some general properties of the radiation from the SASE FEL operating in the linear regime. Indeed, in this case we still deal with Gaussian statistics. As a result, the probability distribution of the instantaneous radiation power should be the negative exponential distribution. Here one should realize clearly that the notion of instantaneous power refers to a certain moment of time and a certain $\hat{z}$ coordinate, and that the analysis must be performed over an ensemble of pulses. Also, the finite-time integrals of the instantaneous power and the integrated spectral density (measured after the monochromator) should fluctuate in accordance with the gamma distribution.

The first-order time correlation function (6.56) depends on both t and t'. In Sect. 6.2.1 when analyzing the time correlations in the case of a long rectangular bunch, we considered the bunch as a part of an infinite current with time-independent statistical characteristics. In other words, we assumed the random process to be stationary. In this case the correlation function g_1 depends only on $t-t'$ and is proportional to the Fourier transform of the spectral density, $\langle|\bar{E}(\Delta\omega)|^2\rangle$. In the case of a gradient profile of the electron bunch this property is not valid. Nevertheless, there exist convenient characteristics of temporal coherence, such as the effective correlation function:

$$g_1^{(\mathrm{eff})}(\tau) = \frac{\int\limits_{-\infty}^{\infty} \langle \tilde{E}(\bar{t}+\tau/2)\tilde{E}^*(\bar{t}-\tau/2)\rangle \mathrm{d}\bar{t}}{\int\limits_{-\infty}^{\infty} \langle|\tilde{E}(\bar{t})|^2\rangle \mathrm{d}\bar{t}}, \tag{6.98}$$

where $\bar{t} = (t+t')/2$ and $\tau = t - t'$. It is natural to define the coherence time for the nonstationary process as

$$\tau_c = \int_{-\infty}^{\infty} d\tau |g_1^{(\text{eff})}(\tau)|^2 \ . \tag{6.99}$$

It is straightforward to show that $g_1^{(\text{eff})}(\tau)$ and $\langle|\bar{E}(\Delta\omega)|^2\rangle$ are connected by the Fourier transform. Using Parseval's theorem, we can write for a narrow-band signal:

$$\frac{\langle|\bar{E}(\Delta\omega)|^2\rangle}{\int_{-\infty}^{\infty} \langle|\bar{E}(\Delta\omega)|^2\rangle d(\Delta\omega)} = \frac{\int_{-\infty}^{\infty} dt \int_{-\infty}^{\infty} dt' \langle \tilde{E}(t)\tilde{E}^*(t')\rangle e^{i\Delta\omega(t-t')}}{2\pi \int_{-\infty}^{\infty} \langle|\tilde{E}(t)|^2\rangle dt} \ ,$$

where $\Delta\omega = \omega - \omega_0$. Going over from variables (t,t') to variables $(\bar{t},\tau)$, and using (6.98), we rewrite the latter expression as

$$\frac{\langle|\bar{E}(\Delta\omega)|^2\rangle}{\int_{-\infty}^{\infty} \langle|\bar{E}(\Delta\omega)|^2\rangle d(\Delta\omega)} = \frac{1}{2\pi} \int_{-\infty}^{\infty} d\tau g_1^{(\text{eff})}(\tau) e^{i\Delta\omega\tau} \ . \tag{6.100}$$

Thus, we can conclude that the correlation function $g_1^{(\text{eff})}(\tau)$ effectively describes the case of a long rectangular bunch producing the same spectrum as that of a bunch with a gradient profile.

In the general case, the spectral correlation function (6.31) is a function of two independent variables, ω and ω'. For a narrow-band signal it is more convenient to operate with the variables $\Delta\omega = \omega - \omega_0$ and $\Delta\omega' = \omega' - \omega_0$. Introducing the effective spectral correlation function,

$$g_1^{(\text{eff})}(\Omega) = \frac{\int_{-\infty}^{\infty} \langle \bar{E}(\Delta\bar{\omega} + \Omega/2)\bar{E}^*(\Delta\bar{\omega} - \Omega/2)\rangle d\Delta\bar{\omega}}{\int_{-\infty}^{\infty} \langle|\bar{E}(\Delta\bar{\omega})|^2\rangle d\Delta\bar{\omega}} \ , \tag{6.101}$$

where $\Delta\bar{\omega} = (\Delta\omega + \Delta\omega')/2$ and $\Omega = \Delta\omega - \Delta\omega'$, we obtain a relation similar to (6.100):

$$\frac{\langle|\tilde{E}(t)|^2\rangle}{\int_{-\infty}^{\infty} \langle|\tilde{E}(t)|^2\rangle dt} = \frac{1}{2\pi} \int_{-\infty}^{\infty} d\Omega g_1^{(\text{eff})}(\Omega) e^{-i\Omega t} \ . \tag{6.102}$$

Therefore, the envelope of the radiation power is the Fourier transform of the effective spectral correlation function. The interval of spectral coherence may be defined as

$$\Delta\omega_c = \int_{-\infty}^{\infty} d\Omega |g_1^{(\text{eff})}(\Omega)|^2 \ . \tag{6.103}$$

Let us present some results of numerical simulations of the SASE FEL with a Gaussian electron bunch. The time structure of a sample radiation pulse and the radiation power averaged over an ensemble are presented in Fig. 6.24 for different bunch lengths. It is seen that for values of $\hat{\sigma}_b$ of about unity, the radiation pulse is visibly shifted off the center of the electron bunch due to the slippage effect. In Fig. 6.25 the number of modes M is plotted versus the longitudinal coordinate for different bunch lengths. We use the following definition of the parameter M:

$$M = \frac{1}{\sigma_{\mathcal{E}}^2} \, ,$$

where $\sigma_{\mathcal{E}}^2 = \langle (\mathcal{E} - \langle \mathcal{E} \rangle)^2 \rangle / \langle \mathcal{E} \rangle^2$ is the relative energy dispersion in the radiation pulse.

6.3 Three-Dimensional Simulations of SASE FEL

Till now we have investigated the SASE FEL in the framework of the one-dimensional model. In particular, this model assumes the input shot noise and the output radiation to have full transverse coherence. In reality the fluctuations of the electron beam current density are uncorrelated in the transverse dimension. Using the notion of the beam radiation modes, we can say that many beam radiation modes are excited when the electron beam enters the undulator. Information on transverse coherence formation can be obtained with three-dimensional, time-dependent simulation code. Analysis of simulation results allows us to find the region of parameters when the transverse coherence of the radiation is settled and we can apply the results of the one-dimensional treatment described above.

6.3.1 Numerical Simulation Algorithm

In this section we describe a three-dimensional, time-dependent algorithm for simulations of the SASE FEL. The main feature of the code is that the radiation fields are calculated using rigorous solution of the electrodynamic problem (see Sect. 4.5.1). Time-dependent effects are included in a similar way to that described in Sect. 6.2.2.

The radiation field and the beam current density are presented in the form:

$$E_y = \tilde{E}(\boldsymbol{r}_\perp, z, t) e^{i\omega_0(z/c-t)} + \text{C.C.} \, ,$$
$$j_y = v_y(z) j_z(z,t)/c \, ,$$
$$j_z = -j_0(\boldsymbol{r}_\perp, z, t) + \tilde{j}_a(\boldsymbol{r}_\perp, z, t) e^{i\psi'} + \text{C.C.} \, ,$$

where $-j_0$ is the ensemble averaged current density. In the case of an axisymmetric beam profile it can be factorized as follows:

$$j_0(\boldsymbol{r}_\perp, z, t) = j_{\max} S(s) S_\perp(r), \quad (6.104)$$

where $j_{\max}$ is the maximal current density, r is the radial coordinate of the cylindrical system (r, φ, z), $s = z - \bar{v}_z t$ is the coordinate introduced in Sect. 6.2.2, and $S(s)$ and $S_\perp(r)$ are functions describing the longitudinal and the transverse profile of the electron bunch, respectively. To be specific, we write down all the expressions for the case of a Gaussian longitudinal and transverse distribution:

$$S(s) = \exp[-s^2/(2\sigma_b^2)], \qquad S_\perp(r) = \exp[-r^2/(2\sigma_r^2)],$$

$$j_{\max} = \frac{I_{\max}}{2\pi\sigma_r^2}, \qquad I_{\max} = -\frac{Nec}{(2\pi)^{1/2}\sigma_b},$$

where N is the total number of particles in the bunch.

We solve the electrodynamic problem using the paraxial approximation. In this case the wave equation may be written in the following form:

$$\left[\nabla_\perp + 2\mathrm{i}\frac{\omega_0}{c}\left(\frac{\partial}{\partial z} + \frac{1}{c}\frac{\partial}{\partial t}\right)\right]\tilde{E}(\boldsymbol{r}_\perp, z, t)$$
$$= -\frac{2\pi\theta_\ell \omega_0 A_{JJ}}{c^2}\tilde{j}_a(\boldsymbol{r}_\perp, z, t). \quad (6.105)$$

The solution of this equation is

$$\tilde{E}(\boldsymbol{r}_\perp, z, t) = \frac{\theta_\ell \omega_0 A_{JJ}}{2c^2}\int_0^z \frac{dz'}{z - z'}\int d\boldsymbol{r}'_\perp \tilde{j}_a\left(\boldsymbol{r}'_\perp, z', t - \frac{z - z'}{c}\right)$$
$$\times \exp\left[\frac{\mathrm{i}\omega_0 |\boldsymbol{r}_\perp - \boldsymbol{r}'_\perp|^2}{2c(z - z')}\right]. \quad (6.106)$$

The complex amplitudes, $\tilde{E}$ and $\tilde{j}_a$, are expanded in a Fourier series in the angle φ,

$$\tilde{E} = \sum_{n=-\infty}^\infty \tilde{E}^{(n)}(r, z, t)\mathrm{e}^{-\mathrm{i}n\varphi}, \qquad \tilde{j}_a = \sum_{n=-\infty}^\infty \tilde{j}_a^{(n)}(r, z, t)\mathrm{e}^{-\mathrm{i}n\varphi}.$$

Then we get from (6.106) the expression for the Fourier harmonics:

$$\tilde{E}^{(n)}(r, z, t) = \frac{\pi\theta_\ell \omega_0 A_{JJ}}{c^2}\mathrm{e}^{-\mathrm{i}n\pi/2}\int_0^z \frac{dz'}{z-z'}\int_0^\infty dr' r'\tilde{j}_a^{(n)}\left(r', z', t - \frac{z-z'}{c}\right)$$
$$\times J_n\left(\frac{\omega_0 rr'}{c(z-z')}\right)\exp\left[\frac{\mathrm{i}\omega_0(r^2 + r'^2)}{2c(z-z')}\right]. \quad (6.107)$$

Prior to the detailed analysis of start-up from noise (i.e. the self-consistent solution of (6.105) and the equations of particle motion under the shot noise initial conditions at the undulator entrance), it is relevant to discuss the region of applicability of the paraxial wave equation (6.105). The paraxial approximation assumes complex amplitude $\tilde{E}(\boldsymbol{r}_\perp, z, t)$ to be a slowly varying function on the scale of the radiation wavelength. When we consider start-up from noise, the beam current, $\tilde{j}_a(\mathbf{r}_\perp, z, t)$, is not a slowly varying function. The first limitation on the problem parameters is similar to that appearing in the one-dimensional model (see Sec. 6.2), i.e., the undulator should be sufficiently long, $k_w z \gg 1$. When the latter condition is fulfilled, we still cannot expect correct results in the three-dimensional case. Indeed, the incoherent undulator radiation has a wide continuous spectrum. When $k_w z \gg 1$, (6.105) correctly describes the fields in the narrow frequency band near the resonance frequency only, $\Delta\omega/\omega_0 \ll 1$. In terms of the far field zone, it gives correct results only for that part of the incoherent undulator radiation which is concentrated within the angle $\Delta\theta \ll 1/\gamma_z$ near the z axis.

When the FEL amplifier starts from the shot noise, a lot of transverse radiation modes are excited at the beginning of the amplification process; the radiation spectrum and the angular distribution in the far zone are relatively large. During the amplification process, the number of transverse radiation modes decreases, and the contribution of the coherent radiation into the total radiation power is increased. Also, the angular distribution of the radiation intensity in the far zone decreases. When it becomes much less than $1/\gamma_z$, we obtain a correct quantitative description of the amplification process starting from the shot noise.

One more relation, connecting the field and the current density, should come from the solution of the dynamical problem. When the space charge field can be neglected, the equations of motion may be written in the form:

$$\frac{dP}{dz} = -\frac{i}{2} A_{JJ} e \theta_\ell \tilde{E} e^{i\psi'} + \text{C.C.},$$

$$\frac{d\psi'}{dz} = \omega P/(c\gamma_\ell^2 \mathcal{E}_0).$$

In the same way as was done in Chap. 4, we perform the normalization procedure:

$$\hat{z} = \Gamma z, \qquad \hat{P} = P/(\rho\mathcal{E}_0), \qquad \hat{r} = r/(\sqrt{2}\sigma_r),$$

$$B = 2\Gamma\omega_0\sigma_r^2/c, \qquad \rho = \Gamma\gamma_\ell^2 c/\omega_0,$$

$$\Gamma = \left[I_{\max} A_{JJ}^2 \omega_0^2 \theta_\ell^2 /(2I_A c^2 \gamma_\ell^2 \gamma)\right]^{1/2}.$$

In addition, we use the following normalization (see Sects. 6.2.2 and 6.2.3 for comparison): $\hat{s} = \rho\omega_0 s/\bar{v}_z$, $\Delta\hat{z} = \Gamma\lambda_w$, $\hat{\sigma}_b = \rho\omega_0\sigma_b/c$, $\hat{a}_1^{(n)} = -\tilde{j}_a^{(n)}/j_0$, $\hat{E}^{(n)} = \tilde{E}^{(n)}/E_0$, and $E_0 = c\mathcal{E}_0\gamma_\ell^2\Gamma^2/(e\theta_\ell\omega_0 A_{JJ})$. As a result, the self-consistent equations can be written in the form:

$$\frac{d\hat{P}}{d\hat{z}} = -\frac{i}{2}\left[\sum_n \hat{E}^{(n)}(\hat{r}, \hat{z}, \hat{s}) e^{i(\psi' - n\varphi)}\right] + \text{C.C.},$$

$$\frac{d\psi'}{d\hat{z}} = \hat{P}, \tag{6.108}$$

$$\hat{E}^{(n)}(\hat{r}, \hat{z}, \hat{s}) = -2\mathrm{e}^{-\mathrm{i}n\pi/2} \int_0^{\hat{z}} \frac{\mathrm{d}\hat{z}'}{\hat{z} - \hat{z}'} \int_0^{\infty} \mathrm{d}\hat{r}'\hat{r}' \exp(-\hat{r}'^2)$$

$$\times \exp\left\{\frac{-[\hat{s} - (\hat{z} - \hat{z}')/2]^2}{2\hat{\sigma}_\mathrm{b}^2}\right\} \hat{a}_1^{(n)}\left(\hat{r}', \hat{z}', \hat{s} - (\hat{z} - \hat{z}')/2\right)$$

$$\times J_n\left(\frac{B\hat{r}\hat{r}'}{\hat{z} - \hat{z}'}\right) \exp\left[\frac{\mathrm{i}B(\hat{r}^2 + \hat{r}'^2)}{2(\hat{z} - \hat{z}')}\right]. \tag{6.109}$$

To perform the simulations, we divide the electron beam into a large number of elementary volumes. The size of the divisions of the electron beam in the longitudinal direction should typically be chosen equal to the radiation wavelength. The number of azimuthal harmonics for calculations of the radiation field, N_φ, defines the number of azimuthal divisions of the electron beam. Typically, it should be by an order of magnitude larger than N_φ. Finally, the radial mesh should be chosen. The simulations are performed with a macroparticle method. The number of macroparticles in each volume is equal to N_m. The equations of motion (6.108) are integrated with the integration step equal to $\Delta\hat{z} = \Gamma\lambda_\mathrm{w}$. At each integration step we calculate the bunching, $\hat{a}_1$, in each elementary volume:

$$\hat{a}_1 = \frac{1}{N_\mathrm{m}} \sum_{m=1}^{N_\mathrm{m}} \mathrm{e}^{-\mathrm{i}\psi'_m}.$$

These values are used to calculate the azimuthal harmonics, $\hat{a}_1^{(n)}$. The radiation field of the nth azimuthal harmonic in the discrete representation is calculated using the rigorous solution (6.109). At the next integration step, the sum of the azimuthal harmonics of the field is substituted into the equations of macroparticles motion in each volume, etc. As a result, one can trace the evolution of the radiation field and the particle distribution when the electron beam passes the undulator.

We have shown in Sect. 4.5.1 that there is a reliable technique for the precise numerical calculation of integrals of the type (6.109). When performing time-dependent simulations, we must recalculate the field at every space point at each integration step, so fast calculation of the integral (6.109) becomes crucial for the simulation code. This problem can be solved in the following way. As we mentioned, in the simulations we divide the electron beam into a large number of elementary volumes with a fixed mesh in the radius and azimuthal angle. The integration step is also fixed. At a fixed value of $\hat{z}$, the integral (6.109) is calculated as described in Sec. 4.5.1. Namely, we let the value of the bunching be constant within each elementary volume, and reduce the integration to the sum of the amplitudes $\hat{a}_1^{(n)}(\hat{r}', \hat{z}', \hat{s} - (\hat{z} - \hat{z}')/2)$ multiplied by the transverse area $\mathrm{d}\hat{r}'\hat{r}'$, by a factor of the current density profile,

$$\exp[-\hat{r}'^2] \times \exp\left[\frac{-[\hat{s} - (\hat{z} - \hat{z}')/2]^2}{2\hat{\sigma}_b^2}\right],$$

and by the integral over $\Delta\hat{z}$:

$$\int_{\hat{z}_k}^{\hat{z}_k+\Delta\hat{z}} \frac{\mathrm{d}\hat{z}'}{\hat{z}-\hat{z}'} J_n\left(\frac{B\hat{r}\hat{r}'}{\hat{z}-\hat{z}'}\right) \exp\left[\frac{iB(\hat{r}^2+\hat{r}'^2)}{2(\hat{z}-\hat{z}')}\right], \tag{6.110}$$

where $\hat{z}_k = \Delta\hat{z} \times k$. Since we use a fixed radial mesh and fixed integration step, we immediately find that the number of different combinations of $\hat{r}$, $\hat{r}'$ and $\hat{z} - \hat{z}'$ entering the integrand in (6.110) is limited. The integrals for each different set of these combinations are calculated only once before running the SASE simulation code and are stored in an array. During the simulation of the FEL process, the elements of the array are weighted by actual current sources given by the product of the value of the density profile function and of the bunching. Such an approach allows one to reduce significantly the required CPU time.

One can also obtain from the integrals (6.109) that the radiation field at each point is defined by the sources located closer than the slippage distance and it is not necessary to keep in memory all the current sources. The procedure of the simulations begins from the tail slice of the electron bunch and the procedure of integration is performed over the whole undulator length. Then the equations of motion for the second slice are integrated taking into account the radiation field from the first slice, etc. As a result, the self-consistent FEL equations can be integrated for an electron bunch of any length. The memory requirements for the code are rather moderate.

When performing the simulation of the linear mode of SASE FEL operation, we can solve directly the equation for the azimuthal harmonics of the bunching instead of solving (6.85):

$$\frac{\mathrm{d}^2 \hat{a}_1^{(n)}}{\mathrm{d}\hat{z}^2} = -\frac{1}{2}\hat{E}^{(n)},$$

which significantly increases the speed of calculations. The space charge field and the energy spread in the beam can be included in the code in the same way as described in Sect. 4.5.1.

The initial shot noise in the electron beam is simulated according to the algorithm presented in Sect. 6.2.2. Since the actual number of particles per elementary volume, N_v, is large, the bunching in each box is the sum of a large number of random phasors with fixed amplitudes and uniformly distributed on $(0, 2\pi)$ phases. Using the central limit theorem, we can conclude that the phases of the bunching parameters are also distributed uniformly and the squared modulus of the amplitudes, $|\hat{a}_1|^2$, are distributed by the negative exponential distribution:

$$p(|\hat{a}_1|^2) = \frac{1}{\langle|\hat{a}_1|^2\rangle} \exp\left(-\frac{|\hat{a}_1|^2}{\langle|\hat{a}_1|^2\rangle}\right), \tag{6.111}$$

where $\langle |\hat{a}_1|^2 \rangle = 1/N_v$. So, a negative exponential random generator with a mean value of $1/N_v$ is used to extract the values of $|\hat{a}_1|^2$ for each volume. The phases of $\hat{a}_1$ are produced by a random number generator for the uniform distribution from 0 to 2π. These values are directly used as input parameters for the linear simulation code. In the nonlinear simulation code the macroparticles are distributed in such a way that the resulting bunching corresponds to the target value of $\hat{a}_1$ in each elementary volume.

The output of the code are the arrays for the field values in the Fresnel diffraction zone. Post-processor programs are used to extract additional information for the field distribution in the far diffraction zone, for the spectrum, for the time, space and spectral correlation functions, and for the probability distributions of the radiation power and the radiation energy.

6.3.2 Transverse Coherence

In Figs. 6.26 and 6.27 we illustrate the formation of transverse coherence. The transverse distribution of the electron beam current density is Gaussian and the value of the diffraction parameter is $B = 1$. It is clearly seen how an initially irregular intensity distribution transforms to one corresponding to the fundamental TEM_{00} mode.

The transverse coherence properties are described in terms of the transverse correlation functions. The first-order transverse correlation function is defined as

$$\gamma_1(\boldsymbol{r}_\perp, \boldsymbol{r}'_\perp, z, t) = \frac{\langle \tilde{E}(\boldsymbol{r}_\perp, z, t) \tilde{E}(\boldsymbol{r}'_\perp, z, t) \rangle}{\left[\langle |\tilde{E}(\boldsymbol{r}_\perp, z, t)|^2 \rangle \langle |\tilde{E}(\boldsymbol{r}'_\perp, z, t)|^2 \rangle \right]^{1/2}} ,$$

where $\tilde{E}$ is the slowly varying amplitude of the amplified wave:

$$E_y = \tilde{E}(\boldsymbol{r}_\perp, z, t) e^{i\omega_0(z/c - t)} + \text{C.C.} . \qquad (6.112)$$

When studying the time correlation functions, we have assumed the statistical process to be stationary. Though this model is very idealized, making use of it is justified by the fact that in many real cases the radiation pulse is much longer than the coherence length. In other words, partial coherence is an essential feature of SASE FEL radiation in the time domain. In the following we will continue to use the model of a stationary process, thus assuming that γ_1 does not depend on time. We can also use ergodicity if needed and can average over time instead of averaging over an ensemble.

The space analogue of a stationary statistical process is a model of a statistically homogeneous field. For such a field the correlation function (6.112) depends only on the remainder $\boldsymbol{\rho} = \boldsymbol{r}_\perp - \boldsymbol{r}'_\perp$:

$$\gamma_1(\boldsymbol{r}_\perp, \boldsymbol{r}'_\perp, z) = \gamma_1(\boldsymbol{\rho}, z) .$$

It is natural to define the area of coherence as (see the definition of coherence time (6.60)):

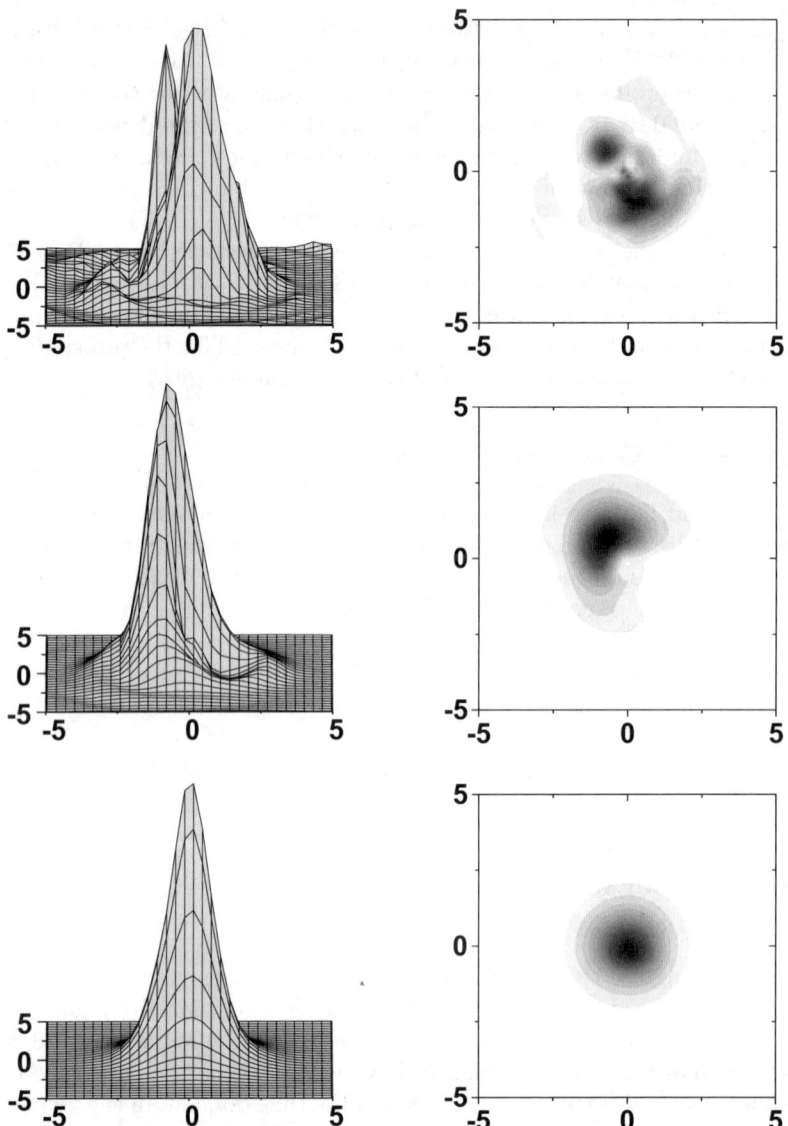

Fig. 6.26. Distributions of the radiation intensity across one slice of the radiation pulse at different undulator lengths, $\hat{z} = 5$, $\hat{z} = 10$, and $\hat{z} = 15$ (upper, middle, and lower plots, respectively). The plots in the right column are gray scale projections of 3-D plots on the $x-y$ plane. The coordinates are normalized to $2^{1/2}\sigma_{\mathrm{r}}$. Here $B = 1$, $\hat{\Lambda}_{\mathrm{p}}^2 \to 0$, and $\hat{\Lambda}_{\mathrm{T}}^2 = 0$. Calculations have been performed with linear simulation code

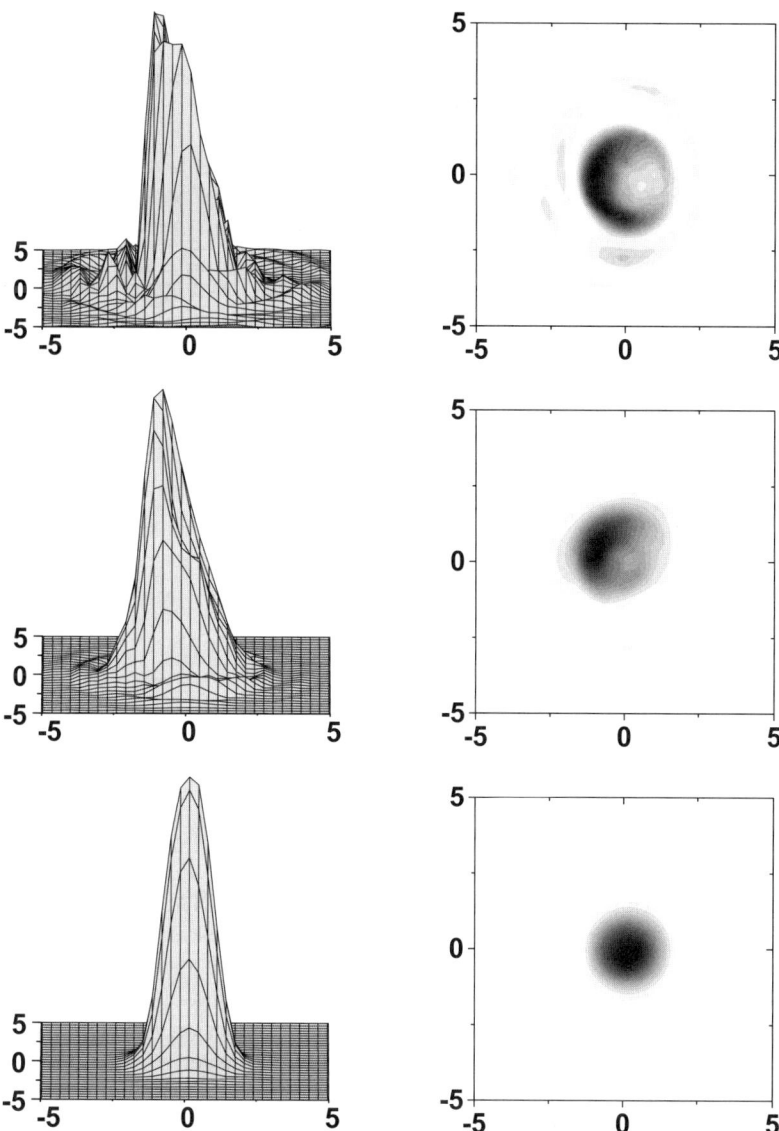

Fig. 6.27. Angular distributions of the radiation intensity for one slice of the radiation pulse at different undulator lengths, $\hat{z} = 5$, $\hat{z} = 10$, and $\hat{z} = 15$ (upper, middle, and lower plots, respectively). The plots in the right column are gray scale projections of 3-D plots on the $\theta_x - \theta_y$ plane. Angles are normalized to $(2^{1/2}\sigma_r\omega_0/c)^{-1}$. Here $B = 1$, $\hat{\Lambda}_P^2 \to 0$, and $\hat{\Lambda}_T^2 = 0$. Calculations have been performed with linear simulation code

$$S_{\mathrm{c}}(z) = \int \mathrm{d}\boldsymbol{\rho} |\gamma_1(\boldsymbol{\rho}, z)|^2 .$$

This definition has a simple physical interpretation: the number of transverse modes, M, within the area of $S \gg S_{\mathrm{c}}$, is equal to:

$$M = \frac{1}{\sigma_{\mathrm{w}}^2} = \frac{S}{S_{\mathrm{c}}} ,$$

where σ_{w} is the normalized dispersion of the instantaneous radiation power passing through the area of S.

Now let us consider the case when a statistically homogeneous field is also isotropic. In this case the correlation function depends on the modulus $\rho = |\boldsymbol{\rho}|$:

$$\gamma_1(\boldsymbol{\rho}, z) = \gamma_1(\rho, z) ,$$

and the expression for the coherence area may be written as

$$S_{\mathrm{c}}(z) = 2\pi \int_0^\infty |\gamma_1(\rho, z)|^2 \rho \, \mathrm{d}\rho .$$

Using the relation $S_{\mathrm{c}} = \pi r_{\mathrm{c}}^2$, it is convenient to introduce the notion of the radius of coherence r_{c}:

$$r_{\mathrm{c}}(z) = \left[2 \int_0^\infty |\gamma_1(\rho, z)|^2 \rho \, \mathrm{d}\rho \right]^{1/2} . \qquad (6.113)$$

The number of modes within a circular aperture of radius $a \gg r_{\mathrm{c}}$ is equal to

$$M = \frac{1}{\sigma_{\mathrm{w}}^2} = \frac{a^2}{r_{\mathrm{c}}^2} .$$

We should mention here that the radiation from a real SASE FEL cannot be properly described in the framework of the model of a statistically homogeneous field. Indeed, the parameters of a SASE FEL are usually chosen in such a way that transverse coherence is achieved at the end of the amplification process. This happens due to transverse mode selection (see Chap. 4), i.e. only one mode survives in the end. So, the finite transverse size of the radiation field makes good sense. That is why in the following we will consider a statistically inhomogeneous field.

We already know that the time correlation function can be used for calculation of the frequency spectrum. Let us now show that the angular spectrum (which defines the divergence of the optical beam) can be reconstructed from the transverse correlation function. The Fourier transform of the complex amplitude $\tilde{E}(\boldsymbol{r}_\perp, z, t)$ over the transverse coordinates is

$$A(\boldsymbol{k}_\perp, z, t) = \int \tilde{E}(\boldsymbol{r}_\perp, z, t) \exp(-\mathrm{i}\boldsymbol{k}_\perp \boldsymbol{r}_\perp) \mathrm{d}\boldsymbol{r}_\perp .$$

Then the angular spectrum for a stationary random process can be written as

$$h(\boldsymbol{k}_\perp, z)$$
$$= \frac{\langle |A(\boldsymbol{k}_\perp, z)|^2 \rangle}{\int \langle |A(\boldsymbol{k}_\perp, z)|^2 \rangle \mathrm{d}\boldsymbol{k}_\perp}$$
$$= \frac{\int \int \langle \tilde{E}(\boldsymbol{r}_\perp, z) \tilde{E}^*(\boldsymbol{r}'_\perp, z) \rangle \exp\left[-\mathrm{i}\boldsymbol{k}_\perp (\boldsymbol{r}_\perp - \boldsymbol{r}'_\perp)\right] \mathrm{d}\boldsymbol{r}_\perp \mathrm{d}\boldsymbol{r}'_\perp}{(2\pi)^2 \int \langle |\tilde{E}(\boldsymbol{r}_\perp, z)|^2 \rangle \mathrm{d}\boldsymbol{r}_\perp} \; . \quad (6.114)$$

Using the notation $\boldsymbol{\rho} = \boldsymbol{r}_\perp - \boldsymbol{r}'_\perp$ and $\boldsymbol{R} = (\boldsymbol{r}_\perp + \boldsymbol{r}'_\perp)/2$, we can rewrite the latter expression as

$$h(\boldsymbol{k}_\perp, z) = \frac{\int \int \langle \tilde{E}(\boldsymbol{R} + \boldsymbol{\rho}/2, z) \tilde{E}^*(\boldsymbol{R} - \boldsymbol{\rho}/2, z) \rangle \exp\left(-\mathrm{i}\boldsymbol{k}_\perp \boldsymbol{\rho}\right) \mathrm{d}\boldsymbol{\rho} \mathrm{d}\boldsymbol{R}}{(2\pi)^2 \int \langle |\tilde{E}(\boldsymbol{R}, z)|^2 \rangle \mathrm{d}\boldsymbol{R}} \; .$$

Then we can introduce the definition of effective transverse correlation function (see the end of Sect. 6.2.3 for comparison):

$$\gamma_1^{(\mathrm{eff})}(\boldsymbol{\rho}, z) = \frac{\int \langle \tilde{E}(\boldsymbol{R} + \boldsymbol{\rho}/2, z) \tilde{E}^*(\boldsymbol{R} - \boldsymbol{\rho}/2, z) \rangle \mathrm{d}\boldsymbol{R}}{\int \langle |\tilde{E}(\boldsymbol{R}, z)|^2 \rangle \mathrm{d}\boldsymbol{R}} \; . \quad (6.115)$$

The angular spectrum and the effective correlation function are connected by the Fourier transform

$$h(\boldsymbol{k}_\perp, z) = \frac{1}{(2\pi)^2} \int \gamma_1^{(\mathrm{eff})}(\boldsymbol{\rho}, z) \exp\left(-\mathrm{i}\boldsymbol{k}_\perp \boldsymbol{\rho}\right) \mathrm{d}\boldsymbol{\rho} \; .$$

Thus, the averaged intensity distribution in the far zone $h(\boldsymbol{k}_\perp, z)$ is totally defined by the effective transverse correlation function. On the other hand, if one knows $h(\boldsymbol{k}_\perp, z)$, then $\gamma_1^{(\mathrm{eff})}(\boldsymbol{\rho}, z)$ can be calculated by means of the inverse Fourier transformation. It should be noted that the correlation function $\gamma_1^{(\mathrm{eff})}$ effectively describes a homogeneous partially coherent source which would generate the same angular distribution of radiation intensity as the actual source.

In what follows we assume the driving electron beam of the SASE FEL to be axisymmetric. In this case the statistical process is isotropic (but not homogeneous): the effective correlation function depends on the modulus $\rho = |\boldsymbol{\rho}|$ and the angular spectrum depends on the modulus $k_\perp = |\boldsymbol{k}_\perp|$. Thus, we have a pair of transformations:

$$h(k_\perp, z) = \frac{1}{2\pi} \int_0^\infty J_0(k_\perp \rho) \gamma_1^{(\mathrm{eff})}(\rho, z) \rho \mathrm{d}\rho \; , \quad (6.116)$$

$$\gamma_1^{(\mathrm{eff})}(\rho, z) = 2\pi \int_0^\infty J_0(\rho k_\perp) h(k_\perp, z) k_\perp \mathrm{d}k_\perp \; . \quad (6.117)$$

Let us introduce the effective radius of coherence for an isotropic inhomogeneous field, generalizing (6.113) as follows

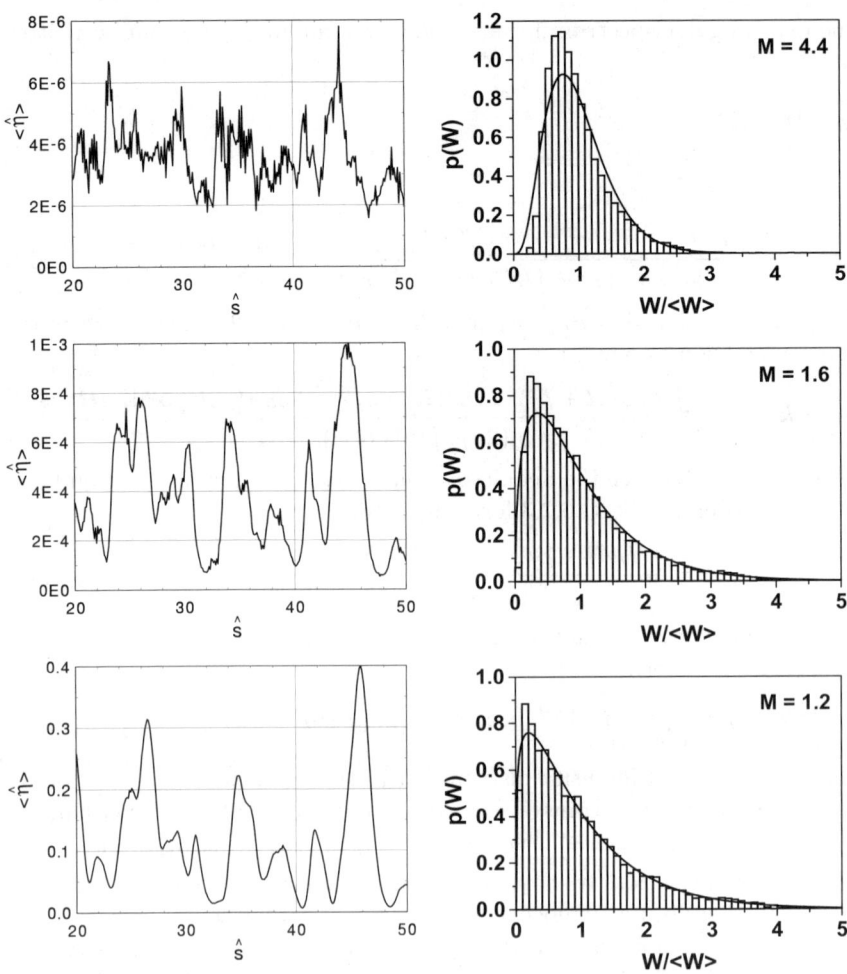

Fig. 6.28. Normalized power in one radiation pulse versus $\hat{s} = \rho\omega_0(z/\bar{v}_z - t)$ (left column) and the probability distributions of the instantaneous radiation power (right column) at different reduced lengths of the undulator of $\hat{z} = 5$, 10, and 15 (upper, middle, and lower plots, respectively). The solid curves in the right column represent the gamma distribution. Here $B = 1$, $\hat{\Lambda}_p^2 \to 0$, and $\hat{\Lambda}_T^2 = 0$. Calculations have been performed with linear simulation code

$$r_c(z) = \left[2 \int_0^\infty |\gamma_1^{(\text{eff})}(\rho, z)|^2 \rho \, d\rho \right]^{1/2} . \tag{6.118}$$

For illustration we consider a fully coherent Gaussian beam, having waist at given z (see (4.144)):

$$\tilde{E} \propto \exp(-r^2/w^2) .$$

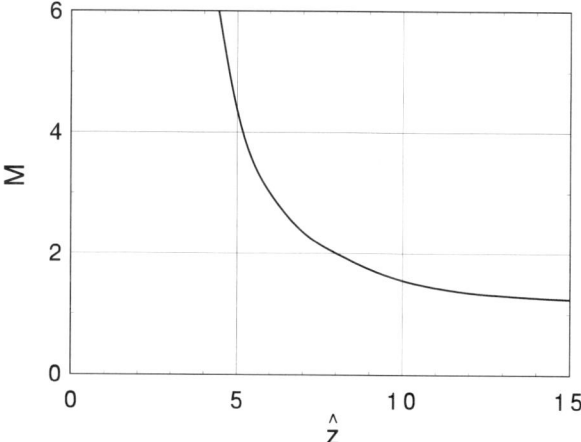

Fig. 6.29. Number of transverse modes, M, versus the reduced undulator length. Here $B = 1$, $\hat{\Lambda}_p^2 \to 0$ and $\hat{\Lambda}_T^2 = 0$. Calculations have been performed with linear simulation code

Then we easily calculate the correlation function

$$\gamma_1^{(\mathrm{eff})}(\rho) = \exp[-\rho^2/(2w^2)]$$

and the radius of coherence $r_c = w$. It is worth noting that the Gaussian beam and the statistically homogeneous isotropic field, correlated over the distance w, have the same width of angular intensity distribution in the far zone.

Now let us consider qualitatively the results of the numerical simulations presented in Fig. 6.26. Let a be the characteristic transverse size of the radiation from the SASE FEL. When $r_c \ll a$, the radiation is partially coherent. This case is shown in the upper plot in Fig. 6.26. Here r_c may be estimated as the typical size of speckles. The gain is not large enough in this case. In the high-gain regime there is almost full transverse coherence, i.e. $r_c \simeq a$ (see the lower plots in Fig. 6.26).

To describe the formation of transverse coherence quantitatively, we should define the degree of coherence. One possible definition can be made as follows. After statistical analysis of the numerical results we find $r_c(z)$. Then we find the radius of coherence r_c^0 for the fully coherent radiation which is represented by the ground TEM_{00} beam radiation mode. In the linear regime the field distribution of this mode can be found by the solution of an eigenvalue equation (see Chap. 4). For a Gaussian density distribution in the electron beam one can use the multilayer approximation method described in Sect. 4.4.5. The degree of coherence, ζ, may be defined as

$$\zeta(z) = \left(\frac{r_c(z)}{r_c^0}\right)^2 .$$

Another possible way to define the degree of coherence is based on the statistical analysis of fluctuations of the instantaneous power

$$W \propto \int |\tilde{E}(\mathbf{r}_\perp, z, t)|^2 \mathrm{d}\mathbf{r}_\perp \ .$$

Since in the linear regime we deal with a Gaussian random process, the power density fluctuates in accordance with the negative exponential distribution and its relative width is equal to 1. If there is full transverse coherence then the same refers to the power. If the radiation is partially coherent, then we have a more general law for power fluctuations, namely the gamma distribution (see Fig. 6.28). The parameter $M = 1/\sigma_w^2$ of this distribution can be considered as the number of transverse modes. Then the degree of coherence in the linear regime, ζ, may be defined as follows

$$\zeta(z) = \frac{1}{M(z)} = \sigma_w^2(z) \ .$$

In Fig. 6.29 we present the dependence of the number of transverse modes on the undulator length for the specific value $B = 1$ of the diffraction parameter.

6.4 SASE FEL: Experiment and Theory

So, we approach the end of the book. Our main goal was to study in a clear form the main physical properties of the free electron laser. Sometimes the price for this clarity was the simplification of the processes with respect to actual FEL devices. In some sense we studied an ideal FEL. A reasonable question (which we are asked frequently) is how this knowledge can be applied to the calculation of practical devices. Our answer is that the theoretical basis, presented in this book, forms a firm ground for practical work. When analyzing practical situations, one should thoroughly study all the effects which might influence the operation of the FEL. Calculation of the dimensionless FEL parameters is a reliable starting point for the investigation, since their values immediately give the power of the corresponding physical effects (i.e. diffraction, space charge, etc.). At the next stage of the analysis it is necessary to obtain a quantitative description of the FEL device with numerical simulation code. As a rule, the codes have limitations on the region of applicability of the output results, thus it is always necessary to check if the numerical code can be applied to the simulation of a specific device. A clear understanding of the interdependence of the FEL parameters (provided by the application of similarity techniques) will be of help in the interpretation of the numerical results.

In this section we apply the FEL theory developed in this book to the calculation of the experiment on the high-gain SASE FEL performed by the UCLA/LANL group (see Table 6.1). A feature of this work is that the experimental conditions (quality of the undulator field, parameters of the electron

6.4 SASE FEL: Experiment and Theory

Table 6.1. Parameters of the UCLA/LANL SASE FEL

Electron beam	
Energy [MeV]	18
Charge per micropulse [nC]	0.3–2.2
Transverse spot size (HWHM) [μm]	120–145
Energy spread (rms) [%]	0.25
Pulse duration (HWHM) [ps]	3–5.5
Undulator	
Type	planar
Period [cm]	2.05
Number of periods	98
Undulator parameter K	1
Beta-function [cm]	20
FEL	
Radiation wavelength [μm]	12
Radiation pulse energy at 2.2 nC [nJ]	32

beam and radiation) are well determined and this offers a unique opportunity for a quantitative comparison between theory and experiment. Following our study, the reader will obtain an understanding of typical problems arising during the analysis of a practical device.

6.4.1 Region of Physical Parameters

We begin our study with the analysis of the physical effects influencing the operation of the UCLA/LANL FEL amplifier. Let us calculate dimensionless parameters of the experiment. The physical parameters of the electron beam are as follows. The axial profile of the bunch current is assumed to be Gaussian:

$$I(s) = \frac{Qc}{\sqrt{2\pi}\sigma_z} \exp\left(-\frac{s^2}{2\sigma_z^2}\right),$$

where Q is the bunch charge and $s = z - v_z t$. Transverse distribution of the beam current density is also supposed to be Gaussian:

$$j_0(s, r) = \frac{I(s)}{2\pi\sigma_r^2} \exp\left(-\frac{r^2}{2\sigma_r^2}\right).$$

The HWHM values of the longitudinal and transverse beam sizes, $\Delta_{z,r}^{\text{HWHM}}$, measured in the experiment, are fitted well by

$$\Delta_{z,r}^{\text{HWHM}} = \sqrt{a^2 + (bQ)^2}.$$

The parameters for the spot size are $a = 120$ μm, $b = 38$ μm$\times$nC^{-1}, and for the pulse length, $a = 0.9$ mm, $b = 0.66$ mm$\times$nC^{-1}. The values of $\sigma_{z,r}$ and $\Delta_{z,r}^{\text{HWHM}}$ are related by $\sigma_{z,r} = \Delta_{z,r}^{\text{HWHM}}/\sqrt{2\ln 2}$.

Let us calculate the parameters of the UCLA/LANL FEL for the value $Q = 2.2$ nC of the bunch charge. The longitudinal and transverse distributions are described by $\sigma_z = 1.45$ mm and $\sigma_r = 125$ µm, respectively. The gain parameter, Γ, defines the scale of the field gain and is equal to

$$\Gamma = \left[I_0 A_{JJ}^2 \omega^2 \theta_\ell^2/(2I_A c^2 \gamma_\ell^2 \gamma)\right]^{1/2} = (17.8 \text{ cm})^{-1}, \qquad (6.119)$$

where $I_0 = Qc/(\sqrt{2\pi}\sigma_z)$. Calculation of the dimensionless parameters gives the following results:

$$B = 2\Gamma \sigma_r^2 \omega/c = 0.091,$$
$$\hat{\Lambda}_p^2 = \Lambda_p^2/\Gamma^2 = 4c^2(\theta_\ell \sigma_r \omega A_{JJ})^{-2} = 1.4,$$
$$\hat{\Lambda}_T^2 = \Lambda_T^2/\Gamma^2 = \langle(\Delta\mathcal{E})^2\rangle/(\mathcal{E}_0^2 \rho^2) = 0.074,$$
$$\rho = c\gamma_\ell^2 \Gamma/\omega = 0.0092. \qquad (6.120)$$

Analysis of the values of the dimensionless parameters shows that the main physical effects defining the operation of the FEL amplifier are the diffraction effects and the space charge effects. Numerical solution of the corresponding eigenvalue equation (see Sect. 4.3.6) confirms this simple physical consideration. It is seen from the plot in Fig. 6.30 that the field gain length should be about 25 cm and is influenced mainly by the diffraction effects and the space charge effects.

The next problem to be studied is that of estimating the slippage effect. If we take the data for the bunch length from Table 6.1, we find that the kinematic slippage is comparable with the bunch length. The actual slippage

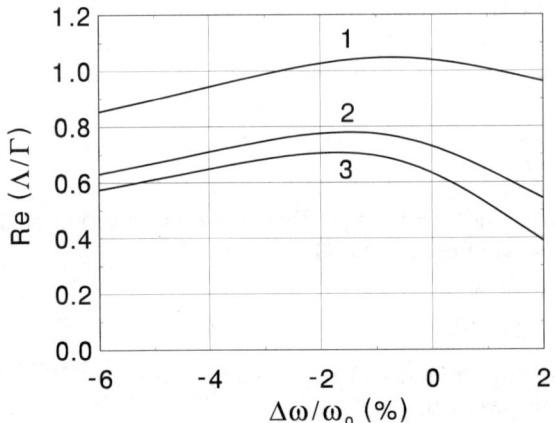

Fig. 6.30. Field gain versus frequency deviation from the resonance value. Calculations have been performed in the steady-state approximation. Curve 1 is calculated with only diffraction effects taken into account, curve 2 also includes space charge effects, and curve 3 is calculated taking into account all the effects (diffraction, space charge, and energy spread)

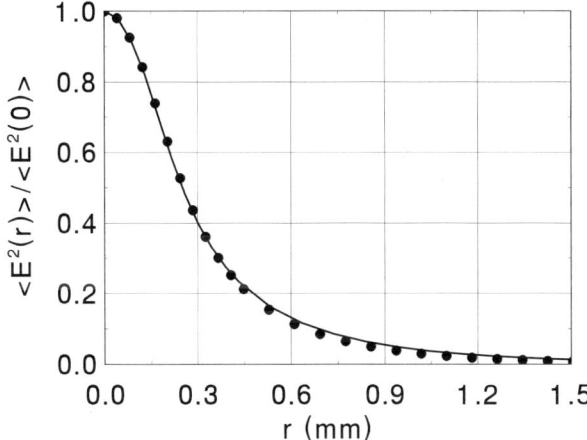

Fig. 6.31. Radial distribution of the radiation intensity at the undulator exit averaged over a large number of shots. The circles present the corresponding distribution for the FEL amplifier operating in the steady-state regime when tuning on the maximum of the gain. The charge in the bunch is 2.2 nC

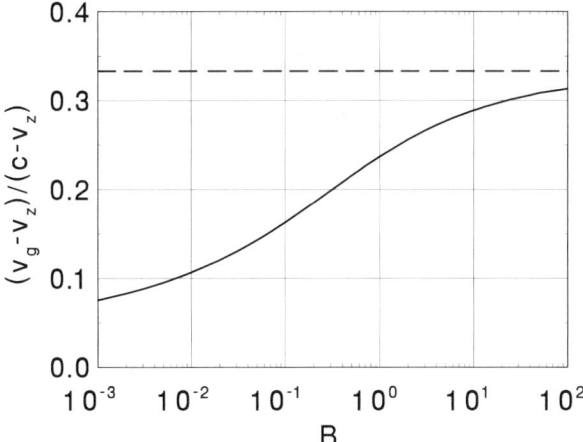

Fig. 6.32. Ratio of the slippage rate of the radiation wavepacket and kinematic slippage rate, $(v_g - \bar{v}_z)/(c - \bar{v}_z)$, as a function of the diffraction parameter B. The dotted line represents the one-dimensional asymptote (6.28)

of the radiation should be less, because the group velocity of the radiation in the electron beam is less than the velocity of light (see Sect. 6.2.1). The group velocity of the amplified wave is given by $v_g^{-1} = \mathrm{d}k/\mathrm{d}\omega = c^{-1} - (\mathrm{d}\,\mathrm{Im}\,\hat{\Lambda}/\mathrm{d}\hat{C})/(2c\gamma_\ell^2)$. Using the solution of the eigenvalue equation (4.156), we can calculate the group velocity. In Fig. 6.32 we present the ratio of the slippage rate of the radiation wavepacket and the kinematic slippage rate, $(v_g - \bar{v}_z)/(c - \bar{v}_z)$, as a function of the diffraction parameter B. In the case under study $B \simeq 0.1$, and we conclude that the slippage effect will be suppressed by a factor of about 6.

The next effect which we should estimate is the influence of the waveguide walls on FEL amplifier operation (see Chap. 5). Radial distribution of the radiation intensity at the undulator exit is presented in Fig. 6.31. It is seen that the radiation spot size is much less than the radius of the vacuum chamber in the experiment, $R \simeq 5$ mm. The merit of the influence of the waveguide effects is the waveguide diffraction parameter, $\Omega = \Gamma \omega R^2/c$. The radius of the vacuum chamber of $R \simeq 5$ mm corresponds to a value of $\Omega \simeq 70$. Remembering the results of Chap. 5, we come to the conclusion that we can neglect the influence of the waveguide walls and treat the UCLA/LANL experiment as an FEL amplifier with an open electron beam.

It is also worth making some remarks on the use of experimental data for the energy spread and for the emittance when calculating the FEL parameters. In particular, in the case under study the electron bunches are produced by RF accelerators and have short pulse duration, of the order of a few picoseconds. This means that the experimental data for the energy spread and the emittance refer to the values averaged over the pulse duration, since it is impossible to resolve instantaneous values within such a short pulse duration. On the other hand, the values of the energy spread and the emittance which should be used for the FEL calculations refer to the instantaneous values (or, slice values). As a rule, the measured energy spread is made up of the instantaneous energy spread and the drift of the mean energy along the bunch. The energy drift can take its origin from the gradient of the accelerating field along the bunch, wake fields, coherent synchrotron radiation, etc. The instantaneous energy spread significantly reduces the field growth rate and the FEL efficiency when its value is comparable to ρ. The merit of the influence of the energy drift on the gain is $(\rho \mathcal{E}_0/\tau_c)^{-1} \mathrm{d}\mathcal{E}/\mathrm{d}t$. When this value is small, the energy drift does not influence the gain and leads to a widening of the spectrum bandwidth only. In the general case, one cannot perform a correct analysis of the experiment when the values of the instantaneous energy spread and energy drift are not separated in the experimental data. Fortunately, in the case of the UCLA/LANL SASE FEL the value of the total energy spread is much less than ρ, and we can proceed with the analysis without going into details about its origin.

In the general case, the problem of the slice and averaged emittance should also be carefully studied. If the value of the slice emittance changes significantly along the bunch, one should keep in mind several effects. The first effect is the axial variation of the beam size. The second effect is that the longitudinal velocity spread changes along the electron bunch. Also, the variation of the beam size influences the longitudinal space charge field. In the case under study the effects of a possible change of the slice emittance are significantly suppressed. First, the longitudinal velocity spread due to the emittance is small (see Sect. 4.6). Second, the diffraction parameter of this

FEL amplifier is small, $B \simeq 0.1$. This is the asymptote of a thin electron beam when the change of the beam size does not influence the characteristics of the FEL amplifier (see Chap. 4 for more details). In Fig. 6.31 we present the transverse distribution of the radiation intensity at the exit of the undulator. It is seen that the spot size of the radiation (which corresponds to the fundamental TEM_{00} mode) is significantly larger than the beam spot size, thus the change of the latter parameter does not cause a significant change of the beam radiation mode.

It is relevant to note that in practical FEL devices there are always imperfections of the beam trajectory caused by undulator field errors. In the case under study these errors are small. In addition, their harmful influence is significantly suppressed due to the small value of the diffraction parameter.

Thus, our analysis shows that the UCLA/LANL SASE FEL experiment falls in the region of applicability of the FEL theory described in this book. Moreover, all the fundamental effects should define the operation of this device, namely the FEL mechanism itself, diffraction effects and space charge effects. In addition, the slippage effect should be significantly suppressed. All the fundamental effects can be studied in the clearest way, since all the non-fundamental effects (energy spread, betatron oscillations, undulator errors, etc.) do not influence the operation of the FEL amplifier.

6.4.2 Numerical Analysis of the Experiment

Thus, the analysis of the physical parameters of the experiment shows that the UCLA/LANL SASE FEL parameters can be simulated correctly using the three-dimensional, time-dependent algorithm described in Sect. 6.3 taking into account diffraction, the space charge field, the slippage, the axial and transverse nonuniformity of the electron bunch, and start-up from shot noise. To calculate characteristics of the FEL amplifier, we performed several thousand statistically independent runs. The input data for the numerical simulation code are the values of the undulator period and the undulator parameter, the values of the energy and the bunch charge, the value of the rms bunch length, and the rms bunch radius.

The first step of the numerical investigations is to define in which mode (linear or nonlinear) the FEL operates. In Fig. 6.33 we present the dependence of the energy in the radiation pulse as a function of the undulator length. Calculations have been performed over 100 statistically independent runs with nonlinear, three-dimensional, time-dependent simulation code (see Sect. 6.3). It is seen that saturation in the UCLA/LANL SASE FEL should occur at an undulator length of about 4 m. Thus, we conclude that for the actual undulator length of 2 m, the UCLA/LANL SASE FEL operates in the high-gain linear regime.

Since we now know that the UCLA/LANL SASE FEL amplifier operates in the linear regime, we can perform all the simulations with the linear simulation code described in Sect. 6.3. The linear simulation code is much

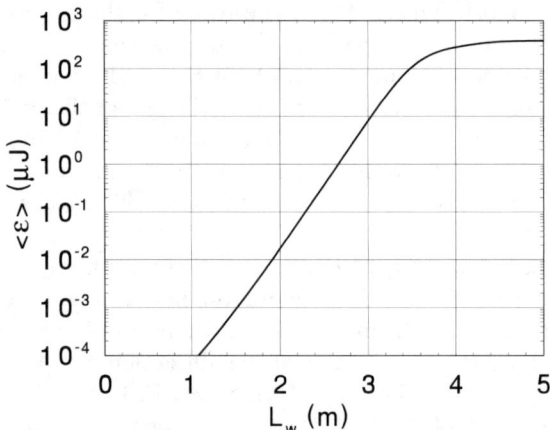

Fig. 6.33. The dependence of the averaged energy in the radiation pulse versus the length of the undulator. Simulations have been performed with nonlinear simulation code. The charge in the bunch is 2.2 nC

faster than the nonlinear code and saves a lot of time. Figure 6.34 shows the typical time structure of the radiation pulse at the undulator exit. Averaging over one thousand independent shots gives the radiation pulse shape, which is plotted in Fig. 6.35. The envelope of the axial beam profile is also shown in this figure. It is seen that the slippage of the radiation is significantly less than the kinematic slippage. This is in good agreement with the estimate presented in the previous section.

In Fig. 6.36 we present the experimental and simulation results for the energy in the radiation pulse for different values of charge. It is seen that the numerical and experimental results agree rather well. It is important to note that we compare absolute values of the radiation energy. Even a small deviation of the numerical model of the FEL process from that of the actual process will result in a significant difference in the results, since we deal with an exponentially growing amplification process with power gain of about 10^5. The results presented in Fig. 6.36 agree to within an accuracy of less than one gain length.

So, the plots in Fig. 6.36 convince us that the physical model of the SASE FEL provides an adequate quantitative description of the actual experiment. Unfortunately, we cannot perform a more detailed comparison of the absolute value of the radiation energy. The reason is that each experiment can be performed with finite accuracy. For instance, the details of the longitudinal profile may influence the output radiation energy. In Fig. 6.36 we present the results for two different models of the axial distribution of the beam current. Also, in the UCLA/LANL experiment the values of $\Delta_{z,r}^{\mathrm{HWHM}}$ were measured with uncertainties of about $\pm 5\%$. Figure 6.37 illustrates that within

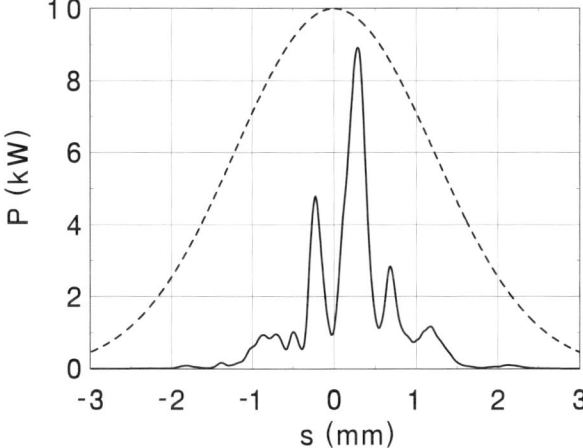

Fig. 6.34. Typical profile of the radiation pulse at the undulator exit. The charge in the electron bunch is 2.2 nC. The dashed line represents the axial profile of the beam current

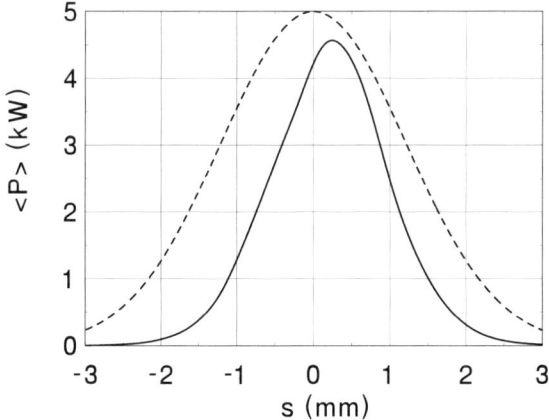

Fig. 6.35. Radiation pulse profile at the undulator exit averaged over 1000 statistically independent runs. The charge in the electron bunch is 2.2 nC. The dashed line represents the axial profile of the beam current

these experimental uncertainties, experimental and theoretical results are in perfect agreement.

It is interesting to explain the unusual behavior of the radiation energy on the value of the transverse beam size: the radiation energy is increased with an increase of the transverse beam size. This effect is connected with the fact mentioned above that the UCLA/LANL FEL amplifier operates in the regime where there is a strong influence of space charge effects. In the

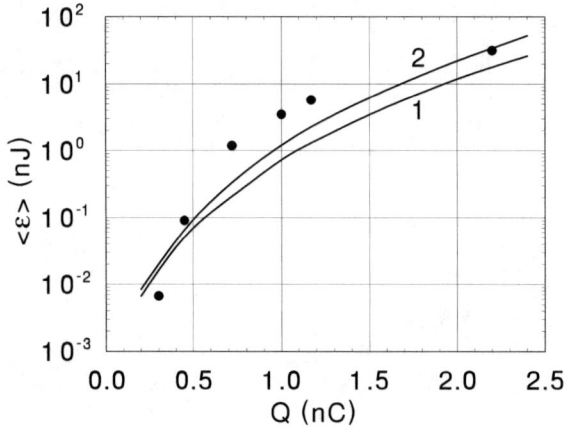

Fig. 6.36. Dependence of the averaged energy in the radiation pulse versus the bunch charge. Curves 1 and 2 correspond to Gaussian and parabolic axial beam profiles. The circles are experimental results

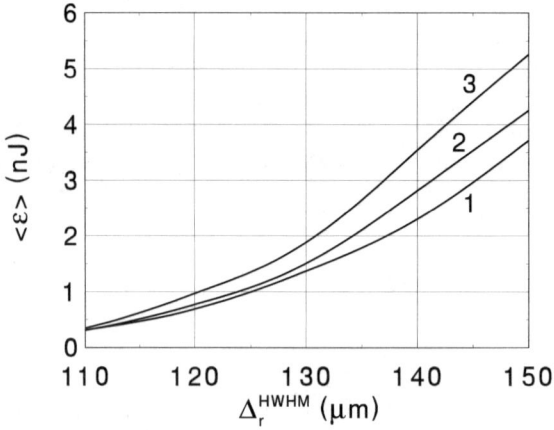

Fig. 6.37. Dependence of the averaged energy in the radiation pulse on the transverse bunch size. Curves 1, 2 and 3 correspond to the values of the longitudinal HWHM bunch size Δ_z^{HWHM} of 1.17 mm, 1.11 mm and 1.05 mm, respectively. The bunch charge is 1 nC

case under study the diffraction parameter B is much less than unity, and increasing the beam size results only in a logarithmic decrease of the field gain due to diffraction effects (see Sect. 4.4). On the other hand, there is a strong influence of the space charge fields. Increasing the transverse beam size leads to a quadratic decrease of the space charge parameter which results in an increase of the field gain. In the region of the parameters traced in Fig. 6.37 this effect dominates the diffraction effects. As a result, the field gain and the

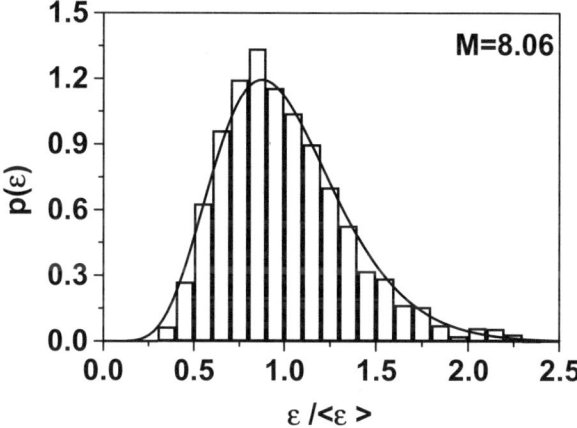

Fig. 6.38. Probability distribution of the energy in the radiation pulse for a bunch charge of 2.2 nC calculated over 2400 statistically independent runs. The solid curve represents the gamma distribution with $M = 8.06$

energy in the radiation pulse grow with an increase of the transverse size of the electron beam. Calculations show that this tendency will take place up to the value $B \simeq 0.3$ of diffraction parameter. Above this point the space charge effect becomes a small perturbation to the FEL process and an increase of the transverse beam size will lead to a decrease of the field gain. In other words this means that the maximal gain in the UCLA/LANL SASE FEL should occur at a larger value of the focusing beta function (or, at a larger value of the emittance) than that used in the experiment.

An important result of the UCLA/LANL experiment consists in the measurement of the fluctuations of the output energy in the radiation pulse. The present theoretical background allows us to perform a quantitative comparison of the experimental and theoretical results. The output radiation field from the SASE FEL can be presented in the form (see Sect. 6.3):

$$E_y = \tilde{E}(z, \boldsymbol{r}_\perp, t) \exp\left[i\omega_0(z/c - t)\right] + \text{C.C.} \,,$$

where the complex amplitude $\tilde{E}(z, \boldsymbol{r}_\perp, t)$ is a slowly varying function. Since the UCLA/LANL SASE FEL operates in the linear regime, we can state immediately that $\tilde{E}$ is a complex Gaussian random process. The theoretical analysis, performed above, states that the finite-time integrals of the radiation energy should fluctuate in accordance with the gamma distribution in this case. A quantitative answer to the parameter of the distribution, M, can be found only from numerical simulations.

Experimental results of the UCLA/LANL SASE FEL shows that the distribution of the energy in the radiation pulses is quite close to the gamma distribution. At the value $Q = 2.2$ nC of the bunch charge, the value of the fluctuations is about $\sigma \simeq 37\%$ which corresponds to the value of the parameter of the gamma distribution, $M \simeq 8$. When performing numerical simulations of the fluctuations of the output energy of an actual device, one should take into account that these fluctuations arise not only from the shot noise in the electron beam, but the shot-to-shot fluctuations of the beam parameters can also contribute to the fluctuations. The UCLA/LANL team performed measurements of the fluctuations of the beam parameters: $\pm 0.75\%$ for the bunch charge, $\pm 5.5\%$ for the transverse beam size, and $\pm 6\%$ for the bunch length. To calculate the probability distribution of the radiation energy, we performed 2400 simulation runs with statistically independent shot noise in the electron beam and fluctuations of the beam parameters. The results of the simulations are shown in Fig. 6.38. It is seen that the probability distribution of the radiation energy follows the gamma distribution with $\sigma \simeq 35\%$ and $M = 8.06$. This result is in good agreement with experiment.

6.5 Suggested Bibliography

The first studies of the high-gain FEL amplifier starting from shot noise were reported in [6.1,6.2]. The first proposals for an X-ray FEL, based on the SASE principle, were published in [6.3,6.4]. A rigorous averaged solution of the one-dimensional linear theory of the SASE FEL was reported in [6.5,6.6]. Analytical considerations of the initial-value problem of the three-dimensional theory of the SASE FEL are given in [6.7,6.8]. Approach for time-dependent numerical simulations of SASE FEL has been developed in [6.9,6.10]. Realization of this approach allowed to obtain some statistical properties of the radiation from SASE FEL operating in linear and nonlinear regimes [6.10–6.12]. A comprehensive study of the statistical properties of the radiation from the SASE FEL is presented in [6.13]. For a general discussion of methods of statistical optics we suggest reading the books [6.14, 6.15]. The first successful demonstration of the high-gain SASE FEL operation in the optical wavelength range was achieved by Pellegrini and co-workers in 1997 [6.16].

Appendices

A.1 The Extended Hamiltonian Formalism

In the framework of the Hamiltonian formalism a system of material points is described by n pairs of canonical variables

$$x_1, x_2, \ldots, x_n$$
$$p_1, p_2, \ldots, p_n$$

with the Hamiltonian function

$$\mathcal{H}(x_1, x_2, \ldots, x_n, p_1, p_2, \ldots, p_n, t) \ .$$

The canonical equations of motion have the form:

$$\mathrm{d}x_j/\mathrm{d}t = \partial \mathcal{H}/\partial p_j \ , \qquad \mathrm{d}p_j/\mathrm{d}t = -\partial \mathcal{H}/\partial x_j \ , \qquad j = 1, \ldots, n \ .$$

In some cases it is convenient to generalize the Hamiltonian formalism by introducing the $(n+1)$th coordinate x_0 coinciding with the independent variable t. In this case the Hamiltonian may be written in the form:

$$\mathcal{H} = \mathcal{H}(x_0, x_1, \ldots, x_n, p_1, p_2, \ldots, p_n) \ .$$

To provide the symmetry, a new $(n+1)$th canonical variable p_0, canonically conjugated with x_0 is introduced. The new Hamiltonian H of $(2n+2)$ variables is of the form:

$$H(x_0, x_1, \ldots, x_n, p_0, p_1, \ldots, p_n) = \mathcal{H}(x_0, x_1, \ldots, x_n, p_1, p_2, \ldots, p_n) + p_0 \ .$$

which leads to the extended system of the canonical equations:

$$\mathrm{d}x_j/\mathrm{d}t = \partial H/\partial p_j \ , \qquad \mathrm{d}p_j/\mathrm{d}t = -\partial H/\partial x_j \ , \qquad j = 0, 1, \ldots, n \ .$$

At $j = 0$ we have:

$$\mathrm{d}x_0/\mathrm{d}t = 1 \ , \qquad \mathrm{d}p_0/\mathrm{d}t = -\partial H/\partial x_0 = -\partial \mathcal{H}/\partial x_0 = -\partial \mathcal{H}/\partial t \ .$$

Let us assume that the variables x_0 and p_0 satisfy the following initial conditions at $t = 0$:

$$x_0(0) = 0 \ , \qquad p_0(0) = -\mathcal{H}(x_0(0), \ldots, p_0(0), \ldots, 0) \ .$$

In this case we get the solutions $x_0(t) = t$ and $p_0(t) = -\mathcal{H}(t)$. Hence, the Hamiltonian H at any time t is equal to zero, $H(t) = 0$, i.e. it is an integral of the motion.

Let us now consider the transformation of the variables given by the following expressions:

$$x_0 = x_0(\bar{x}_0, \bar{x}_1, \ldots, \bar{x}_n, \bar{p}_0, \bar{p}_1, \ldots, \bar{p}_n)$$
$$\ldots$$
$$x_n = x_n(\bar{x}_0, \bar{x}_1, \ldots, \bar{x}_n, \bar{p}_0, \bar{p}_1, \ldots, \bar{p}_n)$$
$$p_0 = p_0(\bar{x}_0, \bar{x}_1, \ldots, \bar{x}_n, \bar{p}_0, \bar{p}_1, \ldots, \bar{p}_n)$$
$$\ldots$$
$$p_n = p_n(\bar{x}_0, \bar{x}_1, \ldots, \bar{x}_n, \bar{p}_0, \bar{p}_1, \ldots, \bar{p}_n) \quad \text{(A.1.1)}$$

which leads to the following Hamiltonian $\bar{H}$:

$$\bar{H}(\bar{x}_0, \bar{x}_1, \ldots, \bar{x}_n, \bar{p}_0, \bar{p}_1, \ldots, \bar{p}_n)$$
$$= H(x_0(\bar{x}_0, \ldots, \bar{x}_n, \bar{p}_0, \ldots, \bar{p}_n), \ldots, p_n(\bar{x}_0, \ldots, \bar{x}_n, \bar{p}_0, \ldots, \bar{p}_n)) . \quad \text{(A.1.2)}$$

The derivatives of the new variables with respect to time are of the form:

$$\begin{aligned}
\mathrm{d}\bar{x}_j/\mathrm{d}t &= \sum_{l=0}^{n}\left[\frac{\partial \bar{x}_j}{\partial x_l}\frac{\mathrm{d}x_l}{\mathrm{d}t} + \frac{\partial \bar{x}_j}{\partial p_l}\frac{\mathrm{d}p_l}{\mathrm{d}t}\right] \\
&= \sum_{l=0}^{n}\left[\frac{\partial \bar{x}_j}{\partial x_l}\frac{\partial H}{\partial p_l} - \frac{\partial \bar{x}_j}{\partial p_l}\frac{\partial H}{\partial x_l}\right] \\
&= \sum_{l=0}^{n}\left\{\frac{\partial \bar{x}_j}{\partial x_l}\sum_{k=0}^{n}\left[\frac{\partial \bar{H}}{\partial \bar{x}_k}\frac{\partial \bar{x}_k}{\partial p_l} + \frac{\partial \bar{H}}{\partial \bar{p}_k}\frac{\partial \bar{p}_k}{\partial p_l}\right]\right. \\
&\quad \left. - \frac{\partial \bar{x}_j}{\partial p_l}\sum_{k=0}^{n}\left[\frac{\partial \bar{H}}{\partial \bar{x}_k}\frac{\partial \bar{x}_k}{\partial x_l} + \frac{\partial \bar{H}}{\partial \bar{p}_k}\frac{\partial \bar{p}_k}{\partial x_l}\right]\right\} \\
&= \sum_{k=0}^{n}\left\{\frac{\partial \bar{H}}{\partial \bar{x}_k}\sum_{l=0}^{n}\left[\frac{\partial \bar{x}_k}{\partial p_l}\frac{\partial \bar{x}_j}{\partial x_l} - \frac{\partial \bar{x}_j}{\partial p_l}\frac{\partial \bar{x}_k}{\partial x_l}\right]\right. \\
&\quad \left. + \frac{\partial \bar{H}}{\partial \bar{p}_k}\sum_{l=0}^{n}\left[\frac{\partial \bar{p}_k}{\partial p_l}\frac{\partial \bar{x}_j}{\partial x_l} - \frac{\partial \bar{x}_j}{\partial p_l}\frac{\partial \bar{p}_k}{\partial x_l}\right]\right\} \quad \text{(A.1.3)}
\end{aligned}$$

and

$$\begin{aligned}
\mathrm{d}\bar{p}_j/\mathrm{d}t &= -\sum_{k=0}^{n}\left\{\frac{\partial \bar{H}}{\partial \bar{p}_k}\sum_{l=0}^{n}\left[\frac{\partial \bar{p}_j}{\partial p_l}\frac{\partial \bar{p}_k}{\partial x_l} - \frac{\partial \bar{p}_j}{\partial x_l}\frac{\partial \bar{p}_k}{\partial p_l}\right]\right. \\
&\quad \left. + \frac{\partial \bar{H}}{\partial \bar{x}_k}\sum_{l=0}^{n}\left[\frac{\partial \bar{p}_j}{\partial p_l}\frac{\partial \bar{x}_k}{\partial x_l} - \frac{\partial \bar{x}_k}{\partial p_l}\frac{\partial \bar{p}_j}{\partial x_l}\right]\right\} . \quad \text{(A.1.4)}
\end{aligned}$$

To reduce (A.1.3) and (A.1.4) we introduce the Poisson brackets $[f, g]$:

$$[f, g] = \sum_{l=0}^{n}[(\partial f/\partial p_l)(\partial g/\partial x_l) - (\partial g/\partial p_l)(\partial f/\partial x_l)] .$$

A.1 The Extended Hamiltonian Formalism

As a result, (A.1.3) and (A.1.4) become:

$$d\bar{x}_j/dt = \sum_{k=0}^{n}\{(\partial\bar{H}/\partial\bar{x}_k)[\bar{x}_k,\bar{x}_j] + (\partial\bar{H}/\partial\bar{p}_k)[\bar{p}_k,\bar{x}_j]\},$$

$$d\bar{p}_j/dt = -\sum_{k=0}^{n}\{(\partial\bar{H}/\partial\bar{p}_k)[\bar{p}_j,\bar{p}_k] + (\partial\bar{H}/\partial\bar{x}_k)[\bar{p}_j,\bar{x}_k]\}. \qquad (A.1.5)$$

Using these expressions one can find that the conditions

$$[\bar{x}_k,\bar{x}_j] = 0, \qquad [\bar{p}_k,\bar{x}_j] = \delta_{kj}, \qquad [\bar{p}_k,\bar{p}_j] = 0. \qquad (A.1.6)$$

are necessary and sufficient for the transformation (A.1.1) to be canonical, and the variables $\bar{x}_j$ and $\bar{p}_j$ satisfy the canonical equations with Hamiltonian $\bar{H}$:

$$d\bar{x}_j/dt = \partial\bar{H}/\partial\bar{p}_j, \qquad d\bar{p}_j/dt = -\partial\bar{H}/\partial\bar{x}_j, \qquad j=0,1,\ldots,n. \qquad (A.1.7)$$

Let us consider, for illustration, the point transformation:

$$x_j = F_j(\bar{x}_0,\ldots,\bar{x}_n), \qquad j=0,\ldots,n. \qquad (A.1.8)$$

We wish to go over to the new system of $(2n+2)$ variables $(\bar{p}_j,\bar{x}_j)$ which are independent functions of p_j and x_j and satisfy the condition:

$$\sum_{j=0}^{n}\bar{p}_j d\bar{x}_j = \sum_{j=0}^{n} p_j dx_j. \qquad (A.1.9)$$

One can easily find the formulae of this transformation:

$$\bar{p}_j = \sum_{l=0}^{n} p_l \partial F_l/\partial\bar{x}_j. \qquad (A.1.10)$$

Calculating the Poisson brackets (A.1.6) one can easily find that (A.1.8) and (A.1.10) determine the canonical transformation.

Let us now go over to a new independent variable $\bar{x}_0$ in place of t. The expressions (A.1.7) lead to the following equations:

$$d\bar{x}_j/d\bar{x}_0 = (d\bar{x}_j/dt)(dt/d\bar{x}_0) = (\partial\bar{H}/\partial\bar{p}_j)(\partial\bar{H}/\partial\bar{p}_0)^{-1} = -[\partial\bar{p}_0/\partial\bar{p}_j]_{\bar{H}},$$

$$d\bar{p}_j/d\bar{x}_0 = (d\bar{p}_j/dt)(dt/d\bar{x}_0) = -(\partial\bar{H}/\partial\bar{x}_j)(\partial\bar{H}/\partial\bar{p}_0)^{-1} = [\partial\bar{p}_0/\partial\bar{x}_j]_{\bar{H}}.$$

Here the symbol $(\ldots)_{\bar{H}}$ means that the corresponding derivative is calculated at the constant value of $\bar{H}(\bar{x}_0,\ldots,\bar{x}_n,\bar{p}_0,\ldots,\bar{p}_n)$. From the condition $\bar{H} = 0$ we get:

$$\bar{p}_0(\bar{x}_0,\ldots,\bar{x}_n,\bar{p}_0,\ldots,\bar{p}_n) = -\mathcal{H}(\bar{x}_0,\ldots,\bar{x}_n,\bar{p}_0,\ldots,\bar{p}_n).$$

Resolving the latter equation with respect to $\bar{p}_0$ we obtain:

$$\bar{p}_0 = -\tilde{\mathcal{H}}(\bar{x}_0,\ldots,\bar{x}_n,\bar{p}_1,\ldots,\bar{p}_n).$$

As a result, the equation of motion may be written in the canonical form:

$$d\bar{x}_j/d\bar{x}_0 = \partial\tilde{\mathcal{H}}/\partial\bar{p}_j, \qquad d\bar{p}_j/d\bar{x}_0 = -\partial\tilde{\mathcal{H}}/\partial\bar{x}_j, \qquad j=1,\ldots,n.$$

A.2 Longitudinal Space Charge Field of a Modulated Electron Beam with Finite Transverse Size

Let us consider a monoenergetic electron beam moving with velocity v_z along the z axis. The beam is modulated at frequency ω which corresponds to the modulation wavelength $\lambda_e = 2\pi v_z/\omega$. In the frame of reference of the electron beam, the scalar potential is given by:

$$\phi'(z', \mathbf{r}_\perp, t') = \int d\mathbf{r}_\perp^{(s)} \int_{-\infty}^{\infty} d\zeta \frac{\rho_1'(\zeta, \mathbf{r}_\perp^{(s)}, t') \cos(k_z'\zeta + \delta_1')}{\sqrt{|\mathbf{r}_\perp - \mathbf{r}_\perp^{(s)}|^2 + (z'-\zeta)^2}},$$

where $k_z' = 2\pi/\lambda_e' = 2\pi/(\gamma_z \lambda_e)$. The amplitude and phase, ρ_1' and δ', of the first harmonic of the charge density are slowly varying functions of the z' coordinate:

$$\partial \rho_1'/\partial z' \ll \rho_1'/\lambda_e', \quad \partial \delta_1'/\partial z' \ll \delta_1'/\lambda_e',$$

and we neglect their change when calculating the integral over ζ. Replacing ζ by $\xi + z'$, we write:

$$\phi' = \cos(k_z' z' + \delta_1') \int d\mathbf{r}_\perp^{(s)} \rho_1'(z', \mathbf{r}_\perp^{(s)}, t') \int_{-\infty}^{\infty} d\xi \frac{\cos(k_z'\xi)}{\sqrt{|\mathbf{r}_\perp - \mathbf{r}_\perp^{(s)}|^2 + \xi^2}}.$$

Using the integral representation of the Bessel function of the second kind, we rewrite the expression for ϕ' in the form:

$$\phi' = \cos(k_z' z' + \delta_1') \int d\mathbf{r}_\perp^{(s)} \rho_1'(z', \mathbf{r}_\perp^{(s)}, t') K_0\left(k_z' |\mathbf{r}_\perp - \mathbf{r}_\perp^{(s)}|\right).$$

Then we remember that the behavior of $K_0\left(k_z'|\mathbf{r}_\perp - \mathbf{r}_\perp^{(s)}|\right)$ for $(k_z' r_b)^2 \gg 1$ approaches the behavior of the delta function (here r_b is the typical transverse size of the beam). As a result, we obtain an asymptotic expression for ϕ':

$$\phi' = \frac{4\pi}{(k_z')^2} \rho_1'(z', \mathbf{r}_\perp, t') \cos(k_z' z' + \delta_1').$$

The longitudinal component of the electric field is equal to:

$$E_z = -\frac{\partial \phi'}{\partial z'} \simeq \frac{4\pi}{k_z'} \rho_1'(z', \mathbf{r}_\perp, t') \sin(k_z' z' + \delta_1').$$

Since E_z is invariant with respect to the Lorentz transformation, this expression is also valid in the laboratory frame of reference. The variables in the moving and laboratory frame of references are related by:

$$z' = \gamma_z(z - v_z t), \quad \rho_1' \lambda_e' = \rho_1 \lambda_e.$$

Thus, the expression for E_z in terms of the laboratory frame of reference takes the form:

$$E_z = 2\rho_1 \lambda_e \sin[2\pi \lambda_e^{-1}(z - v_z t) + \delta_1'].$$

Taking into account that $2\pi v_z/\lambda_e = \omega$, we write:
$$2\pi \lambda_e^{-1}(z - v_z t) + \delta'_1 = \psi - Cz + \delta'_1 = \psi + \psi_1 ,$$
where $C = [k_w - \omega/v_z + \omega/c]$ is the detuning. Finally, we can write (when $v_z \simeq c$):
$$E_z = 2\rho_1 \lambda_e \sin(\psi + \psi_1) \simeq \frac{4\pi}{\omega} j_1 \sin(\psi + \psi_1) .$$
Since the complex amplitude of the first harmonic of the beam current density is defined as
$$j_1 \cos(\psi + \psi_1) = \tilde{j}_1 e^{i\psi} + \text{C.C.} ,$$
we can write the following expression for the complex amplitude of the space charge field:
$$\tilde{E}_z = -\mathrm{i} \frac{4\pi}{\omega} \tilde{j}_1(z, \boldsymbol{r}_\perp) . \tag{A.2.1}$$
Finally we can state that when the transverse size of the electron beam is large, $r_b^2 \gg c^2 \gamma_z^2/\omega^2$, the longitudinal space charge field of the electron beam is defined by the local density modulation.

A.3 Green's Function for a Homogeneous Waveguide

Let us consider a homogeneous waveguide infinite in the z direction. We calculate the Green's function in the following way. First, we find the Green's function for a cavity formed of a piece of the waveguide of length l_z and plane side walls perpendicular to the z axis. Then we let $l_z \to \infty$ and obtain the Green's function for an infinite homogeneous waveguide. The electromagnetic field in the cavity can be expanded in a series of eigenfunctions $\boldsymbol{F}_\nu$:
$$\boldsymbol{A}_\omega = \sum_\nu C_\nu \boldsymbol{F}_\nu(\boldsymbol{r}) . \tag{A.3.1}$$
The eigenfunctions of the cavity, $\boldsymbol{F}_\nu$, are the solutions of the Helmholtz equation:
$$\nabla^2 \boldsymbol{F}_\nu + k_\nu^2 \boldsymbol{F}_\nu = 0 , \tag{A.3.2}$$
satisfying the following conditions:
$$\nabla \cdot \boldsymbol{F}_\nu = 0 , \tag{A.3.3}$$
$$(\boldsymbol{n} \times \boldsymbol{F}_\nu)|_S = 0 . \tag{A.3.4}$$
Here k_ν are the eigenvalues. The eigenfunctions $\boldsymbol{F}_\nu$ are orthogonal, i.e.
$$\int \mathrm{d}\boldsymbol{r} \, \boldsymbol{F}_{\nu_1} \cdot \boldsymbol{F}_{\nu_2} = 0 \quad \text{for} \quad \nu_1 \neq \nu_2 \tag{A.3.5}$$
under integration over the resonator volume. Indeed, let us rewrite the Helmholtz equation (A.3.2) in the form:

$$\nabla \times (\nabla \times \boldsymbol{F}_\nu) - k_\nu^2 \boldsymbol{F}_\nu = 0 , \tag{A.3.6}$$

using the relations

$$\nabla \times (\nabla \times \boldsymbol{F}_\nu) = \nabla(\nabla \cdot \boldsymbol{F}_\nu) - \nabla^2 \boldsymbol{F}_\nu$$

and (A.3.3). Next, we write down two equations (A.3.6) for two different eigenfunctions with indices ν_1 and ν_2. After taking the scalar product of the first equation with $\boldsymbol{F}_{\nu_2}$ and the second equation with $\boldsymbol{F}_{\nu_1}$, we subtract these equations:

$$(k_{\nu_2}^2 - k_{\nu_1}^2) \boldsymbol{F}_{\nu_1} \cdot \boldsymbol{F}_{\nu_2} = \boldsymbol{F}_{\nu_1} \cdot (\nabla \times (\nabla \times \boldsymbol{F}_{\nu_2}))$$
$$- \boldsymbol{F}_{\nu_2} \cdot (\nabla \times (\nabla \times \boldsymbol{F}_{\nu_1})) . \tag{A.3.7}$$

Using the formula

$$\nabla \cdot (\boldsymbol{a} \times \boldsymbol{b}) = \boldsymbol{b} \cdot (\nabla \times \boldsymbol{a}) - \boldsymbol{a} \cdot (\nabla \times \boldsymbol{b})$$

and the Gauss divergence theorem, we obtain from (A.3.7):

$$(k_{\nu_2}^2 - k_{\nu_1}^2) \int d\boldsymbol{r}\, \boldsymbol{F}_{\nu_1} \cdot \boldsymbol{F}_{\nu_2} = \int d\sigma \, \{(\boldsymbol{F}_{\nu_1} \times (\nabla \times \boldsymbol{F}_{\nu_2})) \cdot \boldsymbol{n}$$
$$- (\boldsymbol{F}_{\nu_2} \times (\nabla \times \boldsymbol{F}_{\nu_1})) \cdot \boldsymbol{n}\} , \tag{A.3.8}$$

where the integration is performed over the cavity walls. Remembering that

$$\boldsymbol{c} \cdot (\boldsymbol{a} \times (\nabla \times \boldsymbol{b})) = (\nabla \times \boldsymbol{b}) \cdot (\boldsymbol{c} \times \boldsymbol{a}) ,$$

we rewrite the integral in the right-hand side of (A.3.8) as

$$\int d\sigma \, \{(\boldsymbol{n} \times \boldsymbol{F}_{\nu_1}) \cdot (\nabla \times \boldsymbol{F}_{\nu_2}) - (\boldsymbol{n} \times \boldsymbol{F}_{\nu_2}) \cdot (\nabla \times \boldsymbol{F}_{\nu_1})\} . \tag{A.3.9}$$

It follows from the boundary conditions (A.3.4) that the integrand in (A.3.9) is equal to zero. Finally, we obtain the orthogonality condition for the eigenfunctions:

$$(k_{\nu_2}^2 - k_{\nu_1}^2) \int d\boldsymbol{r}\, \boldsymbol{F}_{\nu_1} \cdot \boldsymbol{F}_{\nu_2} = 0 .$$

It is convenient to normalize the eigenfunction as

$$\int d\boldsymbol{r}\, \boldsymbol{F}_\nu \cdot \boldsymbol{F}_\nu = 1 .$$

Now let us turn to the calculation of the Green's function. Substituting (A.3.1) for $\boldsymbol{A}_\omega$ into (5.12), we obtain

$$\sum_\nu C_\nu \left(\frac{\omega^2}{c^2} - k_\nu^2\right) \boldsymbol{F}_\nu = -\frac{4\pi}{c} \boldsymbol{J}_\omega .$$

We multiply this equation by $\boldsymbol{F}_\nu(\boldsymbol{r})$ and integrate over the resonator volume. Remembering that the eigenfunctions are orthogonal and normalized to unity, we find the value of the expansion factor C_ν:

A.3 Green's Function for a Homogeneous Waveguide   431

$$C_\nu = -\frac{4\pi}{c} \int d\mathbf{r} \frac{\mathbf{J}_\omega \cdot \mathbf{F}_\nu}{\omega^2/c^2 - k_\nu^2} \ .$$

So, (A.3.1) can be written down in the form (5.14) with the tensor Green's function

$$G_\omega^{\alpha,\beta}(\mathbf{r},\, \mathbf{r}') = -\frac{4\pi}{c} \sum_\nu \frac{F_\nu^\alpha(\mathbf{r}) F_\nu^\beta(\mathbf{r}')}{\omega^2/c^2 - k_\nu^2} \ . \tag{A.3.10}$$

The condition $\phi|_S = 0$ results in

$$\int d\mathbf{r} (\boldsymbol{\nabla}\phi \cdot \mathbf{F}_\nu) = \int d\mathbf{r} \left\{ \boldsymbol{\nabla} \cdot (\phi \mathbf{F}_\nu) - \phi \boldsymbol{\nabla} \cdot \mathbf{F}_\nu \right\} = 0 \ .$$

Thus,

$$\int d\mathbf{r} \mathbf{J}_\omega \cdot \mathbf{F}_\nu = \int d\mathbf{r} \mathbf{j}_\omega \cdot \mathbf{F}_\nu \ ,$$

and (5.14) takes the form:

$$A_\omega^\alpha(\mathbf{r}) = \sum_\beta \int G_\omega^{\alpha\beta}(\mathbf{r},\, \mathbf{r}') j_\omega^\beta(\mathbf{r}') d\mathbf{r}' \ . \tag{A.3.11}$$

The gauge condition (A.3.3) reduces the number of independent variables, so the eigenfunction $\mathbf{F}_\nu$ is completely defined by two independent variables at each space point. It is well known that if $\boldsymbol{\nabla} \cdot \mathbf{F}_\nu = 0$, $\mathbf{F}_\nu$ can be represented in the following form:

$$\mathbf{F}_\nu = \boldsymbol{\nabla} \times \left[\mathbf{e}_z \Psi_\nu^{\mathrm{TE}}(\mathbf{r}) \right] + \boldsymbol{\nabla} \times \left(\boldsymbol{\nabla} \times \left[\mathbf{e}_z \Psi_\nu^{\mathrm{TM}}(\mathbf{r}) \right] \right) \ ,$$

where Ψ^{TE} and Ψ^{TM} are independent scalar functions. The vector

$$\mathbf{M}_\nu = \boldsymbol{\nabla} \times \left[\mathbf{e}_z \Psi_\nu^{\mathrm{TE}}(\mathbf{r}) \right]$$

is perpendicular to the z axis. In the case under study we consider the cavity with plane side walls perpendicular to the z axis. This means that $\mathbf{M}_\nu$ takes zero values at $z = 0$ and $z = l_z$. So, $\mathbf{M}_\nu$ can be represented in the form:

$$\mathbf{M}_\nu = A_\nu^{\mathrm{TE}} \boldsymbol{\nabla} \times \left[\mathbf{e}_z \psi_\nu^{\mathrm{TE}}(\mathbf{r}_\perp) \sin(k_z z) \right] \ , \tag{A.3.12}$$

where

$$k_z = m_z \pi / l_z \ , \quad m_z = 1,\, 2,\, \ldots$$

is the longitudinal wavenumber and A_ν^{TE} is the normalizing coefficient. The vector

$$\mathbf{N}_\nu = \boldsymbol{\nabla} \times \left(\boldsymbol{\nabla} \times \left[\mathbf{e}_z \Psi_\nu^{\mathrm{TM}}(\mathbf{r}) \right] \right)$$

can be written in the following form:

$$\mathbf{N}_\nu = \boldsymbol{\nabla}_\perp \frac{\partial \Psi_\nu^{\mathrm{TM}}}{\partial z} - \mathbf{e}_z \boldsymbol{\nabla}_\perp^2 \Psi_\nu^{\mathrm{TM}} \ .$$

The transverse component of the vector $\mathbf{N}_\nu$ is equal to zero on the side walls, so

$$\frac{\partial \Psi_\nu^{\text{TM}}}{\partial z} = 0 \quad \text{at} \quad z = 0, \, l_z .$$

As a result, we obtain the following expression for the vector $\boldsymbol{N}_\nu$:

$$\boldsymbol{N}_\nu = A_\nu^{\text{TM}} \boldsymbol{\nabla} \times \boldsymbol{\nabla} \times \left[\boldsymbol{e}_z \psi_\nu^{\text{TM}}(\boldsymbol{r}_\perp) \cos(k_z z) \right] . \tag{A.3.13}$$

Since $\boldsymbol{M}_\nu$ and $\boldsymbol{N}_\nu$ are solutions of the Helmholtz equation, the functions ψ_ν^{TE} and ψ_ν^{TM} are also solutions of the Helmholtz equation:

$$\boldsymbol{\nabla}_\perp^2 \psi_\nu(\boldsymbol{r}_\perp) + k_\perp^2 \psi_\nu(\boldsymbol{r}_\perp) = 0 , \tag{A.3.14}$$

where $k_\perp^2 = k_\nu^2 - k_z^2$. The boundary conditions for the functions ψ_ν^{TE} and ψ_ν^{TM} are given by the requirements that the tangential components of $\boldsymbol{M}_\nu$ and $\boldsymbol{N}_\nu$ be equal to zero on the side walls of the resonator. The condition

$$(\boldsymbol{n} \times \boldsymbol{M}_\nu)|_S = 0$$

leads to the boundary conditions for ψ_ν^{TE}:

$$\boldsymbol{n} \cdot \boldsymbol{\nabla}_\perp \psi_\nu^{\text{TE}}(\boldsymbol{r}_\perp)|_S = 0 , \tag{A.3.15}$$

and the condition $\boldsymbol{N}_\nu|_S = 0$ leads to

$$\boldsymbol{\nabla}_\perp^2 \psi_\nu^{\text{TM}}(\boldsymbol{r}_\perp)|_S = 0 .$$

Remembering that the function $\psi_\nu^{\text{TM}}(\boldsymbol{r}_\perp)$ is a solution of the Helmholtz equation, we get

$$\psi_\nu^{\text{TM}}(\boldsymbol{r}_\perp)|_S = 0 . \tag{A.3.16}$$

The eigenfunctions $\psi_\nu(\boldsymbol{r}_\perp)$ are orthogonal, i.e.

$$\int d\boldsymbol{r}_\perp \psi_{\nu_1} \psi_{\nu_2} = 0 \quad \text{for} \quad \nu_1 \neq \nu_2 .$$

It is convenient to normalize these functions as

$$\int |\boldsymbol{\nabla}_\perp \psi_\nu|^2 d\boldsymbol{r}_\perp = 1 . \tag{A.3.17}$$

For this normalization the eigenfunctions ψ_ν are dimensionless and

$$\int d\boldsymbol{r} \boldsymbol{M}_\nu \cdot \boldsymbol{M}_{\nu'} = \delta_{\nu\nu'} , \quad \int d\boldsymbol{r} \boldsymbol{N}_\nu \cdot \boldsymbol{N}_{\nu'} = \delta_{\nu\nu'} .$$

Finally, the eigenvectors $\boldsymbol{M}_\nu$ and $\boldsymbol{N}_\nu$ can be written in the form:

$$\boldsymbol{M}_\nu = \sqrt{\frac{2}{l_z}} \boldsymbol{\nabla} \times \left[\boldsymbol{e}_z \psi_\nu^{\text{TE}}(\boldsymbol{r}_\perp) \sin(k_z z) \right] , \tag{A.3.18a}$$

$$\boldsymbol{N}_\nu = \frac{1}{k_\nu} \sqrt{\frac{2}{l_z}} \boldsymbol{\nabla} \times \left(\boldsymbol{\nabla} \times \left[\boldsymbol{e}_z \psi_\nu^{\text{TM}}(\boldsymbol{r}_\perp) \cos(k_z z) \right] \right) . \tag{A.3.18b}$$

Substituting (A.3.18) into (A.3.10), we find the following expression for the Green's function:

$$G_\omega(\mathbf{r},\mathbf{r}')$$
$$= -\frac{8\pi}{cl_z}\sum_\mu \frac{\sin(k_z z)\sin(k_z z')}{\omega^2/c^2 - (k_\mu^{TE})^2}\left[\mathbf{e}_x \frac{\partial \psi_\mu^{TE}(\mathbf{r}_\perp)}{\partial y} - \mathbf{e}_y \frac{\partial \psi_\mu^{TE}(\mathbf{r}_\perp)}{\partial x}\right]$$
$$\otimes \left[\mathbf{e}_x \frac{\partial \psi_\mu^{TE}(\mathbf{r}'_\perp)}{\partial y'} - \mathbf{e}_y \frac{\partial \psi_\mu^{TE}(\mathbf{r}'_\perp)}{\partial x'}\right]$$
$$- \frac{8\pi}{cl_z}\sum_\nu \frac{1}{(k_\nu^{TM})^2(\omega^2/c^2 - (k_\nu^{TM})^2)}$$
$$\times \left[k_z \sin(k_z z)\nabla_\perp \psi_\nu^{TM}(\mathbf{r}_\perp) + \mathbf{e}_z \cos(k_z z)\nabla_\perp^2 \psi_\nu^{TM}(\mathbf{r}_\perp)\right]$$
$$\otimes \left[k_z \sin(k_z z')\nabla_\perp \psi_\nu^{TM}(\mathbf{r}'_\perp) + \mathbf{e}_z \cos(k_z z')\nabla_\perp^2 \psi_\nu^{TM}(\mathbf{r}'_\perp)\right], \quad (A.3.19)$$

where the symbol $\otimes$ denotes the direct product of vectors.

The expression for the Green's function can be significantly simplified for the case of an overmoded waveguide when

$$c^2 k_\perp^2 / \omega^2 \ll 1 \; .$$

Using the paraxial approximation, we can omit all the terms proportional to $k_\perp^2/k_\nu^2$ in (A.3.19), and set k_z/k_ν equal to unity. As a result, the expression for G_ω takes the form:

$$G_\omega(\mathbf{r},\mathbf{r}') = -\frac{8\pi}{cl_z}\sum_\mu \frac{\sin(k_z z)\sin(k_z z')}{\omega^2/c^2 - (k_\mu^{TE})^2}$$
$$\times \left[\mathbf{e}_x \frac{\partial \psi_\mu^{TE}(\mathbf{r}_\perp)}{\partial y} - \mathbf{e}_y \frac{\partial \psi_\mu^{TE}(\mathbf{r}_\perp)}{\partial x}\right]$$
$$\otimes \left[\mathbf{e}_x \frac{\partial \psi_\mu^{TE}(\mathbf{r}'_\perp)}{\partial y'} - \mathbf{e}_y \frac{\partial \psi_\mu^{TE}(\mathbf{r}'_\perp)}{\partial x'}\right]$$
$$- \frac{8\pi}{cl_z}\sum_\nu \frac{\sin(k_z z)\sin(k_z z')}{\omega^2/c^2 - (k_\nu^{TM})^2}$$
$$\times \left[\mathbf{e}_x \frac{\partial \psi_\mu^{TM}(\mathbf{r}_\perp)}{\partial x} + \mathbf{e}_y \frac{\partial \psi_\mu^{TM}(\mathbf{r}_\perp)}{\partial y}\right]$$
$$\otimes \left[\mathbf{e}_x \frac{\partial \psi_\mu^{TM}(\mathbf{r}'_\perp)}{\partial x'} + \mathbf{e}_y \frac{\partial \psi_\mu^{TM}(\mathbf{r}'_\perp)}{\partial y'}\right]. \quad (A.3.20)$$

Let us consider the sum over m_z in (A.3.20):

$$\sum_{m_z=0}^\infty \frac{\sin(k_z z)\sin(k_z z')}{\omega^2/c^2 - k_z^2 - k_\perp^2} = \frac{1}{2}\sum_{m_z=0}^\infty \frac{\cos[k_z(z-z')] - \cos[k_z(z+z')]}{\omega^2/c^2 - k_z^2 - k_\perp^2} \; .$$

In the limit of an infinite waveguide, as $l_z \to \infty$, the values of m_z/l_z tend to be continuous and the sum over discrete values of m_z can be replaced by the integration,

$$\sum_{m_z=0}^{\infty} (\ldots) \to \frac{l_z}{\pi} \int_0^{\infty} dk_z (\ldots) \quad \text{as} \quad l_z \to \infty .$$

So, the calculation of the sum over m_z in the expression for G_ω is reduced to the integral

$$I_z = \int_0^{\infty} \frac{\cos[k_z(z \pm z')]}{\omega^2/c^2 - k_z^2 - k_\perp^2} dk_z . \tag{A.3.21}$$

The integrand in (A.3.21) has a singularity at the point

$$k_z = \sqrt{\omega^2/c^2 - k_\perp^2} ,$$

and a path of integration around this pole should be specified when calculating the integral (A.3.21). Here we should take into account the physical considerations. Till now we have assumed the walls of the waveguide to have infinitely high conductivity. In reality the conductivity of the walls has a finite value, and the electromagnetic wave propagating in the waveguide decays exponentially:

$$E(z, \mathbf{r}_\perp, t) \propto \exp(-\epsilon z) \cos(k_z z - \omega t) = \exp[i(k_z + i\epsilon)z - i\omega t] + \text{C.C.}$$

This formula gives us a way to modify (A.3.21), in order to take into account the finite conductivity of the walls. Namely, we should add a small imaginary term to the longitudinal wavenumber k_z

$$k_z \to k_z + i\epsilon , \quad \epsilon > 0 , \quad k_z > 0 .$$

At the end of the calculations we should let this imaginary contribution tend to zero. This trick allows us to reduce (A.3.21) to integration along a contour. Using Cauchy's residue theorem, we obtain:

$$\frac{\pi}{l_z} \sum_{m_z=0}^{\infty} \frac{\cos[k_z(z-z')] - \cos[k_z(z+z')]}{\omega^2/c^2 - k_z^2 - k_\perp^2}$$

$$\to \frac{\pi c}{\omega} \sin\left[\left(\frac{\omega}{c} - \frac{ck_\perp^2}{2\omega}\right)|z - z'|\right] . \tag{A.3.22}$$

The term proportional to $\cos[k_z(z+z')]$ corresponds to the "reflection" of the wave from the end walls of the cavity and tends to zero for an infinitely long cavity. Using accepted approximations, we keep only the terms corresponding to the wave propagating in the positive direction of the z axis in (A.3.20):

$$\frac{\pi}{l_z} \sum_{m_z=0}^{\infty} \frac{\cos[k_z(z-z')]}{\omega^2/c^2 - k_z^2 - k_\perp^2} \to -\frac{i\pi c}{2\omega} \exp\left[i\left(\frac{\omega}{c} - \frac{ck_\perp^2}{2\omega}\right)|z - z'|\right] .$$

Using this approximation, we finally write down the expression for the Green's function of an overmoded homogeneous waveguide:

$$G_\omega = \frac{2\pi i}{\omega} \sum_\mu \exp\left\{i\left[\frac{\omega}{c} - \frac{c(k_\perp^{TE})_\mu^2}{2\omega}\right]|z - z'|\right\}$$

$$\times \left[e_x \frac{\partial \psi_\mu^{TE}(r_\perp)}{\partial y} - e_y \frac{\partial \psi_\mu^{TE}(r_\perp)}{\partial x}\right]$$

$$\otimes \left[e_x \frac{\partial \psi_\mu^{TE}(r'_\perp)}{\partial y'} - e_y \frac{\partial \psi_\mu^{TE}(r'_\perp)}{\partial x'}\right]$$

$$+ \frac{2\pi i}{\omega} \sum_\nu \exp\left\{i\left[\frac{\omega}{c} - \frac{c(k_\perp^{TM})_\nu^2}{2\omega}\right]|z - z'|\right\}$$

$$\times \left[e_x \frac{\partial \psi_\nu^{TM}(r_\perp)}{\partial x} + e_y \frac{\partial \psi_\nu^{TM}(r_\perp)}{\partial y}\right]$$

$$\otimes \left[e_x \frac{\partial \psi_\nu^{TM}(r'_\perp)}{\partial x'} + e_y \frac{\partial \psi_\nu^{TM}(r'_\perp)}{\partial y'}\right] . \tag{A.3.23}$$

A.4 Eigenfunctions of a Passive Circular Waveguide

It is natural to present the eigenfunctions of a circular waveguide in a polar coordinate system. The function $\psi(r_\perp)$ is the solution of the equation:

$$\frac{\partial^2 \psi}{\partial r^2} + \frac{1}{r}\frac{\partial \psi}{\partial r} + \frac{1}{r^2}\frac{\partial^2 \psi}{\partial \varphi^2} + k_\perp^2 \psi = 0 . \tag{A.4.1}$$

We seek the solution in the form:

$$\psi = u(r) \begin{pmatrix} \sin(m\varphi) \\ \cos(m\varphi) \end{pmatrix} , \tag{A.4.2}$$

where m is an integer. Substituting (A.4.2) into (A.4.1), we get the Bessel equation for $u(r)$:

$$\frac{d^2 u(r)}{dr^2} + \frac{1}{r}\frac{du(r)}{dr} + \left(k_\perp^2 - \frac{m^2}{r^2}\right)u(r) = 0 . \tag{A.4.3}$$

The solution of this equation is given by superposition of Bessel functions of the first and second kind, J_m and N_m:

$$u(r) = A J_m(k_\perp r) + D N_m(k_\perp r) .$$

The field amplitude at the waveguide axis must have a finite value, so we should let $D = 0$. As a result, the expression for function ψ takes the form:

$$\psi = A J_m(k_\perp r) \begin{pmatrix} \sin(m\varphi) \\ \cos(m\varphi) \end{pmatrix} , \tag{A.4.4}$$

which is valid for both ψ^{TE} and ψ^{TM} contributions. The eigenvalue equations for $k_\perp$ follow from the boundary conditions (5.41). For a TE-wave we have:

$$J'_m(k_\perp^{TE} R) = 0 ,$$

where the prime means differentiation over the argument of the Bessel function. For a TM-wave we have:
$$J_m(k_\perp^{\text{TM}} R) = 0 \ .$$

The coefficient A appearing in (A.4.4) is defined by the normalization condition:
$$\int |\boldsymbol{\nabla}_\perp \psi|^2 d\boldsymbol{r}_\perp = 1 \ .$$

Using (A.4.4) we find:
$$\int_0^{2\pi} d\varphi \int_0^R r dr \left\{ \left[\frac{\partial \psi^{\text{TE}}}{\partial r}\right]^2 + \frac{1}{r^2}\left[\frac{\partial \psi^{\text{TE}}}{\partial \varphi}\right]^2 \right\}$$
$$= (A^{\text{TE}})^2 \pi \int_0^R r dr \left\{ \frac{m^2}{r^2} J_m^2(\mu_{mk}r/R) + \left[\frac{d}{dr}J_m(\mu_{mk}r/R)\right]^2 \right\}$$
$$= (A^{\text{TE}})^2 \pi \int_0^R r dr \left\{ \frac{m^2}{r^2} J_m^2(\mu_{mk}r/R) \right.$$
$$\left. + \left[\frac{\mu_{mk}}{R} J_{m-1}(\mu_{mk}r/R) - \frac{m}{r} J_m(\mu_{mk}r/R)\right]^2 \right\} = 1 \ . \quad (A.4.5)$$

It follows from the latter expression that the coefficient A^{TE} is
$$A^{\text{TE}} = \left\{ \frac{\pi \mu_{mk}^2}{2} \left[J_{m-1}^2(\mu_{mk}) - J_{m-2}(\mu_{mk}) J_m(\mu_{mk})\right] \right.$$
$$\left. - \pi m J_m^2(\mu_{mk}) \right\}^{-1/2} \ . \quad (A.4.6)$$

When deriving the expression for A^{TE}, we used the following integral relations for the Bessel functions:
$$\int_0^R J_{m-1}^2(\mu_{mk}r/R) r dr = \frac{R^2}{2} \left[J_{m-1}^2(\mu_{mk}) - J_{m-2}(\mu_{mk}) J_m(\mu_{mk})\right] \ ,$$
$$\int_0^R J_m^2(\mu_{mk}r/R) r^{-1} dr = \frac{1}{2m}\left[1 + J_0^2(\mu_{mk}) + J_m^2(\mu_{mk}) - \sum_{n=0}^{m-1} 2 J_n^2(\mu_{mk})\right] \ ,$$
$$\int_0^R J_m(\mu_{mk}r/R) J_{m-1}(\mu_{mk}r/R) dr$$
$$= \frac{R}{2\mu_{mk}}\left[1 - J_0^2(\mu_{mk}) - \sum_{n=1}^{m-1} 2 J_n^2(\mu_{mk})\right] \ . \quad (A.4.7)$$

Taking into account that $J'_m(\mu_{mk}) = 0$ and using the relations

$$J_{m-2}(\zeta) + J_m(\zeta) = \frac{2(m-1)}{\zeta} J_{m-1}(\zeta)$$

$$J_{m-1}(\mu_{mk}) = \frac{m}{\mu_{mk}} J_m(\mu_{mk}) ,$$

we find

$$J_{m-2}(\mu_{mk}) = J_m(\mu_{mk}) \left[\frac{2m(m-1)}{\mu_{mk}^2} - 1 \right] .$$

As a result, (A.4.6) is reduced to

$$A_{mk}^{\text{TE}} = \frac{\sqrt{2}}{\sqrt{\pi(\mu_{mk}^2 - m^2)} J_m(\mu_{mk})} \quad \text{for} \quad m > 0 .$$

For $m = 0$ we have:

$$A_{0k}^{\text{TE}} = \frac{1}{\sqrt{\pi} \mu_{0k} J_0(\mu_{0k})} .$$

Using a similar procedure, we find the expression for A^{TM}:

$$A_{mk}^{\text{TM}} = \frac{\sqrt{2}}{\sqrt{\pi} \nu_{mk} J_{m-1}(\nu_{mk})} \quad \text{for} \quad m > 0 ,$$

$$A_{0k}^{\text{TM}} = \frac{1}{\sqrt{\pi} \nu_{0k} J_1(\nu_{0k})} .$$

A.5 Calculation of the Sums in (5.119)

Let us consider the sum:

$$\int_0^\infty d\zeta \exp(-\hat{\Lambda}\zeta) \left\{ \sum_{n=1}^\infty \frac{\mu_{1n}^2}{(\mu_{1n}^2 - 1) J_1^2(\mu_{1n})} \right.$$

$$\left. \times J_0(\mu_{1n} r/R) J_0(\mu_{1n} r'/R) \exp\left[-\frac{i\mu_{1n}^2 \zeta}{2\Omega} \right] \right\} , \quad (A.5.1)$$

where μ_{1n} are positive-valued solutions of the equation $J'_1(\mu) = 0$. Using the relations

$$J''_1(\mu_{1n}) = \frac{1 - \mu_{1n}^2}{\mu_{1n}^2} J_1(\mu_{1n}) ,$$

$$J_1(\mu_{1n}) = -i I_1(i\mu_{1n}) , \quad J_0(\mu_{1n} r/R) = I_0(i\mu_{1n} r/R) ,$$

$$I'_1(i\mu_{1n}) K_1(i\mu_{1n}) - I_1(i\mu_{1n}) K'_1(i\mu_{1n}) = \frac{1}{i\mu_{1n}} ,$$

$$J''_1(\mu_{1n}) = i I''_1(i\mu_{1n}) = i I''_1 \left(\sqrt{-2i\Omega\lambda} \right) |_{\lambda = -i\mu_{1n}^2/(2\Omega)} ,$$

we rewrite (A.5.1) as

$$\int_0^\infty d\zeta \exp(-\hat{\Lambda}\zeta) \Bigg\{ \sum_{n=1}^\infty \frac{\mu_{1n}^2}{(\mu_{1n}^2-1)J_1^2(\mu_{1n})}$$

$$\times J_0(\mu_{1n}r/R)J_0(\mu_{1n}r'/R) \exp\left[-\frac{i\mu_{1n}^2\zeta}{2\Omega}\right]\Bigg\}$$

$$= i\Omega \int_0^\infty d\zeta \exp(-\hat{\Lambda}\zeta) \Bigg\{ \sum_{n=1}^\infty \exp\left[-\frac{i\mu_{1n}^2\zeta}{2\Omega}\right]$$

$$\times I_0(i\mu_{1n}r/R)I_0(i\mu_{1n}r'/R)K_1'(i\mu_{1n})$$

$$\times \left[\frac{d}{d\lambda}I_1'\left(\sqrt{-2i\Omega\lambda}\right)\right]^{-1}_{\lambda=-i\mu_{1n}^2/(2\Omega)}\Bigg\}, \qquad (A.5.2)$$

where the prime means differentiation over the argument of the Bessel function.

Let us consider the integral along the closed contour O_1 in the complex λ plane consisting of the line $(-i\infty+\gamma', i\infty+\gamma')$ and a semicircle of infinite radius in the left half-plane:

$$u = \int_{O_1} d\lambda \, \exp(\lambda\zeta) \frac{I_0\left(\sqrt{-2i\Omega\lambda}r'/R\right)}{I_1'\left(\sqrt{-2i\Omega\lambda}\right)}$$

$$\times \left[I_0\left(\sqrt{-2i\Omega\lambda}r/R\right)K_1'\left(\sqrt{-2i\Omega\lambda}\right)\right.$$

$$\left.+ K_0\left(\sqrt{-2i\Omega\lambda}r/R\right)I_1'\left(\sqrt{-2i\Omega\lambda}\right)\right], \qquad (A.5.3)$$

where we assume that $r > r'$. In the opposite case one should exchange r by r' and vice versa. The integrand in (A.5.3) is a single-valued function of λ, since

$$K_0(\zeta) = -I_0(\zeta)\left[\ln\left(\frac{\zeta}{2}\right)+\gamma_E\right] + \left(\frac{\zeta}{2}\right)^2 + \ldots,$$

and

$$I_0''(\zeta) = I_1'(\zeta), \quad K_0''(\zeta) = -K_1'(\zeta),$$

$$I_0''(\zeta)K_0'(\zeta) - I_0(\zeta)K_0''(\zeta) = \frac{1}{\zeta^2}.$$

When $\gamma' > 0$, the integral along the closed contour O_1 is equal to the product of the factor $2\pi i$ with the sum of residues taken at the points

$$\lambda = \lambda_n = -\frac{i\mu_{1n}^2}{2\Omega}, \quad n = 1, 2, \ldots$$

As a result, we obtain the following expression for the function u:

$$u = 2\pi \mathrm{i} \sum_{n=1}^{\infty} \exp\left[-\frac{\mathrm{i}\mu_{1n}^2 \zeta}{2\Omega}\right] I_0(\mathrm{i}\mu_{1n}r/R) I_0(\mathrm{i}\mu_{1n}r'/R) K_1'(\mathrm{i}\mu_{1n})$$

$$\times \left[\frac{\mathrm{d}}{\mathrm{d}\lambda} I_1'\left(\sqrt{-2\mathrm{i}\Omega\lambda}\right)\right]^{-1}_{\lambda = -\mathrm{i}\mu_{1n}^2/(2\Omega)}. \tag{A.5.4}$$

Since the integrand in (A.5.2) satisfies the conditions of Jordan's lemma, the contribution of the integral along the semicircle is negligibly small, and the integral along the contour O_1 can be replaced by the integral along the line $(-\mathrm{i}\infty + \gamma',\ \mathrm{i}\infty + \gamma')$. Using (A.5.2) and (A.5.4), we obtain (for $r > r'$):

$$\int_0^\infty \mathrm{d}\zeta \exp(-\hat{\Lambda}\zeta) \left\{ \sum_{n=1}^\infty \frac{\mu_{1n}^2}{(\mu_{1n}^2 - 1) J_1^2(\mu_{1n})} \right.$$

$$\left. \times J_0(\mu_{1n}r/R) J_0(\mu_{1n}r'/R) \exp\left[-\frac{\mathrm{i}\mu_{1n}^2 \zeta}{2\Omega}\right] \right\}$$

$$= \frac{\Omega}{2\pi} \int_0^\infty \mathrm{d}\zeta \exp(-\hat{\Lambda}\zeta) \int_{-\mathrm{i}\infty+\gamma'}^{\mathrm{i}\infty+\gamma'} \mathrm{d}\lambda \exp(\lambda\zeta)$$

$$\times \frac{I_0\left(\sqrt{-2\mathrm{i}\Omega\lambda}\, r'/R\right)}{I_1'\left(\sqrt{-2\mathrm{i}\Omega\lambda}\right)} \left[I_0\left(\sqrt{-2\mathrm{i}\Omega\lambda}\, r/R\right) K_1'\left(\sqrt{-2\mathrm{i}\Omega\lambda}\right)\right.$$

$$\left. + K_0\left(\sqrt{-2\mathrm{i}\Omega\lambda}\, r/R\right) I_1'\left(\sqrt{-2\mathrm{i}\Omega\lambda}\right)\right]. \tag{A.5.5}$$

Using the inverse Laplace transformation, we can write (for $\operatorname{Re} p > 0$ and $\gamma' > 0$):

$$f(p) = \frac{1}{2\pi\mathrm{i}} \int_0^\infty \mathrm{d}\zeta \exp(-p\zeta) \int_{-\mathrm{i}\infty+\gamma'}^{\mathrm{i}\infty+\gamma'} \mathrm{d}\lambda \exp(\lambda\zeta) f(\lambda).$$

As a result, (A.5.5) is reduced to

$$\int_0^\infty \mathrm{d}\zeta \exp(-\hat{\Lambda}\zeta) \left\{ \sum_{n=1}^\infty \frac{\mu_{1n}^2}{(\mu_{1n}^2 - 1) J_1^2(\mu_{1n})} \right.$$

$$\left. \times J_0(\mu_{1n}r/R) J_0(\mu_{1n}r'/R) \exp\left[-\frac{\mathrm{i}\mu_{1n}^2 \zeta}{2\Omega}\right] \right\}$$

$$= \mathrm{i}\Omega I_0(g\hat{r}) I_0(g\hat{r}') \frac{K_1'\left(g\sqrt{\Omega/B}\right)}{I_1'\left(g\sqrt{\Omega/B}\right)}$$

$$+ \mathrm{i}\Omega \begin{pmatrix} K_0(g\hat{r}) I_0(g\hat{r}') & \text{for } \hat{r} > \hat{r}' \\ I_0(g\hat{r}) K_0(g\hat{r}') & \text{for } \hat{r}' > \hat{r} \end{pmatrix}, \tag{A.5.6}$$

where
$$g^2 = -2\mathrm{i}B\hat{A}, \quad B/\Omega = r_0^2/R^2.$$

Let calculate the sum in (5.119) over the roots of the equation
$$J_1(\nu_{1n}) = 0.$$

Taking into account that
$$J_1'(\nu_{1n}) = J_0(\nu_{1n}) = I_0(\mathrm{i}\nu_{1n}) = I_1'(\mathrm{i}\nu_{1n}),$$
$$I_1'(\mathrm{i}\nu_{1n})K_1(\mathrm{i}\nu_{1n}) - I_1(\mathrm{i}\nu_{1n})K_1'(\mathrm{i}\nu_{1n}) = \frac{1}{\mathrm{i}\nu_{1n}},$$
$$I_1'(\mathrm{i}\nu_{1n}) = I_1'\left(\sqrt{-2\mathrm{i}\Omega\lambda}\right)\big|_{\lambda=-\mathrm{i}\nu_{1n}^2/(2\Omega)},$$

we find:
$$\int_0^\infty d\zeta\, \exp(-\hat{A}\zeta)\left\{\sum_{n=1}^\infty \frac{1}{J_0^2(\nu_{1n})}\right.$$
$$\left.\times J_0(\nu_{1n}r/R)J_0(\nu_{1n}r'/R)\exp\left[-\frac{\mathrm{i}\nu_{1n}^2\zeta}{2\Omega}\right]\right\}$$
$$= \mathrm{i}\Omega\int_0^\infty d\zeta\, \exp(-\hat{A}\zeta)\left\{\sum_{n=1}^\infty \exp\left[-\frac{\mathrm{i}\nu_{1n}^2\zeta}{2\Omega}\right]\right.$$
$$\times I_0(\mathrm{i}\nu_{1n}r/R)I_0(\mathrm{i}\nu_{1n}r'/R)K_1(\mathrm{i}\nu_{1n})$$
$$\left.\times \left[\frac{d}{d\lambda}I_1\left(\sqrt{-2\mathrm{i}\Omega\lambda}\right)\right]^{-1}_{\lambda=-\mathrm{i}\nu_{1n}^2/(2\Omega)}\right\}. \tag{A.5.7}$$

The sum on the right-hand side of the latter expression is replaced by the integral along the closed contour O_1 in the complex λ plane, consisting of the line $(-\mathrm{i}\infty + \gamma', \mathrm{i}\infty + \gamma')$ and a semicircle of infinite radius in the left half-plane (for $r > r'$):

$$u = \int_{O_1} d\lambda\, \exp(\lambda\zeta)\frac{I_0\left(\sqrt{-2\mathrm{i}\Omega\lambda}r'/R\right)}{I_1\left(\sqrt{-2\mathrm{i}\Omega\lambda}\right)}$$
$$\times \left[I_0\left(\sqrt{-2\mathrm{i}\Omega\lambda}r/R\right)K_1\left(\sqrt{-2\mathrm{i}\Omega\lambda}\right)\right.$$
$$\left.+ K_0\left(\sqrt{-2\mathrm{i}\Omega\lambda}r/R\right)I_1\left(\sqrt{-2\mathrm{i}\Omega\lambda}\right)\right]$$
$$= 2\pi\mathrm{i}\sum_{n=1}^\infty \exp\left[-\frac{\mathrm{i}\nu_{1n}^2\zeta}{2\Omega}\right]I_0(\mathrm{i}\nu_{1n}r/R)I_0(\mathrm{i}\nu_{1n}r'/R)K_1(\mathrm{i}\nu_{1n})$$
$$\times \left[\frac{d}{d\lambda}I_1\left(\sqrt{-2\mathrm{i}\Omega\lambda}\right)\right]^{-1}_{\lambda=-\mathrm{i}\nu_{1n}^2/(2\Omega)}. \tag{A.5.8}$$

This procedure is justified, since the integrand in (A.5.8) is a single-valued function of λ. The contribution of the integral along the semicircle is negligibly small, so the integral along the contour O_1 in (A.5.8) is reduced to the integral along the line $(-i\infty + \gamma', i\infty + \gamma')$. Using (A.5.7) and (A.5.8), and applying the inverse Laplace transformation, we obtain:

$$\int_0^\infty d\zeta\, \exp(-\hat{\Lambda}\zeta)\left\{\sum_{n=1}^\infty \frac{1}{J_0^2(\nu_{1n})} J_0(\nu_{1n}r/R) J_0(\nu_{1n}r'/R) \exp\left[-\frac{i\nu_{1n}^2\zeta}{2\Omega}\right]\right\}$$

$$= i\Omega I_0(g\hat{r}) I_0(g\hat{r}') \frac{K_1\left(g\sqrt{\Omega/B}\right)}{I_1\left(g\sqrt{\Omega/B}\right)}$$

$$+ i\Omega \begin{pmatrix} K_0(g\hat{r})I_0(g\hat{r}') & \text{for} & \hat{r} > \hat{r}' \\ I_0(g\hat{r})K_0(g\hat{r}') & \text{for} & \hat{r}' > \hat{r} \end{pmatrix}. \qquad (A.5.9)$$

A.6 List of Symbols

a	Parameter of the Epstein profile
$\hat{a}_1$	Reduced complex amplitude of the first harmonic of the longitudinal component of the beam current density
$\hat{a}_j$	Bunching in the jth box
$\hat{a}_n$	Reduced complex amplitude of the nth harmonic of the longitudinal component of the beam current density
A	Input coupling factor
A_{JJ}	Factor specifying the interaction of an electron with the electromagnetic field in a planar undulator
$\mathbf{A}_\text{w}$	Vector potential of the undulator field
$\mathbf{A}_\omega(\mathbf{r})$	Complex amplitude of the vector potential of an electromagnetic wave
$\mathbf{A}_\perp$	Vector potential of an electromagnetic wave
B	Diffraction parameter
b_0, b_1, b_2	Coefficients of undulator tapering
$\hat{b}_0, \hat{b}_1, \hat{b}_2$	Tapering parameters (reduced coefficients of undulator tapering)
C	Detuning
$\hat{C}$	Detuning parameter (reduced detuning)
$\hat{C}_0$	Reduced detuning at the beginning of a tapered undulator
$\hat{C}_\text{m}, \hat{C}_\text{max}$	Optimal reduced detuning
$\hat{C}_m$	Reduced detuning of the mth longitudinal mode of an optical resonator

c	Velocity of light
a, b, c	Parameters of the hypergeometric function
d	Half-width of a sheet electron beam with stepped profile
d	Boundary of a sheet electron beam with bounded parabolic profile
d_1	Parameter of a sheet electron beam with bounded parabolic profile
d	Length of the drift space
$\hat{d}$	Reduced length of the drift space
E	Amplitude of the electric field
E_0	Parameter of saturation field amplitude
E_g	Amplitude of the electric field of a Gaussian laser beam
$\boldsymbol{E}$	Vector of the electric field
$\boldsymbol{E}_\omega$	Complex amplitude of the electric field vector
$\tilde{E}$	Complex amplitude of the electric field
$\hat{E}^{(j)}$	Normalized field in the jth box
$\tilde{E}_m$	Complex amplitude of the electric field of the mth mode of an optical resonator
$\tilde{E}_{\text{ext}}$	Complex amplitude of the electric field of the input wave
$\tilde{E}_i$	Complex amplitude of the electric field of the radiated wave
$\tilde{E}_{x,y,z}$	Complex amplitude of the electric field components in Cartesian coordinates
$\tilde{E}_{r,\varphi,z}$	Complex amplitude of the electric field components in cylindrical coordinates
$\tilde{E}_{r,z}^{(n)}$	nth coefficient of the Fourier series in the angle φ
$\bar{E}(\omega, z)$	Fourier transform of the electric field $E(t, z)$
$\bar{E}(p)$	Laplace transform of the complex amplitude $\tilde{E}(z)$
$\bar{E}^{(n)}(p, r)$	Laplace transform of the Fourier coefficient $\tilde{E}^{(n)}$
$\mathcal{E}$	Energy
$\mathcal{E}_0$	Nominal energy of an electron
$-e$	Charge of the electron
$\boldsymbol{e}_{x,y,z}$	Unit vectors of Cartesian coordinates
$F(P)$	Distribution function of the canonical momentum
$\hat{F}(\hat{P})$	Distribution function of the reduced canonical momentum
$F(t)$	Profile function of an electron bunch
$\bar{F}(\omega)$	Fourier transform of an electron bunch profile function
$F_{r,\varphi}^{(m)}(\hat{r})$	Active waveguide radial eigenfunction with mth azimuthal number
$\boldsymbol{F}_\nu(\boldsymbol{r})$	Vector eigenfunction
$\boldsymbol{F}$	Force
f	Distribution function in phase space
f_0	Unperturbed distribution function

A.6 List of Symbols

$\tilde{f}_1$	Complex amplitude of the first harmonic of a distribution function
G	Power gain
$\hat{G}$	Green's function
$G_\omega^{\alpha\beta}$	Tensor Green's function
g	Gain per pass in an FEL oscillator
g_s	Small-signal gain
$\hat{g}_s$	Reduced small-signal gain
$g_{1,2}$	First and second order correlation functions
$g_1^{(\text{eff})}$	First-order effective correlation function
H	Hamiltonian
H_A	Spectral Green's function of an FEL amplifier
H_m	Profile of a monochromator line
H_ℓ	Amplitude of the magnetic field of a planar undulator
H_w	Amplitude of the magnetic field of a helical undulator
$\boldsymbol{H}$	Vector of the magnetic field
$\boldsymbol{H}_w$	Vector of the undulator magnetic field
$\boldsymbol{H}_\omega(\boldsymbol{r})$	Complex amplitude of the magnetic field vector
$h(\boldsymbol{k}_\perp, z)$	Angular spectrum
$h(\hat{C})$	Normalized spectral density of the radiation energy
I_0	Beam current
I_A	Alfven current
I_w	Current of helical undulator winding
$\bar{I}(\omega)$	Fourier transform of the beam current $I(t)$
$I(\theta)$	Intensity of radiation in the far zone
j_1	Amplitude of the first harmonic of the longitudinal component of the beam current density
j_n	Amplitude of the nth harmonic of the longitudinal component of the beam current density
$\boldsymbol{j}$	Current density vector
$\boldsymbol{j}_\perp$	Transverse current density vector
$j_{x,y,z}$	Components of the current density vector in Cartesian coordinates
j_0	Unperturbed current density
$(j_0)_{\text{eff}}$	Effective value of the beam current density
$\bar{j}_0$	Unperturbed linear current density of a sheet electron beam
$\boldsymbol{j}_\omega(\boldsymbol{r})$	Complex amplitude of the current density vector
$\tilde{j}_1$	Complex amplitude of the first harmonic of the longitudinal component of the beam current density
$\tilde{j}_n$	Complex amplitude of the nth harmonic of the longitudinal component of the beam current density

$\tilde{j}_a$	Slowly varying complex amplitude of the first harmonic of the longitudinal component of the beam current density in a planar undulator
$\hat{j}_1$	Reduced amplitude of the first harmonic of the longitudinal component of the beam current density
$\hat{j}_n$	Reduced amplitude of the nth harmonic of the longitudinal component of the beam current density
K	Undulator parameter
K_{th}	Threshold coefficient for an FEL oscillator
$\boldsymbol{k}$	Wave vector of an electromagnetic wave
$\boldsymbol{k}_\perp$	Transverse wave vector of an electromagnetic wave
$k_\perp, |\boldsymbol{k}_\perp|$	Modulus of the transverse wave vector of an electromagnetic wave
k_z	Longitudinal wavenumber of an electromagnetic wave
k_m	Longitudinal wavenumber of the mth mode of an optical resonator
k_w	Wavenumber of the undulator
k_ν	Eigenvalue of a passive resonator
k_μ^{TE}	Eigenvalues of TE-modes of a passive waveguide
k_μ^{TM}	Eigenvalues of TM-modes of a passive waveguide
k_1	Reduced parameter of a bounded parabolic profile
L	Base of an optical resonator
l_b	Bunch length
$l_{x,y}$	Transverse dimensions of a rectangular waveguide
l_w	Undulator length
l_g	Gain length
l_1, l_2	Length of two sections of an undulator
$\hat{l}_1, \hat{l}_2$	Reduced length of two sections of an undulator
$\hat{l}_1, \hat{l}_2$	Reduced length of two sections of an undulator
$\hat{l}_b$	Normalized bunch length
$\hat{l}_w$	Normalized length of undulator
M	Transition matrix
M	Number of modes
M	Parameter of the gamma distribution
m	Azimuthal harmonic number
m	Longitudinal mode number in an optical resonator
m_e	Rest mass of the electron
N	Number of electrons in a bunch
N	Number of macroparticles within the interval $(0, 2\pi)$
N_c	Number of cooperating electrons
N_b	Number of boxes per electron bunch
N_m	Number of macroparticles per box

A.6 List of Symbols   445

N_v	Number of particles per elementary volume of the electron bunch
N_w	Number of undulator periods
N_λ	Number of electrons per wavelength
n	Number of round-trips in an FEL oscillator
$\hat{n}$	Reduced number of round-trips in an FEL oscillator
n'	Refractive index
n_0	Unperturbed particle density of an electron beam
$\boldsymbol{n}$	Unit vector orthogonal to the inner wall of a waveguide
P	Canonical momentum
$\hat{P}$	Reduced canonical momentum
p_z	Longitudinal component of the canonical momentum
p_0	Canonical momentum conjugated with time t
p	Argument of the Laplace transform
$p(\mathcal{E}), p(W)$	Probability density distributions
Q	Parameter specifying the interaction of an electron with the electromagnetic field in a planar undulator
R	Radius of a circular waveguide
R_w	Radius of helical undulator winding
r	Radial variable
$\hat{r}$	Reduced radial variable
r_0	Radius of an axisymmetric electron beam with stepped profile
r_0	Boundary of an axisymmetric electron beam with bounded parabolic profile
r_1	Parameter of an axisymmetric electron beam with bounded parabolic profile
r_b	Typical transverse size of an electron beam
r_c	Radius of coherence
r_w	Radius of electron rotation in undulator
S	Transverse area of an electron beam
$S(\hat{x}), S(\hat{r})$	Transverse density profile functions
s	Axial coordinate of a particle within the electron bunch
$\hat{s}$	Normalized axial coordinate of a particle within the electron bunch
T	Duration of a rectangular pulse
T	Transmission coefficient of resonator mirrors
$T(\hat{z})$	Tapering function
t	Time
U	Complex amplitude of the effective potential of particle wave interaction
U_ext	Complex amplitude of the effective potential of particle interaction with an external wave

U_i	Complex amplitude of the effective potential of particle interaction with a radiated wave
$\hat{U}$	Normalized complex amplitude of the effective potential
$\hat{U}_c$	Normalized complex amplitude of the effective potential of the space charge field
u	Amplitude of the effective potential
u_j	Amplitude coefficient of the jth radiation mode
$\hat{u}$	Normalized amplitude of the effective potential
$\hat{u}_{ext}$	Normalized amplitude of the effective potential of particle interaction with an external wave
$\hat{u}^{(j)}$	Normalized amplitude of the effective potential in an optical resonator after the jth round-trip
v	Velocity of an electron
v_z	Longitudinal velocity of an electron
v_{ph}	Phase velocity of an electromagnetic wave
v_g	Group velocity of an electromagnetic wave
$\tilde{v}$	Complex amplitude of electron transverse velocity in a helical undulator
$\boldsymbol{v}$	Vector of electron velocity
$\boldsymbol{v}_\perp$	Transverse vector of electron velocity
W	Power of electromagnetic radiation
W_b	Electron beam power
W_{ext}	Power of the input signal
W_{out}	Power of the output radiation
W_{sh}	Effective power of shot noise
W_0	Saturation power parameter
$\hat{W}_{ext}$	Reduced power of the input signal
w	Waist of a Gaussian laser beam
$\hat{w}$	Reduced waist of a Gaussian laser beam
X	Bunching parameter in the theory of the klystron
x	Cartesian coordinate
$\hat{x}$	Reduced Cartesian coordinate
y	Cartesian coordinate
$\hat{y}$	Reduced Cartesian coordinate
z	Cartesian coordinate
z	Axial coordinate in cylindrical system
z_i	Coordinate of the beginning of undulator tapering
z_0	Longitudinal position of the Gaussian laser beam waist
$\hat{z}$	Reduced Cartesian coordinate
$\hat{z}_0$	Reduced longitudinal position of the Gaussian laser beam waist
α	Relative power losses per resonator round-trip
$\hat{\alpha}$	Reduced parameter of resonator losses

A.6 List of Symbols

β	Efficiency parameter in FEL oscillator theory
β	Beta function
Γ	Gain factor in FEL amplifier theory
Γ	Absorption coefficient of resonator mirrors
γ	Relativistic factor
γ_z	Longitudinal relativistic factor in a helical undulator
γ_ℓ	Longitudinal relativistic factor in a planar undulator
γ_1	Transverse correlation function
$\gamma_1^{(\mathrm{eff})}$	Effective transverse correlation function
γ_{E}	Euler's constant
$\Delta\mathcal{E}$	Energy deviation
$\Delta\omega$	Frequency width of a monochromator with a rectangular line
$\delta\Lambda$	Perturbation of an eigenvalue
$\delta\psi$	Beam-wave phase shift in drift space
$\delta\bar{\psi}$	Nonmultiple phase shift in drift space
ϵ	Emittance
ϵ	Permittivity
η	Efficiency
$\hat{\eta}$	Reduced efficiency
θ	Observation angle
$\hat{\theta}$	Reduced observation angle
θ_{s}	Angle of electron rotation in a helical undulator
θ_ℓ	Maximal angle of electron oscillation in a planar undulator
Λ	Eigenvalue
$\hat{\Lambda}$	Reduced eigenvalue
$\hat{\Lambda}_0$	Zeroth approximation for the reduced eigenvalue in a waveguide
Λ_{p}	Longitudinal plasma wavenumber
$\hat{\Lambda}_{\mathrm{p}}^2$	Space charge parameter
$\hat{\Lambda}_{\mathrm{T}}^2$	Energy spread parameter
λ	Argument of the Laplace transform
λ	Radiation wavelength
λ_{e}	Wavelength of beam density modulation
λ_{e}'	Wavelength of beam density modulation in the rest frame
λ_{w}	Undulator period
λ_j	jth root of an eigenvalue equation
μ_{mn}	nth root of the equation $J_m'(\mu) = 0$
ν_{mn}	nth root of the equation $J_m(\nu) = 0$
$\boldsymbol{\Pi}$	Pointing's vector
Π_z	Longitudinal component of Pointing's vector
Π_m	Density of power loss of the mth resonator mode
ρ	Efficiency parameter in FEL amplifier theory
ρ_{e}	Charge density of an electron beam

σ	Conductivity
σ_T	rms duration of a Gaussian bunch
$\sigma_{\mathcal{E}}$	Relative rms fluctuations of the energy in a radiation pulse
σ_A	rms frequency width of a Gaussian amplification line
σ_b	rms bunch length
$\hat{\sigma}_b$	Normalized rms bunch length
σ_m	rms frequency width of a Gaussian monochromator line
σ_w	rms dispersion of the instantaneous radiation power
τ	Gain factor in FEL oscillator theory
τ_c	Coherence time
$\hat{\tau}_c$	Normalized coherence time
$\Phi(\boldsymbol{r})$	Eigenfunction
$\Phi_n(\boldsymbol{r})$	Radial part of an eigenfunction with azimuthal number n
ϕ	Scalar potential
$\phi_\omega(\boldsymbol{r})$	Complex amplitude of the scalar potential of an electromagnetic wave
φ	Azimuthal angle
$\tilde{\chi}$	Gauge transformation function
χ_m	Susceptibility for the mth longitudinal mode of a resonator
ψ	Variable related to energy-phase variables
ψ_0	Phase of the effective potential of particle wave interaction
ψ_1	Phase of the first harmonic of the longitudinal component of the beam current density
ψ_n	Phase of the nth harmonic of the longitudinal component of the beam current density
$\psi^{\Rightarrow}$	Phase ψ before drift space
$\psi^{\Leftarrow}$	Phase ψ after drift space
ψ_j	Phase of the jth macroparticle
ψ_μ^{TE}	TE-mode eigenfunction of a passive waveguide
ψ_ν^{TM}	TM-mode eigenfunction of a passive waveguide
Ω	Waveguide diffraction parameter
ω	Frequency of the amplified electromagnetic wave
ω_0	Resonance frequency
ω_m	Frequency of the mth longitudinal mode of a resonator

Suggested Further Reading

Books

1. T.C. Marshall, *Free-Electron Lasers* (Macmillan, New York, 1985)
2. C.A. Brau, *Free-Electron Lasers* (Academic Press, Boston, 1990)
3. P. Luchini and H. Motz, *Undulators and Free-Electron Lasers* (Clarendon Press, Oxford, 1990)
4. W.B. Colson, C. Pellegrini and A. Renieri (Eds), *Free Electron Lasers*, Laser Handbook, Vol. 6 (North Holland, Amsterdam, 1990)
5. G. Dattoli, A. Renieri and A. Torre, *Lectures on the Free Electron Laser Theory and Related Topics* (World Scientific, 1993)
6. H.P. Freund and T.M. Antonsen, *Principles of Free Electron Lasers* (Chapman & Hall, New York, 1996)

Review papers

7. C.W. Roberson and P. Sprangle, Phys. Fluids B **1**(1989)3
8. R. Bonifacio et al., La Rivista del Nuovo Cimento, Vol. 13, No. 9 (1990)
9. H.P. Freund and R.K. Parker, in *The 1991 Yearbook of the Encyclopedia of Physical Science and Technology*, p. 49 (Academic Press, San Diego, 1991)
10. R. Bonifacio et al., La Rivista del Nuovo Cimento, Vol. 15, No. 11 (1992)
11. E.L. Saldin, E.A. Schneidmiller and M.V. Yurkov, Physics Reports **260**(1995)187

Conference Proceedings

12. S.F. Jacobs, M. Sargent and M.O. Scully (Eds), *Novel Sources of Coherent Radiation*, The Physics of Quantum Electronics, Vol. 5, (Addison Wesley, Reading, Massachusetts, 1978)
13. S.F. Jacobs, et al. (Eds), *Free-Electron Generators of Coherent Radiation*, The Physics of Quantum Electronics, Vol. 7, (Addison Wesley, Reading, Massachusetts, 1980)
14. S.F. Jacobs et al. (Eds), *Free-Electron Generators of Coherent Radiation*, The Physics of Quantum Electronics, Vol. 8 and 9, (Addison Wesley, Reading, Massachusetts, 1982)
15. D.A.G. Deacon and M. Billardon (Eds), *Proceedings of the Bendorf Free Electron Laser Conference*, Journal de Physique Colloque **C1-44**(1983)

16. C.A. Brau et al. (Eds), *Free-Electron Generators of Coherent Radiation*, SPIE Proceedings **453**(1984)
17. J.M.J Madey and A.Renieri (Eds), *Free Electron Lasers*, Nucl. Instrum. and Methods A **237**(1985)
18. E.T. Scharlemann and D. Prosnitz (Eds), *Free Electron Lasers*, Nucl. Instrum. and Methods A **250**(1986)
19. M.W. Poole (Editor), *Free Electron Lasers*, Nucl. Instrum. and Methods A **259**(1987)
20. P. Sprangle, C.M. Tang and J. Walsh (Eds), *Free Electron Lasers*, Nucl. Instrum. and Methods A **272**(1988)
21. A. Gover and V.L. Granatstein (Eds), *Free Electron Lasers*, Nucl. Instrum. and Methods A **A285**(1989)
22. L.R. Ellias and I. Kimel (Eds), *Free Electron Lasers*, Nucl. Instrum. and Methods A **296**(1990)
23. J.M. Buzzi and J.M. Ortega (Eds), *Free Electron Lasers*, Nucl. Instrum. and Methods A **304**(1991)
24. J.C. Goldstein and B.E. Newnam (Eds), *Free Electron Lasers*, Nucl. Instrum. and Methods A **318**(1992)
25. C. Yamanaka and K. Mima (Eds), *Free Electron Lasers*, Nucl. Instrum. and Methods A **331**(1993)
26. P.W. van Amersfoort, P.J.M. van der Slot and W.J. Witteman (Eds), *Free Electron Lasers*, Nucl. Instrum. and Methods A **341**(1994)
27. T.I. Smith, H.A. Schwettman and R.L. Swent (Eds), *Free Electron Lasers*, Nucl. Instrum. and Methods A **358**(1995)
28. I. Ben-Zvi and S. Krinsky (Eds), *Free Electron Lasers*, Nucl. Instrum. and Methods A **375**(1996)
29. G. Dattoli and A. Renieri (Eds), *Free Electron Lasers 1996*, Nucl. Instrum. and Methods A **393**(1997)
30. J. Xie and X. Du (Eds), *Free Electron Lasers 1997*, Nucl. Instrum. and Methods A **407**(1998)
31. G.R. Neil and S. Benson (Eds), *Free Electron Lasers 1998*, Nucl. Instrum. and Methods A **429**(1999)
32. G. Ingelman and L. Jönsson (Eds), *Future Electron Accelerators and Free Electron Lasers*, Nucl. Instrum. and Methods A **398**(1997)
33. R. Bonifacio, L. De Salvo Souza and C. Pellegrini (Eds), *High Gain, High Power Free Electron Laser: Physics and Applications to TeV Particle Acceleration*, (North Holland, Amsterdam, 1989)
34. R. Bonifacio and W.A. Barletta (Eds), *Towards X-Ray Free Electron Lasers: Workshop on Single Pass, High Gain FELs Starting from Noise, Aiming at Coherent X-rays*, AIP Conference Proceedings **413**(1997)
35. P.G. O'Shea and H.E. Bennett (Eds), *Free-Electron Laser Challenges*, SPIE Proceedings **2988**(1997)
36. A.K. Freund, H.P. Freund and M.R. Howells (Eds), *Time Structure of X-Ray Sources and its Applications*, SPIE Proceedings **3451**(1998)

Special Issues

37. A. Szöke (Editor), *Special Issue on Free-Electron Lasers*, IEEE Journal of Quantum Electronics, Vol. QE-17, No. 8 (1981)
38. L.R. Ellias and W.B. Colson (Eds), *Special Issue on Free-Electron Lasers*, IEEE Journal of Quantum Electronics, Vol. QE-19, No. 3 (1983)
39. V.L. Granatstein and C.W. Roberson (Eds), *Special Issue on Free-Electron Lasers*, IEEE Journal of Quantum Electronics, Vol. QE-21, No. 7 (1985)
40. C.A. Brau and B.E. Newnam (Eds), *Special Issue on Free-Electron Lasers*, IEEE Journal of Quantum Electronics, Vol. QE-23, No. 9 (1987)
41. G. Bekefi and R.H. Pantell (Eds), *Special Issue on Free-Electron Lasers*, IEEE Journal of Quantum Electronics, Vol. QE-27, No. 12 (1991)

References

Chapter 1

1.1 L.R. Elias et. al., Phys. Rev. Lett. **36**(1976)717
1.2 D.A.G. Deacon et. al., Phys. Rev. Lett. **38**(1977)892
1.3 C.A. Brau, *Free-Electron Lasers* (Academic Press, New York, 1990)
1.4 C. Yamanaka, Nucl. Instrum. and Methods A **318**(1992)1
1.5 G.R. Neil et. al., Nucl. Instrum. and Methods A **358**(1995)159
1.6 R. Brinkmann et al. (Eds), "Conceptual Design of 500 GeV e^+e^- Linear Collider with Integrated X-ray Facility", DESY 1997-048, ECFA 1997-182, Hamburg, May 1987
1.7 "A VUV Free Electron Laser at the TESLA Test Facility: Conceptual Design Report", DESY Print TESLA-FEL 95-03, Hamburg, DESY, 1995
1.8 J. Rossbach, Nucl. Instrum. and Methods A **375**(1996)269
1.9 "Linac Coherent Light Sorce (LCLS) Design Study Report", The LCLS Design Study Group, Stanford Linear Accelerator Center (SLAC) Report No. SLAC-R-521, 1998

Chapter 2

2.1 N.M. Kroll and W.A. McMullin, Phys. Rev. A **17**(1978)300
2.2 I. Bernstein and J.L. Hirshfield, Phys. Rev. A **20**(1979)1661
2.3 D.B. McDermot and T.C. Marshall, Phys. Quantum Electron. **7**(1980)509
2.4 E. Jerby and A. Gover, IEEE J. Quantum Electron. **QE-21**(1985)1041
2.5 E.L. Saldin, E.A. Schneidmiller and M.V. Yurkov, Nucl. Instrum. and Methods A **313**(1992)555
2.6 E.L. Saldin, E.A. Schneidmiller and M.V. Yurkov, Opt. Commun. **103**(1993)205
2.7 A. Lin and J.M. Dawson, Phys. Rev. Lett. **42**(1979)1670
2.8 P. Sprangle, C.M. Tang and W.M. Manheimer, Phys. Rev. Lett. **43**(1979)1932
2.9 N.M. Kroll, P. Morton and M.N. Rosenbluth, IEEE J. Quantum Electron. **QE-17**(1981)1436
2.10 T.J. Orzechovski, et. al., Phys. Rev. Lett. **57**(1986)2172.
2.11 R. Bonifacio, C. Pellegrini and L. Narducci, Opt. Commun. **50**(1984)373
2.12 R. Bonifacio et. al., Riv. Nuovo Cimento Vol.13, No.9 (1990)
2.13 E.L. Saldin, E.A. Schneidmiller and M.V. Yurkov, Phys. Rep. **260**(1995)187
2.14 E.L. Stiefel and G. Scheifele, *Linear and Regular Celestial Mechanics* (Springer, Berlin, Heidelberg, 1971)
2.15 G. Ecker, *Theory of Fully Ionized Plasmas* (Academic Press, New York, 1972)

Chapter 3

3.1 J.M.J. Madey, J. Appl. Phys. **42**(1971)1906
3.2 F.A. Hopf, et al., Opt. Commun. **18**(1976)413
3.3 W.B. Colson, Phys. Lett. A **64**(1977)190
3.4 C.M. Tang and P. Sprangle, J. Appl. Phys. **53**(1981)831
3.5 W.B. Colson and P. Elleaume, Appl. Phys. **B29**(1982)101
3.6 I. Schnitzer and A. Gover, Nucl. Instrum. and Methods A **237**(1985)124
3.7 J. Blau and W.B. Colson, Nucl. Instrum. and Methods A **259**(1987)198
3.8 G. Dattoli, et al., Nucl. Instrum. and Methods A **285**(1989)108
3.9 W.B. Colson, IEEE J. Quantum Electron. **QE-17**(1981)1417
3.10 B.W.J. McNeil, Nucl. Instrum. and Methods A **296**(1990)388
3.11 E.L. Saldin, E.A. Schneidmiller and M.V. Yurkov, Opt. Commun. **102**(1993)360
3.12 N.A. Vinokurov, Proc. 10th Int. Conf. on High Energy Charged Particle Accelerators (Serpukhov, 1977), Vol. 2, p. 454
3.13 E.L. Saldin, E.A. Schneidmiller and M.V. Yurkov, Opt. Commun. **103**(1993)297
3.14 P. Sprangle, C.M. Tang and I. Bernstein, Phys. Rev. A **28**(1983)2300
3.15 G. Shvets and J.S. Wurtele, Nucl. Instrum. and Methods A **393**(1997)273
3.16 A.H. Ho, R.H. Pantell and J. Feinstein, IEEE J. Quantum Electron. **QE-23**(1987)1545
3.17 H. Leboutet, Proc. Particle Accelerator Conf. (San Francisco, 1991), Vol. 5, p. 2763
3.18 E.L. Saldin, E.A. Schneidmiller and M.V. Yurkov, Phys. Lett. A **185**(1994)469
3.19 E.L. Saldin, E.A. Schneidmiller and M.V. Yurkov, Opt. Commun. **107**(1994)507
3.20 N.M. Kroll and M.N. Rosenbluth, Physics of Quantum Electr., Vol. 7 (Addison-Wesley, Reading, MA), p. 89, 1980
3.21 W.B. Colson and R.A. Freedman, Opt. Commun. **46**(1983)37
3.22 J.E. Sollid et al., Nucl. Instrum. and Methods A **285**(1989)147
3.23 A. Al-Abawi et al., Opt. Commun. **30**(1979)235
3.24 G. Dattoli, A. Marino and A. Renieri, Opt. Commun. **35**(1980)407
3.25 P. Elleaume, IEEE J. Quantum Electron. **QE-21**(1985)1012
3.26 G. Dattoli, et al., Phys. Rev. A **37**(1988)4334
3.27 S. Ishii, G. Shvets and J.S. Wurtele, Nucl. Instrum. and Methods A **358**(1995)489
3.28 N. Piovella, et al., Phys. Rev. **E52**(1995)5470
3.29 A. Renieri, Nuovo Cim. **53B**(1979)160
3.30 G. Dattoli and A. Renieri, Nuovo Cim. **59B**(1980)1
3.31 P. Elleaume, J. de Physique **45**(1984)997
3.32 G. Dattoli, L. Giannessi and A. Renieri, Nucl. Instrum. and Methods A **358**(1995)338

Chapter 4

4.1 A. M. Kondratenko and E. L. Saldin, Part. Acc. **10**(1980)207
4.2 G.T. Moore, Opt. Commun. **52**(1984)46
4.3 G.T. Moore, Nucl. Instrum. and Methods A **250**(1986)381
4.4 Kwang-Je Kim, Phys. Rev. Lett. **57**(1986)1871
4.5 S. Krinsky and L.H. Yu, Phys. Rev. A **35**(1987)3406

4.6 M. Xie, D.A.G. Deacon and J.M.J. Madey, Phys. Rev. A **41**(1990)1662
4.7 E.L. Saldin, E.A. Schneidmiller and M.V. Yurkov, Nucl. Instrum. and Methods A **307**(1991)531
4.8 E.L. Saldin, E.A. Schneidmiller and M.V. Yurkov, Opt. Commun. **87**(1992)69
4.9 E.L. Saldin, E.A. Schneidmiller and M.V. Yurkov, Opt. Commun. **97**(1993)272
4.10 L.H. Yu, S. Krinsky and R.L. Glukstern, Nucl. Instrum. and Methods A **304**(1991)516
4.11 Y.H. Chin, K.-J. Kim and M. Xie, Nucl. Instrum. and Methods A **318**(1992)481
4.12 C.-M. Tang and P. Sprangle, IEEE J. Quantum Electron. **QE-21**(1985)970
4.13 E.T. Scharlemann and W.M. Fawley, Proc. SPIE 642 (1986)2
4.14 T.M. Tran and J.S. Wurtele, Comput. Phys. Commun. **54**(1989)263
4.15 J.C. Goldstein, T.F. Wang, B.E. Newnam, and B.D. McVey, Proc. 1987 Particle Accelerators Conf., Washington, DC, USA, p. 202
4.16 E.L. Saldin, E.A. Schneidmiller and M.V. Yurkov, Opt. Commun. **95**(1993)141
4.17 H.P. Freund, Phys. Rev. E **52**(1995)5401

Chapter 5

5.1 H.P. Freund and T.M. Antonsen, *Principles of Free-electron Lasers* (Chapman & Hall, New York, 1996)
5.2 E. Jerby and A. Gover, Phys. Rev. Lett. **63**(1989)864
5.3 E.L. Saldin, E.A. Schneidmiller and M.V. Yurkov, Opt. Commun. **85**(1991)117
5.4 E.L. Saldin, E.A. Schneidmiller and M.V. Yurkov, Nucl. Instrum. and Methods A **317**(1992)581
5.5 Y. Pinhasi and A. Gover, Phys. Rev. E **51**(1995)2472
5.6 A. Amir, I. Boscolo and L.R. Elias, Phys. Rev. A **32**(1985)2860
5.7 P.M. Morse and H. Feshbach, *Methods of Theoretical Physics*, Part I (McGraw Hill, New York, 1953)
5.8 L.M. Brekovskikh, *Waves in Layered Media* (Academic Press, New York, 1960)

Chapter 6

6.1 A.M. Kondratenko and E.L. Saldin, Part. Acc. **10**(1980)207
6.2 H. Haus, IEEE J. Quantum Electron. **QE-17**(1981)1427
6.3 Ya.S. Derbenev, A.M. Kondratenko and E.L. Saldin, Nucl. Instrum. and Methods **193**(1982)415
6.4 J.B. Murphy and C. Pellegrini, Nucl. Instrum. and Methods A **237**(1985)159
6.5 K.J. Kim, Nucl. Instrum. and Methods A **250**(1986)396
6.6 J.M. Wang and L.H. Yu, Nucl. Instrum. and Methods A **250**(1986)484
6.7 K.J. Kim, Phys. Rev. Lett. **57**(1986)1871
6.8 S. Krinsky and L.H. Yu, Phys. Rev. A **35**(1987)3406
6.9 W.B. Colson, Review in: W.B. Colson et al. (Eds), *Laser Handbook, Vol. 6: Free Electron lasers* (North-Holland, Amsterdam, 1990), p. 115
6.10 R. Bonifacio, et al., Phys. Rev. Lett. **73**(1994)70 .
6.11 P. Pierini and W. Fawley, Nucl. Instrum. and Methods A **375**(1996)332
6.12 E.L. Saldin, E.A. Schneidmiller and M.V. Yurkov, Nucl. Instrum. and Methods A **393**(1997)157

6.13 E.L. Saldin, E.A. Schneidmiller and M.V. Yurkov, Opt. Commun. **148**(1998)383
6.14 J. Goodman, *Statistical Optics* (Wiley, New York, 1985)
6.15 L. Mandel and E. Wolf, *Optical Coherence and Quantum Optics* (Cambridge University Press, 1995)
6.16 M. Hogan et al., Phys. Rev. Lett. **81**(1998)4867

Index

Absorption coefficient   90, 109
Active
- medium   3, 156, 264
- waveguide   285, 317

Alfven current   20, 92, 170, 215
Amplification bandwidth   26, 28, 56, 233, 250, 253
Amplitude characteristic   57, 72, 78
Angular distribution   179, 180, 201, 216, 217, 242, 249, 253, 404, 411, 413
Approximation
- Fraunhofer   216, 247
- linear   18, 38, 39, 56, 87, 160, 164, 168, 243, 265, 278
- multilayer   158, 195, 209, 213, 238, 239, 278, 309
- paraxial   155, 157, 163, 164, 193, 229, 231, 266, 270, 273, 277, 300, 324, 433
- single-mode   271, 291, 293, 296
- slowly varying amplitude   20, 42, 139, 163, 266

Arbitrary gradient profile   158, 209, 238, 243, 278, 307
Asymptotic
- Bessel functions   219, 222, 238, 296, 323
- high gain   15, 26, 32, 37, 47–49, 52, 57, 61, 155, 171, 191–193, 195, 235, 245, 278, 300, 313, 330, 361, 374–376
- thin electron beam   183, 184, 187, 203–205, 208, 209, 221, 222, 238
- wide electron beam   181, 183, 186, 192, 203, 208, 218, 237

Axisymmetric
- electron beam   158, 214, 215, 225, 234, 238, 246, 248, 276, 278, 303
- radiation mode   216, 223, 244

Bandwidth   26, 28, 56, 69, 89, 91, 135, 136, 233, 250, 253, 360, 361, 367, 369, 373

Beam
- axisymmetric   158, 214, 215, 225, 234, 238, 246, 248, 276–278, 285, 288, 297, 303, 320
- cold   24, 38, 43, 91, 111, 113, 115, 118, 138, 324, 362
- sheet   158, 168–170, 188, 195
- thin   183, 184, 187, 203–205, 208, 209, 221, 222, 238
- wide   181, 183, 186, 192, 203, 208, 218, 237

Bessel equation   214, 236, 279, 304, 327, 435
Bessel functions   44, 67, 111, 123, 124, 215, 216, 219, 227, 238, 248, 258, 287, 296, 301, 307, 311, 317, 318, 323, 435, 436
- asymptotic forms   219, 222, 296, 301, 317, 323
- modified   67, 215, 222, 311

Beta function   259, 415, 423
Betatron oscillations   259–261
Boundary conditions   87, 89, 135, 136, 156, 172, 190, 198, 228, 264, 268–270, 277, 280, 305, 308, 430, 432, 435
- Leontovich's   88, 89, 135, 136, 265, 344, 346, 348, 349, 351

Bunching parameter   382, 383, 405, 406

Canonical
- coordinate   16, 246
- equations   16, 100, 425, 427
- momentum   16, 17, 21, 246
- transformation   16, 427
- variables   16, 425, 427

Cartesian coordinates   168, 195, 265, 320, 322
Cauchy's residue theorem   23, 32, 94, 140, 191

Central limit theorem   358, 366, 384, 406
Circular polarization   89, 135, 230, 264, 301
Circular waveguide   264, 276, 277, 288, 320
Coefficient
- absorption   90, 109
- Fourier   226, 310
- power gain   25, 26, 28, 55, 56, 193, 194, 229, 231, 232, 244, 247, 256, 316, 330
- threshold   143, 149
- transmission   90, 109
Coherence
- area   355, 407, 410
- degree of   402, 413, 414
- longitudinal   355
- partial   355, 407, 413, 414
- radius   402, 410, 411, 413
- spectral interval   366, 369, 373, 401
- time   149, 151, 355, 374, 376, 391, 401
- transverse   355, 396, 402, 407, 410, 413, 414
- volume   355
Cold electron beam   24, 38, 43, 91, 111, 113, 115, 118, 138, 324, 362
Complex plane   23, 24, 31, 191, 229
Condition
- boundary   87, 89, 135, 136, 156, 172, 190, 198, 228, 264, 268–270, 277, 280, 305, 308, 430, 432, 435
- initial   14, 15, 21, 38–41, 52, 87, 99, 115, 139, 151, 188, 225, 245, 265, 384
- resonance   6, 10, 66, 68, 263
Conductivity   265, 342–344, 348
Confluent hypergeometric function   206, 207, 209, 304
Constant
- Euler's   222, 368
Conventional laser   1, 3
Correlation function
- spectral   366, 367
- time   148, 374, 376
- transverse   407, 410, 411, 413
Coulomb gauge   268–270
Coupling factor   26, 30, 31, 37, 231
Cubic equation   24, 30, 31, 181, 203, 295, 362
Current density   18, 19, 43, 51, 85, 90, 98, 135, 137, 138, 141, 145, 146, 157, 161, 163, 164, 167, 169, 170, 188, 210, 225, 238, 243, 248, 266, 270, 276, 278, 285, 303, 307, 357, 360, 361
- distribution   194, 205, 213, 234, 235, 246, 278
- first harmonic   19, 39, 50, 51, 90, 94, 98, 99, 115, 163, 164, 169
Cylindrical coordinates   214, 309, 322

Damping factor   90, 142
- reduced   92, 99, 101, 104, 106, 109, 132, 143, 149
Decibel   55
Degeneracy parameter   355
Delta function   38, 51, 60, 91, 98, 140, 141, 164, 268, 357
Density
- current   18, 19, 39, 41, 43, 48, 50, 51, 85, 90, 94, 98, 99, 115, 135, 137–139, 141, 145, 146, 157, 158, 161, 163, 164, 167, 169, 170, 188, 194–196, 205, 210, 213, 225, 234, 235, 238, 243, 244, 246, 248, 264, 266, 270, 276, 278, 284, 285, 303, 307, 328, 357, 360, 361
- modulation   14, 18, 41, 43, 52, 88, 112, 113, 123, 135, 138, 146, 161, 225, 264, 284, 361
- power flow   52, 136, 140, 142, 272, 300, 302
- probability   358, 365, 369, 376, 384, 389, 414
- spectral   356, 374
Detuning   17, 41, 42, 50, 67, 79, 90, 138, 160
- parameter   20, 25, 26, 28–31, 35–37, 49, 56, 60, 61, 67–69, 73, 75–77, 80, 92, 94, 96–98, 100, 110, 117, 120, 123, 127, 130, 134, 143, 144, 170, 181, 201, 207, 215, 222, 223, 243, 249, 250, 253, 288, 291, 298, 330, 362
Differential equation
- Bessel   214, 236, 279, 304, 327
- hypergeometric   196
Diffraction
- angle   164
- effects   3, 83, 155, 159, 218, 255, 256
- grating   150, 151
- limit   3
- parameter   14, 170, 178, 180, 181, 192–194, 201, 208, 215–218, 221–223, 230, 231, 243, 246, 248–250, 253, 255, 256, 288, 290, 291, 300, 303
- waveguide parameter   280, 288, 291, 298, 300

Dimensionless
- equations   2, 195, 234, 309
- parameters   15, 49, 66, 180, 214, 329

Dispersion section   88, 111, 112, 117, 122, 124

Distribution
- current   155, 213
- current density   194, 195, 205, 234, 235, 239, 246, 278
- energy   31, 72, 101, 106, 161, 215, 367
- field   85, 156, 178, 201, 207, 216, 230, 241, 245, 247, 249, 253–256, 306
- function   18, 21, 23, 24, 31, 32, 38, 43, 47, 91, 92, 94, 113, 115, 116, 160, 168, 171, 195, 265, 268, 278
- probability density   356, 358, 365, 369, 372, 376, 384, 389, 414

Domain
- frequency   360, 365, 376
- time   356, 361, 362, 370

Drift space   63, 107, 111, 113, 116, 117, 119, 134

Effective correlation function
- spectral   401, 411
- time   400

Effective initial intracavity power   88, 91, 147

Effective potential of interaction   17, 42, 50, 53, 75, 90, 97, 132, 149, 160, 167, 246, 247, 249, 265, 268, 270, 276, 321, 328

Effective power of shot noise   388, 389

Efficiency   15, 52, 66, 67, 69, 72, 74–78, 80, 82, 88, 97, 99, 101, 111, 113, 124, 128, 130, 151, 152, 247, 253, 255
- optimization   68, 73, 77, 80, 99, 101, 106, 123, 124, 130, 134
- parameter   20, 45, 49, 51, 75, 82, 93, 97, 98, 100, 143, 171, 215, 246, 279, 360
- practical   108, 109
- reduced   49, 52, 53, 56, 60, 61, 78, 97, 100, 101, 104, 106, 108–110, 123, 124, 126, 132, 134, 249, 250, 332

Eigenfunction   156, 157, 174, 175, 187, 199, 201, 206, 207, 209–211, 216, 236, 238, 239, 288, 300, 302, 429–432, 435

Eigenvalue equation   14, 24, 35–37, 41, 48, 59, 158, 174, 175, 177, 181, 183, 187, 192, 199, 201, 203, 205, 207–209, 212, 215, 218, 220, 223–225, 236, 239, 278, 280, 291, 293, 307, 314, 315, 362, 435

Electron
- charge   5, 140, 357
- energy   6, 17, 42, 49, 52, 66, 71, 90, 100, 159, 167, 275
- – deviation   90, 100
- – nominal   17, 42, 48, 49, 160
- mass   5
- velocity   5, 6, 11, 41, 42, 50, 89, 159, 168, 263, 265

Electron beam profile
- arbitrary gradient   158, 195, 209, 234, 238, 243, 278, 307
- Epstein   158, 194, 196, 210, 213
- Gaussian   213, 239
- parabolic   158, 205, 234, 235, 239, 278, 303, 304, 309
- stepped   158, 170, 207, 214, 225, 236, 246, 248, 278

Emittance   259, 418, 421

Energy spread   14, 24, 38, 48, 53, 61, 62, 72, 73, 88, 91, 92, 99, 101, 104, 106, 112, 116, 121, 126, 159, 164, 169, 181, 183, 184, 203, 223, 243, 245, 248, 299, 360
- Gaussian   21, 32, 33, 46, 92, 171, 172, 192, 195, 215, 217, 278, 304, 315
- Lorentzian   31
- parameter   15, 21, 35–37, 47, 49, 62, 88, 93, 94, 106, 117, 121, 122, 171, 195, 215, 217, 223, 235, 243, 246, 249, 250, 279, 288

Ensemble   63, 106, 140, 246, 248, 357, 362

Epstein profile   158, 194, 196, 209, 210, 213

Euler's constant   222, 368

Even modes   174, 177–181, 183, 184, 186, 193, 199–201, 203, 204, 207, 209, 212

Extended Hamiltonian formalism   16, 425

External monochromatization   128, 130, 150, 151

Factor
- damping   88, 90, 142
- filling   135
- input coupling   26, 30, 31, 37, 231
- trapping   68, 73

Far zone   179, 180, 201, 203, 216, 241, 245

Feedback   87, 135
Fiber, optical   155, 156
Field
- amplitude distribution   156, 178, 201, 207, 216, 230, 241, 245, 247, 249, 253–256, 306
- growth rate   26, 28, 30, 36, 56, 59, 62, 73, 146, 155, 181, 183, 186, 193, 201, 203, 205, 207, 217, 222, 223, 229, 243, 248, 253, 256, 288, 290, 298, 306
- space charge   15–18, 28, 29, 35, 50, 59–61, 73, 74, 88, 94, 104, 155, 159, 161, 222, 245, 247, 299
Filling factor   135
First-order correlation   141, 358, 359, 366, 372
- function   148, 366, 374, 385
Fluctuations of radiation energy   367, 369, 376, 389
Focusing of master oscillator radiation   230, 244
Fourier
- coefficients   226, 310
- harmonic   60, 141, 358–362, 365
- series   59, 169, 226, 310
- transform   137, 141, 166, 358, 359, 374, 385
Fredholm equation   176
Frequency
- deviation   27, 253
- domain   145, 360, 365, 370, 376
- lasing   100, 113, 128, 130, 134, 151
Fresnel number   83
Function
- Bessel   44, 67, 111, 123, 124, 215, 216, 219, 222, 227, 236, 238, 248, 258, 287, 296, 301, 307, 311, 317, 318, 323, 327, 435, 436
- delta   38, 51, 60, 91, 98, 140, 141, 268
- error   33, 368
- gamma   199, 369
- Green's   59, 188–190, 226–228, 245, 270, 310, 360, 362, 365, 429–434
- hypergeometric   198, 238
-- confluent   206, 207, 209, 304
- integral cosine   248, 368
- integral sine   248, 368
- spherical   198–200
Fundamental mode   216, 217, 222, 223, 238, 241, 243, 253, 254

Gain
- bandwidth   28

- coefficient   25, 26, 28, 55, 56, 59, 193, 194, 229, 231, 232, 244, 247, 256, 316, 317, 330
- length   51, 66, 69, 349, 355, 360
- parameter   20, 49, 92, 98, 99, 109, 110, 113, 115, 170, 181, 203, 209, 215, 220, 223, 225, 246, 291
Gamma distribution   356, 369, 376, 400, 414, 423, 424
Gamma function   199, 369
Gauge   17
- Coulomb   268–270
- transformation   16
Gaussian
- distribution function   32, 62, 63, 106, 248, 370, 376
- energy spread   21, 32, 33, 46, 61, 92, 106, 171, 172, 192, 195, 215, 217, 278, 304, 315
- laser beam   158, 230, 231, 244, 245
- line of monochromator   368, 369
- mode   145
- profile of electron beam   158, 213, 239, 240
- statistics   355, 357
Green's function   59, 188–190, 226–228, 245, 270, 310, 360, 362, 365, 429–434
Ground mode   91, 145, 146
Group velocity   365, 387, 416
Growth rate   26, 28, 30, 31, 36, 37, 56, 59, 60, 62, 73, 146, 155, 181, 183, 186, 193, 201, 203, 205, 207, 217, 222, 223, 229, 230, 243, 248, 253, 256, 288, 290, 291, 298, 299, 306

Hamiltonian   15–17, 42, 49, 50, 60, 74, 246, 425–427
- extended formalism   16, 425
Helical
- trajectory   4, 5, 89, 159, 265
- undulator   4, 15, 41, 45, 48, 67, 89, 113, 116, 135, 156, 157, 214, 246, 263, 265, 276, 277, 309, 320
Helmholtz equation   269, 270, 429, 432
High gain
- asymptotic   15, 25, 26, 28, 32, 33, 35, 37, 47–49, 52, 53, 57, 60, 61, 155, 156, 171, 175, 191–193, 195, 214, 229, 235, 245, 278, 300, 303, 313, 316, 330, 361, 362, 374–376, 388
Homogeneous waveguide   263, 276, 429, 434

Hypergeometric function   198, 238
- confluent   206, 207, 209, 304

Index of refraction   135, 137, 343
Initial conditions   14, 15, 21, 38–41, 52, 87, 99, 115, 139, 151, 161, 188, 225, 245, 265
Initial-value problem   14, 22, 28, 32, 38, 39, 59, 90, 147, 156, 158, 187, 188, 192, 225, 243, 278, 309, 362
Input coupling factor   26, 30, 31, 37, 231
Instantaneous power distribution   372, 389
Integral equation   163, 164, 175, 177
Integro-differential equation   14, 20, 22, 37, 163, 170, 188, 244, 266, 329
Inverse Laplace transformation   22, 46, 94, 140, 191, 229, 313

Jordan's lemma   23, 24, 31, 32, 94, 191

Kinetic energy of electron   17
Kinetic equation   19, 40, 43, 113, 161, 164, 243
Kummer equation   206, 236

Landau method   23
Laplace transform   14, 22, 32, 45, 46, 94, 139, 140, 188, 191, 226, 229, 310, 313
Laser
- conventional   1, 3
- master   230, 244, 245
- X-ray   1, 11, 155, 263, 353, 355, 384, 385, 424
Lasing threshold   100, 113, 143–147
Leontovich's boundary conditions   88, 89, 135, 136, 265, 344, 346, 348, 349, 351
Limit
- diffraction   3
- high gain   25, 26, 28, 32, 33, 35, 37, 47, 52, 57, 61, 155, 156, 171, 191–193, 195, 235, 245, 278, 300, 303, 313, 316, 330, 361, 362, 374–376
Line of monochromator
- Gaussian   368, 369
- rectangular   368, 369
Linear approximation   18, 38, 39, 56, 87, 160, 164, 168, 243, 265, 278
Liouville's equation   18
Littrow mounted grating   150, 151
Lorentz force   5

Lorentzian energy spread   31

Macroparticles   51, 59, 61–63, 69, 96, 98, 104, 106, 158, 246, 248, 249, 384
Master laser   230, 244, 245
Maxwell's equations   14, 18, 19, 39, 43, 59, 96, 135, 137, 155, 157, 164, 169, 243, 268, 277
Method
- Green's function   59, 188, 226, 245, 310
- Landau   23
- multilayer approximation   158, 195, 209, 213, 214, 238, 239, 278, 309
Model
- one-dimensional   13, 15, 19, 20, 42, 43, 48, 51, 59, 82, 87, 99, 135, 137, 141, 145, 147, 155, 181, 183, 186, 192, 193, 203, 209, 218, 221, 223, 225, 245, 253, 256, 356, 361, 365
Modes
- degeneration   264, 302, 307
- even   174, 177–181, 183, 184, 186, 187, 193, 199–201, 203, 204, 207, 209, 212
- fundamental   216, 217, 222, 223, 238, 241, 243, 253, 254
- Gaussian   145
- ground   91, 145, 146
- longitudinal   87, 88, 135, 136, 140, 143, 145, 146, 148, 149
- odd   177, 183, 199, 203, 205, 207, 212
- rotating   285, 287, 328
- TE   271, 288, 293, 303, 328, 330
- TEM   145, 217, 221, 223, 225, 230, 238, 241, 243, 253, 254, 256, 298, 302
- TM   271, 288, 293, 328, 330
Monochromator line
- Gaussian   368, 369
- rectangular   368, 369

Negative exponential distribution   356, 358, 365, 366, 369, 372, 376, 384, 389, 400, 406, 414
Nominal energy   7, 17, 42, 48–50, 74, 90, 100, 138, 159, 160, 246
Number of transverse modes   404, 410, 414
Numerical simulation   14, 48, 51, 55, 59, 61–63, 66, 67, 72, 74–77, 80, 82, 88, 96, 98, 104, 106, 107, 110, 117, 122, 126, 133, 148, 149, 152, 164, 244–246, 248–250, 255, 258, 332, 357, 362, 366, 372, 384, 385

One-dimensional model 13, 15, 19, 20, 42, 43, 48, 51, 59, 82, 87, 99, 135, 137, 141, 145, 147, 155, 181, 183, 186, 192, 193, 203, 209, 218, 221, 223, 225, 245, 253, 256, 356, 361, 365
Optical
- fiber 155, 156, 158
- guiding 155
- klystron 88, 111, 112, 118, 121, 122, 126
- resonator 88, 135
- waveguide 156, 157, 196, 209

Parabolic equation 178
Parameter
- detuning 15, 20, 25–31, 35, 36, 39, 49, 56, 60, 61, 67–69, 73, 75–77, 79, 80, 92, 96, 97, 100, 110, 111, 117, 119, 120, 123, 127, 130, 134, 143, 144, 170, 181, 195, 201, 207, 215, 222, 223, 243, 249, 250, 253, 258, 278, 288, 291, 298, 330, 362
- diffraction 14, 170, 178, 180, 181, 192–195, 201, 208, 214–218, 220, 222, 223, 230–232, 243, 246, 248–250, 253, 255, 256, 258, 278, 288, 290, 291, 300, 303
- efficiency 20, 45, 51, 66, 75, 82, 93, 98, 100, 111, 143, 171, 195, 215, 240, 246, 258, 279, 360
- energy spread 15, 21, 35–37, 47, 49, 61, 62, 73, 88, 93, 97, 106, 111, 117, 121, 122, 171, 186, 195, 215, 223, 235, 240, 243, 246, 249, 250, 258, 279, 288, 360
- gain 20, 49, 92, 98, 99, 109–111, 113, 115, 170, 181, 195, 203, 209, 214, 215, 220, 223, 225, 240, 246, 258, 278, 291
- saturation field amplitude 49, 53, 99, 111, 362
- space charge 15, 28–31, 35, 39, 49, 60, 61, 73, 88, 96, 97, 104, 106, 111, 170, 195, 214, 222, 240, 243, 246, 250, 258, 288
- undulator 67, 75, 80, 82, 145, 258
- waveguide diffraction 280, 288, 290, 291, 296, 298, 300
Paraxial approximation 155, 157, 163, 164, 193, 229, 231, 266, 270, 273, 277, 300, 324, 433
Parseval's theorem 362
Partial coherence 355, 407, 413, 414
Partial waves 23–26, 31

Pendulum equation 8
Period of undulator 6, 47, 51, 67, 79, 81, 82, 116, 263
Permittivity 343
Phase
- velocity 181–183, 263, 288
Planar undulator 4, 41, 45, 47, 63, 82, 111, 116, 168, 170, 258, 361
Plane wave 14, 166, 193
Poisson
- brackets 426, 427
- equation 161
Potential
- effective 17, 42, 50, 53, 66, 75, 90, 97, 132, 149, 160, 167, 246, 247, 249, 265, 268, 270, 275, 276, 321, 322, 328
- scalar 16, 268, 269, 428
- vector 16, 17, 19, 20, 50, 268, 269
Power
- balance 51, 52, 100, 164, 168, 272, 276
- gain coefficient 25, 26, 28, 55, 56, 193, 194, 229, 231, 232, 244, 247, 256, 316, 330
Poynting's vector 136, 272, 273, 300, 362
Prebuncher 88, 111, 113, 126, 128, 133, 134, 151, 152
Probability density distribution 356, 358, 365, 369, 372, 376, 384, 389, 414
Profile of electron beam
- arbitrary gradient 158, 195, 209, 234, 238, 243, 278, 303, 307
- Epstein 158, 194, 196, 209, 210, 213, 214
- Gaussian 158, 210, 213, 239, 240
- parabolic 158, 205, 207, 209, 234, 235, 239, 278, 303, 304, 306, 307, 309
- stepped 158, 170, 188, 208, 214, 225, 236, 237, 246, 248, 278, 280

Radius of coherence 402, 410, 411, 413
Rectangular waveguide 264, 340, 341
Reduced
- damping factor 88, 92, 97, 99, 101, 104, 106, 109, 132, 143, 149
- efficiency 49, 52, 53, 56, 60–62, 78, 100, 101, 104, 106, 108–110, 123, 124, 126, 132, 134, 249, 250, 332
- equations 20, 41, 44, 50, 51, 79, 170, 225, 246, 303
Refractive index 135–137, 344
Residue theorem 23, 32, 94, 140, 191

Index   463

Resonance
- condition   6, 10, 66, 68, 263
- frequency   130, 357, 360, 374
Rotation angle of electron   5, 75, 77, 89, 159, 265
Runge-Kutta technique   37, 99, 248, 330

Scalar potential   16, 268, 269, 428
Second-order correlation function   366–368, 376, 385
Self-consistent equations   14, 18, 21, 38, 39, 41, 44, 48, 51, 73, 75, 79, 87, 97, 109, 117, 157–159, 163, 170, 195, 210, 225, 243, 263, 266, 268, 320, 328, 330
Sheet beam   158, 168–170, 188, 195
Shot noise   89, 99, 134, 135, 137, 140, 142, 146, 357, 360, 365, 388
Sideband instability   88, 89, 149, 151, 152, 397
Similarity techniques   2, 14, 15, 48, 66, 88, 97, 159, 245, 288
Slippage   85, 360, 365, 386, 390, 398, 402, 406, 416, 419, 420
Slowly varying amplitude   20, 42, 139, 163, 266, 355, 372, 374
Small-signal gain   88–91, 98, 100, 111–113, 118, 120–122, 128, 130, 134, 143, 151
Space charge
- field   15–18, 28, 35, 50, 59–61, 73, 88, 94, 104, 155, 159, 222, 245, 247, 299
- parameter   15, 28–31, 35, 39, 49, 60, 73, 88, 96, 104, 106, 111, 170, 195, 214, 222, 240, 243, 246, 250, 258, 288
Spectral correlation functions   366, 367
Spectral Green's function   356, 360–362, 365
Spherical
- function   198–200
- wave   216, 247
Spike   362, 365
Spontaneous
- emission   146
- undulator radiation   91, 145, 146
Statistical optics   355, 356, 369, 391, 402, 424
Steady-state
- Green's function   360
- simulation codes   388
- theory   13, 361
Stepped profile   158, 170, 188, 208, 214, 225, 236, 237, 246, 248, 278, 280

Storage ring FEL   113, 153
Superbunching   391
Superconducting accelerator   152
Susceptibility   138, 156, 157, 196

Tapering parameters   68, 76, 80, 116, 128, 130, 258, 336
TE modes   271, 288, 293, 303, 328, 330
Technique
- Laplace transform   14, 22, 32, 45, 139, 310
- Runge-Kutta   37, 99, 248, 330
- similarity   2, 14, 15, 48, 66, 88, 97, 159, 245, 288
TEM modes   145, 217, 221, 223, 225, 230, 238, 241, 243, 253, 254, 256, 298, 302
Theorem
- Cauchy's   23, 32, 94, 140, 191
- central limit   358, 366, 384, 406
- Parseval's   362
Thin electron beam   183, 184, 187, 203–205, 208, 209, 221, 222, 238
Time domain   356, 361, 362, 370
TM modes   271, 288, 293, 328, 330
Transformation
- canonical   16, 427
- Fourier   137, 141, 166, 358, 361, 374, 385
- gauge   16
- Laplace   14, 22, 32, 45, 46, 94, 139, 140, 188, 191, 226, 229, 310, 313
Transition matrix   39
Transmission coefficient   90, 109
Trapping factor   68, 73

Undulator
- helical   4, 5, 15, 41, 45, 48, 89, 113, 116, 135, 156, 157, 214, 246, 263, 265, 276, 277, 290, 309, 320
- magnetic field   5, 15, 16, 27, 42, 48, 67, 79–81, 111, 116, 152, 168, 253, 265
- nonuniform   88, 111, 113, 117, 126
- parameter   5, 67, 75, 79, 80, 82, 116, 145, 258
- period   6, 47, 51, 67, 79, 81, 82, 89, 93, 98, 116, 130, 143, 263
- planar   4, 41, 42, 45, 47, 48, 63, 82, 111, 116, 168, 170, 258, 361, 362
- tapering   15, 66–68, 74, 76, 80–82, 116, 128, 130, 133, 134, 151, 152, 253
- vector potential   16
- wavenumber   5, 48, 78

Variables
- canonical   16, 50, 160, 246, 425, 427
- reduced   143, 329
Vector potential   16, 17, 19, 20, 50, 268, 269
Velocity
- group   365, 387, 416
- phase   181–183, 263, 288
Vlasov equation   14, 18, 38, 87, 155, 157, 160, 265, 266, 268

Waist of Gaussian beam   145, 230, 232
Wave
- partial   23–26, 31
- plane   14, 166, 193
- spherical   216, 247

Wave equation   19, 40, 43, 137, 155, 163, 169, 231, 266, 324
Waveguide
- active   285, 317
- circular   263, 264, 276, 277, 288, 309, 320
- homogeneous   84, 263, 276
- passive   285, 288, 290, 291, 317
- rectangular   264, 340, 341
Wavenumber
- plasma   20, 171, 265
- undulator   5, 48, 69, 78
Wide electron beam   181, 183, 186, 192, 203, 208, 218, 237

X-ray laser   1, 11, 155, 353, 384, 424

Printing: Mercedes-Druck, Berlin
Binding: Stürtz AG, Würzburg